Advances in Bioprocess Technology

Pogaku Ravindra
Editor

Advances in Bioprocess Technology

Springer

Editor
Pogaku Ravindra
University Malaysia Sabah School
of Engineering and Info tech
Kotakinabalu, Malaysia

ISBN 978-3-319-17914-8 ISBN 978-3-319-17915-5 (eBook)
DOI 10.1007/978-3-319-17915-5

Library of Congress Control Number: 2015945338

Springer Cham Heidelberg New York Dordrecht London

Printed on acid-free paper

Springer International Publishing AG Switzerland is part of Springer Science+Business Media (www.springer.com)

To all the noble souls who are relentlessly striving to integrate technology with human touch for sustainable society.

It has become appallingly obvious that our technology has exceeded our humanity.

ALBERT EINSTEIN

Preface

Fuelled by rapid economic growth, bioprocessing of a variety of feedstocks has increased drastically, making bioprocess technology the larger revenue generator in the world within a short span of few decades. There is continuously ever-increasing gap between traditional and bioprocess technology, which is a positive sign for the welfare of a sustainable society. Taking into account the trend of bio economic growth, every nation's strong dependence on traditional process technology cannot be mitigated in the future without developing self-dependent bio-based alternatives, which not only secure the sustainability of its own economy, but also impact the rest of the world for a healthy society. On an average, every country has leading grain production in the world, producing more than 300 MT of grain per year on an average, and correspondingly generating 600–700 MT agricultural residues. Due to the lack of economically viable technologies for their utilization, most agricultural residues are burnt in the field by farmers, which pollutes environments and even presents public hazards such as the disruption of air transportation by smoke clouds in the sky. The abundant of cellulosic waste can be used for bio-based products. Therefore, producing biofuels, bioenergy, and bio-based chemicals through the refinery of lignocellulose biomass has been acknowledged worldwide as an alternative to the oil refinery, since the oil crisis occurred in the 1970s. It has also recently been highlighted again because of global climate change caused by the over-consumption of petroleum-based products, particularly vehicle emissions. Without doubt, the successful development of biomass refineries will help it to sustain its own economy, and in the meantime contribute to the whole world. Driven by these imperatives, governmental funding for R&D of biomass refineries has increased significantly in many nations. This momentum is expected to be maintained for a long term to make such a pathway economically competitive.

I am honored to be invited by the Springer to edit the book entitled "Advances in Bioprocess technology," with a focus on bioprocess technology, biofuels, and bioenergy to present major progress achieved by eminent scientists and challenges to be addressed collectively by international communities.

The bio-based platform is the prerequisite for the bioconversion of lignocellulose biomass, and highly efficient and low-cost cellulose enzymes are the bottleneck. Therefore, all the authors have emphasized and reviewed the bioprocess technology scope for the sustainable world. The book is divided into five parts. The first part is with a theme of advances in biochemical engineering. It comprises five chapters. Chap. 1 is authored by Abu Zahrim et al. It has focused on life cycle review on biomass combustion. Chap. 2 deals with bioleaching of nickel written by Pogaku Ravindra et al. Hari Vuthaluru reviews on ash formation from pulverized fuel combustion in Chap. 3. Waste management methods are reviewed by Faheem in Chap. 4 and Chap. 5 makes a comprehensive study of free fatty acid modelling in palm oil refinery by Ravindra et al. In Part II, with the title of Biomass and Bionergy, there are ten chapters. Ravindra et al. authored Chap. 6 entitled Production of Biogas from Palm Oil Mill Effluent. Danquah authored Chap. 7, a very relevant topic, on the process analysis of microalgae biomass. Chapter 8 is reviewed by Gangagni Rao on biogas from poultry. Chapter 9, on biogas production, is reviewed by Chan et al. The detailed review on bioenergy is well reviewed by See Ram et al. in Chap. 10. Optimization of catalytic coal gasification for hydrogen is highlighted by Suzana Yusup in Chap. 11. Further Suzana Yusup et al. wrote Chap. 12 on the effect of process parameters on bio oil yield. Anantha Raman et al. narrated agro residue as fuel in Chap. 13. A novel method for biogas as clean fuel is described by Vijay Kumar et al. in Chap. 14. Sarma et al. have described the thermochemical processing of biomass in Chap. 15. Part III under the title Bioprocess Technology has three chapters. Chapter 16 is written by Azlina Kamruddin et al., with a focus on dynamic enzymatic kinetic resolution of NSAID. This is followed by catgut waste utilization for protease production in Chap. 17 by Jegan et al., and Chap. 18 discusses about membrane processes for microalgae by Rosalam et al. Part IV emphasizes on food biotechnology. It has three chapters. Chapter 20, authored by Inge Russell et al., narrates innovation in alcohol beverage production. Chapter 21 is written on starter cultures technology for fermented foods by Ravindra et al., and Chap. 22 highlights specialized studies on liquid core capsules for lactobacilli fermentation. This chapter is authored by Boon Beng Lee et al. The final Part V contains policy and regulations in Chap. 23. The framework of policy regulations for bioprocess is well explained by Sripathi Kulkarni Rao et al.

We expect this special volume to be a window for international colleagues to learn the current R&D progress in bioprocess technology, biofuels, and bioenergy in many nations. On behalf of Springer and Co, I express my sincere thanks to all authors and reviewers for their dedication, contributions, and valuable comments.

Professor Devinder Mahajan, Chemical & Molecular Engineering Materials Science Engineering Dept. Stony Brook University, NY, USA for his encouragement.

Gregory Baer and **Merry Stuber** of Springer Science+Business Media for their delicate coordination.

Also, I greatly appreciate the generosity of my University Malaysia Sabah (UMS), my colleagues and graduates at Faculty of Engineering and my family for all their support.

Kotakinabalu, Malaysia Dr. Pogaku Ravindra

About the Editor

Pogaku Ravindra has diverse and intense, yet rewarding experiences in teaching, research, industry, executive and administrative fields spanning over 35 years.

Professor Ravindra was visiting scientist at Cornell University and visiting professor at Pennsylvania State University. He has an expertise in the area of bioprocess technology for high value products and bioenergy. At present, his research group is focused on bioprocessing of palm oil waste and bio-derived energy for sustainable development. Over the course of his 35 years, he has published more than 200 articles in journals and proceedings. He has edited 4 books, authored 8 books and 12 book chapters. He has reviewed more than 1000 journal manuscripts for reputed international journals. He has a patent and four copyrights.

Professor Ravindra is bestowed with the national and international prestigious awards. He has received gold and silver medals for his research contributions in the Oil and Gas, Chemical and Bioprocessing fields. Professor Ravindra's LCP model was given an award from International Invention and Innovation Exhibition (ITEX), Malaysia. Professor Ravindra was recipient of distinguished chemical engineer award from the Indian institute of Chemical engineers. He has received the best researcher award from the *International Journal of Science and Technology*.

Professor Ravindra was also a UNESCO consultant on Sustainable energy projects. He has also carried out at least 25 major industry studies. He serves as the editor-in-chief, editorial board member, guest editor and reviewer for multiple referred journals.

Professor Ravindra has delivered invited lectures, plenary talks and keynote address at various National and International institutions, symposia, conferences etc. He is also advisory committee member for international conferences. Professor Ravindra has organized short-term refresher courses, workshops, conferences and seminars.

Professor Pogaku Ravindra's primary research interest is to develop the sustainable process for bioconversion of lignocellulose into renewable energy and bio-chemicals, by bridging the gap between research laboratories and industries. Professor Ravindra's focus is on the Green engineering and technology for sustainable development of the society.

Contents

Part I
Advances in Biochemical Engineering

Chapter 1
A Review on the Empty Fruit Bunch Composting: Life Cycle Analysis and the Effect of Amendment(s)

A.Y. Zahrim, T. Asis, M.A. Hashim, T.M.T.M.A. Al-Mizi, and P. Ravindra

Introduction

Palm oil industry significantly contributes to the national economy in Malaysia and currently accounts for RM53 billion (Kabbashi et al. 2014). The empty fruit bunches (EFB) of oil palm is one of the major wastes from oil palm industry. It is reported that about 3.0 million tons of oil palm empty fruit bunch (EFB) fibers are produced every year (Sajab et al. 2013). The typical physicochemical analysis for EFB is shows in Table 1.1. Due to the fact that EFB are generated daily and every year its disposal becoming a great concern, thus sustainable technology for EFB disposal is vital to be developed. Composting is regarded as a proven technology for processing EFB from the palm oil mill (Zahrim and Asis 2010). Composting is an accelerated bioconversion of organic matter to humic substances known as compost. Furthermore, composting could reduce the volume and initial weight of the fresh EFB by 85 % and 50 % respectively (Saletes et al. 2004).

The compost can be applied as a soil amendment (Yaser et al. 2007) or mulching as well other non-agricultural usage such as biofilter. Recently, indigenous micro-organisms from EFB compost were developed into advanced or multifunctional biofertilizer products (Phua et al. 2012). Among others, the quality of compost depends on the raw material itself (Zahrim and Asis 2010) as well as turning frequency (Tiquia et al. 2002).

A.Y. Zahrim (✉) • P. Ravindra
Faculty of Engineering, Universiti Malaysia Sabah, Sabah, Malaysia
e-mail: zahrim@ums.edu.my

T. Asis • M.A. Hashim • T.M.T.M.A. Al-Mizi (✉)
Prestige Central Management, Sdn. Bhd., Selangnor, Malaysia

P. Ravindra (ed.), *Advances in Bioprocess Technology*,
DOI 10.1007/978-3-319-17915-5_1

Table 1.1 Typical physicochemical analysis of EFB

Parameters	Value	Ref.
Moisture, %	58–65	Zahrim and Asis (2010)
Cellulose, %	46	Saletes et al. (2004)
Lignin, %	16.5	Saletes et al. (2004)
C/N	45–70	Saletes et al. (2004)
N, % on dry	1.1–1.2	Zahrim and Asis (2010)
P_2O_5, % on dry	0.05–2.6	Zahrim and Asis (2010)
K_2O, % on dry	2.4–2.7	Zahrim and Asis (2010)
MgO, % on dry	0.4–0.5	Zahrim and Asis (2010)

Life Cycle Assessment (LCA)

Life cycle assessment (LCA) has been used to evaluate and compare the impacts of different waste disposal scenarios, including composting. By using a common metric, LCA methodology allows for the quantification and comparison of environmental impacts between stages of a product or service throughout its life cycle, including raw material acquisition, processing, distribution, use, and end of life. Several LCA studies have reported that composting is more advantageous, i.e. less environmental impacts, than other organic waste disposal scenarios, such as landfill and incineration (Saer et al. 2013).

Stichnothe and Schuchardt (2010) studied a detailed life cycle model has been used to calculate the environmental impacts of POME and EFB treatment. The authors investigated several options, i.e. (1) dumping EFB and storing POME and ponds, (2) returning EFB to the plantation and POME as before, (3) using EFB and POME for co-composting and returning the produced compost to the plantation, (4) generating biogas from POME and thereafter as in (3). The sensitivity analysis has been carried out in order to estimate the influence of good and poor management practice on the environmental performance. From this study, Stichnothe and Schuchardt (2010) stated that the main contributor to the Global Warming Potential (GWP) is methane from POME and EFB dumping. The GWP of palm oil mill waste treatment can be reduced from 245 kg CO_{2eq} per ton FFB to up to 5 kg CO_{2eq} per ton FFB due to the reduction of methane emissions and nutrient recycling. Co-composting of POME and EFB leads to considerable nutrient recovery, in addition to GWP reduction (Stichnothe and Schuchardt 2010). Recently, Chiew and Shimada (2013) analysed seven technologies for EFB management: ethanol production, methane recovery, briquette production, biofuel for combined heat and power (CHP) plants, composting, medium density fiberboard (MDF) production, and pulp and paper production. The authors reported that the methane recovery and composting are more environmentally friendly than other technologies, as measured by reduction of greenhouse gas emissions (Chiew and Shimada 2013). In another study, Norhasmillah et al. (2013) compared the life cycle inventory (LCI) obtained from three commercial oil palm biomass composting projects in Malaysia which use the open windrow composting system. Interestingly, the authors found that composting saved 65 % of time required for a complete degradation of POME when compared to ponding system, and 89 % of time required for a complete degradation of EFB compared to mulching (Norhasmillah et al. 2013).

The advantages of EFB composting not only reduces environmental burdens; it also leads to net environmental benefit regarding most environmental impact categories, e.g., acidification potential, eutrophication potential, ozone layer depletion potential, etc. due to the avoided emissions from inorganic fertilizer production (Stichnothe and Schuchardt 2010). Yoshizaki et al. (2013) reported that the compost production using shredded empty fruit bunch (EFB) and POME anaerobic sludge obtained from the anaerobic digester is equivalent of 579 tonnes, 151 tonnes and 761 tonnes per year of nitrogen, phosphorus and potassium respectively (Yoshizaki et al. 2013).

A Review on the Effect of Different Amendment(s)

Rapid decomposition of empty fruit bunch (EFB) can be obtained by adding suitable material(s) such as animal waste and palm oil mill effluent (Table 1.2). The effects of composting EFB alone, EFB-poultry layer deep-litter-urea, and EFB-poultry broiler floor-litter-urea were studied by Thambirajah and Kuthubutheen (1989). The initial C: N ratios of the three mixtures were 40:1, 33:1 and 26:1, respectively. After 8 weeks of composting the C: N ratios of the mixtures were 26:1, 17:1 and 16:1, respectively (Thambirajah and Kuthubutheen 1989). The composting of EFB alone, EFB-goat dung, EFB-cow dung and EFB-chicken manure were studied by Thambirajah et al. (1995). The initial C:N ratios (52:1, 35:1, 48:1, 47:1) for the four compost heaps were significantly reduced to 24:1, 14:1, 18:1 and 12:1, respectively, after 60 days of composting (Thambirajah et al. 1995). In this study, the maximum heap temperature of 70 °C was maintained for 3 days during composting phase. Both mesophilic and thermophilic bacteria showed consistent activity throughout the process, whereas fungal activity was completely suppressed during the peak heating phase (Thambirajah et al. 1995). Saletes et al. (2004) have added urea and/or ripe compost to amend the initial C/N of EFB. They reported that after 70 days, the compost could be considered matured. However, almost 50 % of the phosphorus, 70 % of the potassium, 45 % of the magnesium and between 10 % and 20 % of the calcium theoretically applied were lost during composting period. The authors suggested that the better distribution of the effluent applications, combined with a system to

Table 1.2 Mass of each raw material in the compost heap

	Banana skin, kg	EFB, kg	Total initial compost, kg	% Banana skin
H0	0	100.0	100	0.0
H5	5.3	100.0	105.3	5.0
H10	11.1	100.0	111.1	10.0

recover the leachings, should substantially reduce these losses, while maintaining suitable humidity for microbial degradation (Saletes et al. 2004).

Schuchardt et al. (2005) reported that the conventional ponding system of palm oil mill effluent (POME) treatment is not only contribute highest pollution of the environment also the system with the lowest profit. The authors recommended utilisation of nutrients from POME and EFB for composting process (Schuchardt et al. 2005). The investigation of co-composting EFB with partially treated palm oil mill effluent (POME) was carried out by Baharuddin et al. (2009). The temperature was increased up to 58.5 °C at day three of treatment, after that fluctuated between 50 and 62 °C and then decreased in the latter stage of the process. The pH of the system (7.75–8.10) did not vary significantly during the treatment period while moisture content was reduced horn 65–75 % to about 60 % at the end of the treatment. The initial C/N ratio of 45 was significantly reduced to 12 after 60 days of composting. The final cured compost contained a considerable amount of nutrients (carbon, nitrogen, phosphorus, potassium, calcium, magnesium, sulfur and iron) and trace amounts of manganese, zinc, copper. In addition, very low levels of heavy metals were detected in the compost. The number of bacteria involved in the composting process was decreased at the end of the composting period. The compost product may useful in palm oil plantation as fertilizer and soil amendment (Baharuddin et al. 2009).

The co-composting of pressed-shredded EFB and palm oil mill effluent (POME) anaerobic sludge from 500 m^3 closed anaerobic methane digested tank was investigated by Baharuddin et al. (2010). High nitrogen and nutrients content were observed in the POME anaerobic sludge. The sludge was subjected to the pressed-shredded EFB to accelerate the co-composting treatment. The co-composting treatment was completed in a short time within 40 days with a final C/N ratio of 12.4. The co-composting process exhibited a higher temperature (60–67 °C) in the thermophilic phase followed by curing phase after 4 weeks of treatment. Meanwhile, pH of the composting pile (8.1–8.6) was almost constant during the process and moisture content was reduced from 64.5 % (initial treatment) to 52.0 % (final matured compost). The use of pressed-shredded EFB as a main carbon source and bulking agent contributed to the optimum oxygen level in the composting piles (10–15 %). The biodegradation of composting materials is shown by the reduction of cellulose (34.0 %) and hemicellulose (27.0 %) content towards the end of treatment. In addition, considerable amount of nutrients and low level of heavy metals were detected in the final matured compost (Baharuddin et al. 2010). Baharuddin et al. (2011) reported that shredding-pressing treatment on EFB gave better results in removing the debris and silica bodies as compared to only shredding treatment. With the aspiration of reducing the composting yard area and energy consumption, the production of non-shredded EFB-POME compost in industrial scale windrow was reported by Zahrim and Asis (2010). The authors reported that the total composting time including preparation of the windrow was about 40–45 days and the windrow was turned every 10 days. Compost quality i.e. moisture content, pH, nitrogen and other nutrients, was evaluated during 6 months operations, and it was found that the nutrient in compost is acceptable for soil conditioner (Zahrim and Asis 2010). A study on EFB-POME- non-food cassava starch was carried out by Mohammad et al. (2015). The authors reported that the total period of composting was 35–40 days and found that the 2-mm particle size of EFB was most suitable to grow microbes, gave the highest protein of 85 g kg^{-1} and degraded fastest (lowest C/N ratio of 16). In addition, substrate ratio 1:3 (EFB and POME) and pH 5.0 were found to be favorable for mature compost (Mohammad et al. 2015).

Decanter cake (DC) is generated by palm oil milling plant from three-phase CPO purification. The production rate of DC is about 4–5 wt% of fresh fruit bunch processed and consuming a lot of space. Moreover when dried, the DC could becoming fire hazard and contribute toward increasing the amount of suspended particles in the vicinities of mills (Dewayanto et al. 2014). Composting EFB-decanter cake (DC)-palm oil mill effluent (POME) was reported by Yahya et al. (2010). The addition of decanter cake slurry has accelerated the composting process of the EFB. The C/N ratio after 51 days for the mature compost with the decanter cake slurry was 18.65 while that of the matured compost without the decanter cake slurry remained high at 28.96. The compost formed from the addition of decanter cake (DC) to EFB and POME had 46.4 % nitrogen, 17.9 % phosphorus, 17.7 % potassium and 23.1 % calcium more than that without decanter cake (Yahya et al. 2010). Nutongkaew et al. (2014a) investigated the composting of EFB-palm oil mill sludge (POMS)-DC. The compost appeared dark brown in color, crumbly, attained an ambient temperature and had the C/N ratio of 11:1 after 40 days fermentation, indicating the maturity of the compost. The authors also reported that the compost quality complied with the national compost standard set by the Ministry of Agriculture, Thailand (Nutongkaew et al. 2014a). In another study, Nutongkaew et al. (2014b) investigated a composting study on EFB-DC-palm oil mill biogas sludge (POMS)-palm oil fuel ash (POFA). The authors found that the compost piles turned dark brown and attained an ambient temperature after 40 days incubation. The pH values were stable in the range of 6.9–7.8 throughout the process whereas the moisture content tended to decrease till the end with the final value around 30 %. After 60 day's incubation, the mixture ratio of POMS:PEFB: DC at 2:1:1 with the addition of biogas effluent gave the highest quality of the compost. Its nitrogen content was 31.75 % higher than the other treatments that may be a result of growth of ink cap mushroom (*Coprinus* sp.). This is the first report on the occurrence of this mushroom during composting. In addition, its nutrients (3.26 % N, 0.84 % P and 2.03 % K) were higher than the level of the Organic Fertilizer Standard (Nutongkaew et al. 2014b). However, in another study Kananam et al. (2011) reported that the use of decanter sludge did not have an effect on any biochemical conditions of either aerobic or anaerobic EFB composting. Moreover, the oil palm EFB compost with decanter sludge in an aerobic condition completed within 30 days whereas compost in the anaerobic condition failed to complete composting within 90 days. By adding red soil to the compost pile, it does not affected the composting time, but it reduced the odour generated from the pile (Kananam et al. 2011). In another study, Kabbashi et al. (2014) studied composting of EFB-POME-DC-sawdust. The maturity of the composting could be reflected by the best C/N ratio obtained was run 6 which are 16.51, pH, and germination index result of 154 %. This simple technology urge to enhance the productivity and sustainability of the Malaysian palm oil milling industry by improving the local isolated fungal strains and increasing composting utilization (Kabbashi et al. 2014).

Addition of recycled paper mill sludge (RPMS) to EFB for the production of compost was investigated by Rosazlin et al. (2011). Then, the EFB-RPMS compost mixtures were evaluated for physical, chemical, phytotoxicity and short term plant growth effects. These composts mixtures had no toxicity effects on plants, had 100 % seed germination, high in nutrient contents, low in C/N ratio and had fine particle size of <18 mm. The concentrations of heavy metals were also within the recommended level of the Council of European Communities (CEC) for compost (Rosazlin et al. 2011). Composting oil palm wastes (EFB-frond-trunk) with sewage sludge was carried out by Kala et al. (2009). Shredded oil palm wastes were mixed with sewage sludge in three different ratios (1:0, 3:1 and 4:1 ratio). Oil palm wastes with sewage sludge at 4:1 ratio was found to be the most optimum compost as potting media for ornamental plants because of its texture suitable for potting media, not stringent or stiff, had high nutrient contents (2.05 % N, 0.640 % P, 1.39 % K, 0.705 % Ca, 0.229 % Mg), pH 6.2 and low C/N ratio, 19 (Kala et al. 2009). The efficiency of EFB-frond- poultry litter composting was investigated by Vakili et al. (2012). From the study, the 1:3 ratio of EFB-frond and poultry litter had the lowest C/N ratio, TOC and the highest value of TKN, 18, 27 % and 1.48 %), respectively (Vakili et al. 2012).

An investigation of several organic materials in Malaysia as additives/amendments for composting EFB was carried out by Chai et al. (2013). The authors found that the organic waste materials with a C/N ratio of <30 can be applied as a nitrogen source in EFB co-composting. The outcome of this study suggested that the percentage of EFB ranged between 50 % and 60 %, which is considered as the ideal mixing ratio in EFB co-composting (Chai et al. 2013). Talib et al. (2014) the composting of EFB-rabbit manure using forced-aeration system and reported that aeration rate of 0.26 L min^{-1} dry matter^{-1} provided enough oxygen level (10 %) for the rest of composting period, showing 40.5 % of OM reduction that is better than other aeration rates (Talib et al. 2014). From the above review, selection and dosage of amendments are critical for the enhancement of EFB composting. In the next section, a new amendment for EFB composting i.e. banana skins is discussed.

Case Study: Addition of Banana Skin as Amendment

As one of the most consumed fruits in the world, banana is a very common fruit. The main banana residue is the fruit skin, which accounts for 30–40 % of the total fruit weight. It was reported that several tons of banana peels are produced daily in small-medium food processing industry, marketplaces, household garbage and restaurants (Mohammed and Chong 2013). In Malaysia, banana skins have not been fully utilized for production of useful by-products.

Co-composting of banana skins with empty fruit bunch could reduce waste management problems and conserve plant nutrients. Nasreen and Qazi (2012) investigated composting of banana skins in glass jars. The authors found that the seed germination indices for the compost is 63 %, indicated the conversion of the wastes into value added phytotoxin free fertilizer, which can escalate the agricultural output (Nasreen and Qazi 2012). In another study, Kalemelawa et al. (2012) evaluated the efficacy of aerobic and anaerobic composting of inoculated banana peels, and assess the agronomic value of banana peel-based compost. The study suggested that the final composts contained high K (>100 g kg^{-1}) and TN (>2 %), indicating high potential as a source of K and N fertilizer (Kalemelawa et al. 2012). Recently, a co-composting banana stem-swine manure-eucalyptus bark was carried out by Deng et al. (2014). The authors reported that when C/N ratio of the composting material was 25–27, the heaps of compost were the highest in temperature, reaching up to 56 °C or higher and maintained the high temperature for 10–11 days, respectively, and they were also higher in content of nutrients, and the pot experiment also shows that composts had a certain growth-promoting effect on banana seedlings (Deng et al. 2014).

The aim of this study was to evaluate the effect of amount of banana skin on the heap temperature and final nutrient of EFB-POME compost. Results from this study could give valuable insight on the effect of banana skin as an amendment for enhancing composting performance.

Methodology

The palm oil mill effluent (POME) from an anaerobic digestion pond No. 1 and empty fruit bunch were collected from Merotai Palm Oil Mill, Tawau, Sabah. Banana skin was collected from various small food stall around Tawau, Sabah. One hundred kilograms of EFB was mixed with different percentage of banana skins (BS) as indicated in Table 1.2. The 28 m^3/h POME (from day 0 to day 35) was sprayed carefully so that the banana peel does leached from the compost heap. The composting process was performed over the course of 45 days. Turning of the compost was carried out at day 5, 10, 15, 20, 25, 30 and 35. This study was carried out at Merotai Composting Plant, Tawau, Sabah. All experiments were carried out in duplicate.

Physicochemical Analysis

During the whole co-composting process, the temperatures at three points along the length of the middle of composting mixtures (5 cm, 10 cm deep and at the core of the compost heap) were measured almost daily. The three readings per composting mixture were averaged. Ambient temperature was also recorded. Sampling for nutrients analysis was made on ten randomly selected points on each compost heap. The chemical analysis was carried out by Sime Darby Research Sdn. Bhd. Method used for moisture content burn at 103 °C. The nitrogen content was analyzed using distillation method (MS 677: Part 1–VIII:1980). The organic matter (OM) content (volatile solids) was determined with a furnace at 550 °C (MS 417: Part 8:1997). The total P was estimated using spectrophotometric molybdovanadophosphate (MS417: Part 4:1994). The potassium and magnesium were determined using atomic absorption spectroscopy (AAS) (MS417: Part 5 and Part 6:1994).

Result and Discussion

Temperature Profile

The temperature monitoring is a very simple way to follow the progress of composting (Yaser et al. 2007). In this study, the oxygen content for all heaps i.e. H0, H5 and H10, is maintained around 19–20 %, which is necessary for maximum biodegradation (Baharuddin et al. 2009). The trends of the temperatures in the H0, H5 and H10 were not similar indicating different effect of percentage of BS addition. Stentiford (1996) suggested that temperatures higher than 55 °C maximized sanitation, those between 45 and 55 °C maximized the biodegradation rates, and between 35 and 40 °C maximized microbial diversity in the composting process. Throughout the experimental period, the ambient temperature ranged from 24 to 29 °C. During the composting period, at day 3, the H10 achieved the maximum temperature of 47.5 °C while the H0 and H5 only achieved temperature of 36.0 and 40.5 °C (Fig. 1.1). The sharp increase in temperature to greater than 45 °C for H10 in

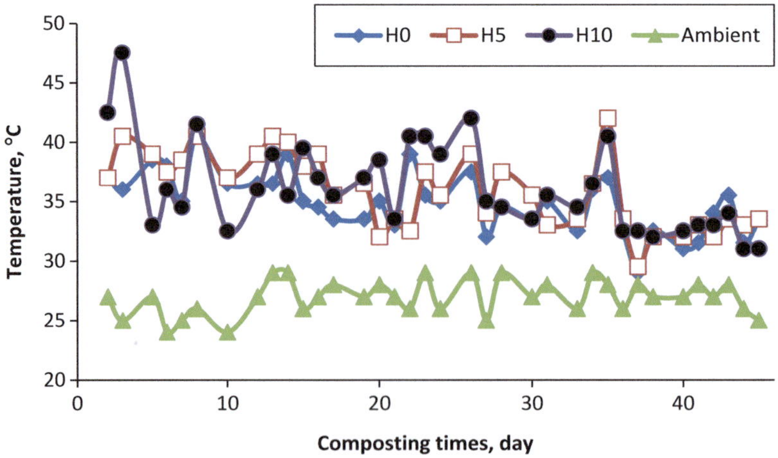

Fig. 1.1 Temperature during the composting EFB-POME-BS as affected by different percentage of BS. Ambient air temperature is also indicated and was the same for all treatments

the first 3 days of composting reflects the rapid initiation of the composting process (Placha et al. 2013; Stentiford 1996). Both H0 and H5 did not achieve temperature for maximum biodegradation rates. The maximum temperature for the H0 is 40.5 °C and was achieved at day 8 while the H5's maximum temperature is 42 °C (day 35). However, it can be seen also that the large drop in temperature also experienced by the H10 might be due to the limited ability of BS to adsorb heat.

From Fig. 1.1, the temperature in H10 can be raised again by turning the heap. After biodegradation process, the structure of compost e.g. porosity, may change. The turning operation improves the aeration of the compost materials by increasing its porosity. Therefore, the availability of more air within the composting materials favors the renewal of microbial activities which indirectly increase the decomposition process of the composting materials (Yahya et al. 2010). In this study, the effect of turning was very significant especially for H10. It can be seen that at day 5, the temperature drop to 33 °C, but after turning process, the temperature is rise again to 41.5 °C. Similar trend also found after each turning process until after day 35 (i.e. last turning was carried out), the temperature for the H10 is around 34 °C due to lack of carbon that has been fully utilized by the microorganisms and consequently, minimize the metabolism and the heat production (Zahrim and Asis 2010).

Moisture Content, pH and Nutrients

The final physicochemical compositions of the final compost are shown in Table 1.3. From the table, it could be seen that the value for moisture content, pH, phosphorus, and potassium of the compost material with banana skin slurry are greater than un-amended compost (control). In comparison with others studies (Table 1.3), the potassium in this study is greater than others, indicating positive improvement by adding banana skin as amendments.

Increasing of the final moisture content due to addition of banana skin might be due to the ability of banana skin to adsorb water. Due to the fact that the protein content in banana is around 8–11 % (Happi Emaga et al. 2007) while only 2–3 % for EFB (Abdullah et al. 2009), as expected, the final pH (protein degradation) for both H5 and H10 were higher than the control.

The addition of banana skin contributes little effect on the final N content based on R^2 value in Eq. (1.1). The best equation to describe the correlation for mass of banana skin (x) and nitrogen content is as follows:

$$\begin{aligned}\text{Nitrogen content, }\% &= 0.066\,(\%\text{ banana skin}) + 1.5463,\\ &\quad\text{Coefficient correlation, } R^2 = 0.60\end{aligned} \tag{1.1}$$

From Table 1.3, it is suggested the addition of sludge from anaerobic digestion system to improve the N content (Yaser et al. 2007; Baharuddin et al. 2010;

Table 1.3 Nutrients content in EFB compost for several studies

Amendment(s) material	Composting time, days	Moisture content, %	pH	C/N	N, %	P, %	K, %	Ref.
Partially treated POME	45	75	6.8	27	1.6	0.2	2.7	This study
Partially treated POME + 5 % banana skin	45	79	8.3	29	1.5	1.4	3.4	This study
Partially treated POME + 10 % banana skin	45	79	8.5	27	1.6	1.7	3.0	This study
Anaerobic digestion sludge + decanter cake	60	60	7.8	13	3.3	0.9	2.0	Nutongkaew et al. (2014b)
Fresh POME + decanter cake	52	58	8.8	19	1.7	1.0	2.5	Yahya et al. (2010)
Anaerobic digestion sludge	40	52	8.1	12	2.3	1.4	2.8	Baharuddin et al. (2010)
Fresh POME	45	55	7.9	20	1.9	0.6	2.0	Zahrim and Asis (2010)
Partially treated POME	60	61	8.1	13	2.2	1.3	2.8	Baharuddin et al. (2009)
POME + Wheat flour	60	ND	5.6	20	ND	ND	ND	Kabbashi et al. (2007)
Without any amendment material	60	65	8.5	24	1.7	ND	ND	Thambirajah et al. (1995)
Goat dung	60	65	8.5	14	2.5	ND	ND	Thambirajah et al. (1995)
Cow dung	60	65	8.5	18	1.9	ND	ND	Thambirajah et al. (1995)
Chicken manure	60	65	8.5	12	2.0	ND	ND	Thambirajah et al. (1995)
Sewage sludge + oil palm trunk, frond	90	60	6.2	19	2.1	0.6	1.4	Kala et al. (2009)

ND = not determined

Nutongkaew et al. 2014b). While the best equation to describe the correlation for mass of banana skin (x) and K_2O is given as:

$$\begin{aligned}\%K_2O &= 0.0277(\%\text{banana skin}) + 2.9018,\\ \text{Coefficient correlation, } R^2 &= 0.13\end{aligned} \tag{1.2}$$

The banana skin increases the potassium content from 2.7 % to around 3.0 %, however the addition of more than 5 % BS did not increases the K content might be due to the solubilisation of K in compost leachate as shown in Eq. (1.2). However, the addition of banana skin highly affected the final P_2O_5 and MgO (data not shown) as shown in Eqs. (1.3) and (1.4).

$$\begin{aligned}\%\,P_2O_5 &= 0.1401\,(\%\text{ banana skin}) + 0.3962,\\ \text{Coefficient correlation, } R^2 &= 0.88\end{aligned} \tag{1.3}$$

$$\begin{aligned}\%\,MgO &= 0.1599\,(\%\text{ banana skin}) + 0.2802,\\ \text{Coefficient correlation, } R^2 &= 0.99\end{aligned} \tag{1.4}$$

Conclusion

(1) LCA study by various investigators confirmed that composting is more really environmentally friendly based on the greenhouse gas reduction measurement.
(2) The EFB composting with suitable amendments give acceptable quality of compost and accelerate the process to less than 60 days.
(3) In the case study, the addition of banana skin could enhance rapid EFB decomposition and increase nutrients such as P and K.

Acknowledgement A.Y. Zahrim would like to thank to Universiti Malaysia Sabah and Prestige Central Management Sdn. Bhd. for giving permission in conducting 1 year industrial internship.

References

Abdullah, S. S. S., Hassan, M. A., Shirai, Y., Funaoka, M., Shinano, T., & Idris, A. (2009). Effect of solvent pretreatment on lignophenol production from oil palm empty fruit bunch fibres. *Journal of Oil Palm Research, 21*, 700–709.

Baharuddin, A. S., Hock, L. S., Yusof, M. Z. M., Rahman, N. A. A., Shah, U. K. M., Hassan, M. A., et al. (2010). Effects of palm oil mill effluent (POME) anaerobic sludge from 500 m^3 of closed anaerobic methane digested tank on pressed-shredded empty fruit bunch (EFB) composting process. *African Journal of Biotechnology, 9*, 2427–2436.

Baharuddin, A. S., Rahman, N. A. A., Shah, U. K. M., Hassan, M. A., Wakisaka, M., & Shirai, Y. (2011). Evaluation of pressed shredded empty fruit bunch (EFB)-palm oil mill effluent (POME) anaerobic sludge based compost using Fourier transform infrared (FTIR) and nuclear magnetic resonance (NMR) analysis. *African Journal of Biotechnology, 10*, 8082–8089.

Baharuddin, A. S., Wakisaka, M., Shirai, Y., Abd-Aziz, S., Abdul-Rahman, N. A., & Hassan, M. A. (2009). Co-composting of empty fruit bunches and partially treated palm oil mill effluents in pilot scale. *International Journal of Agricultural Research, 4*, 69–78.

Chai, E. W., H'ng, P. S., Peng, S. H., Wan-Azha, W. M., Chin, K. L., Chow, M. J., et al. (2013). Compost feedstock characteristics and ratio modelling for organic waste materials co-composting in Malaysia. *Environmental Technology, 34*, 2859–2866.

Chiew, Y. L., & Shimada, S. (2013). Current state and environmental impact assessment for utilizing oil palm empty fruit bunches for fuel, fiber and fertilizer—A case study of Malaysia. *Biomass and Bioenergy, 51*, 109–124.

Deng, X. K., Dong, C. M., Xin, K., Zhao, N., Yang, J. M., & Ruan, Y. Z. (2014). Comparative study on effects of composting of banana stem, eucalyptus bark and swine manure relative to ratio of ingredients. *Journal of Ecology and Rural Environment, 30*, 658–663.

Dewayanto, N., Isha, R., & Nordin, M. R. (2014). Use of palm oil decanter cake as a new substrate for the production of bio-oil by vacuum pyrolysis. *Energy Conversion and Management, 86*, 226–232.

Happi Emaga, T., Andrianaivo, R. H., Wathelet, B., Tchango, J. T., & Paquot, M. (2007). Effects of the stage of maturation and varieties on the chemical composition of banana and plantain peels. *Food Chemistry, 103*, 590–600.

Kabbashi, N. A., Alam, Z., & Ainuddin, M. (2007). Bio-composting process development by SSF for utilization agro-industrial wastes. In *Third Kuala Lumpur international conference on biomedical engineering, 2006* (Vol. 15, pp. 464–468).

Kabbashi, N. A., Noruldin, N., & Mirghani, M. E. (2014). Studies of locally effective microorganisms (EM) to enhance the degradation of empty fruit bunches, pome and cake during composting process. *Journal of Pure and Applied Microbiology, 8*, 815–819.

Kala, D. R., Rosenani, A. B., Fauziah, C. I., & Thohirah, L. A. (2009). Composting oil palm wastes and sewage sludge for use in potting media of ornamental plants. *Malaysian Journal of Soil Science, 13*, 77–91.

Kalemelawa, F., Nishihara, E., Endo, T., Ahmad, Z., Yeasmin, R., Tenywa, M. M., et al. (2012). An evaluation of aerobic and anaerobic composting of banana peels treated with different inoculums for soil nutrient replenishment. *Bioresource Technology, 126*, 375–382.

Kananam, W., Suksaroj, T. T., & Suksaroj, C. (2011). Biochemical changes during oil palm (*Elaeis guineensis*) empty fruit bunches composting with decanter sludge and chicken manure. *ScienceAsia, 37*, 17–23.

Mohammad, N., Alam, M. Z., & Kabashi, N. A. (2015). Optimization of effective composting process of oil palm industrial waste by lignocellulolytic fungi. *Journal of Material Cycles and Waste Management, 17*, 91–98.

Mohammed, R. R., & Chong, M. F. (2013). Treatment and decolorization of biologically treated Palm Oil Mill Effluent (POME) using banana peel as novel biosorbent. *Journal of Environmental Management, 132*, 237–249.

Nasreen, Z., & Qazi, J. I. (2012). Lab scale composting of fruits and vegetable waste at elevated temperature and forced aeration. *Pakistan Journal of Zoology, 44*, 1285–1290.

Norhasmillah, A. H., Puah, C. W., Ibrahim, N. A., Baharuddin, A. S., & Choo, Y. M. (2013). Life cycle inventory of the commercial production of compost from oil palm biomass: A case study. *Environment, Development and Sustainability, 15*, 1663–1670.

Nutongkaew, T., Duangsuwan, W., Prasertsan, S., & Prasertsan, P. (2014a). Effect of inoculum size on production of compost and enzymes from palm oil mill biogas sludge mixed with shredded palm empty fruit bunches and decanter cake. *Songklanakarin Journal of Science and Technology, 36*, 275–281.

Nutongkaew, T., Duangsuwan, W., Prasertsan, S., & Prasertsan, P. (2014b). Physicochemical and biochemical changes during composting of different mixing ratios of biogas sludge with palm oil mill wastes and biogas effluent. *Journal of Material Cycles and Waste Management, 16*, 131–140.

Phua, C. K. H., Abdul Wahid, A. N., & Abdul Rahim, K. (2012). Development of multifunctional biofertilizer formulation from indigenous microorganisms and evaluation of their N2-fixing capabilities on Chinese cabbage using 15N tracer technique. *Pertanika Journal of Tropical Agricultural Science, 35*, 673–679.

Placha, D., Raclavska, H., Kucerova, M., & Kucharova, J. (2013). Volatile fatty acid evolution in biomass mixture composts prepared in open and closed bioreactors. *Waste Management, 33*, 1104–1112.

Rosazlin, A., Fauziah, C. I., Wan Rasidah, K., Rosenani, A. B., & Kala, D. R. (2011). Assessment on the quality of Recycled Paper Mill Sludge mixed with oil palm empty fruit bunch compost. *Malaysian Journal of Soil Science, 15*, 49–62.

Saer, A., Lansing, S., Davitt, N. H., & Graves, R. E. (2013). Life cycle assessment of a food waste composting system: environmental impact hotspots. *Journal of Cleaner Production, 52*, 234–244.

Sajab, M. S., Chia, C. H., Zakaria, S., & Khiew, P. S. (2013). Cationic and anionic modifications of oil palm empty fruit bunch fibers for the removal of dyes from aqueous solutions. *Bioresource Technology, 128*, 571–577.

Saletes, S., Siregar, F. A., Caliman, J. P., & Liwang, T. (2004). Ligno-cellulose composting: Case study on monitoring oil palm residuals. *Compost Science and Utilization, 12*, 372–382.

Schuchardt, F., Wulfert, K., & Darnoko, D. (2005). New process for combined treatment of waste (EFB) and waste water (POME) from palm oil mills—Technical, economical and ecological aspects. *Landbauforschung Volkenrode, 55*, 47–60.

Stentiford, E. T. (1996). *Composting control: Principles and practice*. London, UK: Chapman & Hall.

Stichnothe, H., & Schuchardt, F. (2010). Comparison of different treatment options for palm oil production waste on a life cycle basis. *International Journal of Life Cycle Assessment, 15*, 907–915.

Talib, A. T., Mokhtar, M. N., Baharuddin, A. S., & Sulaiman, A. (2014). Effects of aeration rate on degradation process of oil palm empty fruit bunch with kinetic-dynamic modeling. *Bioresource Technology, 169*, 428–438.

Thambirajah, J. J., & Kuthubutheen, A. J. (1989). Composting of palm press fiber. *Biological Wastes, 27*, 257–269.

Thambirajah, J. J., Zulkali, M. D., & Hashim, M. A. (1995). Microbiological and biochemical changes during the composting of oil palm empty fruit bunches—Effect of nitrogen supplementation on the substrate. *Bioresource Technology, 52*, 133–144.

Tiquia, S. M., Richard, T. L., & Honeyman, M. S. (2002). Carbon, nutrient, and mass loss during composting. *Nutrient Cycling in Agroecosystems, 62*, 15–24.

Vakili, M., Haque, A. A. M., Hosseini, S. M., & Salamatinia, B. (2012). Evaluation of maturation and stability some indexes of mixed palm oil biowaste composting process and poultry litter. *World Applied Sciences Journal, 19*, 234–240.

Yahya, A., Sye, C. P., Ishola, T. A., & Suryanto, H. (2010). Effect of adding palm oil mill decanter cake slurry with regular turning operation on the composting process and quality of compost from oil palm empty fruit bunches. *Bioresource Technology, 101*, 8736–8741.

Yaser, A. Z., Abd Rahman, R., & Kalil, M. S. (2007). Co-composting of palm oil mill sludge-sawdust. *Pakistan Journal of Biological Science, 10*, 4473–4478.

Yoshizaki, T., Shirai, Y., Hassan, M. A., Baharuddin, A. S., Raja Abdullah, N. M., Sulaiman, A., et al. (2013). Improved economic viability of integrated biogas energy and compost production for sustainable palm oil mill management. *Journal of Cleaner Production, 44*, 1–7.

Zahrim, A. Y., & Asis, T. (2010). Production of non shredded empty fruit bunch semi-compost. *The Journal of the Institution of Engineers, Malaysia, 71*, 11–17.

Chapter 2
Effect of Adaptation of *Acidothiobacillus ferrooxidans* on Ferrous Oxidation and Nickel Leaching Efficiency

Pogaku Ravindra, Bharathi Kodli, and V. Prasarnth Raaj Veera Rao

Introduction

It is known that microorganisms possess ability to tolerate the presence of particular element in growth medium through development of adaptation of cells over a prolonged period (Elzeky and Attia 1995). In natural environment, the metal ions are accumulated in to solutions and confer metal resistance due to acclimatization of the bacteria to that environment. This property was utilized to enhance the metal tolerance as well as the leaching ability of microorganisms (Garcia and da'Silva 1991; Ballester et al. 1990; Bharathi et al. 2004).

Bacterial adaptation to ores and floatation concentrates prior to bioleaching was found to play an important role in enhancing the leaching rates (Li and Ke 2001; Das et al. 1997). Adaptation to leaching environment enables the bacteria to function more efficiently in high concentrations of metals and low pH thereby enhancing the leaching efficacy of the bacterium. Sulfide mineral leaching involves the role of lixiviant Fe^{3+} ion produced from the oxidation of pyrite, which acts as the typical energy source of chemolithotrophs such as *Acidothiobacillus ferrooxidans* (*Tf*).

Tf is the preferred organism for leaching of sulfide concentrates within mesophilic temperatures (Mason and Rice 2002; Adamov et al. 1997; Torma 1997) with an unique character to derive energy from oxidation of Fe^{2+}, generating heat energy that is required for chemical reactions of sulfides. The barophilic nature of *Tf* renders the organism to get adapted to versatile mining conditions leading to enhanced leaching efficacy (Natarajan and Iwasaki 1983). Mineralogy of ores and

P. Ravindra (✉) • V.P.R. Veera Rao
Energy Research Unit, Faculty of Engineering, University Malaysia Sabah, 88400 Kota Kinabalu, Sabah, Malaysia
e-mail: ravindra@ums.edu.my; dr_ravindra@hotmail.com

B. Kodli
University College of Technology, Osmania University, Hyderabad, India
e-mail: bharathi_kodali@yahoo.com

P. Ravindra (ed.), *Advances in Bioprocess Technology*,
DOI 10.1007/978-3-319-17915-5_2

flotation concentrates plays a role in selective leaching of metals (Valix et al. 2001). Oxidation of pentlandite mineral phase is essential to selectively leach nickel (Giaveno and Donati 2001; Mehta et al. 1997, 1999).

UCIL (Uranium Corporation of India Limited), India, produces tons of copper flotation concentrate which is dumped out as waste since this concentrate is unworthy for efficient copper metallurgy due to high nickel content in it. Processing of this concentrate by conventional technologies has emerged as uneconomical and environmentally unfriendly (Bandhyopadhyay 2003). Recently, mineral biotechnology shot in to prominence in mineral engineering for the utilization of wastes and low-grade ores (Kar et al. 2003; Bosecker 2001).

Bacterial oxidation process has been established in gold and copper industries. But, application in nickel extraction from ore wastes and lower grade nickel sulfide resources is sparse (Bharathi et al. 2000, 2002; Acevado 2002). Hence, the need to improve the bacterial growth rates and enhance the efficiency of *Tf* strains to achieve better yields of nickel during leaching process should be focused (Bharathi et al. 2004).

Though the technique of adaptation of *Tf* prior to leaching is attractive, much attention has not been paid by researchers regarding ferrous oxidation ability and selective leaching of nickel by adapted cells of *Tf*. Hence, comparative studies were conducted to leach nickel by unadapted strains and strains adapted to synthetic salts of copper, nickel and copper flotation concentrate.

Materials and Method

Copper Flotation Concentrate

The representative samples of copper flotation concentrate were obtained from UCIL, Jaduguda, Chhattisgarh, India. The finely ground concentrate samples were subjected to elemental analysis using AAS. The chemical composition of the concentrate sample and different mineral phases were analyzed by X-ray diffraction spectroscopy (X-RD) (Tables 2.1 and 2.2).

Microorganism

Tf-44 and *Tf*-231 were obtained from Agharkar Research Institute, Pune, India. *Tf* strains were activated and regularly subcultured on modified 9K medium.

Table 2.1 Chemical composition of copper flotation concentrate obtained by X-RD analysis

Sample	Elements	% Composition
Copper flotation concentrate	S	31.5
	Fe	29.5
	Cu	21.4
	Ni	2.73
	Mo	1.38

Table 2.2 Presence of different mineral phases of the copper concentrate sample. Prior to and after leaching as analyzed by X-RD analysis

Sample	Major phases	Minor phases
Copper flotation concentrate (prior to leaching)	Chalcopyrite ($CuFeS_2$) Pentlandite [$(Fe,Ni)_9S_8$]	Pyrite (FeS_2), Pyrrhotite (FeS), Violarite [$(NiFe)_3S_4$]
Copper flotation concentrate (leach residue)	Chalcopyrite ($CuFeS_2$)	Pentlandite [$(FeNi)_9S_8$] Hydronium jarosite [$H_3OFe_2(SO_4)_3(OH)_6$]

The composition of the medium in g L^{-1}; ammonium sulfate-2.0, magnesium sulfate-0.5, dipotassium hydrogen orthophosphate-0.025, ferrous sulfate-40 was adjusted to pH value of 2.5 (Paknikar and Agate 1995).

Bacterial Activation and Growth

Tf cultures were maintained in 250 ml standard Erlenmeyer shake flasks on a biological incubator shaker (model: Remi-RIS-24) at 120 rpm (revolutions per minute), pH-2.5 and temperature 30 °C. For inoculation 10 % v/v cells were harvested from late exponential phase of growth. The cells were harvested by centrifugation by refrigerated centrifuge at 4 °C (Hitachi make, model-Mikro 22R) followed by resuspension in mineral salts solution. The growth studies were conducted in terms of substrate utilization (ferrous oxidation) of *Tf* strains by dichromate titration using sodium diphenylamino sulfonate as indicator (Jeffery et al. 1989; Vogel 1978).

Preadaptation Studies

For 0.3 g of powdered concentrate sample, 0 ml of 105 tartaric acid solution and 6 ml of concentrated nitric acid (HNO_3) were added and allowed to stand at room temperature over night for digestion. The solution prepared was then heated on a sand bath for 2–3 h to reduce al sulphur compounds to elemental sulphur. The undissolved residue was washed with deionized water and filtered. The concentration of the elements of interest against standard and blank containing similar acid concentrations was analyzed by Atomic Absorption Spectroscopy (Ravindra and Bharathi 2009).

Adaptation to Metals

Tf cultures were serially subcultured in $M9K^-$ medium (without ferrous sulphate) in the presence of copper and nickel salts by gradually increasing their concentration up to 20 g L^{-1}.

Adaptation to Copper Flotation Concentrate

Tf strains were preadapted to copper concentrate by continuously growing in $M9K^-$ medium (flotation concentrate in the absence of chemical supplementation of ferrous). The concentration of flotation concentrate was gradually increased from 1 g L^{-1} to 10 g L^{-1}.

Efficacy of Adaptation

The ferrous oxidizing ability was determined after every subculture. The efficacy of adaptation was tested by the oxidizing ability of adapted *Tf* strains in fresh M9K medium (Selvi et al. 1998).

Bacterial Leaching Technique

Bacterial leaching in batch operation at lab scale level was carried in 250 ml and 500 ml Erlenmeyer flasks. Aeration of the pulp in the flasks was achieved by agitation on a rotatory incubator shaker at a frequency of 140–150 rpm in order to prevent the sedimentation of solids from pulp. The ratio of the volume of the reactor to the pulp was 1:2.5. The pH, E_h of the pulp and concentration of Fe^{2+}, Ni^{2+} and Cu^{2+} in the liquid phase of the pulp was checked intermittently at fixed interval of 24 h. The temperature of shaker was kept at 30–32 °C and the pH was set to 2.3. The pH of leaching suspensions was monitored periodically and adjusted, till the pH stabilized to 2.3 using 1 N H_2SO_4. Once the pH was stabilized at set pH, medium was inoculated with the bacterial culture of exponential phase. The loss of medium due to evaporation was compensated by adding distilled water. Total contents of each flask were filtered through Whatman no. 1 filter paper. Sterile controls were kept in parallel using 1 % thymol as bactericide. The liquid filtrate was used for the analysis of Fe^{2+}, Ni^{2+} and Cu^{2+}.

Leachability Studies

Leaching experiments were conducted under optimal conditions (Bharathi and Ravindra 2006). Bacterial leaching abilities of unadapted and adapted strains of *Tf* were compared by conducting separate bioleaching tests. Uninoculated control experiments were also carried out in order to determine the contributions from chemical leaching due to aerial oxidation.

Analytical Methodology

Various metals present in the leach liquor and leach residue were analyzed by Atomic absorption spectroscopy. Different mineral phases of the concentrate were analyzed by X-ray diffraction spectroscopy.

Results and Discussion

Adaptation of Tf *Strains to Copper and Nickel*

The development of membrane-associated enzyme protecting system, which enables the cells to secure energy provision through ferrous iron oxidation, is metal specific and hence the lag phase depends on specific metal cation (Gehrke et al. 1998, 2001). Therefore, the bacterial strains tolerant to a particular metal are preferred. Figure 2.1a, b show the effect of *Tf* strains adapted to Ni^{2+}, Cu^{2+} and mixture of Ni^{2+} and Cu^{2+} at a concentration of 20 g L^{-1} each on iron utilization ability. Ni^{2+} adapted strains could sustain the ferrous oxidizing ability. But, strains adapted to copper and mixture of Ni and Cu ions lost their ability to oxidize iron when grown in modified 9K medium due to their acclimatization to the adapted environment.

Adaptation of Tf *Strains to Copper Flotation Concentrate*

Figure 2.2a, b show the ferrous oxidizing ability of concentrate adapted cells of *Tf* strains in fresh M9K growth medium. Time taken for total oxidation of ferrous (T_{ox}) was 6 days initially in the case of unadapted strain and it was increased to 9 days after subculturing three times. At the end of sixth subculture, T_{ox} could not be achieved even after 11 days of incubation. The efficiency of oxidation decreased with increased time period of adaptation. After nine subcultures, *Tf* strains failed to utilize synthetic ferrous sulfate. The loss of ferrous oxidizing ability of adapted strains indicates the efficacy of adaptation of bacterial cells to the new environment as observed by Selvi et al. (1998).

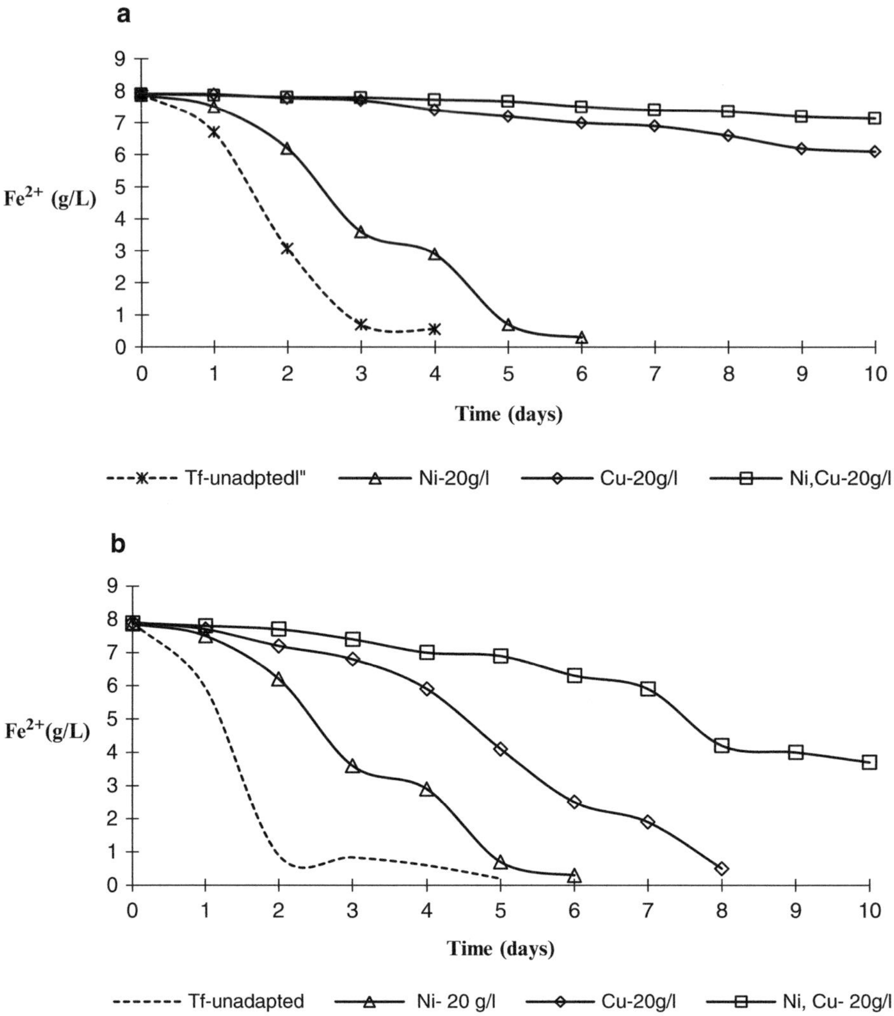

Fig. 2.1 Effect of adaptation of *Tf* onto Cu^{2+} and Ni^{2+} in M9K medium. Cultures used: (**a**) *Tf*-44, (**b**) *Tf*-234

Nickel Leaching by Metal/Concentrate Adapted Strains

Metal adapted strains have shown variable results on efficacy of Ni leaching. Table 2.3 shows the percent leachabilities of Ni obtained by Cu^{2+} tolerant, Ni^{2+} tolerant and strains tolerant to Cu^{2+} and Ni^{2+}. The % recoveries of Ni by strains adapted to copper and mixture of Cu^{2+} and Ni^{2+} were less compared to Ni^{2+} adapted strains. The percent recovery of nickel by Ni^{2+} adapted strains was improved by 10 % over that achieved by unadapted strains. Nickel recovery was significantly increased by *Tf* strains adapted to concentrate.

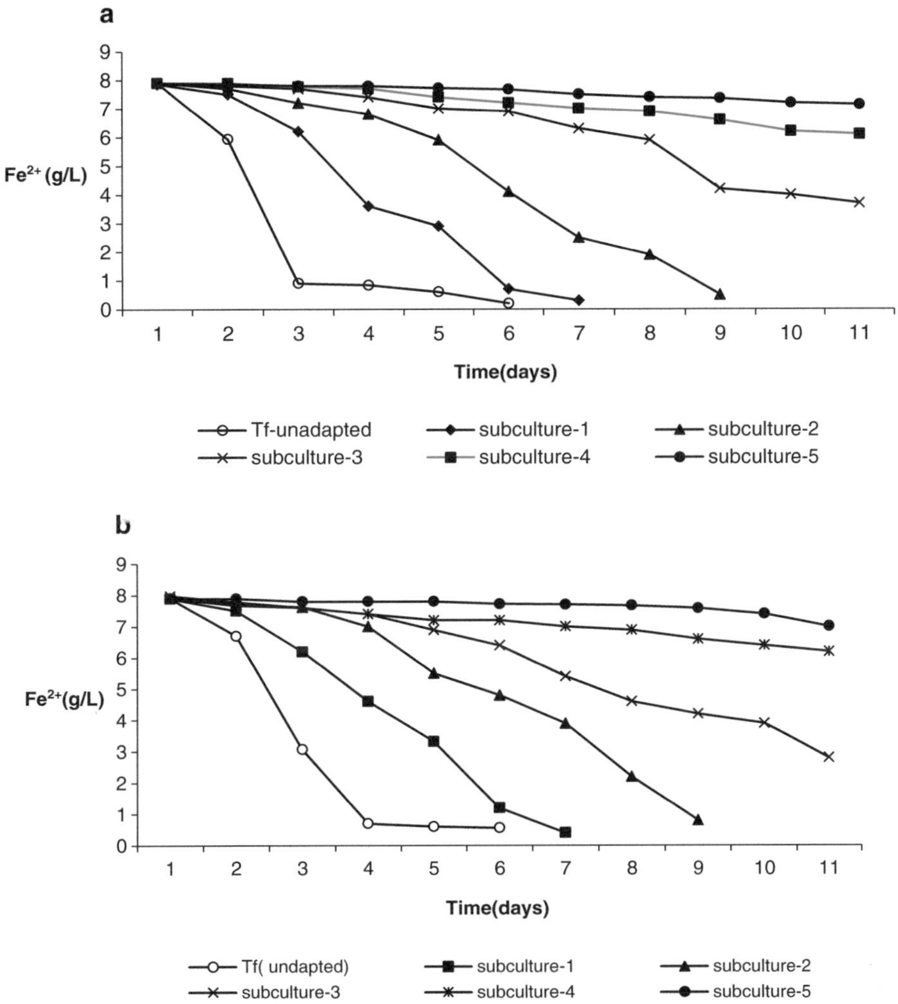

Fig. 2.2 Loss of ferrous oxidizing ability of *Tf* strains with preadaptation to copper flotation concentrate in M9K medium. Cultures used: (**a**) *Tf*-44, (**b**) *Tf*-231

Table 2.3 Percent recoveries of nickel by metal adapted and concentrate adapted strains

	% Ni leachability	
Type of *Tf* strains	*Tf*-44	*Tf*-231
Cu^{2+} adapted strains	12	10
Cu^{2+} + Ni^{2+}adapted strains	25	20
Ni^{2+} adapted strains	65	57
Concentrate adapted strains	83.4	80

Comparison of Nickel Leachabilities by Different Strains

Figure 2.3 shows the percent recovery of nickel by different strains of *Tf*. The recovery of nickel from concentrate was 85 % by strains adapted to concentrate. The percent recovery by sterile control was only 5 % showing that 80 % of the nickel was leached by bacterial action (Table 2.3). The percent leachability of controls containing bactericide 1 % thymol/Hg_2Cl_2 was lower than any bacterial system. Both the *Tf* strains adapted to high concentrations of nickel have shown improved Ni leachabilities compared to unadapted as well as strains adapted to copper. Better leachabilities were obtained by concentrate adapted strains of *Tf*. This is due to the increased tolerance of bacterial cells to high concentration of metals which affected the physiology of *Tf* cells leading to enhanced leachabilities as studied by Natarajan et al. (1994) and Puskas et al. (1980). These results coincide with the fact that the efficiency of microbiological metal recovery depends on the ability of the microorganisms to tolerate or become adapted to high concentrations of soluble compounds in the leach liquor as observed by Leduce et al. (1997) and Kai et al. (1995).

An improvement of nickel leachability by 40 % was achieved by *Tf* strains adapted to the copper flotation concentrate under optimized conditions within in a period of 15 days of residence time of leaching. Bacterial adaptation has successfully eliminated lag period of cells or initial pH rise during bioleaching process owing to increase in the secretion of iron and sulfur enzymes by the bacterial cells during adaptation as observed by Li and Ke (2001). The improvement in Ni leachabilities that were obtained by adapted strains could also be attributed to the rapid development of membrane-associated enzyme protecting system as reported by Mason and Rice (2002). Selective leaching of nickel was possible due to the development of galvanic couple between the mineral phases present in the concentrate. Upon

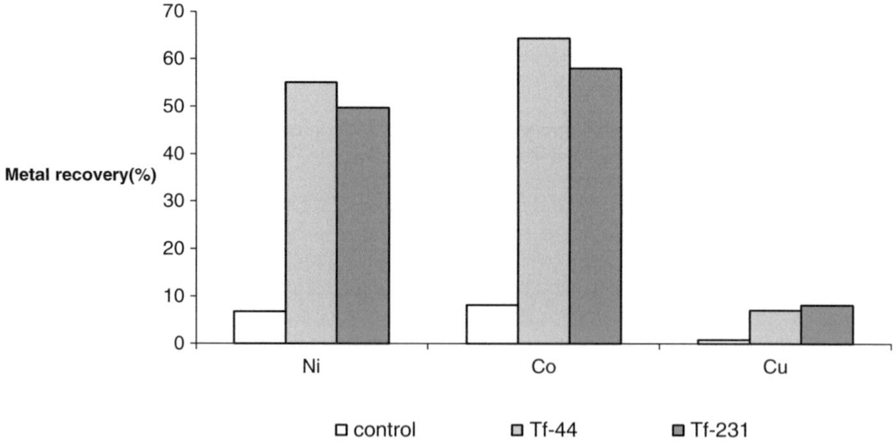

Fig. 2.3 Comparison of leachability of nickel by different strains of *Tf*. Cultures used: (**a**) *Tf*-44, (**b**) *Tf*-231

adaptation through sub culturing, Ni leaching performance of the organisms increased significantly due to preferential leaching of pentlandite from a predominantly chalcopyrite concentrate (Bharathi et al. 2008) (Table 2.2).

Conclusions

Adaptation of *Tf* cells to high concentration of nickel and flotation concentrate has resulted in increased metal tolerance levels and selective leachability of nickel. The efficacy of *Tf* strains to utilize ferrous from concentrate was enhanced with adaptation of cells to nickel, copper and copper flotation concentrate. Better nickel leachabilities were obtained by nickel adapted strains, in comparison to unadapted strains. Therefore, it can be concluded that the acclimatization of cells to the leaching environment played key role in enhancing ferrous recovery and nickel leaching performance of *Tf* strains. The order of efficiency of strains could be termed as concentrate adapted strains > Ni^{2+} adapted strains > mixture of Cu^{2+} and Ni^{2+} adapted strains.

References

Acevado, F. (2002). Present and future of biotechnology in developing countries. *Electronic Journal of Biotechnology* (online). IISN-0717-3458. www.ejb.org/content/vol3/issue3/full/4/4/index/html.

Adamov, E. V., Karavaiko, G. I., Koreshkov, N. G., Lankov, B. Y. U., & Krylova, L. N. (1997). Use of *Acidothiobacillus ferrooxidans* for bacterial oxidation of pyrrhotite concentrates. *Prikladnaya Biokhimiya i Mikrobiologiya, 33*(2), 189–193.

Ballester, A., Gonzalez, F., Blazanez, M. L., & Miller, J. L. (1990). The influence of various ions in the bioleaching of metal sulphides. *Hydrometallurgy, 23*, 221–235.

Bandhyopadhyay, A. (2003). Environmental issues in mineral processing. In V. N. Misra, G. Yadav, & K. S. Rao (Eds.), *Mineral processing and engineering* (pp. 202–206). India: Indian Institute of Chemical Engineers.

Bharathi, K., Lakshmi Narasu, M., Ravindra, P., & Bhagvanth Rao, M. (2004). Effect of biochemical reactions on enhancement rate of bioleaching. *Chemical Engineering Science, 59* (22–23), 5069–5073.

Bharathi, K., & Ravindra, P. (2006). Optimization of bacterial oxidation process parameters for selective leaching of nickel by *Acidohiobacillus ferrooxidans. International Journal of Chemical Reactor Engineering, 4*, A1. http://www.bepress.com/ijcre/vol4/A1.

Bharathi, K., Lakshmi Narasu, M., & Ravindra, P. (2008). Role of galvanic interaction in selective leaching of nickel from copper flotation concentrate. *Advances in Natural and Applied Sciences, 2*(2), 68–72.

Bharathi, K., Bhagvanth Rao, M., & Ravindra, P. (2000). New trends in mineral bioprocessing—An over view. In *Proceedings of CHEMCON, Dec, Calcutta, India.* BIO 81–84.

Bharathi, K., Suresh, P., Bhagvanth Rao, M., Polasa, H., & Ravindra, P. (2002). Comparative studies of nickel leachability with different Strains of *Thiobacillus ferrooxidans*. In *Proceedings of CHEMCON, Dec, Hyderabad, India.*

Bosecker, K. (2001). Microbial leaching in environmental clean up programmes. *Hydrometallurgy, 59*, 245–248.
Das, A., Modak, J. M., & Natarajan, K. A. (1997). Studies on multi metal tolerance of *Acidothiobacillus ferrooxidans. Minerals Engineering, 10*(7), 743–749.
Elzeky, M., & Attia, Y. A. (1995). Effect of bacterial adaptation on kinetics and mechanisms of bioleaching ferrous sulfides. *The Chemical Engineering Journal, 56*, B115–B124.
Garcia, O., Jr., & da'Silva, L. L. (1991). Differences in growth and iron oxidation among *Acidothiobacillus ferrooxidans* cultures in the presence of some toxic metals. *Biotechnology Letters, 13*(8), 567–570.
Gehrke, T., Hallmann, R., Kinzler, K., & Sand, W. (2001). The EPS of *Acidothiobacillus ferrooxidans*—A model for structure–function relationships of attached bacteria and their physiology. *Water Science and Technology, 43*(6), 159–167.
Gehrke, T., Telegdi, J., Thierry, D., & Sand, W. (1998). Importance of extracellular polymeric substances from *Acidothiobacillus ferrooxidans* for bioleaching. *Applied Environmental Microbiology, 64*(7), 2743–2747.
Giaveno, A., & Donati, E. (2001). Bioleaching of heazelwoodite by *Thiobacillus* spp. *Process Biochemistry, 36*, 955–962.
Jeffery, G. H., Bassett, J., Mendham, J., & Denny, R. C. (1989). Titrimetric analysis. In *Vogel's textbook of quantitative analysis* (5th ed., pp. 376–377). London, UK: ELBS-Longman.
Kai, T., Nishi, M., & Takahashi, T. (1995). Adaptation of *Acidothiobacillus ferrooxidans* to nickel ion and bacterial oxidation of nickel sulfide. *Biotechnology Letters, 17*(2), 229–232.
Kar, R. N., Sukla, K. M., & Misra, V. N. (2003). Perspectives of mineral biotechnology. In V. N. Misra, G. Yadav, & K. S. Rao (Eds.), *Mineral processing and engineering* (pp. 163–171). India: Indian Institute of Chemical Engineers Publications.
Leduce, L. G., Ferroni, G. D., & Trevors, J. T. (1997). Resistance to heavy metals in different strains of *Acidothiobacillus ferrooxidans. World Journal of Microbiology and Biotechnology, 13*(4), 453–455.
Li, H. L., & Ke, J. J. (2001). Influence of Cu^{2+} and Mg^{2+} on the growth and activity of Ni^{2+} adapted *Acidothiobacillus ferrooxidans. Hydrometallurgy, 61*, 151–156.
Mason, L. J., & Rice, N. M. (2002). The adaptation of *Acidothiobacillus ferrooxidans* for the treatment of nickel-iron sulfide concentrates. *Minerals Engineering, 15*, 759–808.
Mehta, K. D., Pandey, B. D., & Premchand. (1999). Bio-assisted leaching of copper, nickel and cobalt from converter slag. *Materials Transactions, JIM, 40*(3), 214–221.
Mehta, K. D., Pandey, B. D., & Premchand. (1997). Bioleaching of copper, nickel and cobalt from copper converter slag by *Acidothiobacillus ferrooxidans. NML Technical Journal, 39*, 59–70.
Natarajan, K. A., & Iwasaki, I. (1983). Environmental leaching behaviour of copper-nickel bearing Duluth Gabbro and floatation tailings. *Hydrometallurgy, 10*, 329–342.
Natarajan, K. A., Sudeesha, K., & Rao, G. R. (1994). Stability of copper tolerance in *Acidothiobacillus ferrooxidans. Antonie Van Leeuwenhoek, 66*(4), 303–306.
Paknikar, K. M., & Agate, A. D. (1995). Laboratory manual. In *International workshop on metal–microbe interactions and their applications. United Nations Environment Programme* (pp. 24–36).
Puskas, A., Pap, G., Hollo, J., & Lakatos, T. (1980). Effects of dissolved metal ions on the physiology of *Acidothiobacillus ferrooxidans*. In *Proceedings of international conference on use of microorganisms in hydrometallurgy* (pp. 189–196). Hungary: Hungarian Academy of Sciences.
Ravindra, P., & Bharathi, K. (2009). *Bacterial leaching of low-grade nickel ores*. EMS Publications. ISBN 978-967-5224-37-9.
Selvi, S. C., Modak, J. M., & Natarajan, K. A. (1998). Electrobioleaching of sphalerite floatation concentrate. *Minerals Engineering, 11*(8), 783–788.
Torma, A. E. (1997). The role of *Acidothiobacillus ferrooxidans* in hydrometallurgical process. *Advances in Biochemical Engineering, 6*, 1–31.
Valix, M., Tang, J. Y., & Cheung, W. H. (2001). The effects of mineralogy on the biological leaching of nickel lateritic ores. *Minerals Engineering, 14*(12), 1629–1635.
Vogel, A. I. (1978). *Textbook of quantitative inorganic analysis*. London, UK: Longman Publishers. 354.

Chapter 3
A Review on Ash Formation During Pulverized Fuel Combustion: State of Art and Future Research Needs

Kalpit V. Shah, Mariusz K. Cieplik, and Hari B. Vuthaluru

Introduction

Pulverized coal and a variety of biomass fuels are used as a feed in the power station boilers, where a large amount of thermal energy is generated because of the exothermic reaction taking place during the combustion of fed hydrocarbon which is later converted to electrical energy by several other means. The mineral matter present in quite significant proportions alongside with the hydrocarbons usually fragments, devolatilize (evaporates) and subsequently partly condenses during combustion. This inorganic, mineral residue after combustion, commonly called ash, travels towards the smokestacks carried by the flue gas, may lead to various operational problems such as slagging, fouling, corrosion and erosion of heat exchanging, internal boiler and flu gas duct surfaces etc.

Extensive studies on ash formation during combustion have been conducted World-wide. As a result, theories on ash formation mechanisms have been formulated and described in detail by several researchers (Livingston 2007; Sarofim and Helbe 1994; Baxter 1993; Sarofim et al. 1977; Van Lith 2005). It is evident from several experimental investigations that solid fuel particles undergo various physical transformations during combustion, as shown in Fig. 3.1. The important physical

K.V. Shah (✉)
Priority Research Centre for Energy, Faculty of Engineering and Built Environment, Chemical Engineering, School of Engineering, The University of Newcastle, Callaghan, NSW 2308, Australia
e-mail: kalpit.shah@newcastle.edu.au

M.K. Cieplik
Energy Research Centre of Netherlands, P.O. Box 1, 1755 ZG Petten, The Netherlands

H.B. Vuthaluru
School of Chemical and Petroleum Engineering, Curtin University of Technology, GPO Box U 1987 Perth, WA 6845, Australia
e-mail: h.vuthaluru@curtin.edu.au

P. Ravindra (ed.), *Advances in Bioprocess Technology*,
DOI 10.1007/978-3-319-17915-5_3

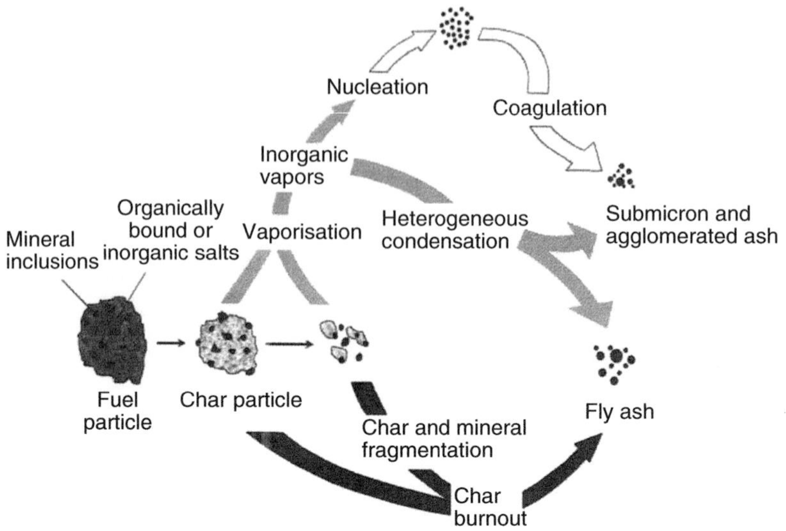

Fig. 3.1 Physical transformations involved for ash formation during coal/biomass combustion (Sarofim and Helbe 1994)

transformations are fragmentation and/or coalescence and vaporization. It is postulated that the fragmentation/coalescence of the ash/char particles along with chemical oxidation and physical devolatilization, will lead to coarse ash formation. The vaporized minerals chemically react with other gas-borne matter, and may condense homogeneously or heterogeneously to form submicron aerosols and fine ash particles. The physical and chemical transformations during thermal conversion of solid fuel are time-dependent and very difficult to understand as a continuous process.

These physical and chemical transformations of minerals depend on several fuel characteristics i.e. fuel, fixed carbon, volatile matter, total ash content and mineral matter elemental composition, mineralogy (either included or excluded especially for coal), char reactivity, char morphology, density, particle size etc. The fuel characteristics will be different for different fuels according to their age, formation history and handling. This paper highlights the effects of all the above fuel characteristics on ash transformations during combustion.

The mineral transformations can also be significantly influenced by several operating conditions i.e. mode of combustion, temperature, pressure, heating rate, residence time, reaction kinetics of various mineral gaseous, slag and solid species etc. Currently, a broad range of technologies is available for the combustion and co-firing of coal and biomass. These include: Atmospheric Fluidized Bed (AFBC), Pressurized Fluidized Bed (PFBC), Pulverized Fuel (PF) and Grate Fired (GF) combustors. All the technologies have their own advantages and limitations (European Biomass Industry Association). The route of mineral transformations will be similar in nearly all the mentioned options, but the extent and criticality will be different for each technology due to differences in operating conditions. The present paper reviews mainly the efforts made to identify the effect of the mentioned operating parameters on ash transformations during pulverized fuel (PF) combustion.

Experiments ranging from lab-scale-combustion simulators to pilot- and plant-scale furnaces under laminar- through turbulent flow conditions, are usually designed and analyzed to understand the ash formation processes during combustion. To date, several methods/sub models/models have been employed to study and identify the effect of different fuel characteristics and operating parameters on ash formation. The present work also briefly reviews some of the basic analytical methods used to measure various parameters responsible for ash formation. It also highlights the modeling efforts undertaken to date, ranging from the simple calculations to advanced numerical simulations for predicting the ash transformations during PF combustion. As there appears to be a lack of a comprehensive literature review to date covering all of this basic information related to ash transformations, such synthesized information may give an overview on the updates in the concerned field. Furthermore it also gives some insight on the future research needs in this area.

Parameters Responsible for Mineral Transformations During PF Combustion

Fuel Mineral Matter Composition and Their Association

Coal and Biomass (or their blend) can be subjected to different ash formation mechanisms during pulverized fuel (PF) combustion, as the fuel mineral matter composition and their association varies greatly in different fuels. The mineral matter in the fuel may be present in the form of free ions, salts, organically bound or as excluded minerals. The lignite and woody biomass contain a major fraction of volatile compounds (and less excluded minerals) compared to Bituminous or anthracite coals. Alkalis in low rank coals and woody biomass, remain primarily in included minerals as free ions, salts and organically associated inorganic elements and start vaporizing at lower temperatures. Even before reaching the char burnout, these vaporized species will chemically react and will condense, nucleate and coagulate on each other or onto the furnace surfaces, to produce submicron ash. Other elements such as calcium and magnesium partly devolatilize, fragment or coalesce (Baxter 1993; Schurmann et al. 2007). Thy (2000) found that if alkali metals occur as network-modifying and charge balancing cations in highly depolymerized melts, such as typical for wood ash, they are easily evaporated during prolonged heating and subsequently deposit onto the heat exchanger surfaces. However, if the melt is highly polymerized such as in the case for rice straw, where alkali metals occur as network modifying cations, they are strongly retained in the polymerized network. During diffusion-limited char combustion, the interior of the particle becomes hot and fuel rich. The non-volatile oxides (e.g. Al_2O_3, SiO_2, MgO, CaO, and Fe_2O_3) can be reduced to more volatile suboxides or even down to elements, and partly vaporized. These reduced species re-oxidize while passing through the boundary layer surrounding the char particle, becomes instantaneously highly supersaturated which make them nucleate homogeneously (Kramlich et al. 1995).

Ash melting behavior is affected by the elemental composition of ash (alkali metals, phosphorous, chlorine, silicon and calcium species), as well as the chemical concentration of the compounds which can alter reaction kinetics of the fuel combustion. Commonly analyzed ash-forming elements are silicon (Si), aluminum (Al), iron (Fe), calcium (Ca), magnesium (Mg), manganese (Mn), sodium (Na), potassium (K), phosphorus (P), sulfur (S) and chlorine (Cl).

Baxter (1992) studied three different ranks of coal (high-volatile bituminous, sub-bituminous, and lignite) and observed that for high-volatile bituminous coals more than 100 fly ash particles were formed from a single 80 μm (initial diameter) char particle, whereas only ten fly ash particles are produced from single 20 μm (initial diameter) char particle. However, regardless of its initial size, fragmentation of lignite particles was far less extensive, with less than five fragments from a single char particle.

The volatile inorganic matter content is one of the most important parameter in coal and biomass as far as submicron particulate formation is concerned. Buhre et al. (2005) observed that formation of submicron aerosol ash particles during coal combustion is mainly due to condensation of evaporated species and not due to the fragmentation.

Mineralogy

Mineralogy of coal and biomass can also play a critical role in various physical and chemical transformations. Physically, the inorganics can be present as included and excluded minerals in the fuel especially for coal. Excluded minerals present in biomass are mainly a result of the contamination with soil during the harvest or handling while presence in coal is due to mining or handling. It is quite obvious that the amount of excluded minerals in most of the biomass fuels will be significantly lower than the in coals, of which deposits are inherently in close contact with rocks and soil. Included minerals in the biomass are the inorganics required for plant growth; and as such they are still present in coals even after millennia of peatification and coalification (Terttalisia 1999), however their physical and chemical form may be altered by the said geological processes.

Included minerals have a higher tendency to remain in the char during combustion. Due to exothermic reactions occurring in the char during combustion, the included mineral matter can reach very high temperatures (above the temperature of the surrounding flue gas). As included minerals are situated close to each other, reactions between them can easily take place. Included minerals may contain more volatile inorganic matter than excluded minerals. The volatile minerals from the included and excluded minerals will be vaporized in the early stage of combustion. The vaporized minerals will condense later on to produce sub-micron ash. During char burnout, the included minerals may either appear as molten particle on a reducing char surface or as a lattice network in char particle. As char burnout proceeds, the minerals may coalesce within a single particle or fragment into several smaller entities. The extent of fragmentation or coalescence depends on several operating parameters and fuel characteristics. This subject has been studied in detail by many research groups, combining both experimental as well as modeling work. Wilemski et al. (1992) and Kang (1991) tried to validate experimental results with no and a full coalescence limit. Morone (1989) in her Ph.D. thesis reported that a partial coalescence is likely to occur in real life systems. Helble and Sarofim (1989) observed very small number of fragments being created during devolatilization. Wilemski and Srinivasachar (1993) later on validated his shrinking core model with partial coalescence limit. Wigley et al. (1997) stated that coal particles containing included mineral matter will have a greater specific heat capacity than particles consisting of organic material alone, hence included particles would be expected to heat up and combust more slowly. Included minerals may fuse and coat the surface of burning char particles, reducing the rate of char combustion. On the other hand, the included mineral matter may catalyze char combustion. The difference in thermal expansion between included minerals and their organic matrix may cause localized thermal stress, thus, leading to an increased char fragmentation. Agglomeration may occur when particles collide or when they meet on a deposit surface on a boiler wall or tube. Mitchell (1997) observed attrition, breakage and percolative-type fragmentation of included minerals during the devolatilization stage. Excluded minerals (especially in the case of coal) on the other hand will reach lower temperatures than included minerals, and they will not be influenced by locally reducing environment. The transformations occurring in excluded minerals and the behavior with respect to the ash deposition may therefore be significantly different from included minerals. Excluded minerals can either be carried through the combustion system with their original structure intact or they can melt and fragment. Decombe et al. (1999), Yan et al. (2001a) and many others concluded that excluded minerals always fragment randomly, due to thermal stress. ten Brink et al. (1996) and Yan et al. (2001b) observed that calcite and pyrite as excluded minerals fragment at high temperature and high heating rate conditions while siderite and ankerite grains did not fragment at the same conditions.

Particle Shape, Size and Density

Experimental and theoretical investigations indicate that particle shape, size and density influence particle dynamics, including drying, heating rate and oxidation reaction rate (Baxter et al. 2008). It is generally observed that spherical particles devolatilize quickly compared to other shape particles. Badzioch and Hawksley (1970) found that particle size had no significant effect on the weight loss because the heating rate of the particle was controlled mainly by the heating rate of the carrier gas, so that the large particles heated only at slightly lower rates than the fine particles. Mathews et al. (1997) observed that mineral matter and macerals composition of the char will be different for different particle sizes, which can affect the devolatilization rate. No and Syred (1990) and Decombe et al. (1999) observed that large particles form more fragments than small particles, likely due to larger

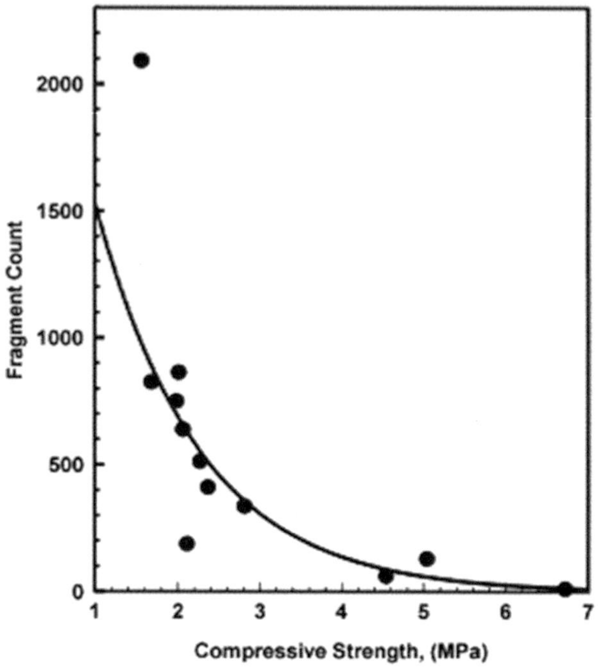

Fig. 3.2 Variation in the extent of the fragmentation of a coal against compressive strength (Dacombe et al. 1999)

internal temperature gradient. Wigley et al. (1997) confirmed that a decrease in char particle size may lead to more complete combustion. Decombe et al. (1999) suggested the relationship of fragmentation extent with compressive strength as shown in Fig. 3.2. However, compressive strength of the coal particle is inversely proportional to the particle size.

The ash transport behavior is affected to a large extent by the size of the particle after combustion. Large ash particles tend to impact onto boiler heat transfer surfaces by inertia, whereas fine ash particles tend to reach wall surfaces by thermophoresis or Brownian motion. For example, a 60 μm ash particle was estimated to reach the deposit surface almost three times faster compared to 30 μm particle primarily due to inertial effect (Yan et al. 2001a).

Liu et al. (2008) studied Chinese bituminous coal with three density fractions. The fragmentation was severe with light density fractions as shown in Fig. 3.3. The median size of each coal fraction was almost the same. The reasons for the above were particle size, mineralogy and swelling ratio. The light fraction and the medium fraction of the coal contained mostly included minerals, and the heavy fraction contained largely excluded minerals.

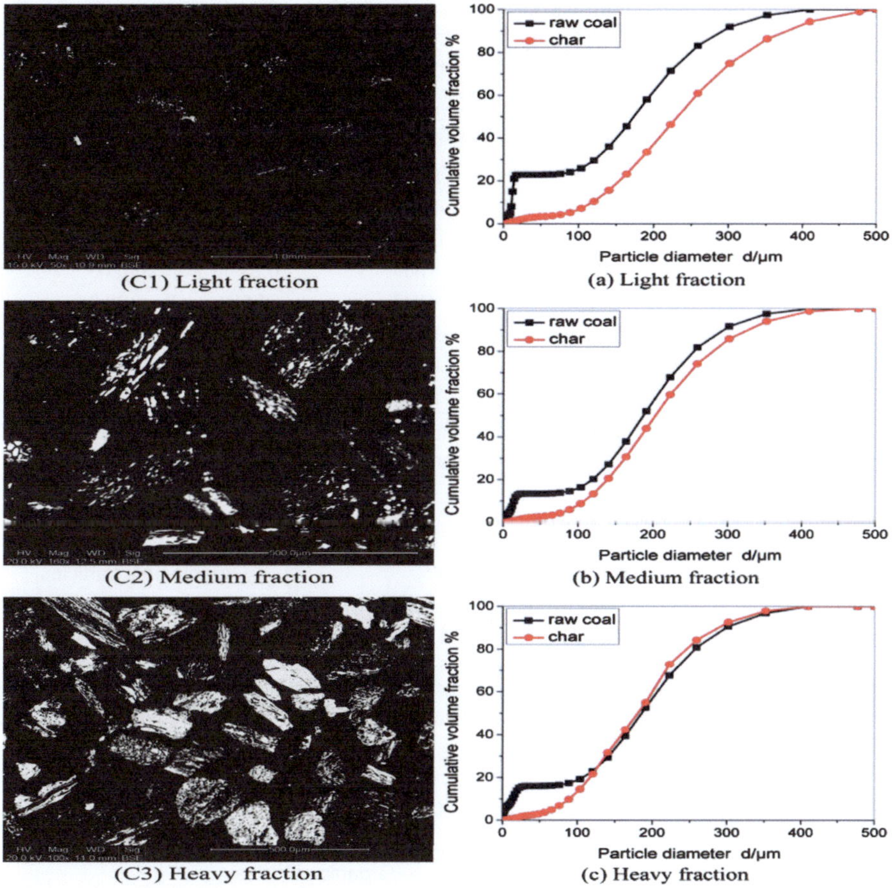

Fig. 3.3 BSE images with comparison of PSD (by volume) of char and coal (Liu et al. 2008). (**a**) Light fraction. (**b**) Medium fraction. (**c**) Heavy fraction

Fuel Characteristics After Milling

Milling of raw coals or biomass fuels, i.e. fineness of the material after grinding as well as the applied mill technology, has a profound effect on ash formation. It has been observed by several researchers that mineralogy, ash percentage, volatile matter, density and char reactivity will be different for different particle size ranges (PSD). Bridgeman et al. (2007) studied two energy crops (switchgrass and reed canary grass) in terms of their physical and chemical properties in different size fractions after grinding with ball mills at lab scale. The results summarized in Table 3.1 indicate that smaller particles of the two grasses have a significantly higher concentration of inorganic matter as well as the moisture content than larger particles. In contrast the larger-sized fractions had higher carbon content and lower

Table 3.1 Proximate and ultimate analysis of biomass (Bridgemana et al. 2007)

%		RCG < 90 μm	RCG > 90 μm	SW < 90 μm	SW > 90 μm
Moisture		6.45	5.71	8.64	7.93
Ash		6.0	3.62	6.88	3.12
Volatiles		72.62	74.89	70.58	72.57
Fixed carbon		14.92	15.78	13.9	16.37
C		43.56	44.9	42.33	44.32
H		6.1	6.14	5.98	5.99
N		0.47	<0.04	0.23	0.03
O		37.65	39.07	37.58	38.24
CV (kJ/kg)	Measured	17,100	17,700	16,600	17,100
	OLS	17,200	17,700	16,700	17,500
	PLS	17,300	17,800	16,800	17,500

Ultimate, proximate and CV analyses of different size fractions of reed canary grass and switchgrass (Institution 1), where RCG = reed canary grass, and SW = switchgrass

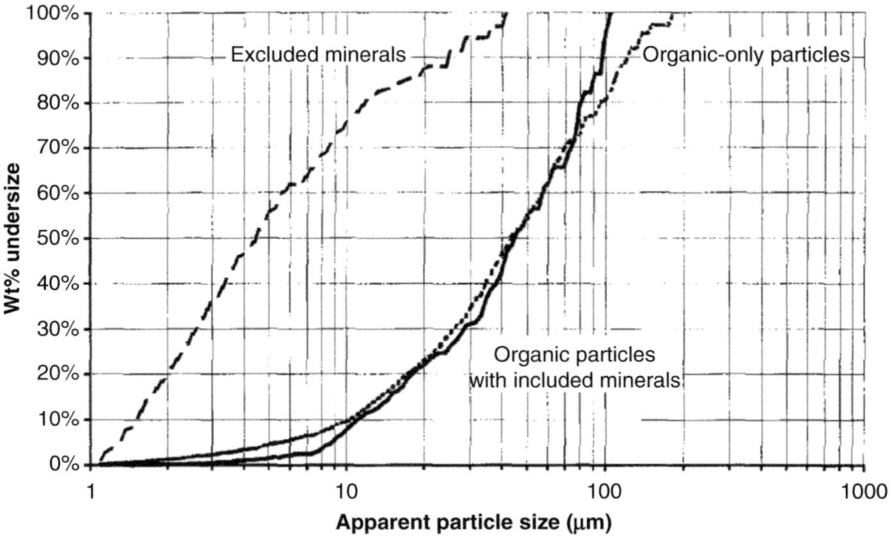

Fig. 3.4 Classifier effect: size distributions for excluded minerals, organic particles with included mineral, and organic-only for typical coal (Wigley et al. 1997)

nitrogen content, with a resulting higher calorific value. The volatile content was also higher in the larger sized fraction.

However, Wigley et al. (1997) stated that the presence of included mineral matter could alter the size distribution of pulverized coal particles leaving the mill classifier and entering the boiler. The mineral inclusions will increase the average density of a coal particle. As classifiers separate particles on a combined size and density basis, denser coal particles would be expected to be slightly finer.

The observations on particle size distribution do indicate that included minerals have slightly finer size distributions than organic-rich particles, and the very largest particles are almost purely organic, as shown in Fig. 3.4.

Char Structure

The change in the internal structure of a char particle is one of the most important issues during coal devolatilization and is closely associated with the coal particle swelling phenomenon, during the plastic stage. The extent to which the pore structure changes is dependent on the fuel type and is strongly affected by the conditions under which the fuel is devolatilized (Yua et al. 2007). Hurt et al. (1991) concluded that CO_2 gasification reactions took place primarily on the surfaces of larger pores during kinetic-diffusion controlled regime. The fragmentation will be increased as char burnout shifts from a diffusion-controlled to a chemically-controlled regime (Liu et al. 2000) as shown in Fig. 3.5. It was observed

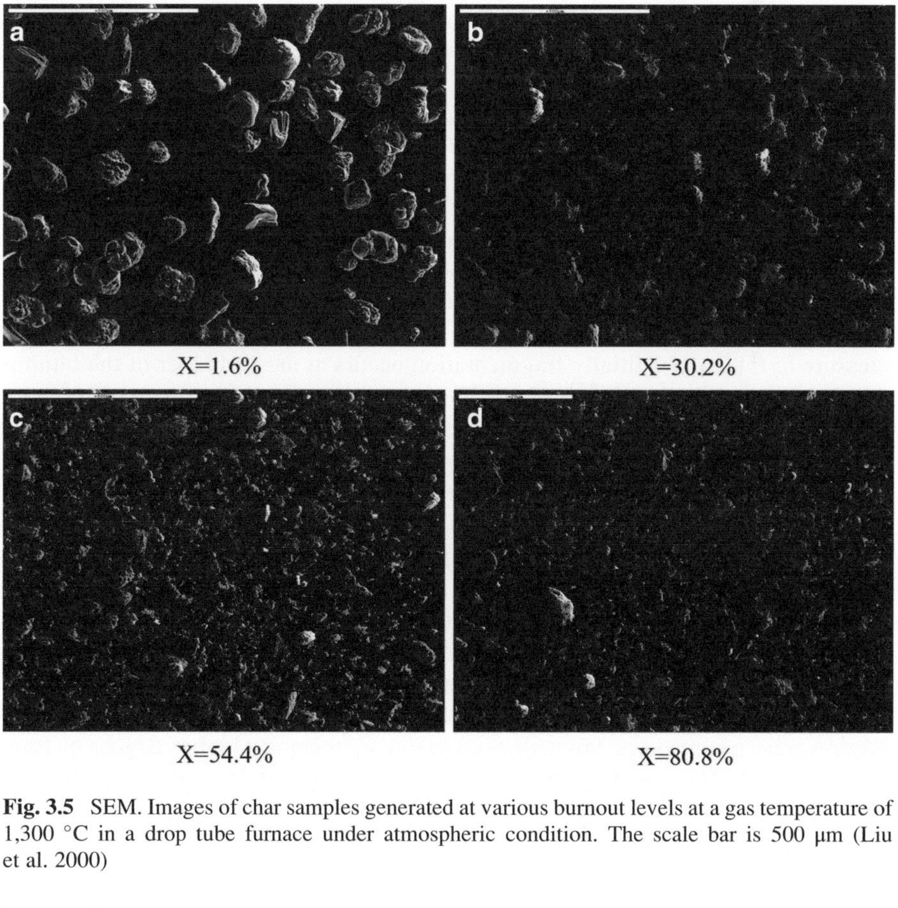

Fig. 3.5 SEM. Images of char samples generated at various burnout levels at a gas temperature of 1,300 °C in a drop tube furnace under atmospheric condition. The scale bar is 500 μm (Liu et al. 2000)

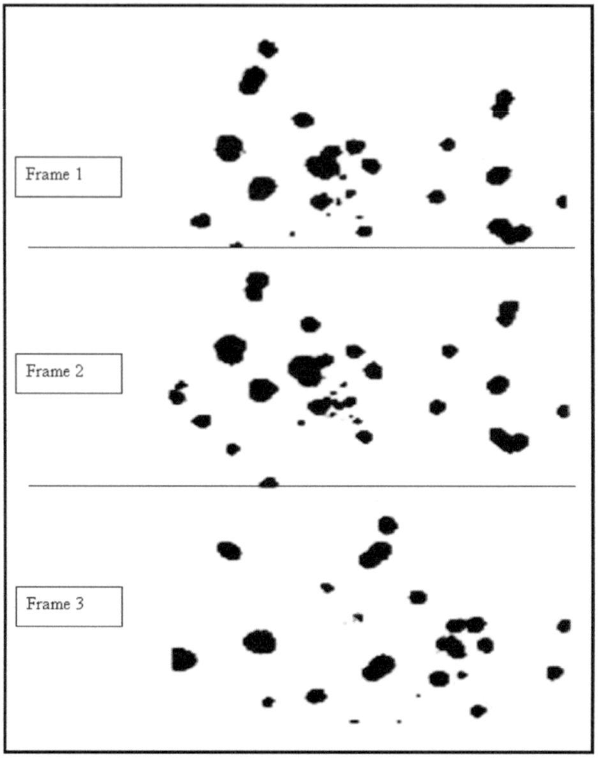

Fig. 3.6 Negative prints of three successive frames from high speed film (approximately 4,000 frames per second) (Helble and Sarofim 1989). (**a**) X = 1.6 %. (**b**) X = 30.2 %. (**c**) X = 54.5 %. (**d**) X = 80.8 %

by many researchers that during the initial heat up and devolatilization in a kinetic-diffusion controlled regime, char particles do not change much in shape and size. As shown in Fig. 3.6, Helble and Sarofim (1989) observed with a high speed camera (approximately 4,000 frames per second) that at 1,250 K and at high oxygen partial pressure (>0.80 atm) initially fragmentation occurs at the perimeter of the bituminous coal char particles. Mitchell (1997) also mentioned attrition-type of behavior during the initial heat up and devolatilization of char particles. He also noted that large amount of aerosols were formed by the attrition of large particles from the peripheral diffusion in regime II which describes the particle burning rate during the char oxidation at high temperatures in which the characteristic rates for pore diffusion and chemical reaction are comparable, making both effects important in determining overall mass loss rates.

Menendez et al. (1993) ranked the most important char characteristics with gradual increase in combustion temperature as follows: (1) The total surface which may be accessible to the reacting gases; (2) Porosity of the char particle; and (3) Char particle size. These parameters are crucial in modeling of PF combustions and gasification. Mitchell observed a significantly higher degree of fragmentation, with less porous chars during the heat-up and devolatilization stages, suggesting that the more open the porous char structure, the lesser will be the extent of fragmentation during heat-up and devolatilization, induced by either thermal stresses or stresses due to build-up of pressure of volatiles in the pore work.

Highly porous char particles can attain the chemically-controlled regime earlier than dense chars, due to a higher extent of both the devolatilization and fragmentation. The fragmentation has also been found to have a significant impact on the chemistry of the final ash particles. Kaiho and Toda (1979) and Ezra and Kantorovich (2001) examined the role of pore structure in the fragmentation of highly-porous char particles and claimed that the reason for local fragmentation under non-uniform oxidation is the increase in the local macro-porosity. Kang et al. (1992) with his experiments concluded that the fragmentation of a macro-porous char can influence the final ash size distribution.

Yua et al. (2007) summarized extensive efforts made in the past decades to classify morphologically complicated char structures. Char structures have been classified on the bases of char morphological parameters including macro-porosity, the wall thickness, particle shape etc. (Yua et al. 2007; Benfell and Bailey 1998). A three-group classification system (Table 3.2) suggested by Benfell and Bailey has been adopted by number of researchers (Yan et al. 2001b; Yua et al. 2007).

The macerals composition of coal plays a dominant role in the morphology of the char during devolatilization. Vitrinite-containing bituminous coal particles commonly produce cenospheric chars while the intertinite produces a char with low porosity.

For softening coals, the formation of different types of char structures is closely associated with their thermoplastic behavior such as fluidity and swelling during heating (Yua et al. 2007). The porosity of the chars from non-plastic coal increases steadily with increasing temperature. Gale et al. (1995) found that the overall porosity and swelling ratio of char increases with increasing heating rates up to 10^3 K s^{-1}, with a further increase in the heating rate above 2×10^4 K s^{-1} resulting in a decreased porosity and swelling, as shown in Fig. 3.7. This is due to the rapid

Table 3.2 Summary of the threefold char structure classification system by Bailey and Benfell

Char groups	Group I 	Group II 	Group III 
Char subtypes	Cenosphere tenuisphere, tenuinetwork	Crassisphere, crassinetwork, mesosphere, mixed porous (mixed dense)	Inertoid, solid, fusinoid (mixed dense)
Char particle shape	Spheroidal	Spheroidal to irregular	Subspheroidal, rectangular or irregular
Porosity	>80 %	>50 %	~50 %
Pore shape	Spheroidal	Variable	Spheroidal to elongate and angular
Wall thickness	<5 pm	Variable	>5 μm
Dominant maceral components	Vitrinite	Vitrinite and inertinite	Inertinite
Swelling ratio	>1.3	<1.0	<0.9

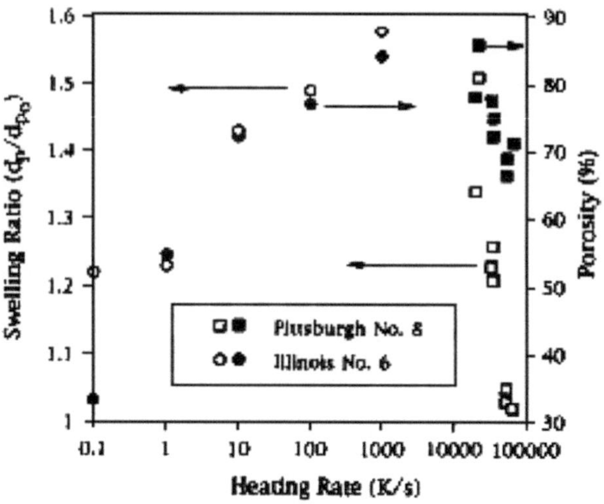

Fig. 3.7 The porosity and swelling ratio as a function of heating rate (Yua et al. 2007)

release of volatile matter than the relaxation time for expanding the char particle. The temperature gradient in a particle at a very fast heating rate may also affect the process.

Other Fuel Characteristics

The ignition temperature of the fuel particle has an indirect relation with ash formation as it is an important parameter for defining the early start of the combustion process. A number of investigations (Gupta 2005) have been devoted to this issue and it was found that the ignition temperature decreases with an increase in particle size, oxygen partial pressure and volatile matter content. And larger particle size and higher volatile content can lead to various ash related problems.

Char reactivity is defined as the mass loss per unit external surface area. The average char reactivity was found to decrease with an increase in burn out levels but was ranging greatly even within the same particle size (Gupta 2005). Koranyi (1989) found within a set of three British bituminous coals that a qualitatively good correlation exists between the char reactivity and its micro-porosity. Ash, moisture and fixed carbon percentages can also be interlinked with other fuel characteristics such as ignition temperature, char reactivity, char morphology, char mineralogy, char structure and particle size.

Relation with Operating Parameters

During the char oxidation at high temperatures, particle burning rates initially lie in the so called zone II burning regime, in which the characteristic rates for pore diffusion and chemical reaction are comparable, making both effects important in determining overall mass loss rates. As burning progresses, particles become smaller and pores become enlarged, decreasing mass transport limitations. Thus, later in the burn-off, a transition is expected from the "zone II" burning regime to the zone I regime, in which chemical reaction rates are dominant in controlling overall mass loss rates (Mitchell 2000). Zone I, can be summarized as a chemically controlled regime and Zone II as a kinetic-diffusion controlled regime. In all the regimes, density and particle size are the most important parameters which change with mass loss rate simultaneously (Mitchell 1997, 2000; Yua et al. 2007). The effect of operating parameters will be different in the two regimes.

Heating rate has a significant effect on ash formation. The effect of heating rate will be different for included and excluded minerals, porous vs. non-porous structures and small vs. large-sized particles. Excluded minerals have more specific heat capacity and will therefore heat up slowly compared to included mineral matter or purely organic particles. Highly porous char that undergoes much more extensive devolatilization during heating will burn out at an earlier stage compared to less porous (solid) char particles; even though they burn at a similar "per carbon site" rate (Wall et al. 2002). Large particles, although experiencing a higher temperature gradient making them susceptible to fragmentation, will heat up later than the small entities.

Temperature and pressure can significantly affect the extent of ash formation, as well as its characteristics. For instance, Erickson et al. (1992) found in his experiments with synthetic coal in a drop tube furnace at temperatures of 900, 1,100, 1,300, 1,500 °C, and at constant heating rate conditions, that at high temperature fly ash formation was dominated by fragmentation, as shown in Fig. 3.8.

Wu et al. (2000) performed combustion experiments employing a bituminous coal with a size fraction of 6390 μm, under oxidizing atmosphere (air) in a drop tube furnace (DTF) and a pressurized drop tube furnace (PDTF) at a gas temperature of 1,300 °C with same heating rate conditions and pressures of 0.1, 0.5, 1.0, and 1.5 MPa. As shown in Fig. 3.9, ash generated at high pressure was found to be much finer than ash generated at low pressure due to the differences in the pressure gradient.

Usually, PF combustion occurs at atmospheric pressure and high temperature with high heating rates (in excess of 10^5 K/s). Though operating parameters such as temperature, pressure, heating rate will be in a relatively narrow, specified range for PF combustion, the effect of operating conditions will be varying significantly for different fuels. The measures of the fuel characteristics such as char reactivity, ash, moisture, volatile matter and fixed carbon percentages, as well as its density and porosity will also be different even within the same fuel with varying particle size. Therefore, each particle size range in single fuel will behave differently under PF combustion conditions. Kinetic reaction rates of the mineral chemical conversions are also highly dependent on several operating parameters (temperature, pressure, residence time etc.) and fuel characteristics (ash contents, mineralogy, particle size etc.).

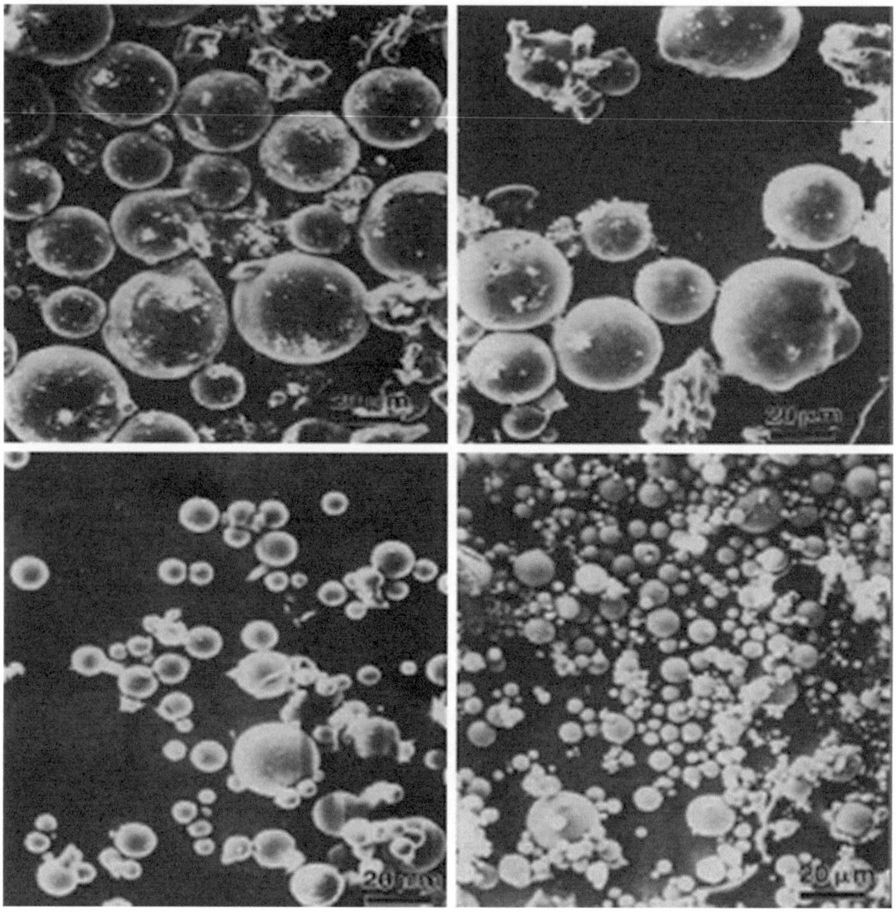

Fig. 3.8 Fly ash particles formed at 900 °C (*left-top*), 1,100 °C (*right-top*), 1,300 °C (*left-bottom*) and 1,500 °C (*right-bottom*) (Erickson et al. 1992)

Prediction of Ash Formation During PF Combustion

Extensive research has been carried out to identify the inorganic behavior during coal, biomass combustion and co-firing and many uncertainties have been clarified. Experiments ranging from lab-scale-combustion simulators to pilot- and plant-scale furnaces under laminar through turbulent flow conditions have been run and analyzed. Many methods/model/sub models starting from simple and traditional ash analysis to advance numerical modeling have been attempted based on the achieved understanding so far.

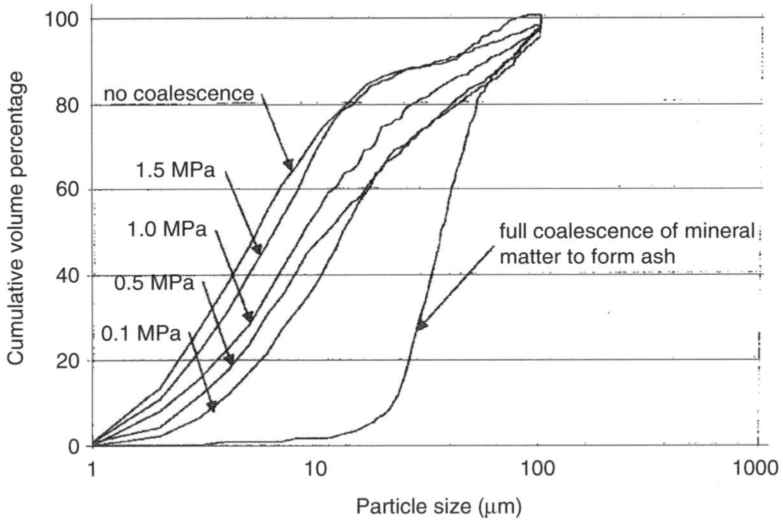

Fig. 3.9 Particle size distribution of ash generated at different pressures (Wu et al. 2000)

Analytical Methods

Proximate and ultimate analyses of the fuel are considered as the most basic and necessary analyses that need to be carried out to understand the combustion characteristics of the fuel and deciding on optimum operating conditions in an installation of a particular design (Speight 2005). A typical proximate analysis includes the moisture, ash, volatile matter, and fixed carbon contents. It also gives an idea about the calorific value of the fuel. The ultimate analysis indicates the various elemental chemical constituents such as carbon, hydrogen, oxygen, sulfur, etc. It is useful in determining the quantity of air required for combustion and the volume and composition of the resulting combustion gases. However these analyses combined are only giving clues for optimal operating conditions and are incapable of measuring the mineralogical composition of the fuel (Speight 2005).

The elemental analysis of a fuel/ash can be performed by means of several traditional methods such as atomic absorption spectroscopy (AAS), graphite furnace atomic absorption spectroscopy (GFAAS), inductively coupled plasma atomic emission spectrometry (ICP-AES), inductively coupled plasma/mass spectrometry (ICP/MS), X-ray fluorescence spectrometry (XRF), glow discharge mass spectrometry (GDMS) and spark source mass spectrometry (SSMS) (Jacobs 2000). Molecular beam mass spectrometry (MBMS) (Dayton et al. 1999) and high pressure mass spectrometry (HPMS) (Wolf 2002) are also found to be used at lab-, pilot- or plant scales to measure the alkali gaseous phase release. Korbee et al. (2006) and Frandsen et al. (2007) used ICP-AES and SEM-EDX for finding the elemental characteristics for a range of coals and biomasses.

However, the traditional elemental analysis is of little use in getting insights into ash formation characteristics. To know about the mineral matter association in the fuel, pH-static leaching and chemical fractionation are often used. Doshi et al. (2009) used these methods to analyze the mineral association in the fuel matrix, classifying the elements present as free ions, salts, organically bound and excluded minerals. This was done by measuring the solubility of inorganics in water, ammonium acetate and HCl. The free ions, salts and organically bound minerals easily devolatilize and are responsible mainly for aerosol ash formation, primarily through condensation processes during combustion. Excluded minerals remain in the solid form and play the primary role in coarse ash formation. However, the pH-static leaching and chemical fractionation methods were found to overestimate the release measured dynamically in lab-scale facilities. This is mainly due to the assumption that the entire 'reactive' fraction (the fraction of the inorganic elements leachable in water or ammonium acetate) is released into the gas phase during combustion. In a real combustion situation, the 'reactive' fraction may interact with the 'less reactive' fraction (leachable in HCl or not leachable at all), thereby decreasing the fraction of the inorganic elements being released to the gas phase (Frandsen et al. 2007).

Thermogravimetric analysis (TGA/STA) is often used to find out the reaction kinetics of the fuel organics alongside with their included and excluded minerals. TGA measures the weight change in the materials as a function of time and temperature (Lu et al. 2007). The measurements provide basic information about the thermal stability of the fuel and its composition. TGA is nowadays one of the most commonly applied thermal techniques used to characterize both char oxidation and devolatilization rates. It is often also used for predicting the total residence time required for a fuel under given operating conditions. Lu et al. (2007) studied the combustion kinetics for various coals and biomass using the TGA method. Vuthaluru (2004) used thermogravimetric analysis to study the pyrolytic behavior of coal and biomass blends. Hurt et al. (2003) did his kinetics study using TGA to investigate the fuel transformations during advanced coal combustion and gasification. da Silva Filho and Milioli (2008) studied the kinetics of Brazilian coal while Zhaosheng et al. (2009) and Miranda et al. (2008) studied rice, wheat straw and olive residue using thermogravimetric analysis.

Traditional ash analyses and chemical methods are time-consuming and often of limited scope. Furthermore, such methods are incapable of providing physical characteristics like the size and shape of coal mineral particles and mineral distribution on a particle basis, which clearly play important roles in understanding the ash formation during combustion to the full extent (Terama et al. 1999). For this purpose, computer controlled scanning electron microscopy (CCSEM) has been extensively used. This technique is relatively expensive, and still not very common and mostly used for research. Terma et al. (1999) studied the ash transformations of coal during PF combustion by extensive use of CCSEM. Chen et al. (2004) investigated the mineral matter composition of fine particulate matter of coal using CCSEM, while Wang et al. (2007) used CCSEM to investigate the interactions of the inherent minerals. Yu et al. (2007) studied six highly heterogeneous Chinese coals while Brunner et al. (2001) studied bark, waste wood and wood chips using the CCSEM technique. The use of CCSEM has increased remarkably in the last two decades. It can be used to find out the mineral matter distribution along with organic matter in different size fractions of the raw fuel as well as ashes from various stages of the combustion process. Many unknown facts and uncertainties are made clear using CCSEM techniques, adding to a better understanding of the combustion process. CCSEM has been also used to identify and characterize ash particles in deposits, in order to gain insights into deposition and fouling characteristics of coals under conditions of various combustion regimes. In recent efforts, QEMSCAN has been used to determine mineral-mineral associations, particle size, mineral compositions, and particle texture in coal and ash samples. The above mentioned procedure is similar as CCSEM analysis, but characterized by an even higher degree of automation (Liu et al. 2005). Vuthaluru and French (2008a, b) have analyzed the ash chemistry and mineralogy of an Indonesian coal at various stages of combustion. Many other coals have been analyzed using QEMSCAN. However, analysis of different biomass using the same is very limited to date.

Chemical transformations can be predicted using thermo-chemical equilibrium calculations based on Gibbs-free energy minimization principle. This method assumes that the chemical equilibrium exists at each time fraction but ignores any intermediate products. As an input, this method requires the elemental mineral matter composition in gas-slag-solid phases. It describes the composition of stable gas-solid species at different temperature conditions. The input elemental composition is provided using traditional ash analyses, possibly extended with the chemical fractionation method. Such an approach was used by Doshi et al. (2009) who applied FACTsage (Thermo-chemical equilibrium software) to predict compositions of gas and solid species for different coal and biomass fuels.

Several empirical indices for several aspects of coal combustion and ash formation have been proposed based on the experiments performed at lab, pilot and plant scale in the function of mineral matter composition. These empirical indices quantify primarily various ash-related issues such as slagging, fouling, corrosion, erosion and aerosol formation etc. rather than lead to an in-depth knowledge of ash formation mechanisms. The well-known empirical indices are ash fusion temperature, base-acid ratio, slagging factor, T250 °C temperature, iron/calcium ratio, iron plus calcium, slagging index, silica percentage etc. (Lockwood). Large discrepancies were observed in the use of the majority of these indices on a wider range of coals, let alone for coals and biomass co-firing. The empirical indices were found to be a rather poor tool for the prediction of ash formation or deposition behavior (Barrosoa et al. 2007). Nonetheless for an isolated set of fuels and under constant firing conditions, where variations in mineral matter are small, mechanisms of ash formation and depositions are consistent and operating and design variables remain constant, a well-developed set of indices may actually predict the ash behavior very well.

Mathematical Modeling

Several mathematical/analytical (sub) models have been developed to predict the stepwise combustion process. As discussed earlier, inorganics may undergo a number of physical and chemical transformations during combustion and these transformations depend on multiple operating parameters and fuel characteristics. Therefore, it is difficult to develop an integrated mathematical model for all the mechanisms, as with more of variables complexity of the models increase (Benson et al. 2002). To make the simulations simpler, the different mechanisms, such as char oxidation, devolatilization, fragmentation, chemical reactions along with gaseous phase nucleation, coagulation, homogeneous and heterogeneous condensation etc. are studied separately, supported by different analytical methods as discussed above, often resulting in the development of several "rival" sub models describing the same system. Also the effects of various operating parameters such as the heating rate, temperature, pressure, and the residence time with different particle sizes, mineralogy and mineral matter composition ranges, as well as each of the chemical and physical transformations are often represented by separate subroutines to avoid simulation complexity. For a more realistic simulation interlinking of the above described sub models is also attempted.

A common approach for the ash formation modeling during PF combustion comprises a dual size ash mode (Wu et al. 1999; Oberberger 2003): (1) Coarse ash and (2) Aerosol. Coarse ash formed during combustion usually participates in slagging, fouling, corrosion and erosion while aerosols contribute to environmental and health hazards, although they can also play a role in fouling at intermediate temperatures. The two important mathematical/numerical models for the prediction of the overall ash formation process are the aerosol ash formation models and the coarse particle size distribution evolution model, which are described in more details in the following section.

Coarse Ash Formation

The critical physical transformations responsible for coarse ash formation considered in most of the studies reported in the literature are fragmentation and/or coalescence. From the experimental and the industrial observations, it was found that fragmentation/coalescence during combustion is a very complex phenomena, particularly when covering a broad range of fuels at different operating conditions. Flagen and Friedlander (1978), Kang et al. (1990), Decombe et al. (1999), Wilemski and Srinivasachar (1993), Mitchell (1997) and Yan et al. (2001b) all modeled fragmentation/coalescence. The different models developed to date include particle break up and/or the coalescence of molten grains (Wilemski and Srinivasachar 1993; Flagen and Friedlander 1978), macroporosity due to the thermal stress (Dacombe et al. 1999), the percolative fragmentation, based on macroporosity, in which oxidation progressively erodes the solid network until the solid phase becomes spatially discontinuous (Yan et al. 2001a, b; Liu et al. 2000; Kang et al. 1990; Edwards and Ghosal 1988; Wigley and Williamson 1998), collision-induced attrition (Salatino et al. 1992, 1993) and pressure-induced fracture and macroporosity due to attrition, breakage and percolation during devolatilization (Mitchell 1997). This section summarizes the gradual development of the individual models from simple theory based calculations to advanced numerical modeling.

Break Up Model

Laboratory work and studies of full scale coal-fired boilers in the early 1980s encouraged Flagen and Friedlander (1978) to model the residual ash formation during coal combustion. The simple breakup model described by Flagen and Friedlander (1978) assumed that char particles containing mineral matter, fragment during combustion. Major assumptions in the model include: (1) all coal particles contain same percentage and amount of mineral matter, independently of size, (2) all coal particles break into exactly the same number of char particles during combustion. The breakup number identified in this model is influenced by the breakup of char during burnout, from shedding at burning char surface and from the fragmentation of discrete included and excluded minerals. Despite several assumptions, the basic breakup model has proven to be a useful engineering and interpretative tool. Building further on Falgen's model, Demle et al. (1982) added the vaporization-condensation model for submicron particle formation and evaluated the performance of the model with actual experiments.

Fragmentation Model Based on Thermally Induced Stress

This mechanism implies that the fragmentation due to thermal stress is the dominant driving force for particle breakage, as there are always internal temperature gradients during particle heating up, especially with large particles. These internal temperature gradients cause significant thermal stress, leading to the production of more tiny fragments than those generated from small parent particles. Decombe et al. (1999) developed a theoretical model for initial fragmentation based on thermally-induced stresses. A transient analysis of these stresses allowed the fragmentation point to be determined. The results suggest a mode of fragmentation where many small particles are produced from the outer region and a few large particles from the inner one. As small particles experience a smaller temperature gradient resulting in much lower stress, fragmentation can be delayed to the char burnout phase (Baxter et al. 2008). Decombe et al. (1999) found the relationship of the fragmentation extent with compressive strength. Furthermore, he observed that the extent of the fragmentation was found to increase with the particle size. To the best of the authors' knowledge, no further model was developed based on compressive strength.

Shrinking Core Model

Wiesmiki et al. (1992) and Wilemski and Srinivasachar (1993) developed a shrinking core model based on the observations that ash particles are produced during combustion by transformations and interactions of the mineral inclusions within the coal particle. The growth behavior of ash particles on the char surface is described in the model by considering redistribution and coalescence processes (ash formation mode). Several redistribution sub models have been used in the shrinking core model. Charon et al. (1990) applied Monte Carlo methods to simulate the random distribution of minerals among a set of coal particles. Barta et al. (1992) developed an analytical model based on Poisson statistics for determining the size and chemical composition distributions for the minerals based on CCSEM data. Later on the same group developed the Random coalescence model to predict the PSD with elemental composition. However, all these models inaccurately predict that no small inorganic particles are present as excluded minerals. Wilemski et al. (1992) used a composite method that combines Poisson statistics for distributing the smallest minerals among the smallest coal particles with a Monte Carlo method for handling all of the larger minerals and coal particles. Ash distributions were predicted with a full and no coalescence ash formation modes (Wilemski et al. 1992). In the no coalescence mode, each mineral inclusion is assumed to produce one ash particle while in the full coalescence mode, the minerals in each coal particle are assumed to coalesce fully, producing one ash particle per coal particle. Based on Kang's (1991) thesis, the char fragmentation mode was also studied for cenospherical chars. In this case, ash particles are formed by the coalescence of inclusions that have high probabilities of encountering each other, as the cenospherical char shell burns away. In the said model, the char fragmentation mode has independent variables such as porosity, swell volume of char particle, and cenosphere shell thickness. These parameters are entered in the model with the use of CCSEM analysis of the coal. Yan et al. (2001b) investigated the implications of the shrinking core model on ash deposition and thermal behavior. Liu et al. (2000) used the shrinking core model to mechanistically predict ash particles size distributions and chemical compositions of the included and excluded minerals, throughout coal combustion. The upgraded mechanistic model of shrinking core method is based on partial coalescence (Morone 1989) of included minerals with three different groups of chars (Benfell and Bailey 1998). The excluded mineral fragmentation in this model is calculated with the Poisson statistical distribution method (Yan et al. 2001b).

Percolation Model

Based on a series of experiments by various researchers during the 1970–1980s, it was found that char oxidation is percolative in nature and char macroporosity is the single most important factor governing char breakup and the resulting residual ash size. Mohanty et al. (1982) was the first to suggest the application of percolation theory to fluid-solid reaction systems, accompanied by pore volume changes. Kerstein and Niksa (1984) applied this theory to char oxidation system, and developed a model to explain the fragmentation of chars in the chemically controlled reaction regime. Reyes and Jensen (1986) applied the same theory to the Bethe lattice char model. Kerstein and Edwards (1987) performed the simulation of char oxidation and fragmentation using the percolation model on lattice. To target the effect of char fragmentation on ash formation, Kang et al. (1988) developed the first stochastic model based on the percolation theory on lattice for the external diffusion controlled regime. Porosity, mineral matter grain size distribution in the coal particle, reaction rate, and ash surface coalescence are the independent variables in the model. Salatino et al. (1992, 1993) stated that the fragmentation process may extend to the entire particle structure (uniform percolation, occurs only in chemical-kinetic-controlled regime, porosity develops uniformly within the particle, simultaneously throughout the particle, until conditions for loss of particle connectedness are reached) or may be restricted to its periphery (peripheral percolation, under diffusion controlled regime, where porosity increases non-uniformly and the fragmentation threshold is reached at particle periphery earlier than within the particle). However, they proposed that peripheral percolation is to be considered for carbon oxidation studies as both chemical kinetic and intraparticle diffusive resistance are considered in this approach. Salatino et al. (1992, 1993) developed discrete, uniform percolation and peripheral percolation model. Most of the percolation-based models were discrete models with one of the main disadvantages of the discrete model being the associated high computational power. Continuous models, however, take less computational power but provide only qualitative results. Salatino et al. (1992) developed the percolative fragmentation model combining discrete and continuous methods where continuity equations are pseudo-stationary, as far as the oxygen concentration is concerned. Yan et al. (2001b) studied the above percolation model in the diffusion-controlled regime. It was further extended and used to predict structural changes such as particle swelling due to coal devolatilization (Kurose et al. 2003, 2007; Suzuki et al. 2002).

Particle Population Balance Model

Most models that include time-dependent relationships for the growth of numbers of particles are based on the work of Dunn-Rankin and coworkers (1987, 1988). Mitchell (1997, 2000), Decombe et al. (1999) and many others found that during the initial conversion stages like heating up and devolatilization, the char particle will be under peripheral kinetic-diffusion controlled regime and later on after significant char burnout, it will be diverted to chemical-kinetic-controlled regime. Mitchell (1997) observed three types of fragmentation behavior during combustion at different stages in his experiments with synthetic chars at different heating rate affecting the attrition, breakage and percolation. During attrition fragmentation, numerous small particles are produced from the surface while parent particle diminishes in size slightly. During breakage fragmentation, only a few large fragments are produced, not much smaller than the parent particle. Percolation fragmentation refers to the transition from a connected solid network to a completely fragmented state. It was observed during the experiments that the degree of fragmentation with less porosity was much higher during the heat-up and devolatilization stages, which suggest that the more open the porous structure, the lesser will be the extent of fragmentation during heat-up and devolatilization induced by either thermal stresses or stresses due to build-up of pressure of volatiles in the pores. Experimental studies showed that both char particle diameter and apparent density change as burning progresses. Based on his experimental observations, Mitchell (2000) adopted the particle population balance model of Dunn-Rankin and coworkers (1987, 1988) to predict particle size distribution (PSD) as a function of time during coal combustion. The particle population model was developed based on a power law expression used to correlate mass, density and diameter changes with burning rate. The burning rate constant in the model is based on an Arrhenius parameter that was obtained for each fuel experimentally. As this approach failed to account for functional variations, later on the intrinsic kinetic-based particle population model was developed that employ power-law-controlled mode of burning, in which particle size and apparent density variations are dependent on the particular physical and chemical characteristics of char (Mitchell 2000). Syred et al. (2007) solved the particle population balance model analytically for fragmentation and tried to incorporate it into CFD modeling. However, fragmentation alone is not a complete way of presenting combustion, as particle should be burnt in a finite time period. Recently, Shah et al. (2010) in their simple approach, have extended Syred's work by solving the particle population balance model analytically for two size class with inclusion of a burning term. The simplified model has then been validated with Polish coal experiments. However, the model is at initial stage of development and needs further improvement in the future.

The above described coarse ash formation models have only occasionally been developed and deployed for biomass fuels or co-firing modeling. Furthermore they predict ash size distributions qualitatively, except a few, which give also quantitative results. Likewise only a few models (Morone 1989; Yan et al. 2001b) predict chemical compositions along with the qualitative or quantitative size distributions.

Ash Release/Aerosol Formation

The second major ash formation mechanism is the generation of submicron aerosols through vaporization and a number of gas-to-particle conversion mechanisms. When the ash size distribution is plotted in terms of number density or particles numbers, the submicron generally peak at around 0.1 μm. Although these particles account only for a small mass fraction, they can present a large fraction of surface area and become the preferred site for the condensation of more volatile oxides and toxic metal components deeper in the boiler. To avoid complexity in the simulations, vaporization and condensation mechanisms are treated separately with dedicated models.

The vaporization is often predicted by a combination of several analyses and sub models. Proximate and ultimate analyses alongside with chemical fractionation techniques are used to decide the volatile matter in the fuel. Additionally, several empirical correlations are used to predict the vaporization mechanism accurately. Recently, CCSEM techniques have also been used to quantify the volatile matter in the fuel. TGA (Thermogravinomatric analysis) has also been used to model the char devolatilization rate. Finally the vaporization is then often predicted using thermo-chemical equilibrium modeling (FACTSage).

Buhere et al. (2005) and several others observed that aerosols are mainly formed due to condensation of released gaseous species rather than fragmentation. There are two competing routes for the condensing vapor. Firstly, the vapor may condense directly onto the internal surface of the furnace forming slag. Alternatively, the vapors may undergo gas-to-particle conversion to form aerosols by either homogeneous nucleation or heterogeneous condensation on existing particles entrained in the flue gas (Doshi et al. 2009). The droplets and aerosols begin to form larger particles through coagulation and agglomeration until finally accumulating as ash particles. The condensation of the aerosols onto the coarse ash fraction developed by fragmentation or coalescence is also possible. Several methods/models ranging from simple calculation to numerical modeling have been developed to predict the gas-to-particle conversion processes.

Doshi et al. (2009) have reported simple calculations based on the aerosol formation of the alkali chlorides and sulfates to model the aerosol ash formation. In the numerical methods, finite element techniques have also been applied with some degree of success (Gelbard and Seinfeld 1978), and the method of moments can be applied if the equations for evolution of the moments of the size distribution can be obtained in a closed form (McGraw 1997). However, Gelbard et al. (1980) found that for problems involving simultaneous nucleation, growth, and coagulation, the methods most widely used are based on a sectional representation of the size distribution. In sectional methods, the size distribution is divided into a number of sections or size classes within which all particles are assumed to have the same properties. Jokiniemi et al. (1994) used plug flow model developed by Im et al. (1985) for aerosol dynamics by simulating alkali species during coal combustion process. However, the use of fixed size sections for problems involving aerosol growth, may lead to numerical diffusion that in turn results in the error of sharp changes in size distributions. Gelbard (1990) introduced the moving sectional method for gas-to-particle conversion. Jacobson and Turco (1995) and Wu and Biswas (1998) modified aspects of coagulation and condensational growth. Later on Christensen and Livbjerg (2000) numerically simulated the Plug flow model with nucleation, growth, coagulation and gas phase reactions using moving sectional method. Zeuthe (2007) in his Ph.D. thesis, validated the one dimensional model of Christensen for aerosol formation from biomass fuels such as straw and household waste with a detailed view at particle composition and particle size distribution. The main particle formation mechanisms included in the aerosol formation plug flow model are nucleation, condensation, coagulation and agglomerations, together with the precipitation mechanisms (diffusion, thermophoresis, inertial impaction and gravitational settling) on to the particle or on the furnace wall. The gas phase is usually modeled with thermodynamic equilibrium model (FACTSage) or with advanced fuel characteristics.

The mentioned simulation models, focused primarily on coal combustion and if biomass combustion was considered, straw was the fuel of choice. Furthermore, if particle formation mechanisms were treated in detail, either alkali metal compounds or heavy metal compounds were considered for particle formation from the gas phase.

Summary

1. It can be inferred from the literature review that ash formation during PF combustion depends on several fuel characteristics and operating parameters.
2. PF combustion is performed normally at atmospheric pressure, at a high temperature with high heating rates. Though operating parameters such as temperature, pressure, heating rate will be in a relatively narrow, specified range for PF combustion, the effect of operating conditions will be varying significantly for different fuels. The measures of the fuel characteristics such as char reactivity, ash, moisture, volatile matter and fixed carbon percentages, as well as its density and porosity will also be in same range for the same fuel with varying particle size. Therefore, each particle size range in a single fuel will behave differently under PF combustion conditions. Kinetic reaction rates of the mineral chemical conversions are also highly dependent on several operating parameters (temperature, pressure, residence time etc.) and fuel characteristics (ash contents, mineralogy, particle size etc.).
3. Several analytical methods/tools/models are available and used so far to quantify ash formation process. Proximate and ultimate analysis of the fuels are the basic and essential primary test which decides several design parameters and gives an idea about the mineral percentage present in the fuel. To know the mineral matter composition in the fuel or residual ash, several traditional ash analysis techniques such as atomic absorption spectrophotometer (AA), graphite furnace atomic absorption (GFAA), inductively coupled plasma optical emission spectrometry (ICP/ICP-OES), inductively coupled plasma/mass spectrometry (ICP/MS), X-ray fluorescence spectrometry (XRF), glow discharge mass spectrometry (GDMS) and spark source mass spectrometry (SSMS) are used. However, the traditional ash analysis methods are time consuming and of limited scope sometimes as they are inadequate of providing the mineral matter distribution with varying particle sizes in the fuel or residual ash. Therefore, these techniques are complimented by more efficient and advanced CCSEM analysis. The use of CCSEM has increased considerably in the last two decades. QEMSCAN has also been used in place of CCSEM to determine mineral-mineral associations, particle size, mineral compositions, and particle texture in the coal and ash samples with higher degree of automation in recent efforts. To decide the mineral matter association in the fuel mineral matrix, pH

dependent leaching and/or chemical fractionation methods are often employed. STA/TGA is performed to decide the overall reaction kinetics which includes residence time, char oxidation and devolatilization rate etc. Thermochemical equilibrium models are used to find out the stable gaseous and solid species at given operating conditions.

4. The mathematical modeling efforts made to predict ash formation are also quite significant. The modeling of ash formation is mainly divided into two parts such as coarse ash formation due to the fragmentation/coalescence and aerosol ash formation due to vaporization followed by condensation. Models mentioned in this paper, focused mostly on coal combustion and if biomass combustion was considered, straw was the most commonly modeled fuel. Furthermore, if particle formation mechanisms were treated in detail, either alkali metal compounds or heavy metal compounds were considered for particle formation from the gas phase.

Conclusions and Future Research Needs

1. Ash formation during PF combustion is a very complex phenomenon, depending on a broad range of variables, either associated with the fuel or the conversion technology.
2. Different fuels will have different physical and chemical properties and therefore will behave differently during combustion. Moreover, within the same fuel, different particle size will have different physical and chemical properties after milling and classifying based on size and density. The investigations on lab/pilot/plant scale with narrow/single size range particles under well-controlled conditions and a greater number and more diverse fuels are very limited in the literature and therefore much needed in future research to predict the ash formation process more precisely.
3. Analytical methods are well advanced to determine the particle sizes and their mineralogy at different residence times. Nevertheless, the particle size reduction rate is difficult to measure or calculate accurately even with the methods available in the literature. This is primarily due to the various size altering processes (such as burning and fragmentation) occurring simultaneously during combustion. The size reduction rate is often assumed in the models or derived inaccurately from the experiments. Innovatively, the use of Particle image velocimetry (PIV) is appreciated for the same in future.
4. The determination of the extent of slagging and fouling phenomena are usually tackled on the industrial scale by quantitative methods such as thermal conductivity and slag thickness measurements, ash deposition probes etc., which are good indicators for local phenomena but are not efficient in bringing the overall chart along with more importantly the reasons and details of the underlying deposition mechanisms. These aspects will therefore need more scientific attention in the future.
5. Aerosol formation creates serious environmental and health hazards while coarse ash creates problems such as erosion and also participates in slagging, fouling,

corrosion etc. At the research level, they are often quantified by expensive and time consuming lab-pilot-plant scale trials. The use and development of ash formation modeling to date is very limited in predicting various ash related problems. The experimental and measurement techniques are more accurate but often time consuming and expensive. Therefore, the development and extensive use of ash formation modeling is highly recommended in the future.
6. Ash formation models are mainly of two types: coarse ash and aerosol formation. They are mostly used separately and for different purposes such as flow, slagging, fouling, corrosion erosion, environmental and health hazards modeling, and to date seldom brought together and interlinked to simulate the overall combustion and ash formation process. The integration of these ash transformations is essential to predict overall ash formation hence, the integration of both of these models is highly appreciated in future to detail the overall ash formation process in detail.
7. The models typically developed for coals are yet to be validated for biomass and co-firing conditions.

Acknowledgement The authors gratefully acknowledge the financial support of the EC through the research project BIOASH—SES6-CT-2003-502679. The authors also want to thank staff from Energy Research Center of the Netherlands, Holland and Curtin University, Australia for their support and assistance.

References

Badzioch, S., & Hawksley, P. G. W. (1970). Kinetics of thermal decomposition of pulverized coal particles. *Industrial Engineering and Chemistry Process Design and Development, 9*, 521–530.

Barrosoa, J., Ballester, J., & Pinaa, A. (2007). Study of coal ash deposition in an entrained flow reactor: Assessment of traditional and alternative slagging indices. *Fuel Processing Technology, 88*, 865–876.

Barta, L. E., Toqan, M. A., Beer, J. M., & Sarofim, A. F. (1992). *Prediction of fly ash size and chemical composition distributions: The random coalescence model, Twenty-fourth symposium (international) on combustion* (pp. 1135–1144). Pittsburgh, PA: The Combustion Institute.

Baxter, L. L. (1992). Char fragmentation and fly ash formation during pulverized-coal combustion. *Combustion and Flame, 90*, 174–184.

Baxter, L. L. (1993). Ash deposition during biomass and coal combustion: a mechanistic approach. *Biomass and Bioenergy, 4*(2), 85–102.

Baxter, H., Lu, H., Ip, E., Scott, J., Foster, J, & Vickers, M. (2008). Effect of particle shape and size on devolatilization of biomass particle. *Fuel*. doi:10.1016/j.fuel.2008.10.023.

Benfell, K. E., & Bailey, J. G. (1998). *Comparison of combustion and high pressure pyrolysis chars from Australian black coals* (pp. 157-162). Eighth Australian Coal Science Conference.

Benson, S. A., Erickson, T. A., Jensen, R. R., & Laumb, J. D. (2002). Transformations model for predicting size and composition of ash during coal combustion. *Fuel Chemistry Division Preprints, 47*(2), 796.

Bridgemana, T. G., Darvell, L. I., Jones, J. M., Williams, P. T., Fahmi, R., Bridgwater, A. V., et al. (2007). Influence of particle size on the analytical and chemical properties of two energy crops. *Fuel, 86*, 60–72.

Brunner, T., Obernberger, I., Jöller M., Arich, A., & Polt, P. (2001) Behavior of ash forming compounds in biomass furnaces – Measurement and analysis of aerosols formed during fixed bed biomass combustion (pp. 75–80). Aerosols from Biomass Combustion (International Seminar), International Energy Agency (IEA) and the Swiss Federal Office of Energy, BioenergyTask 32: Biomass Combustion and Cofiring (2001).

Buhre, B. J. P., Hinkley, J. T., Gupta, R. P., Wall, T. F., & Nelson, P. F. (2005). Submicron ash formation from coal combustion. *Fuel, 84*, 1206–1214.

Charon, O., Sarofim, A. F., & Beer, J. M. (1990). Distribution of mineral matter in pulverized coal. *Progress in Energy and Combustion Science, 16*, 319–326.

Chen, Y., Shah, N., Huggins, F. E., Huffman, G. P., Linak, W. P., & Miller, C. A. (2004). Investigation of primary fine particulate matter from coal combustion by computer-controlled scanning electron microscopy. *Fuel Processing Technology, 85*, 743–761.

Christensen, K. A., & Livbjerg, H. (2000). A plug flow model for chemical reactions and aerosol nucleation and growth in an alkali-containing flue gas. *Aerosol Science and Technology, 33*, 470–489.

da Silva Filho C. G., & Milioli, F. E. (2008). A thermogravimetric analysis of the combustion of a Brazilian mineral coal. *Química Nova 31*, doi:10.1590/S0100-40422008000100021

Dacombe, P., Pourkashanian, M., Williams, A., & Yap, L. (1999). Combustion-induced fragmentation behavior of isolated coal particles. *Fuel, 78*, 1847–1857.

Dayton, D. C., Belle-Oudry, D., & Nordin, A. (1999). Effect of coal minerals on chlorine and alkali metals released during biomass/coal cofiring. *Energy & Fuels, 13*, 1203–1211.

Demle, S., Ensor, D. S., & Ranade, M. B. (1982). Coal combustion aerosol formation mechanisms: A review. *Aerosol Science and Technology, 1*, 119–133.

Doshi, V., Vuthaluru, H. B., Korbee, R., & Kiel, J. H. A. (2009). Development of a modeling approach to predict ash formation during co-firing of coal and biomass. *Fuel Processing Technology, 90*(9), 1148–1156.

Dunn-Rankin, D. (1988). Kinetic model for simulating the evolution of particle size distributions during char combustion. *Combustion Science Technology, 58*, 297–314.

Dunn-Rankin, D., & Kerstein, A. R. (1987). Numerical simulation of particle size distribution evolution during pulverized coal combustion. *Combustion and Flame, 69*, 193–209.

Edwards, B. F., & Ghosal, A. K. (1988). *Model of ash size distribution from coal char oxidation*. Morgantown, WV: Department of Physics, West Virginia University.

Erickson, T. A., Ludlow, D. K., & Benson, S. A. (1992). Fly ash development from sodium, sulphur and silica during coal combustion. *Fuel, 71*, 15–18.

European Biomass Industry Association official website. Retrieved from http://www.eubia.org/333.0.html.

Ezra, Z., & Kantorovich, I. I. (2001). Mutual effects of porosity and reactivity in char oxidation. *Progress in Energy and Combustion Science, 27*, 667–697.

Flagen, R. C., & Friedlander, S. K. (1978). Recent developments. In D. T. Shaw (Ed.), *Aerosol science*. New York, NY: Wiley.

Frandsen, F. J., van Lith, S. C., Korbee, R., Yrjas, P., Backman, R., Obernberger, I., et al. (2007). Quantification of the release of inorganic elements from biofuels. *Fuel Processing Technology, 88*, 1118–1128.

Gale, T. K., Bartholomew, C. H., & Fletcher, T. H. (1995). Decreases in the swelling and porosity of bituminous coals during devolatilization at high heating rates. *Combustion and Flame, 100*, 94–100.

Gelbard, F. (1990). Modeling multicomponent aerosol particle growth by vapor condensation. *Aerosol Science and Technology, 12*, 399–412.

Gelbard, F., & Seinfeld, J. H. (1978). Numerical solution of the dynamic equation for particulate systems. *Journal of Computational Physics, 28*, 357–375.

Gelbard, F., Tambour, Y., & Seinfeld, J. H. (1980). Sectional representation for simulating aerosol dynamics. *Journal of Colloid Interface Science, 76*(2), 541–556.

Gupta, R. P. (2005). Coal research in Newcastle—Past, present and future. *Fuel, 84*, 1176–1188.

Helble, J. J., & Sarofim, A. F. (1989). Influence of char fragmentation on ash particle size distributions. *Combustion and Flame, 76*, 183–196.

Hurt, R. H., Calo, J. C., Fletcher, T., & Sayre, A. (2003). *Fundamental Investigations of Fuel Transformations in Advanced Coal Combustion and Gasification Processes.*

Hurt, R. H., Sarofim, A. F., & Longwell, J. P. (1991). The role of microporous surface area in the gasification of chars from a subbituminous coal. *Fuel, 70*, 1079–1082.

Im, K. H., Ahluwalia, R. K., & Chuang, C. F. (1985). RAFT: A computer model for formation and transport of fission product aerosols in LWR primary systems. *Aerosol Science and Technology, 4*, 125–140.

Jacobs, M. L. (2000). *Instrumentation for elemental analysis of coal ash products*. Golden, CO: Denver Division, Wyoming Analytical Laboratories, Inc. http://www.mcrcc.osmre.gov/PDF/Forums/CCB3/2-2.pdf.

Jacobson, M. Z., & Turco, R. P. (1995). Simulating condensational growth, evaporation, and coagulation of aerosols using a combined moving and stationary grid. *Aerosol Science and Technology, 22*, 73–92.

Jokiniemi, J. K., Lazaridis, M., Lehtinen, K. E. J., & Kauppinen, E. I. (1994). Numerical simulation of vapor-aerosol dynamics in combustion processes. *Journal of Aerosol Science, 25*(3), 429–446.

Kaiho, M., & Toda, Y. (1979). Change in thermoplastic properties of coal under pressure of various gases. *Fuel, 58*, 397–398.

Kang, S. G. (1991). *Fundamental studies of mineral matter transformations during pulverized coal combustion*. PhD thesis.

Kang, S. G., Helble, J. J., Sarofim, A. F., & Beer, J. M. (1988). *Time-resolved evolution of fly ash during pulverized coal combustion. Proceedings twenty second symposium (international) on combustion* (pp. 231–238). Pittsburgh, PA: The Combustion Institute.

Kang, S. G., Kerstein, A. R., Helble, J. J., & Sarofim, A. F. (1990). Simulation of residual ash formation during pulverized coal combustion: Bimodal ash particle size distribution. *Aerosol Science and Technology, 13*, 401–412.

Kang, S. G, Sarofim, A. F., Beer, J. M. (1992). Effect of char structure on residual ash formation during pulverized coal combustion (pp. 1153–1159). 24th Symposium (international) on combustion, The Combustion Institute.

Kerstein, A. R., & Edwards, B. F. (1987). Percolation model for simulation of char oxidation and fragmentation time-histories. *Chemical Engineering Science, 42*(7), 1629–1634.

Kerstein, A. R., & Niksa, A. (1984). *Fragmentation during carbon conversion: predictions and measurements Proceedings of twentieth symposium (international) on combustion* (pp. 941–949). Pittsburgh, PA: The Combustion Institute.

Koranyi, A. D. (1989). The relationship between specific reactivity and the pore structure of coal chars during gasification. *Carbon, 27*, 55–61.

Korbee, R., Lensselink, J., & Cieplik, M. (2006). *Energy Research Centre of the Netherlands, Final report of task 1.3 – Release of ash forming matter in pulverized fuel systems, SES6-CT-2003-502679*. (2006).

Kramlich, J. C., Chenvert, B., & Park, J. (1995). *Suppression of fine ash formation in pulverized coal flames*, DOE grant no. DE-FG22–92PC92548, Quarterly Technical progress report no.10, For U.S. Department of Energy.

Kurose, R., Makino, H., Hashimoto, N., & Suzuki, A. (2007). Application of percolation model to particulate matter formation in pressurized coal combustion. *Powder Technology, 172*(1), 50–56.

Kurose, R., Matsuda, H., Makino, H., & Suzuki, A. (2003). Characteristics of particulate matter generated in pressurized coal combustion for high-efficiency power generation system. *Advanced Powder Technology, 14*(6), 673–694.

Liu, Y., Gupta, R. P., Sharma, A., Wall, T. F., Butcher, A., Miller, G., et al. (2005). Mineral matter-organic matter association characterisation by QEMSCAN and applications in coal utilization. *Fuel, 84*, 1259–1267.

Liu, G., Wu, H., Gupta, R. P., Lucas, J. A., Tate, A. G., & Wall, T. F. (2000). Modeling the fragmentation of non-uniform porous char particles during pulverized coal combustion. *Fuel, 79*, 627–633.

Liu, X., Xu, M., Yao, H., Yu, D., Lv, D., & Zhou, K. (2008). The formation and emission of particulate matter during the combustion of density separated coal fractions. *Energy & Fuels, 22*, 3844–3851.

Livingston, W. R. (2007). *Biomass ash characteristics and behavior in combustion, gasification and pyrolysis systems, Report No: 34/07/005, Project/Sub-Project:78541/SD001*. Crawley: Doosan Babcock Energy Limited.

Lockwood, F. C., Costen, P. G., Siddiqi, M. M., & Harrison, P. J. *Mineral ash transformations, technical report SW7 2BX JOF3-CT95-0024*. The Imperial College of Science, Technology and Medicine; Mechanical Engineering Department; London. ftp://ftp.euro-cleancoal.net/pub/pdf/JOF3-CT95-0024-pdf-files/JOF3-CT95-0024-02%20Lockwood-ICSTM.pdf

Lu, H., Jia, C., Zhang, L., & Su, G. (2007). *The Study on Combustion Characteristics and Kinetics of Coal and Biomass, Challenges of Power Engineering and Environment, Springer, Berlin, Heidelberg(doi.* doi:10.1007/978-3-540-76694-0).

Mathews, J. P., Hatcher, P. G., & Scaroni, A. W. (1997). Particle size dependence of coal volatile matter: is there a nonmaceral- related effect? *Fuel, 76*, 359–362.

McGraw, R. (1997). Description of aerosol dynamics by the quadratum method of moment. *Aerosol Science and Technology, 27*(2), 255–265.

Menendez, R., Vleeskens, J. M., & Marsh, H. (1993). The use of scanning electron microscopy for classification of coal chars during combustion. *Fuel, 72*, 611–617.

Miranda, T., Esteban, A., Rojas, S., Montero, I., & Ruiz, A. (2008). Combustion analysis of different olive residues. *International Journal of Molecular Sciences, 9*, 512–525.

Mitchell, R. E. (1997). Char fragmentation and its effect on unburned carbon during pulverized coal combustion, Contract no.: DE-FG22-92PC92528, For U.S. Department of Energy.

Mitchell, R. (2000). An intrinsic kinetics-based, particle population balance model for char oxidation during pulverized coal combustion. *Proceedings of the Combustion Institute, 28*, 2261–2270.

Mohanty, K. K., Ottino, J. M., & Davis, H. T. (1982). Reaction & and transport in disordered composite media: Introduction of percolation concepts. *Chemical Engineering Science, 37*, 905–924.

Morone, L. S. (1989). *An experimental and modeling study of residual fly ash formation in combustion of a bituminous coal*. PhD thesis, Massachusetts Institute of Technology.

No, S. Y., & Syred, N. (1990). Thermal stress and pressure effects on coal particle fragmentation and burning behavior in a cyclone combustor. *Journal of the Institute of Energy, 63*, 151–159.

Oberberger, I. (2003). Aerosols in fixed-bed biomass combustion, Bio-aerosols, Budapest, 16—17 October 2003. Retrieved from ftp://ftp.cordis.europa.eu/pub/sustdev/docs/energy/bioenergy_c04.pdf.

Reyes, S., & Jensen, K. F. (1986). Percolation concepts in modeling of gas-solid reactions—I. Application to char gasification in the kinetic regime. *Chemical Engineering Science, 41*(2), 333–343.

Salatino, P., Miccio, F., & Massimilla, L. (1992). *Monte Carlo simulation of combustion induced percolative fragmentation of carbons. Twenty-fourth symposium (international) on combustion* (pp. 1145–1151). Pittsburgh, PA: The Combustion Institute.

Salatino, P., Miccio, F., & Massimilla, L. (1993). Combustion and percolative fragmentation of carbons. *Combustion and Flame, 95*, 342–350.

Sarofim, A. F., & Helbe, J. J. (1994). The impact of ash deposition on coal fired plants. In J. Williamson & F. Wigley (Eds.), *Proceedings of the engineering foundation conference (1993), Solihull, England*. London: Taylor & Francis.

Sarofim, A. F., Howard, J. B., & Padia, A. S. (1977). The physical transformation of the mineral matter in pulverized coal under simulated combustion conditions. *Combustion Science and Technology, 16*, 187–204.

Schurmann, H., Monkhouse, P. B., Unterberger, S., & Hein, K. R. G. (2007). In situ parametric study of alkali release in pulverized coal combustion. *Proceedings of the Combustion Institute, 31*, 1913–1920.

Shah, K. V., Cieplik, M. K., Betrand, C. I., van de Kamp, W. L., & Vuthaluru, H. B. (2010). A kinetic-empirical model for particle size distribution evolution during pulverized fuel combustion. *Fuel, 89*(9), 2438–2447.

Speight, J. G. (2005). *Handbook of Coal Analysis (chemical Analysis: A Series of Monographs on Analytical Chemistry and Its Applications), ISBN 0-471-52273-2 (cloth), Wiley-Interscience.*

Suzuki, A., Yamamoto, T., Aoki, H., & Miura, T. (2002). Percolation model for simulation of coal combustion process. *Proceedings of the Combustion Institute, 29*(1), 459–466.

Syred, N., Kurniawan, K., Griffiths, T., Gralton, T., & Ray, R. (2007). Development of fragmentation models for solid fuel combustion and gasification as subroutines for inclusion in CFD codes. *Fuel, 86*, 2221–2231.

ten Brink, H. M., Eenkhoorn, S., & Weeda, M. (1996). The behavior of coal mineral carbonates in a simulated coal flame. *Fuel Processing Technology, 47*, 233–243.

Terama, T., Yamashita, T., & Ando, T. (1999). *Behavior of Inorganic Materials during Pulverized Coal Combustion*. Springer US: Impact of Mineral Impurities in Solid Fuel Combustion.

Terttalisia, L. (1999). *Ash formation in circulating fluidized bed combustion of coal and solid biomass*. PhD thesis, Helsinki University of Technology, Finland

Thy, P., Lesher, C. E., & Jenkins, B. M. (2000). Experimental determination of high-temperature elemental losses from biomass slag. *Fuel, 79*, 693–700.

Van Lith, S. C. (2005) *Release of inorganics elements during wood-firing on grate*. PhD Thesis, CHEC Research Centre, Technical University of Denmark.

Vuthaluru, H. B. (2004). Investigations into the pyrolytic behavior of coal/biomass blends using thermogravimetric analysis. *Bioresource Technology, 92*, 187–195.

Vuthaluru, H. B., & French, D. (2008a). Ash chemistry and mineralogy of an Indonesian coal during combustion: Part 1. Drop-tube observations. *Fuel Processing Technology, 89*(6), 595–607.

Vuthaluru, H. B., & French, D. (2008b). Ash chemistry and mineralogy of an Indonesian coal during combustion: Part II — Pilot scale observations. *Fuel Processing Technology, 89*(6), 608–621.

Wall, T. F., Liu, G. S., Wu, H. W., Roberts, D. G., Benfell, K. E., Gupta, S., et al. (2002). The effect of pressure on coal reactions during pulverized coal combustion and gasification. *Progress in Energy and Combustion Science, 28*, 405–433.

Wang, Q., Zhang, L., Sato, A., Ninomiya, Y., & Yamashita, T. (2007). Interactions among inherent minerals during coal combustion and their impacts on the emission of PM10. 1. Emission of micrometer-sized particles. *Energy and Fuels, 21*(2), 756–765.

Wigley, F., & Williamson, J. (1998). Modeling fly ash generation for pulverized coal combustion. *Energy Combustion and Science, 24*, 337–343.

Wigley, F., Williamson, J., & Gibb, W. H. (1997). The distribution of mineral matter in pulverized coal particles in relation to burnout behavior. *Fuel, 76*(13), 1283–1288.

Wilemski, G., & Srinivasachar, S. (1993). Prediction of ash formation in pulverized coal combustion with mineral distribution and char fragmentation models (pp. 151–164). *Proceeding of the engineering foundation conference*, England.

Wilemski, G., Srinivaschar, S., & Sarofim, A. (1992). *Modeling of mineral matter redistribution and ash formation in pulverized coal combustion* (p. 545). New York: ASME.

Wolf, K. F. (2002) Studies of alkali sorption kinetics for pressurized fluidized bed combustion by high pressure mass spectrometry. Retrieved from www.osti.gov/bridge/servlets/purl/836714.pdf.

Wu, C. Y., & Biswas, P. (1998). Study of numerical diffusion in a discrete-sectional model and its application to aerosol dynamics simulation. *Aerosol Science and Technology, 29*, 359–378.

Wu, H., Bryant, G., & Wall, T. (2000). The effect of pressure on ash formation during pulverized coal combustion. *Energy & Fuels, 14*, 745–750.

Wu, H., Wall, T., Liu, G., & Bryant, G. (1999). Ash liberation from included minerals during combustion of pulverized coal: The relationship with char structure and burnout. *Energy and Fuels, 13*(6), 1197–1202.

Yan, L., Gupta, R., & Wall, T. (2001a). The implication of mineral coalescence behavior on ash formation and deposition during pulverized coal combustion. *Fuel, 80*, 1333–1340.

Yan, L., Gupta, R., & Wall, T. (2001b). Fragmentation behavior of pyrite and calcite during high-pemperature processing and mathematical simulation. *Energy and fuels, 15*, 389–394.

Yu, D., Xu, M., Zhang, L., Yao, H., Wang, Q., & Ninomiya, Y. (2007). Computer-controlled scanning electron microscopy (CCSEM) – Investigation on the heterogeneous nature of mineral matter in six typical Chinese coals. *Energy and Fuels, 21*(2), 468–476.

Yua, J., Lucasb, J. A., & Wall, T. F. (2007). Formation of the structure of chars during devolatilization of pulverized coal and its thermoproperties: A review. *Progress in Energy and Combustion Science, 33*, 135–170.

Zeuthe, J. H. (2007). *The formation of aerosol particles during combustion of biomass and waste*. PhD thesis, The Aerosol Laboratory, Department of Chemical Engineering, Technical University of Denmark, Lyngby.

Zhaosheng, Y., Xiaoqiana, M., & Aoa, A. (2009). Thermogravimetric analysis of rice and wheat straw catalytic combustion in air- and oxygen-enriched atmospheres. *Energy Conversion and Management, 50*, 561–566.

Chapter 4
Waste Management Methods and Sustainablity

S.M. Faheem and M.A. Khan

Introduction

Waste is generated as a result of production and consumption activities and tends to increase with the level of prosperity and economic development of the country. The primary sources of wastes are domestic and industrial activity, e.g. sewage, agriculture and food wastes from processing, textile and leather processing wastes and a wide range of toxic industrial chemical products and by-products (Smith 2009). In the final assessment, wastes represent the end of the technical and economic life of products. Costs for adequately dealing with wastes are escalating, and much attention is presently devoted to efficient and effective waste management, which will include costs of collection, storage, processing and removal of wastes. Wastes are broadly classified as Solid and liquid category.

Solid Waste Management

Solid waste is unwanted solid material generated from combined residential, industrial and commercial activities in a given place usually comprises of non-hazardous refuse that contain less than 70 % water. It can be categorized according to its origin (domestic, industrial, commercial, construction or institutional); according to its contents (organic material, glass, metal, plastic paper etc.);

S.M. Faheem (✉)
School of Life Sciences, Manipal University, P.O. Box 345050,
Academic City, Dubai, United Arab Emirates
e-mail: proffaheem@gmail.com

M.A. Khan
College of Sustainability Sciences and Humanities, Zayed University, P.O. Box 19282,
Dubai, United Arab Emirates

P. Ravindra (ed.), *Advances in Bioprocess Technology*,
DOI 10.1007/978-3-319-17915-5_4

or according to hazard potential (toxic, non-toxic, flammable, radioactive, infectious etc.). Global generation of municipal solid waste alone is projected to rise to 69.3 % by 2025, with average annual increase of +5.3 %. Currently it stands at 1.3 billion tons/year and is likely to be >2.2 billion tons/year by 2025. A major chunk (97 %) of this growth will be contributed by Asia and Africa alone (Halbach 2013). Management of solid waste is necessary to reduce or eliminate adverse effects on the environment and human health and improves quality of life and contributes to economic development. A number of processes are involved in effectively managing waste for a municipality. These include monitoring, collection, transport, processing, recycling and disposal. Viable private sector participation (PSP) also becomes necessary in effective waste management. Appalling attitude of waste generators towards waste management personnel, lack of good transport network and traffic congestions and unwillingness to pay for the services by the waste generator are significant challenges for the operation of PSP. Hence, Knowledge and perceptions on safety measure practices and challenges of PSP operators need to be addressed in order to make PSP operated waste management business viable (Sridhar and Adejumo 2014). Most countries adapt the following traditional and modern approaches of solid waste disposal.

Solid Waste Disposal Methods

Open Dumping

Also referred to as crude tipping, is the indiscriminate dumping of solid waste without environmental and health concerns. In developing countries, most of the municipal solid waste is dumped on land, far from cities, in an uncontrolled manner. Open dumping is un-economical use of the available space. It often produces unpleasant odors and is not a sustainable and eco-friendly practice.

Barging into the Sea

Coastal and riverine communities with inadequate arable lands use canoes and barges to carry solid waste far into rivers and seas for disposal. Dumping of solid waste into water may result in eutrophication which causes destruction of aquatic life due to depletion of oxygen and thus makes the water unsuitable for drinking and domestic use.

Open Air Burning

Involves the ignition of waste matter in the open from which the products of the combustion are emitted directly into the air without passing through a stack of chimney filter. Open burning has been practiced by a number of urban centers because it reduces the volume of refuse received at the dump and therefore extends the life of their dumpsite. These include dioxins, particulate matter, polycyclic aromatic compounds, volatile organic compounds, carbon monoxide, hexachlorobenzene and ash. All of these chemicals pose serious risks to human health. The dioxins are capable of producing a multitude of health problems; they can have adverse effects on reproduction, development, disrupt the hormonal systems or even cause cancer. The polycyclic aromatic compounds and the hexachlorobenzene are considered to be carcinogenic. The particulate matter, apart from creating smoke and haze in the air, can also lead to many respiratory problems such as asthma or bronchitis while carbon monoxide can cause neurological symptoms. Open burning also effects the environment by releasing acidic gases of halo-hydrides and oxides of nitrogen and carbon. Oxides of nitrogen contribute to smog, acid rain, ozone depletion and global warming. Carbon monoxide being a greenhouse gas also produces harmful ozone in reaction with sunlight.

Hog Feeding

It is a method of feeding garbage to domestic animals. This practice is encouraged as it is in compliance with the principle of integrated waste management. It encourages waste sorting and reuse, energy recycling and cost reduction. However the practice is restricted to domestic garbage and requires sorting at source.

Landfills

Landfills are sites where waste is isolated from the environment until it is safe. It is the oldest and most prevalent form of solid waste disposal with open dumps in most countries and sanitary landfills with high degree of planning, engineering and administration in developed nations. Modern landfills are engineered to isolate the waste and resultant leachate from the surrounding environment. A landfill waste is considered safe, when it has completely degraded biologically, chemically and physically. Sanitary landfills are most common in developed countries. A Sanitary landfill also known as controlled tipping is the practice of solid waste disposal in which waste is stacked to decay under controlled conditions. A typical landfill system comprises of a bottom liner to prevent direct contact with the soil, disposal cells approved for waste deposition, a leachate collection system, rain water drain system and methane gas vents. Before a site can be regarded as a sanitary landfill, four basic conditions are to be ensured.

- Full or partial hydro-geological isolation: The land site should naturally contain leachate from the solid waste else additional lining materials should be brought to the site to reduce leakage from the base. Containment of leachate reduces contamination of groundwater and soil.
- Engineering designs should be developed from local geological and hydro-geological investigations and in compliance to disposal and restoration plans.

- Control: trained personnel based at the landfill are deployed for supervise of site during preparation and construction, deposition of waste and the regular operation and maintenance.
- Waste deposition and covering: solid waste should be spread in compacted layers. A small working area which is covered daily helps make the waste less accessible to worms and pests.

Sometimes, Mechanical Biological Treatment (MBT) is undertaken prior to landfilling. Such a pre-treatment is required to make the material more conducive to environment. This is usually done in four stages: waste input and control, mechanical conditioning, biological treatment and emplacement of treated waste at a landfill. The mechanical stage sorts out non-biodegradable and recyclable material. Next, the residual waste is prepared for biological treatment by comminution, mixing and even moistening. The waste is exposed to atmospheric oxygen to induce aerobic decomposition, or anaerobically broken down in the absence of atmospheric oxygen before it is deposited. The MBT process reduces the waste volume, lengthens the useful life of the landfill and also reduces the rate of gas formation thus reducing the danger of landfill fires and the leachate load. Landfills could be identified in various types as follows:

1. Municipal solid waste landfill: uses plastic or synthetic liner to isolate trash from the environment.
2. Sanitary Landfill: uses a clay liner to separate the waste from the environment.
3. Industrial waste landfill: comprises nontoxic waste associated with industrial and manufacturing processes.
4. Construction and demolition waste landfill: consists of the debris generated during demolition, renovation and construction of infrastructural facilities.

Incineration

It is the process where waste is subjected to burning at high temperatures, in the range of 900–1,200 °C, producing steam and ash. This is a controlled combustion process to convert solid combustible wastes into gases and residues in the form of ash. It is a volume reduction process. When incineration is operated correctly, it reduces the volume of waste bulk by 90 %. Incineration however, does not replace the need for land filling but reduces the amount to be thrown in. It is recognized as a practical way of disposing biohazardous waste from hospitals, clinics and other places. Incineration can be moving grate, fixed grate, rotary kiln and fluidized bed types.

Incineration completely destroys toxic materials. Incineration is widely practiced in developing countries while being criticized for destroying not only the raw materials, but also all the energy, water, and other natural resources used to produce it. Some energy can be reclaimed as electricity from the combustion steam which can drive electric generators.

Waste incineration has reached the first goal of waste management i.e. protection of people and the environment, and is also on the verge of meeting the second goal which is resource conservation and material recovery (Brunner and Rechberger 2015). Incineration is the key tool to destroy hazardous organic substances and to concentrate toxic metals in relatively small amounts of filter residues. Modern "Waste to energy" plants guarantee complete mineralization of organic substances including refractory compounds such as persistent organic pollutants. Microorganisms including pathogens are reliably destroyed and waste volume is also decreased in magnitude. Because of sophistication in air pollution control, the resulting emissions are way below emission standards. Technologies ensure immobilization of incineration residues before disposal. Hence, "waste to energy" seem to be an important tool for material recycling and resource conservation, too. Due to complexity in processes and large volumes of data, ingenious solutions to waste management are possible through information and communications technologies (Iona and Gheorgheb 2014).

Composting

Humus like material called compost is formed when microorganisms break down the organic matter in the waste. Composting is carried out aerobically (in presence of oxygen) and anaerobically (in absence of oxygen). During aerobic composting, aerobic microorganisms oxidize organic compounds in the solid waste to carbon dioxide, nitrite and nitrate. The carbon from organic compounds is used as a source of energy while nitrogen is recycled. Heat is generated due to metabolic activity of microbes. During anaerobic process, microorganisms break down organic compounds through a process of reduction releasing a very small amount of energy. The gases evolved are mainly methane and carbon dioxide.

Compost is valuable as manure in agriculture and depends on the quantity and quality of feed materials poured into the composting pit. Compost pits can be underground or over-ground.

Vermicomposting

Vermicomposting is the process of composting using various worms, usually earthworms to create a heterogeneous mixture of decomposing organic waste, bedding materials and vermicast. Vermicomposting is the result of combined activity of microorganisms and earthworms. Microbial decomposition of organic waste occurs through primary decomposition, by the action of extracellular enzymes, and the secondary decomposition again by the action of microorganisms inhabiting the earthworm gut. Ingested substrate materials are subjected to grinding inside worm gut resulting in reduction of particle size. Vermitechnology, a tripartite system which involves biomass, microbes and earthworms is influenced by the biotic factors such as temperature, moisture, aeration etc. Conditions adverse to aerobic decomposition result in mortality of earthworms and subsequently effects vermicomposting. Other unfavorable conditions such as temperature increase, anaerobic conditions, toxicity of decomposition products etc. influence activity of worms. Hence, preprocessing of the waste as well as providing favorable environmental condition is necessary for vermicomposting. The worms that are used in vermicomposting may include both exotic and indigenous varieties. But a local variety is recommended to the extent possible. These worms are known to survive in the moisture range of 20–80 % and the temperature range of 20–40 °C. The worms do not survive in pure organic substrates containing more that 40 % fermentable organic substances. Hence fresh waste is generally mixed with the stabilized waste before it is subjected to vermicomposting. The process of vermicomposting can be carried out in compost beds as well as tanks at both household and community levels.

The rate of compost formation is controlled by the composition and constituents of the substrate material particularly carbon to nitrogen (C/N) ratio, the temperature, the moisture content and air. The C/N ratio is critical for the process to be efficient. The microorganisms require carbon as an energy source and nitrogen for the synthesis of some proteins. Moisture content, otherwise called water activity, significantly influences metabolism of microbes and hence the composting process. A higher temperature eliminates pathogenic organisms. However, temperatures above 75 °C are not favorable to the useful microorganisms. Hence, optimum temperature for the process is around 60 °C. Aeration is also crucial and the quantity of air needs to be properly controlled when composting. Insufficient oxygen kills aerobic species and thus will be replaced by anaerobic microbes. This eventually slows down the process and results in bad odor and evolves inflammable methane gas. Aeration can be achieved by churning the composting material.

Pyrolysis and Gasification

Pyrolysis and gasification are similar processes they both decompose organic waste by exposing it to high temperatures and low amounts of oxygen. Gasification uses a low oxygen environment while the pyrolysis allows no oxygen. These techniques use heat and an oxygen starved environment to convert biomass into other forms. These processes produce a mixture of combustible and non-combustible gases as well as pyroligenous liquid. All of these products have a high heat value and can be utilized. Gasification is advantageous since it allows for the incineration of waste with energy recovery and without the air pollution that is characteristic of other incineration methods.

Pulverization

Solid waste is shredded for several reasons, including volume reduction. Under certain circumstances, shredded refuse can be disposed of in a landfill. If solid waste is to be converted to refuse derived fuel (RDF), shredding and/or pulverizing is an element of the RDF production process. There are four types of shredders used for the shredding or pulverizing of solid waste: hammer mills, drum pulverizes, crushers, and wet pulverizes. Each of the equipment has a variety of designs, advantages, and disadvantages. Primary considerations in selecting a shredder are its capacity, speed, power requirements, maintenance needs, ability to produce the end product desired and, most important, and reliability. These characteristics will differ significantly for various types of solid waste and differing end products. In choosing the type and particular design of a shredder, it is desirable to obtain information on the performance of the shredder in conditions similar to those for which the machine is to be used.

Biogas Technology

When biodegradable organic solid waste is subjected to anaerobic decomposition, a gaseous mixture of Methane and Carbon dioxide, known as Biogas could be produced under favorable conditions. The decomposition of the waste materials is mainly carried out by anaerobic bacteria through fermentation process. The process involves a series of reactions by several kinds of bacteria feeding on the raw organic matter. The anaerobic digestion of the organic waste matter occurs in three different stages: hydrolysis, acidogenesis and methanogenesis.

Most of the organic waste suitable for methanogenesis is made up of cellulose, hemicellulose, lignin etc. which are insoluble in water. In the first step of digestion, these macromolecules are broken down to smaller molecules by enzymes secreted by bacteria. Large amount of carbon dioxide (CO_2) is released and the major end product of hydrolysis is glucose which becomes the substrate for the acid forming bacteria. The acid forming bacteria convert the water soluble substances into volatile acids with acetic acid as a major component. Besides this, some other acids like butyric acid, propionic acid and gasses like CO_2 and H_2 are also produced. The acid forming bacteria during the conversion process utilize the amount of oxygen remaining in the medium and make the environment anaerobic. In the last stage of the biogas generation, the methanogenic bacteria convert the volatile acids formed into Biogas which is a mixture of methane (55–65 %) and carbon dioxide (35–45 %) along with trace amounts of hydrogen, Hydrogen sulfide and ammonia. Biogas or the marsh gas is a combustible gas and can be used for heating, lighting, powering irrigation pump, generating electric power and for local use for cooking purpose. The gas is smokeless, environment friendly and efficient fuel. The by-product is slurry which is environment friendly and is used as organic fertilizer. It can be used to enrich the soil or even the compost.

Waste Reduction Through Reduce, Reuse and Recycle

Waste treatment techniques seek to transform the waste into a form that is more manageable, reduce the volume or reduce the toxicity of the waste thus making the waste easier to dispose of. Treatment methods are selected based on the

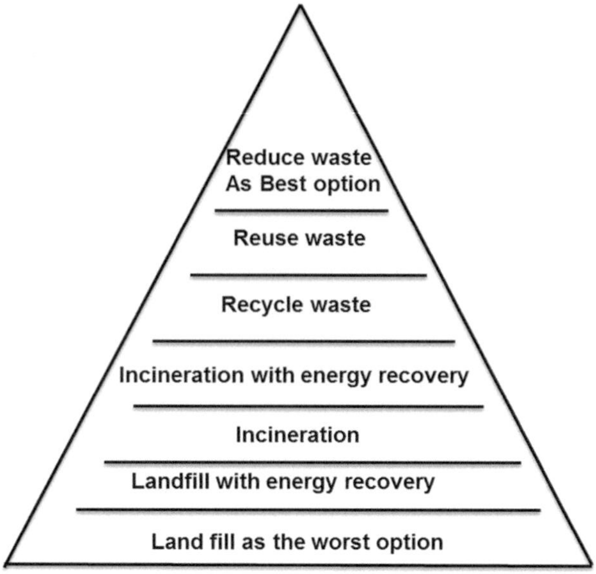

Fig. 4.1 Pyramid of solid waste management hierarchy (www.sustainability-ed.org)

composition, quantity, and form of the waste material. Some waste treatment methods being used today include subjecting the waste to extremely high temperatures, dumping on land or land filling and use of biological processes to treat the waste. It should be noted that treatment and disposal options are chosen as a last resort to management by reducing, reusing and recycling strategies (Fig. 4.1).

Methods of waste reduction, waste reuse and recycling are the preferred options when managing waste. There are many environmental benefits that can be derived from the use of these methods. They reduce or prevent greenhouse gas emissions, reduce the release of pollutants, conserve resources, save energy and reduce the demand for waste treatment technology and landfill space. Therefore it is advisable that these methods be adopted and incorporated as part of the waste management plan. Waste reduction and reuse Waste reduction and reuse of products are both methods of waste prevention. They eliminate the production of waste at the source of usual generation and reduce the demands for large scale treatment and disposal facilities. Methods of waste reduction include manufacturing products with less packaging, encouraging customers to bring their own reusable bags for packaging, encouraging the public to choose reusable products such as cloth napkins and reusable plastic and glass containers, backyard composting and sharing and donating any unwanted items rather than discarding them. All of the methods of waste prevention mentioned require public participation. In order to get the public onboard, training and educational programs need to be undertaken to educate the public about their role in the process. Also the government may need to regulate the types and amount of packaging used by manufacturers and make the reuse of shopping bags mandatory.

Recycling refers to the removal of items from the waste stream to be used as raw materials in the manufacture of new products. Thus from this definition recycling occurs in three phases: first the waste is sorted and recyclables collected, the recyclables are used to create raw materials. These raw materials are then used in the production of new products.

The sorting of recyclables may be done at the source (i.e. within the household or work area) for selective collection by the municipality or to be dropped off by the waste producer at a recycling centers. The pre-sorting at the source requires public participation which may not be forthcoming if there are no benefits to be derived. Also a system of selective collection by the government can be costly. It would require more frequent circulation of trucks within a neighborhood or the importation of more vehicles to facilitate the collection. Another option is to mix the recyclables with the general waste stream for collection and then sorting and recovery of the recyclable materials can be performed by the municipality at a suitable site. The sorting by the municipality has the advantage of eliminating the dependence on the public and ensuring that the recycling does occur. The disadvantage however, is that the value of the recyclable materials is reduced since being mixed in and compacted with other garbage can have adverse effects on the quality of the recyclable material.

Waste from our homes is generally collected by our local authorities through regular waste collection, or by special collections for recycling. Within hot climates such as that of the UAE, the waste should be collected at nights on daily basis to control fly breeding, and the harboring of other pests in the community. Other factors to consider when deciding on frequency of collection are the odors caused by decomposition and the accumulated quantities.

According to one study (Guerrero et al. 2013), solid waste management (SWM) is a challenge for municipal authorities in developing countries mainly due to the increasing generation of waste, burden on the municipal budget, lack of understanding of factors that affect the different stages of waste management and articulation of handling system and functioning. Recent reports on waste management from developing countries show very few articles with quantitative information. Waste management involves a large number of different stakeholders, with different fields of interest. They all play a role in shaping the system of a city, but often it is seen only as a responsibility of the local authorities. Detailed understandings on who the stakeholders are and the responsibilities they have in the structure are necessary steps in order to plan, change or implement waste management systems in cities and establish an efficient and effective system.

Solid waste management is a multi-dimensional issue. Municipalities in general seek for equipment as a path in finding solutions to the diversity of problems they face. An efficient system is not only based in technological solutions but also environmental, socio-cultural, legal, institutional and economic linkages that should be present to enable the overall system to function. Solid waste services incur costs from skilled personnel, equipment, right infrastructure, proper maintenance and operational resources that are not often recovered. The financial support of the central government, the interest of the municipal leaders in waste management issues, the participation of the service users and the proper administration of the funds are essential for a modernized sustainable system. Positive changes may require development of integrated waste management strategies and involvement of universities/research centers in preparing professionals and technicians in waste management.

Solid waste management has become an issue of increasing global concern as urban populations continue to rise and consumption patterns change. The health and environmental implications associated with SWM are rising, particularly in the context of developing countries. While systems analyses largely targeting well-defined, engineered systems have been used to help SWM agencies in industrialized countries, collection and removal dominate the SWM sector in developing countries. In developed countries, public health, environment, resource scarcity, climate change, and public awareness and participation have acted as drivers towards the current paradigm of integrated SWM. However, urbanization, inequality, and economic growth; cultural and socio-economic aspects; policy, governance, and institutional issues; and international influences have complicated SWM in developing countries. There is a need to develop new SWM approaches for developing countries. While the need for a systems approach to SWM has been both explicitly (Seadon 2010) and inexplicitly recognized through many calls for integrated methodologies, there is a lack of literature exploring the actual application of post-normal approaches and complex, adaptive systems thinking to SWM systems in context of developing country. This kind of publicly engaged systems thinking can provide some understanding to create approaches for coping with complexity (Waltner-Toews et al. 2008). It has been widely recognized that it is counterproductive for developing countries to use strategies and policies developed for developed countries. All developed countries have evolved their current systems in a series of steps; developing countries can benefit from that experience, but to expect to move from uncontrolled dumping to a 'modern' waste management system in one great leap is just not realistic" (Wilson 2007, p. 205). There is a need for new approaches emerging from the interface of SWM, post-normal science, and complex-adaptive systems research as the state of SWM systems in many developing regions continues to threaten and degrade the health of the most vulnerable human populations and the ecosystems they are a part of (Marshall and Farahbakhsh 2013).

An integrated sustainable waste management to examine how cities in developing countries have been tackling their solid waste problems is discussed in a report by Wilson et al. (2012). Such a system examines both the physical components (collection, disposal and recycling) and the governance aspects (users and service providers; financial sustainability; institutions supported by proactive policies). Evidence suggests that efficient, effective and affordable systems can be tailored to local needs and conditions, developed with direct involvement of service beneficiaries. Despite the challenges, sustainable solid waste and resources management is feasible for developing countries. About 20–30 wt% of recycling rate is achieved by the informal recycling sector in many developing countries. Integrating existing informal sectors in the formal system presents a huge win–win situation because of huge cost cutting and providing livelihood to large numbers of the urban poor. Although there are still many serious problems and numerous challenges, there are equally many opportunities and examples of city authorities and their communities working together to achieve appropriate and sustainable solutions. The evidence suggests that sustainable solid waste and resources management is a feasible proposition for developing countries.

Sustainability in Solid Waste Management

Sustainability ensures the path of reconciliation for man, nature, and economy. Responsibility for the waste management returns to those who produce it. Governments need to develop policies and regulatory frameworks as well as priorities for prevention, recovery and reducing impact of waste on health and environment. In 2005, management of wastes was estimated to produce about 5 % of the total greenhouse gas emissions (UNEP 2010). Waste recycling and composting systems are known to fulfil the clean development mechanism objectives, stated in Kyoto Protocol (United Nations 1997), better than landfill disposal systems. Composting systems provide twice the reduction in greenhouse gas emissions in comparison to landfill gas combustion projects. Landfill methane emissions are increasing in developing countries because of larger quantities of municipal solid waste from urban areas, economic development and engineered landfills (Couth and Trois 2012). The greenhouse gas emission reductions can be effectively provided over the period of waste production rather than over a generation. They are more sustainable in developing countries and do not leave an environmental hazard for future generations.

Wastewater Treatment

Declining reserves of potable water and increasing unit costs, have forced us to consider recycling of wastewater. Moreover, wastewater if not properly handled and disposed, could be hazardous to human health and environment. In the past twentieth century, modern water supply and sanitation is considered the most significant contribution to public health (Mackenbach 2007) and also, it is stated as one of the greatest engineering achievements of the century by National Academy of Engineering (Constable and Sommerville 2003). Yet we are faced with new challenges today to develop newer and more sustainable water and sanitation systems driven by the exploding growth of population, urbanization coupled by the rising standards of living (Daigger 2007b, 2008a; Wallace 2005). Fortunately today, many technologies do exist but impetus is on refining and integrating them. The challenge is to choose among available options and develop institutional arrangements for implementing them in the most effective ways (Daigger and Crawford 2005).

Table 4.1 Triple bottom line of sustainability goals (Glen T. Daigger 2008)

Sustainability	Goal
Economy	Financially stable utilities with enough resources to maintain the infrastructure
Environment	Locally sustainable water supply. Recharge must exceed net withdrawal
	Energy-neutral system (or positive if possible), with minimal consumption of chemicals
	Nutrient management to minimize dispersal to the aquatic environment
Society	Access to clean water and appropriate sanitation for all

Applications of a variety of new wastewater treatment technologies, such as membrane filtration systems and advanced oxidation, have led to new ways of managing urban water systems and water resources. These new treatment regimes, especially the integration of urban water and waste management systems, promise to dramatically improve the sustainability of our water resources. The requirements for sustainability are consistent with the "triple bottom line" definition of sustainability which includes social, environmental and economic goals (Table 4.1).

However, an integrated closed-loop systems approach for recycling water and waste material, with a combination of decentralized and centralized elements, is necessary in order to meet the environmental sustainability goals (Daigger 2007a; 2008a, b). For example, by segregating water supply into potable and non-potable uses, the net removal of water from the environment for potable uses can be dramatically reduced since the amount of potable urban water consumption is small. The bulk of the domestic and commercial water supply is non-potable water, which can be supplied from a variety of local sources, including recycled water and captured rainwater (Daigger 2008a).

Wastewater Treatment Technologies

Wastewater treatment technologies with environment friendly designs and low-cost sanitation additionally provide benefits from the reuse of water. These technologies use natural aquatic and terrestrial systems classified into three principal types.

Aquatic Systems

Aquatic systems are classified as facultative lagoons, aerated lagoons and hydrograph controlled release (HCR) lagoons. Such lagoon based systems can be supplemented with pre-treatment or post-treatment methods through constructed wetlands, aqua-cultural production systems and sand filtration which are suitable to treat a variety of wastewaters under a wide range of weather conditions.

Facultative lagoons are the most common form of aquatic lagoons currently in use. The water layer near the surface is aerobic while the bottom sludge deposit is anaerobic. The intermediate layer constitutes the facultative zone. Aerated lagoons are smaller and deeper than facultative lagoons. These systems evolved from stabilization ponds when aeration devices were added to counteract odors arising from septic conditions.

Hydrograph controlled release (HCR) lagoons are recently developed in order to counter high effluent solids (exceeding 100 mg/L) common to lagoon systems. In HCR lagoons, wastewater is discharged only during periods when the stream flow is adequate to prevent water quality degradation. When stream conditions prohibit discharge, wastewater is accumulated in a storage lagoon.

Sand filters, Constructed wetlands and aquaculture operations are generally the most successful methods of refining the effluent from lagoons. They provide additional treatment upon secondary treatment.

Sand filter systems use a sand layer and the wastewater is treated biologically by the epiphytic flora associated with the sand particles. The treatment is further achieved by physical filtration of suspended solids and chemical adsorption. The sand filtration systems may differ in the method of application of the wastewater. Intermittent filters are flooded with wastewater and then allowed to drain completely before the next application. In contrast, recirculating filters recirculate the effluent back to the filter.

Constructed wetlands with surface water flow or subsurface water flow systems utilize the roots of aquatic plants as habitat for bacteria and oxygen transfer. Bacteria utilize nutrients from the effluent while plants consume some nitrogen. Typically, these systems are narrow basins planted with aquatic vegetation. The shallow groundwater systems use a gravel or sand medium to provide rooting support for aquatic plants.

Aquaculture systems are basically shallow ponds covered with floating aquatic plants that detain wastewater for a period of time. Usually, aquatic plants such as *Eichhornia crassipes* (water hyacinth) and *Lemna minor* (duckweed) are grown in these wastewater holding basins. Bacteria growing on these plant habitats remove the vast majority of dissolved nutrients.

Terrestrial Systems

Terrestrial systems make use of the nutrients contained in wastewaters; plant growth and soil adsorption convert biologically available nutrients into less-available forms of biomass, which is then harvested for a variety of uses, including methane gas production, alcohol production, or cattle feed supplementation. In addition to wastewater treatment and low maintenance costs, these systems yield additional benefits by providing water for groundwater recharge and irrigation purposes. These systems rely upon physical, chemical, and biological reactions. Terrestrial treatment systems are further classified into Slow-rate overland flow system, Slow-rate subsurface infiltration and Rapid infiltration systems.

Slow-rate overland flow systems are natural systems, where in, vegetation is a critical component. Slow-rate subsurface infiltration systems and rapid infiltration systems are "zero discharge" systems. Slow-rate overland flow systems positively impact the sustainable practices. In addition to treating wastewater, they provide reusable water and nutrients for the purpose of agriculture and reforestation in arid and semi-arid lands. In these systems, wastewater flowing at a controlled rate is treated as it passes through the soil layer by filtration, adsorption, ion exchange, precipitation, microbial action, and plant uptake processes. Vegetation serves to extract nutrients, reduce erosion, maintain soil permeability and ensure maximum contact time.

Overland flow systems are a land application treatment method in which treated effluent is finally discharged to surface water. Wastewater is applied intermittently across the tops of terraces constructed on soils of very low permeability and allowed to flow across the vegetated surface to the run-off collection channel. Treatment, including nitrogen removal, is achieved primarily through sedimentation, filtration, and biochemical activity.

In rapid infiltration systems, most of the applied wastewater percolates through the soil, and the treated effluent drains naturally to surface waters or recharges the groundwater. Wastewater is applied to soils that are highly permeable. The system converts ammonia nitrogen in the wastewater to nitrate nitrogen before discharge. Vegetation is not necessary keeping cost and manpower requirements very low.

Subsurface infiltration systems are designed for very small municipalities. They are usually designed for individual homes (septic tanks) or community clusters.

Mechanical Treatment Systems

Mechanical systems utilize a combination of physical, biological and chemical processes to achieve the treatment objectives. Using essentially natural processes within an artificial environment, mechanical treatment systems require mechanical components to treat wastewater and various types of instrumentation to control parameters such as flow of wastewater. An activated sludge process, which is a suspended-growth system (Fig. 4.2), is an integral process of mechanical treatment systems. Sequencing batch reactors (SBR), oxidation ditches, and extended aeration systems are all variations of the activated sludge process. In contrast, a trickling filter process (TFP) is an attached growth system.

Sustainability of Natural Treatment Systems

The advantages as well as disadvantages of the various natural wastewater treatment technologies are summarized in Table 4.2. Natural treatment systems, discussed above, are capable of producing an effluent quality equal to that of a mechanical

Fig. 4.2 Conventional aeration tank (*left*) and extensive foaming observed in the activated sludge process of Dubai's sewage treatment plant, Dubai-UAE. (*Courtesy*: Dubai Municipality, DSTP Division)

Table 4.2 Advantages and disadvantages of the various wastewater treatment technologies (UNEP source book 1997)

Treatment	Advantages	Disadvantages
Aquatic systems		
Stabilization lagoons	Low operation and maintenance costs Low technical manpower requirement	Requires a large area of land May produce undesirable odors
Aerated lagoons	Requires relatively less area	Requires mechanical devices to aerate the basins High suspended solids in the effluent May produce undesirable odors
Terrestrial systems		
Septic tanks	Suited for households Easy operation and maintenance	Low treatment efficiency Periodic disposal of sludge and septage
Constructed wetlands	Removes up to 70 % of solids and bacteria Low capital cost and less operation and maintenance	Largely experimental and restricted to areas where suitable native plants are available Periodic removal of excess plant material
Mechanical systems		
Filtration systems	Minimal land requirement Low cost and easy operation	Requires mechanical devices
Biological reactors	Requires less area Highly efficient treatment method Suitable for small and large scale operation	Complex technology with high cost Requires highly technical manpower for operation and maintenance High energy requirement
Activated sludge	Requires less area Highly efficient treatment method Suitable for small and large scale operation	High cost Sludge bulking and foaming (Fig. 2, Faheem and Khan 2009) Requires sludge disposal area Technical manpower for operation and maintenance

treatment system (UNEP Source book 1997). The limits established for secondary treatment, defined as biological oxygen demand (BOD) and total suspended solids (TSS) concentrations of less than 30 mg/L are generally met by all systems. All except the lagoons can also produce effluents of advanced treatment standards (BOD and TSS concentrations of less than 20 mg/L). Lagoons and oxidation ditches are suitable where there is plenty of land available and the aquatic systems are best suited to regions where suitable aquatic plants are grown naturally while mechanical systems are the least land intensive among the wastewater treatment methods based on natural processes.

New and improved, aesthetically pleasing and efficient natural treatment systems (NTS) in the form of constructed wetlands, farm irrigation, golf courses and greenhouses are being developed. According to Environment Protection Agency (EPA), a true natural treatment system has minimum dependence on mechanical elements in the wastewater treatment process. Instead, NTS use plants and microorganisms to break down pollutants in wastewater. Natural systems use, rather than dispose, wastewater with minimum use of chemicals and energy. NTS take advantage of natural processes to improve water quality (Kadlec and Knight 1996) and are increasingly being used for water reclamation. These natural systems have the advantage of being able to remove a wide variety of contaminants including pathogens and other toxic substances.

The demand for reliable, efficient and low-cost wastewater treatment systems is increasing around the world, especially in urban areas. There is great deal of importance on adequately treating wastewater on-site before it is released into the environment.

Sustainable Decentralized Wastewater Treatment

Unlike centralized system where sewage is collected at one treatment plant, a Decentralized wastewater treatment system (DWT) refers to the use of onsite treatment of wastewater generated by individual homes, clusters or community. Decentralized wastewater treatment consists of a variety of approaches for collection, treatment and dispersal or reuse of wastewater. Today, Decentralized wastewater treatment is being considered for most communities because of its economic and environmental advantages. DWT system is typically installed near to the point where the wastewater is generated. It can function as a part of permanent infrastructure and can be managed as stand-alone facility or can also be integrated into centralized sewage treatment systems. It effectively and efficiently treats domestic sewage in a green and sustainable way and hence is a smart alternative for communities considering new systems or modifying or replacing existing wastewater treatment systems. When appropriately designed, operated and maintained, DWT systems meet public health and water quality goals much like centralized systems. The U.S. Environmental Protection Agency (EPA) is in the process of developing a number of management models for decentralized wastewater systems owing to the merits it has to offer. The major advantages of Decentralized wastewater treatment are:

1. Provides treatment for both domestic and industrial wastewater.
2. Cost-effective and economical with low capital as well as operation and maintenance costs.
3. Wise use of land and energy.
4. Environment friendly and safe with respect to public health and water quality.
5. Green and sustainable method.

Wastewater from a DWT system stays in a local watershed as it returns to the drain field, dispersing into the underlying soil and eventually recharging groundwater and reentering the watershed. Advanced DWT systems are able to achieve treatment levels comparable to centralized wastewater treatment systems while minimizing levels of phosphates and nitrogen entering into the ground water. Decentralized systems can be designed to meet specific treatment goals and to handle unusual site conditions. Further, decentralized designs including rain gardens and green roofs, water-efficient appliances and landscapings, help beautify cities and towns, enhance water supply, recover energy and nutrients and improve our health and environment.

Most decentralized systems take advantage of gravity flow rather than using energy to pump the wastewater. Additionally, they often incorporate septic tanks at the wastewater source resulting in reduced costs and energy for treatment of septage prior to land dispersal. Decentralized systems mitigate contamination and health risks by using natural treatment properties of the soil which acts as a natural filter and removes harmful bacteria, viruses and nutrient contaminants. Hence DWT can be a sensible solution for communities of any size and demographic. Where they are determined to be a good fit, decentralized systems help communities reach the above stated triple bottom line definitions of sustainability (Table 4.1). However, constraints that hamper acceptance of decentralized wastewater systems arise due to lack of knowledge, regulatory and management barriers and also financial limitations of the community.

Cultural Acceptability

Governments and the private sector fail to recognize the necessity of wastewater treatment and the importance of water quality in improving the quality of life of existing and future generations. The contamination of natural resources is a major impediment to achieving the stated objective of Agenda 21 of environmentally sustainable economic growth and development (Agenda 21 1992). The acceptability of technology depends on the region since some cultures do not accept treatment and reuse of wastewater. It is also essential to determine an appropriate balance between cost and efficiency.

New Technologies for Sustainable Systems

Wastewater treatment technologies are integral to urban water systems. Fortunately today, we are able to reclaim and reuse wastewater. Microfiltration and ultrafiltration membrane technologies for removing particulate matter, nano-filtration and reverse osmosis for removal of dissolved solids are increasingly being employed. Membrane bioreactors (MBR) in which particle removal membranes (Jefferson et al. 2000) are coupled with biological systems are fast becoming essential for water reclamation (Daigger et al. 2005; DiGiano et al. 2004). Further, ultraviolet (UV) radiation, Ozone, and hydrogen peroxide to produce highly reactive hydroxyl radical (OH) are combined in an oxidative treatment process. Activated carbon is still being used additionally in water reclamation processes to remove many toxic compounds and color contaminants through adsorption.

Membrane Filtration Systems

Membrane systems have remarkably improved the treatment in advanced water reclamation. Use of micro- and ultra-filtration membranes provided excellent pre-treatment for Reverse Osmosis application, which can remove a wide range of dissolved solids. Membrane filtration systems have led to the development of Membrane biological reactors (MBR), which are becoming increasingly important in the water reclamation processes. Both aerobic and anaerobic biological treatment methods have been extensively used to treat domestic and industrial wastewater (Visvanathan et al. 2000). Using MBR, complete biological treatment and retention of microorganisms, including viruses, is possible as the solids retention time or solids residence time (SRT) of biological solids is considerably increased. Thus, treatment with MBR produces a highly clarified effluent that can be more easily disinfected and can be ideally used for producing non-potable water. MBR must be followed by RO and UV treatment for the purpose of reclaiming potable water (Tao et al. 2005, 2006). Advances in Reverse Osmosis technology include improved membranes and configurations, more efficient pumping and energy-recovery systems, and the development of process technology, such as membrane distillation (Kim et al. 2008).

Nanotechnology

Developments in material science offer potential new approaches to the challenges of water disinfection, decontamination and desalination to cope with increasing demands for cleaner water. Dramatic improvements are feasible in the near future (Shannon et al. 2008). New membranes with ability to sequester viruses through biomimetic surface chemistries or deactivate pathogens with nanostructured photocatalysts are being investigated for thorough disinfection. Titanium Oxide (TiO_2)

nanowire membranes have been successfully fabricated with the capability of filtering organic contaminants from water with simultaneous photo-catalytic oxidation (Xiao et al. 2010). Carbon nanotube based membranes, which are known to have high water fluxes and salt-rejection coefficients, are promising improved efficiency in the decontamination/disinfection of water. Nanotechnology concepts are being studied for higher performance of membranes with selective transport or rejection characteristics, improved hydraulic conductivity and lesser fouling.

Microbial Fuel Cells

Application of Microbial Fuel Cells represents new generation approach to wastewater treatment with production of sustainable clean energy. Microbial fuel cell is a device that uses bacteria as the catalysts to oxidize organic and inorganic matter and generate electric current by electrochemical technology (Logan et al. 2006). The microbes, for example certain metal reducing bacteria, participate in the electron transfer chain between fuel substrate and the electrode, thereby enhancing the cell current. Electrons produced by such bacteria from these substrates are transferred to the anode and flow to the cathode. Microbial Fuel cells with mixed cultures are capable of generating substantially greater power densities than those with pure cultures (Rabaey et al. 2004). Microorganisms are grown as a biofilm on the electrodes and the electron donor is separated from the electron acceptor by a proton exchange membrane, which establishes an electrical current. Electrical energy is then generated through the oxidation of organic matter.

Although this technology is still in the early stages of development, it has immense potential to produce electricity directly using wastewater.

Electrocoagulation

Electrocoagulation is a water treatment process that uses electric current to remove various contaminants. The technology is commonly used in the oil & gas, construction, and mining industries to treat water containing emulsified oil, petroleum hydrocarbons, suspended solids, heavy metals, and other hard to remove contaminants (Martin 2014). Electrocoagulation is performed by applying unique form of direct current across metal plates (usually Iron or Aluminum) that are submerged in wastewater. Heavy metals, organics, and inorganics are primarily held in water by electrical charges. When electrical current is applied, metal ions from the electrodes split and get into the liquid medium in the form metal oxide or hydroxide nuclei which will attract these contaminant particles to coagulate and precipitate. The precipitated mass can now be easily separated and removed from the water. Electrocoagulation effectively removes dissolved, suspended or emulsified molecules, elements or ions (Butler et al. 2011).

The technology eliminates the need for chemical additives and thus, is environmentally sustainable. Using electrocoagulation, water could be recycled directly from a municipal wastewater treatment plant. Instead of releasing it back into the environment, it can be reused for irrigation of parks, golf courses and other landscapes.

Further Developments

Advanced treatment technology, coupled with wastewater reduction and reuse initiatives, offer hope of minimizing loss of useful water. The cost-effectiveness of existing and new wastewater treatment technologies needs to be improved. New designs of mechanical systems which address this concern and also optimize resource consumption are being introduced. The use of vertical reactor tanks for activated sludge system is being tested. Similar developments are occurring to wisely use aquatic and terrestrial plants and hybrid hydroponic systems as a means of wastewater treatment; however, these technologies are still evolving and will require successful pilot studies prior to being accepted as standard treatment technologies. The simplest approach to liquid waste management is to control the quality of wastewater at its point of treatment and discharge. This places regulation and control at institutional level. As in many developed nations like US, Canada and Europe, the quality of the discharge can be regulated to fit the type of use.

There is a need for legislation and regulations for the control of treatment and use of wastewater for consistent application of technology. Studies need to be undertaken to determine the variations in effluent quality and its effect on the raw water being reused. In addition, education to create awareness on the importance of wastewater treatment and the sustainability aspects of the emerging technologies remains critical at all levels of governance and society.

References

Agenda 21. (1992). *United Nations Conference on Environment & Development*, Rio de Janerio, Brazil, June 3–14, 1992. Retrieved from sustainabledevelopment.un.org/content/documents/Agenda21.pdf.

Brunner, P. H., & Rechberger, H. (2015). Waste to energy – Key element for sustainable waste management. *Waste Management, 37*, 3–12.

Butler, E., Hung, Y. T., Yeh, R. Y. L., & Al Ahmad, M. S. (2011). Electrocoagulation in wastewater treatment. *Water, 3*, 495–525.

Constable, G., & Sommerville, B. (2003). *A century of innovation: Twenty engineering achievements that transformed our lives*. Washington, DC: Joseph Henry Press.

Couth, R., & Trois, C. (2012). Sustainable waste management in Africa through CDM projects. *Waste Management, 32*, 2115–2125.

Daigger GT (2007) Creation of sustainable water resources by water reclamation and reuse. *Proceedings of the 3rd international conference on sustainable water environment: integrated water resources management-new steps* (pp. 79–88). Sapporo, Japan, October, 2007.

Daigger, G. T. (2007b). Wastewater management in the 21st century. *Journal of Environmental Engineering, 133*(7), 671–680.

Daigger, G. T. (2008). AEESP lecture-evolving urban water and residuals management paradigms: Water reclamation and reuse, decentralization, resource recovery. *Proceedings of the Water Environment Federation 81st Annual Conference and Exposition*, Chicago, October, 2008.

Daigger, G.T. (2008). Decentralization: A practitioner's perspective. *Proceedings of the 6th IWA leading edge water and wastewater treatment technology conference*, Zurich, Switzerland, June 2008.

Daigger, G. T. (2008). *New approaches and technologies for wastewater management, The bridge technologies for clean water*. Salt Lake City, UT: National Academy of Engineering, *38*(3).

Daigger, G. T., & Crawford, G. V. (2005). *Wastewater treatment plant of the future-decision analysis approach for increased sustainability, 2nd IWA leading-edge conference on water and wastewater treatment technology. Water and environment management series*. London: IWA Publishing.

Daigger, G. T., Rittmann, B. E., Adham, S., & Andreottola, G. (2005). Are membrane bioreactors ready for widespread application? *Environmental Science and Technology, 39*(19), 399A–406A.

DiGiano, F. A., Andreottola, G., Adham, S., et al. (2004). Membrane bioreactor technology and sustainable water. *Water Environment Research, 76*(3), 195–196.

Faheem, S. M., & Khan, M. A. (2009). *A Study on filamentous bacteria in activated sludge process of sewage treatment plant in Dubai, United Arab Emirates*. London: IWA Publication. *Water Practice and Technology*, *4*(2).

Guerrero, L. A., Maas, G., & Hogland, W. (2013). Solid waste management challenges for cities in developing countries. *Waste Management, 33*, 220–232.

Halbach, T. R. (2013) *International trends in solid waste handling: 2013*.Solid waste management & recycling export roundtable. Department of Soil, Water and Climate, University of Minnesota, Carlson School of Management Minneapolis, Minneapolis, MN, March 14, 2013.

Iona, I., & Gheorgheb, F. F. (2014). The innovator role of technologies in waste management towards the sustainable development. *Procedia Economics and Finance, 8*, 420–428.

Jefferson, B., Laine, A. L., Judd, S. J., & Stephenson, T. (2000). Membrane bioreactors and their role in wastewater reuse. *Waster Science and Technology, 41*(1), 197–204.

Kadlec, R. H., & Knight, R. L. (1996). *Treatment wetlands*. New York, NY: CRC Press-Lewis Publishers.

Kim, I.S., Oh, B.S., & Hyu, H.W. (2008) Moving desalination forward. *Proceedings of the Singapore international water week water convention*. Singapore, June 25–26, 2008.

Logan, B. E., Hamelers, B., Rozendal, R., Schröder, U., Keller, J., Freguia, S., et al. (2006). Microbial fuel cells: Methodology and technology. *Environmental Science and Technology, 40*(17), 5181–5192.

Mackenbach, J. P. (2007). Medical milestones: celebrating key advances since 1840. *BMJ, 334*, s1–s20.

Marshall, R. E., & Farahbakhsh, K. (2013). Systems approaches to integrated solid waste management in developing countries. *Waste Management, 33*, 988–1003.

Martin, L. (2014). Electrocoagulation: A shocking approach to wastewater treatment. www.wateronline.com.

Rabaey, K., Boon, N., Siciliano, S. D., Verhaege, M., & Verstraete, W. (2004). Biofuel cells select for microbial consortia that self-mediate electron transfer. *Applied and Environmental Microbiology, 70*, 5373–5382.

Seadon, J. K. (2010). Sustainable waste management systems. *Journal of Cleaner Production, 18*(16–17), 1639–1651.

Shannon, M. A., Bohn, P. W., Elimelech, M., Georgladis, J. G., Mariñas, B. J., & Mayes, A. M. (2008). Science and technology for water purification in the coming decades. *Nature, 452*(20), 301–310.

Smith, J. E. (2009). *Biotechnology* (5th ed.). New York, NY: Cambridge University Press.

Sridhar, M. K. C., & Adejumo, M. (2014). Health and safety challenges, and perceptions of private sector waste operators in lagos. *Nigeria Health, 6*, 632–640.

Tao, G. H., Kekre, K., Qin, J. J., Oo, M. W., Viswanath, B., & Seah, H. (2006). MBR-RO for high-grade water (NEWater) production from domestic used water. *Water Practice and Technology, 1*(2), 1–8.

Tao, G., Kekre, K., Wei, Z., Lee, T. C., Viswanath, B., & Seah, H. (2005). Membrane bioreactors for water reclamation. *Water Science and Technology, 51*, 431–440.

UNEP. (1997). *Source book of alternative technologies for freshwater augmentation in Latin America and the Caribbean*. Osaka: UNEP-International Environmental Technology Centre. www.oas.org.

UNEP. (2010). *Waste and climate change: Global trends and strategy framework*. United Nations Environmental Program, Division of Technology, Industry and Economics. Retrieved from www.unep/Publications/spc/WasteClimate Change/WasteClimateChange.pdf.

United Nations. (1997). *Kyoto protocol 1997: United nations convention on climate change (UNFCCC)*. UNFCCC. Retrieved December 11, 1997, from http://unfccc.int/kyoto_protocol/items/2830.php.

Visvanathan, C., Ben Aim, R., & Parameshwaran, K. (2000). Membrane separation bioreactors for wastewater treatment. *Critical Reviews in Environmental Science and Technology, 30*(1), 1–48.

Wallace, B. (2005). *Becoming part of the solution: The engineer's guide to sustainable development*. Washington, DC: American Council of Engineering Companies.

Waltner-Toews, D., Kay, J., & Lister, N.-M. E. (2008). *The ecosystem approach: Complexity, uncertainty, and managing for sustainability*. New York, NY: Columbia University Press.

Wilson, D. C. (2007). Development drivers for waste management. *Waste Management & Research, 25*(3), 198–207.

Wilson, D. C., Velis, C. A., & Rodic, L. (2012). Integrated sustainable waste management in developing countries. *Waste and Resource Management, 166*(2), 52–68.

Xiao, Y. T., Xu, S. S., Li, Z. H., An, X. H., Zhou, L., Zhang, Y. L., et al. (2010). Progress of applied research on TiO2 photocatalysis-membrane separation coupling technology in water and wastewater treatments. *Chinese Science Bulletin, 55*(14), 1345–1353.

Chapter 5
Modeling of Free Fatty Acid Content in the Deodorization Process of Palm Oil Refinery Using Six Sigma with Response Surface Methodology

Ravindra Pogaku, Anisuzzaman S.M., and V. Prasarnth Raaj Veera Rao

Introduction

Six Sigma is a statistical concept that measures a process in terms of defects. Achieving six sigma means any processes that are delivering only 3.4 defects per million opportunities. Sigma indicates defects in the outputs of a process, and helps us to understand how far the process deviates from perfection. Defect is a measurable characteristic of the process or its output that is not within the acceptable customer limits (Brue 2002). Six Sigma is about practices that help to eliminate defects and always deliver products and services that meet customer specifications (Staudter et al. 2008).

Motorola was the first large company to implement Six Sigma in the 1980s. In 1983, reliability engineer Bill Smith concluded that if a product was defective and corrected during production, then other defects were probably being missed and later found by customers. The founder of the Motorola Six Sigma Research Institute, Mikel Harry, further refined the methodology, to not only eliminate process waste, but also turn it into growth currency (Brue 2002).

Six Sigma methodologies use a specific problem-solving approach and Six Sigma tools to improve processes and products. This methodology is data-driven with a goal of reducing unacceptable product is no more than three defects per million parts (Aggogeri and Gentili 2008). The real-world application of Six Sigma in most companies is to make a product that satisfies the customer and minimizes supplier losses to the point that it is not cost-effective to pursue tighter quality (Brussee 2010).

R. Pogaku (✉) • A. S.M. • V.P.R. Veera Rao
Energy Research Unit, Faculty of Engineering, University Malaysia Sabah, Kota Kinabalu, Sabah 88400, Malaysia
e-mail: ravindra@ums.edu.my; dr_ravindra@hotmail.com

P. Ravindra (ed.), *Advances in Bioprocess Technology*,
DOI 10.1007/978-3-319-17915-5_5

In the first year of Six Sigma implementation, General Electrics (GE) trained 30,000 employees at a cost of $200 million and got back about $150 million in savings. From 1996 to 1997, GE increased the number of Six Sigma projects from 3,000 to 6,000 and achieved $320 million in productivity gains and profits. By 1998, the company had generated $750 million in Six Sigma savings above and beyond their investment. GE expects to receive $1.5 billion savings the next year (Evans and Lindsay 2014).

In a typical palm oil refinery industry, free fatty acid (FFA) plays an important role. It is desired that the oil palm produced has a lower FFA content. It is not an easy task to control the FFA content in the product as the feedstock Crude Palm Oil (CPO) FFA content varies from 4 to 5 %. Basically, there is currently no effective method on controlling or finding the important parameters such as temperature, pressure in various unit operations that played a significant role in managing the FFA content in the product. Hence, Six Sigma methodology can be applied in order to improve the process and product quality. In addition, it can reduce the cost by reducing customers' rejection possibility of the product. This concept can be considered as user friendly since it is totally based on modeling and simulation approach. This indicates the horizon of green technology to improve the process and product.

Design for Six Sigma methodology was introduced into this study, and it was used as the backbone for the whole modeling process (Hoerl 2001). Six sigma tools such as Supplier Input Process Output Customer (SIPOC) and Process Map were used to identify the possible key variables of palm oil refinery. Before the data can be used for analysis, the data went through pre-processing. Subsequently, key variables that have significant effect on FFA content were determined by using regression function in Minitab software which was developed by three Penn State professors; Barbara F. Ryan, Thomas A. Ryan, Jr., and Brian L. Joiner, in 1972 (Jahirul 2014). Next, response surface methodology (RSM) in Minitab software was used to develop the prediction model. Then, the resulted model went through a validation run to test the ability of the model. Finally, optimization study of the FFA content is proposed by the developed RSM model.

Methodology

Sample Preparation

Plant operation data were collected from FELDA Vegetable Oils Sdn. Bhd. The plant operation data were used in the Measure Phase for simulation by using Minitab 16. Basically, Minitab is statistical software that often used in conjunction with the implementation of Six Sigma for improvement methods or techniques. In this project, Minitab 16 software was used for simulation and modeling purpose. The RSM model was developed using this software as well.

Fig. 5.1 DMADV

Experiment Procedure

This research work is based on "**Design for Six Sigma**" (DFSS), and it can be divided into five phases as shown in Fig. 5.1.

Define Phase

The goals of define phase is to describe the specific problem, identify the project's goal and exact scope. Besides, this phase also determines the key customers (parameters) of the process to be optimized as well as their "Critical to Quality" (CTQ). Lastly, CTQ ensures that the project gains acceptance from the end users.

Project charter is developed in order to depict the problems and project goals briefly and clearly (Staudter et al. 2008). Besides that, project charter focus clearly on a specific process or sub-process for optimization. It also nominates all the important project participants.

SIPOC process diagram has to be developed in order to ensure that there is a shared understanding about the process to be improved. In addition, determines the relationship between end users and suppliers through the relevant process inputs and outputs. It also identifies the key parameters in the process.

The goal of this CTB known as "Critical to Business" tool is to translate end users' needs into specific, measurable customer requirements, which are critical to quality.

Measure Phase

The goal for measure phase is to collect data with respect to the specifications needed for meeting end user requirements. Graphs and charts were used as a descriptive statistic to analyze the key output measurements and identify their specific characteristics. The measurement matrix is also known as CTQ or CTB output matrix (Staudter et al. 2008). The goal for this tool is to ensure that a good output measurement is found for each CTQ/CTB. The goal of this plan is to describe the data collection in an overview, which data is to be collected how, when, and by whom. It is necessary to generate a process map to show all the key process parameters for each unit operation.

Analyze Phase

The main goal of analyze phase is to collect and verify the suspected causes which affects the FFA content of the product, identify the relationship between output, process, and input, and lastly to deduce the root causes to the problem handled. The 156 sets of 17 process parameters (FELDA Vegetable Oils Sdn. Bhd) has undergo some data elimination work to remove any unrelated data. Multiple regressions analyzes the relationship between a continuous, independent variable and a continuous, dependent output. Besides that, it also determines a linear function with the capability of explaining as many data points as possible, and to minimize the data deviation from the function. Normality test transforms data of any distribution into data in normal distribution. Basically, this step is required in order to ensure the reliability of the data before proceeding to the "Design" phase. Stability test ensures the data passes Run Chart and I-MR chart test in order to remove noise and unusual observation from the data. Test and confidence interval for two variances is to determine whether the final model is valid and reliable for the prediction of the free fatty acid content. There are two tests conducted, which are F test and Levene's test.

Design Phase

Basically, the goal of design phase is to design or optimize the current process or generate an alternative process based on the results analyzed by using suitable techniques. RSM model was developed by build-up the regression model in terms of the inputs and outputs (Staudter et al. 2008).

Verify Phase

The goal of verify phase is to verify the potential erosion and to maintain the improved performance which was achieved through implementation of proposal changes. Two sample T-Test was conducted in this phase. If the p-value calculated from two samples T-test indicates >0.05, it is 95 % confident that the prediction model can be used to predict the outcome of the input (Staudter et al. 2008).

Results and Discussion

Define Phase

Suppliers-Inputs-Process-Outputs-Customers (SIPOC) Diagram

Table 5.1 below shows the SIPOC diagram in this project.

Critical to Quality (CTQ) and Critical to Business (CTB) Matrix

The two identified CTQ requirements are free fatty acid content must be less than 0.1 % and predict the free fatty acid content by using response surface methodology. While the major CTB is to lower the FFA content by optimization process conditions, and through the developing of a prediction model which will lower the cost and reduce rejection among customers.

Table 5.1 SIPOC diagram

Suppliers	Bleaching process
Inputs	Bleached palm oil
Process	Deodorizing process
Outputs	Refined bleached de-odorized *palm oil* (RBDPO) with FFA <0.1
Customers	End users

Measure Phase

Measurement Matrix

There are total five measurement outputs that were identified which have strong relation between CTQ and CTB, which are pressure, flow rates, deodorizer vacuum, deodorizer steam sparging, temperatures and FFA content.

Data Collection Plan

There are total of 156 sets of 20 key parameter's data for the deodorizing process are collected from FELDA Vegetable Oil Products Sdn. Bhd. All data were taken each week from 1 January 2011 until 31 December 2013 in deodorizing process, FELDA Vegetable Oils Sdn. Bhd.

Analyze Phase

Preliminary Data Selection

In this study, 156 sets of process parameters had been divided into 104 sets and 52 sets. These 104 sets of data will be used as training set and the remaining 52 sets will be used as validation set.

Multiple Regressions

The response for multiple regressions is FFA content, and the variables are the remaining key parameters. Figure 5.2 shows the Final Outcome from Multiple Regression for 6 Process Parameters.

From Fig. 5.2 above, we can see that these six process parameters have a significant influence on the FFA content. Since the p-values are less than 0.05 and the VIF values are less than 10.

Normality Test

Figure 5.3a shows that the P-Value is 0.639, which are higher than 0.05. This means that the data has passed this normality test in the first residual. It is very seldom that data will pass the normality test for the first residual. However, the passing of the normality test for this does not mean that the data will pass the normality test for the subsequent residual due to this the test has been repeated again after the

Regression Analysis: FFA versus G760T, G760P, TE704, TE705, T750, STEAMB

```
The regression equation is
FFA = - 0.127 - 0.000115 G760T + 0.00174 G760P + 0.000685 TE704 + 0.000393 TE705
      - 0.00167 T750 - 0.00291 STEAMB
```

Predictor	Coef	SE Coef	T	P	VIF
Constant	-0.12650	0.08252	-1.53	0.132	
G760T	-0.00011455	0.00003999	-2.86	0.006	1.926
G760P	0.0017371	0.0008016	2.17	0.035	1.240
TE704	0.0006847	0.0002855	2.40	0.020	1.065
TE705	0.0003927	0.0001318	2.98	0.004	1.268
T750	-0.0016738	0.0005542	-3.02	0.004	5.041
STEAMB	-0.002914	0.001008	-2.89	0.006	3.721

S = 0.00681441   R-Sq = 59.0%   R-Sq(adj) = 54.0%

Analysis of Variance

Source	DF	SS	MS	F	P
Regression	6	0.00333507	0.00055585	11.97	0.000
Residual Error	50	0.00232181	0.00004644		
Total	56	0.00565688			

Source	DF	Seq SS
G760T	1	0.00016952
G760P	1	0.00000000
TE704	1	0.00009812
TE705	1	0.00000020
T750	1	0.00267923
STEAMB	1	0.00038800

Fig. 5.2 Final outcome from multiple regression for six process parameters

unusual observation has been removed from the data in order to pass the stability test in the next section.

Figure 5.3b shows the probability plot of second Residual for the normality test. There are total of 57 numbers of data that follow the normal distribution with P-Value of 0.706 after the removal of unusual observation. P-value should be higher than 0.05 for this normality test using Minitab. According to Minitab if $p > 0.05$ its fail to reject null hypothesis, in other terms accept null hypothesis. Therefore, the second residuals passed the normality test.

Stability Test

Figure 5.4 shows the Run Chart of the second Residual for the stability test. As shown in the figure, all the four P-value was greater than 0.05. This means that total of 57 data had passed the Run Chart test after the removal of unusual observation from the data.

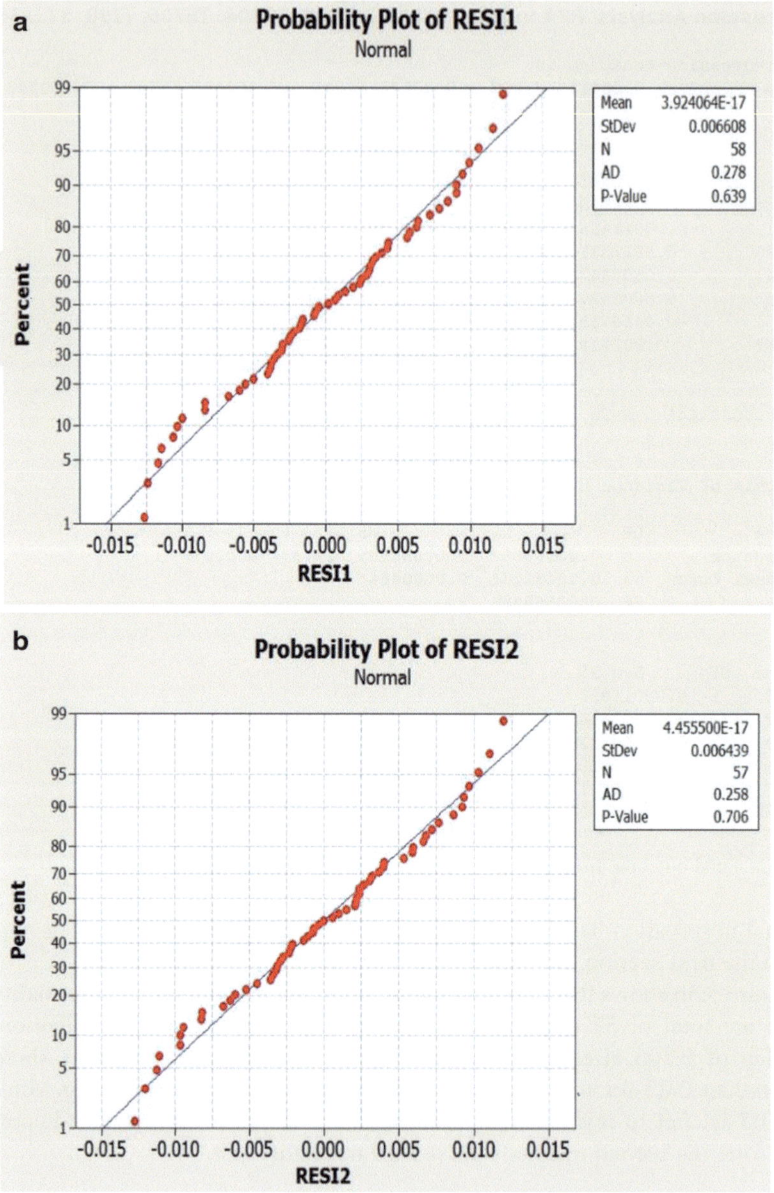

Fig. 5.3 Probability plot of (**a**) first residual (**b**) second residual for the normality test

Figure 5.5 shows the I-MR Chart of second Residual for Stability Test. There are no red dots indicated in this chart, which means the data has passed this test and there is no noise in the data region. Normality test and Run Chart test are repeated again to ensure these test were passed as well.

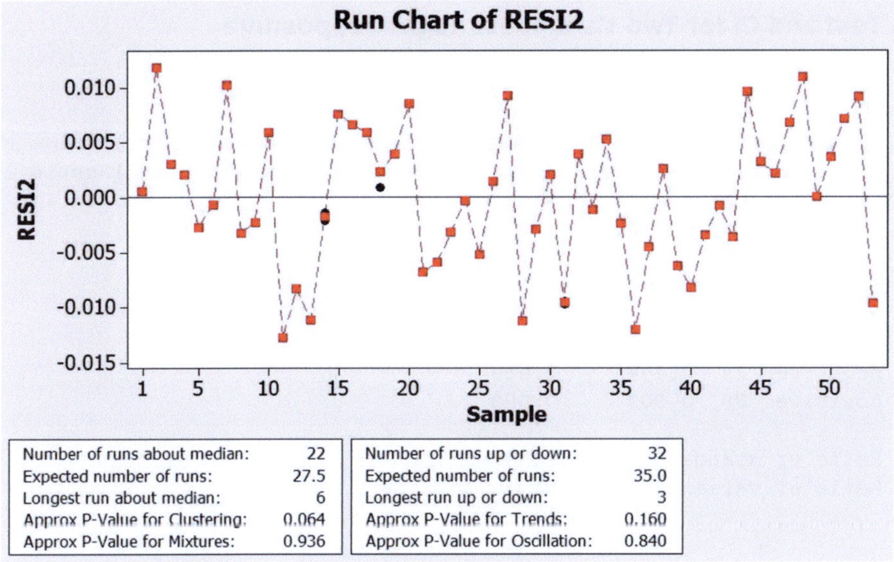

Fig. 5.4 Run Chart Test for second residual for the stability test

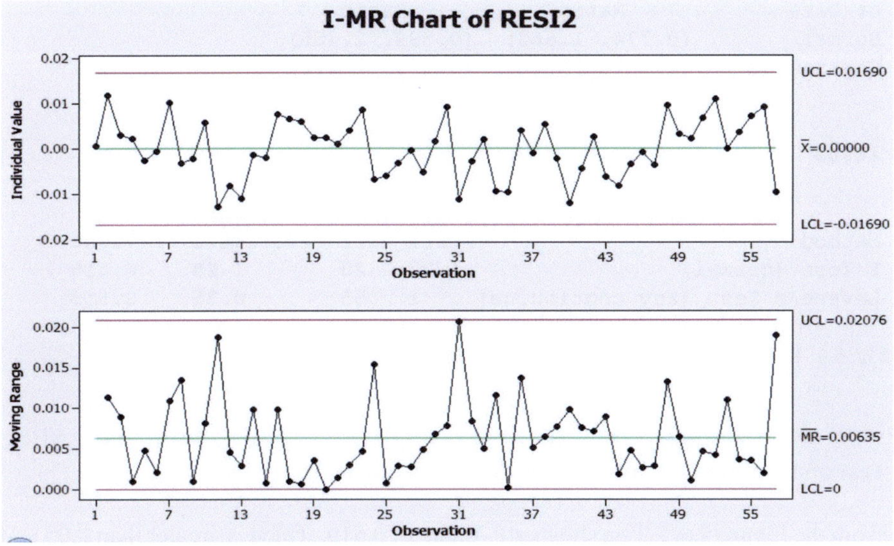

Fig. 5.5 I-MR Chart of second residual for the stability test

```
Test and CI for Two Variances: negative, positive

Method

Null hypothesis          Sigma(negative) / Sigma(positive) = 1
Alternative hypothesis   Sigma(negative) / Sigma(positive) not = 1
Significance level       Alpha = 0.05


Statistics

Variable   N  StDev  Variance
negative  28  0.004     0.000
positive  29  0.003     0.000

Ratio of standard deviations = 1.132
Ratio of variances = 1.281


95% Confidence Intervals

                                      CI for
Distribution   CI for StDev         Variance
of Data            Ratio              Ratio
Normal         (0.774, 1.660)   (0.599, 2.755)
Continuous     (0.752, 1.666)   (0.565, 2.776)


Tests

                                                Test
Method                           DF1  DF2  Statistic  P-Value
F Test (normal)                   27   28       1.28    0.519
Levene's Test (any continuous)     1   55       0.35    0.558
```

Fig. 5.6 Results from test and confidence interval for two variances

Test and Confidence Interval for Two Variances

From the Figure 5.6, F Test shows a P-Value of 0.519, which is greater than 0.05 and Levene's Test shows a P-Value of 0.558, which is greater than 0.05 as well. In addition, there are 95 % confidence intervals for the model. Hence, we can say that the first negative column and the second positive column were equally distributed. Then, the final model is valid and reliable for the prediction of the FFA content. Therefore, the final model can proceed to the "Design" Phase.

Design Phase

Response Surface Methodology for Flue Gas Duct Temperature

Equation below shows the full quadratic response surface model of the FFA. The significant process parameters used in the prediction model including G760T (A1), G760P (B1), TE704 (C1), TE705 (D1), T750 (E1) and STEAMB (F1).

$$
\begin{aligned}
\text{FFA} = & -15.7646 - (0.00291817 \times \text{A1}) + (0.201865 \times \text{B1}) + (0.0941681 \times \text{C1}) \\
& - (0.0833892 \times \text{D1}) - (0.0232187 \times \text{E1}) + (0.165876 \times \text{F1}) \\
& + (0.00000463968 \times \text{A1} \times \text{A1}) + (0.00125797 \times \text{B1} \times \text{B1}) \\
& - (0.0000595113 \times \text{C1} \times \text{C1}) + (0.0000292055 \times \text{D1} \times \text{D1}) \\
& + (0.000200311 \times \text{E1} \times \text{E1}) + (0.0011118 \times \text{F1} \times \text{F1}) \\
& - (0.0000559563 \times \text{A1} \times \text{B1}) + (0.00000981924 \times \text{A1} \times \text{C1}) \\
& - (0.0000132082 \times \text{A1} \times \text{D1}) + (0.00000138712 \times \text{A1} \times \text{E1}) \\
& + (0.000185331 \times \text{A1} \times \text{F1}) - (0.001294 \times \text{B1} \times \text{C1}) \\
& + (0.000286744 \times \text{B1} \times \text{D1}) - (0.0000668125 \times \text{B1} \times \text{E1}) \\
& - (0.00150126 \times \text{B1} \times \text{F1}) + (0.00027595 \times \text{C1} \times \text{D1}) \\
& + (0.000032273 \times \text{C1} \times \text{E1}) - (0.000547342 \times \text{C1} \times \text{F1}) \\
& - (0.00005798 \times \text{D1} \times \text{E1}) - (0.000423172 \times \text{D1} \times \text{F1}) \\
& + (0.000516041 \times \text{E1} \times \text{F1})
\end{aligned}
$$

Training Data Set

Figures 5.7 shows the graphs for training set data, where the data influences or has an effect on the equation. Therefore, a new set of data is required to validate the equation above. For this purpose, we have removed some portion of data, which wasn't being used for purpose of making the equation; these data will be used for validation data sets.

Validation Data Set

Figure 5.8 shows the comparison of actual and predicted value from January 2013 till December 2013. As we can see there is similarities in terms of the shape of the line. The peaks follows when there is ups and downs. The extreme shootout can be explained as this data is pure and haven't been filtered yet. Possibilities that they are caused by maintenance in the factory, broken equipment, emergency shutdowns or breakdowns. This may be the cause for such shootouts to occur.

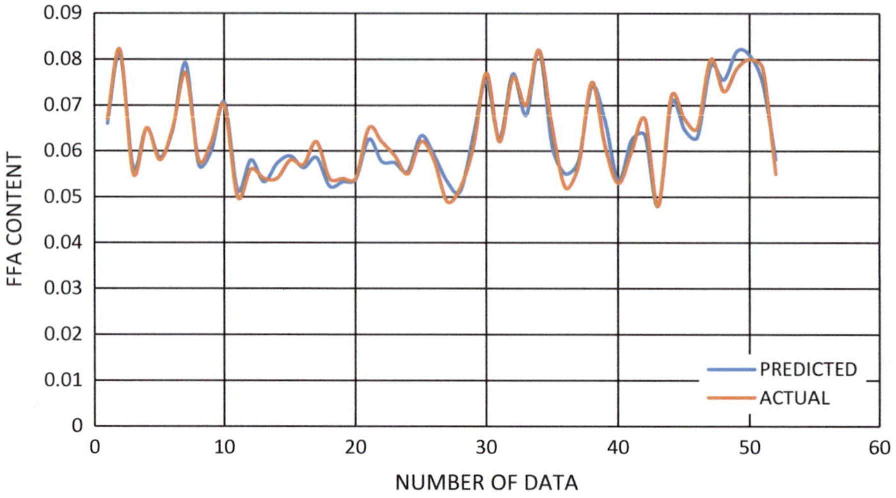

Fig. 5.7 Comparison of predicted and actual FFA content from January 2011 to December 2012

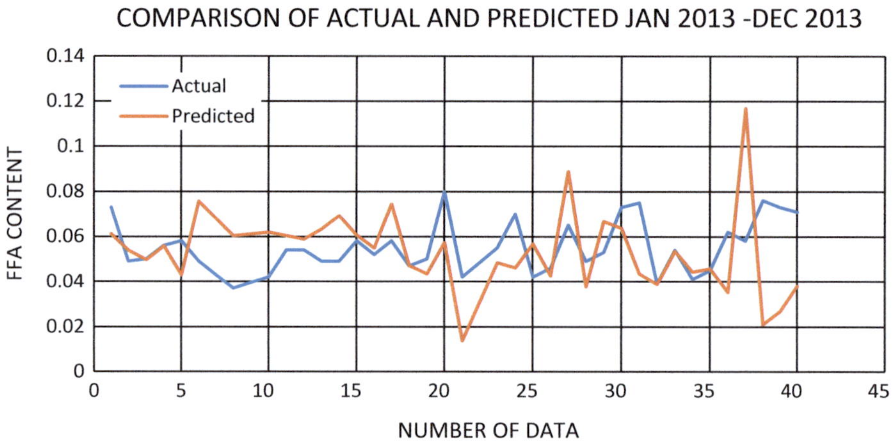

Fig. 5.8 Comparison of actual and predicted from January 2013 to December 2013

Relationship Between Process Parameters and FFA Content

Figure 5.9 shows the contour plots for FFA content versus process parameters. As we can see there is no proportionality and linearity between them. There are only few specific points in order to obtain FFA <0.05. As we can observe, the dark-blue spots shows where the FFA content is <0.050. Figure 5.9a shows that dark blue spot is on when G760P is at 60 bar and G760T is in the range of 330–340 °C. As for Fig. 5.9b, the TE704 range is in between 250–253 °C and at 270 °C, whereas for the

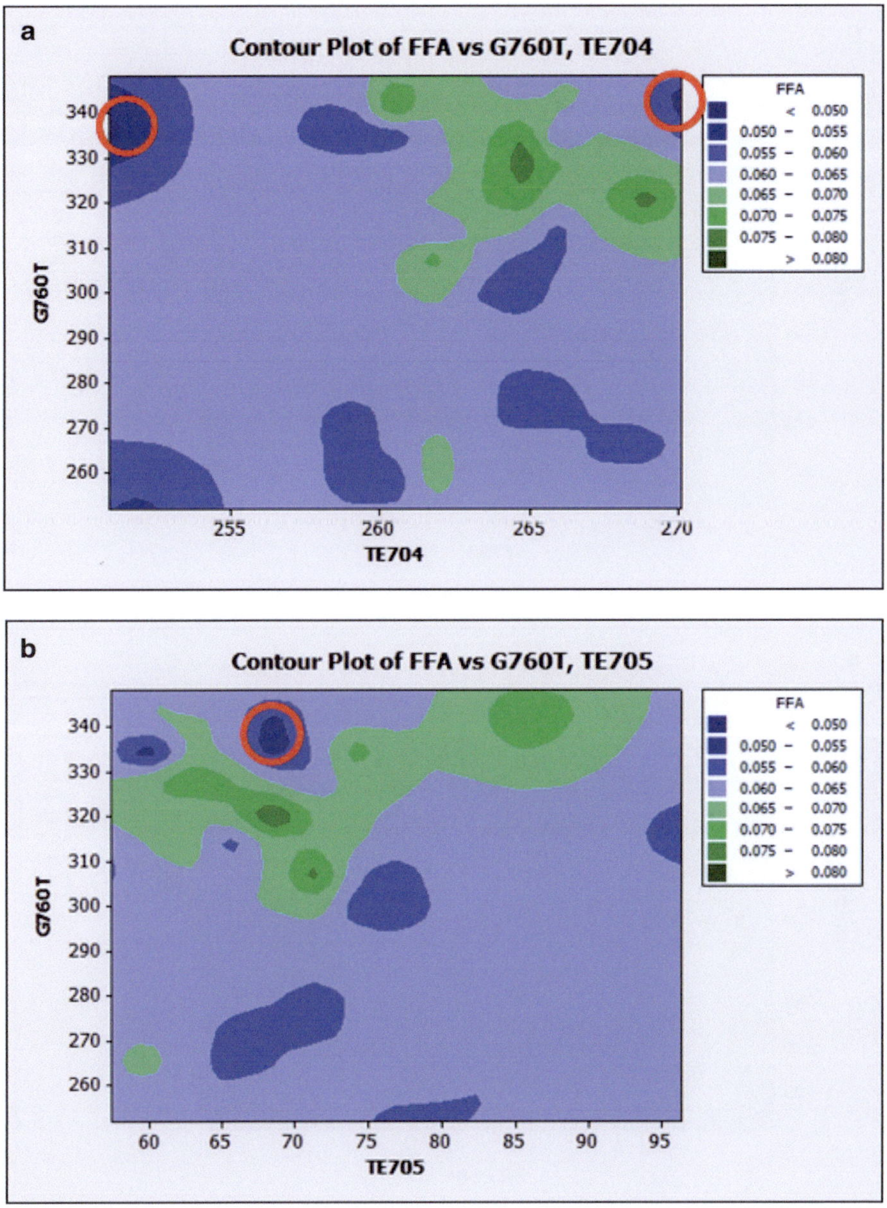

Fig. 5.9 Contour plot (**a**) FFA content versus G760T and TE704 (**b**) FFA content versus G760T and TE705 (**c**) FFA content versus G760T and TE750 (**d**) FFA content versus G760T and STEAMB

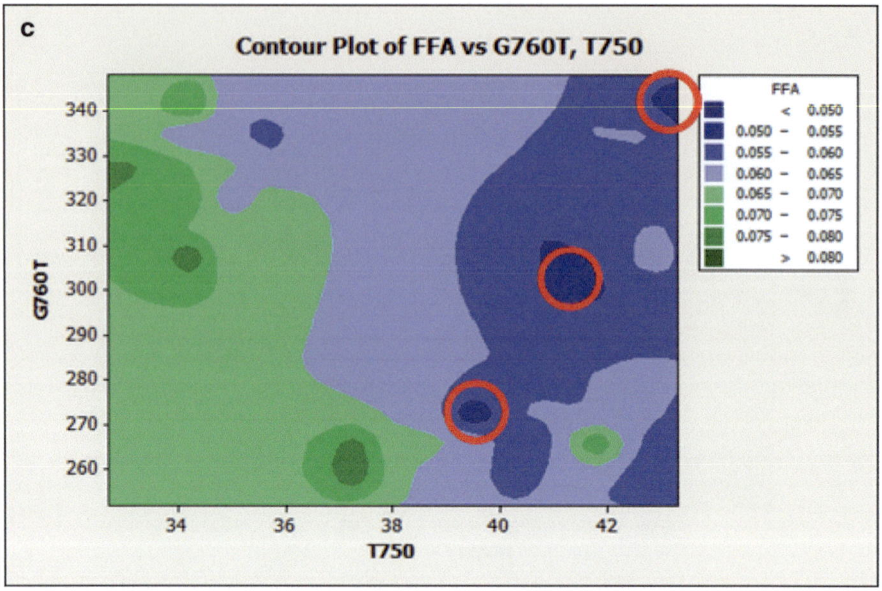

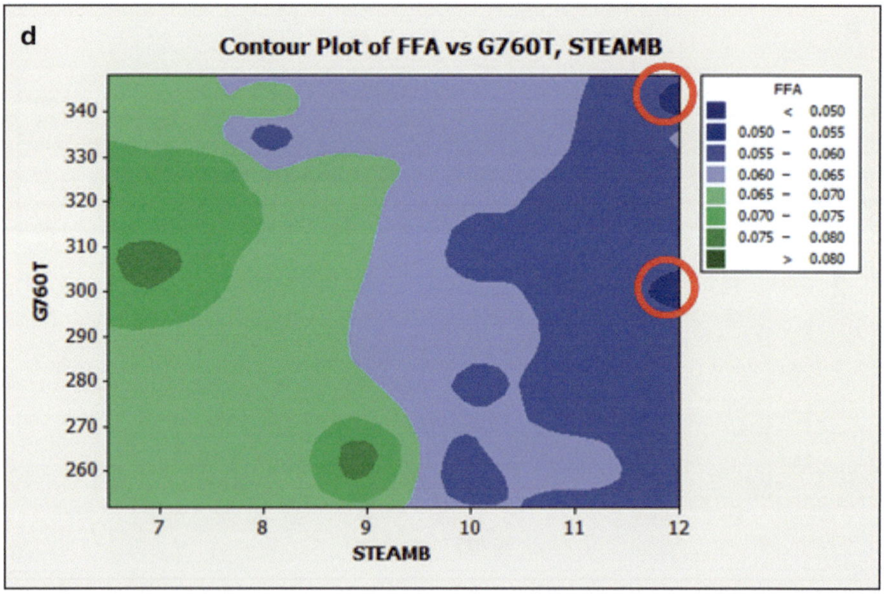

Fig. 5.9 (continued)

G760T is in the range of 330–340 °C. For Fig. 5.9c, TE705 is in the range of 65–70 °C, for G760T the range is still the same. Figure 5.9d is the relation between G760T and TE750, the range for G760T varies this time to 300–310 and 340 °C. While for TE750 the suitable temperature would be at 39, 41 and 43 °C.

Table 5.2 Parameters and their suitable range

Process parameters	Range
G760T	330–340 °C
G760P	60 bar
TE704	250–253, 270 °C
TE705	65–70 °C
TE750	39, 41, 43 °C
STEAMB	11.5–12.0 bar

Two-Sample T-Test and CI: Actual FFA, Predicted FFA

```
Two-sample T for Actual FFA vs Predicted FFA

                N    Mean   StDev  SE Mean
Actual FFA     37  0.0555  0.0116   0.0019
Predicted FFA  37  0.0535  0.0184   0.0030


Difference = mu (Actual FFA) - mu (Predicted FFA)
Estimate for difference:  0.00203
95% CI for difference:  (-0.00514, 0.00920)
T-Test of difference = 0 (vs not =): T-Value = 0.57  P-Value = 0.573  DF = 60
```

Fig. 5.10 Outcome of two sample T-test from Minitab 16

For Fig. 5.9d, STEAMB is in the range of 11.5 bar till 12.0 bar, whereas G760T is at temperature 300 and 340 °C.

Table 5.2 summarizes the parameters and their suitable range in order to obtain FFA <0.05.

Verify Phase

Figure 5.10 shows the outcome of two-sample t-test from Minitab 16. As shown, Minitab gives three results, which are Mean, standard deviation (StDev) and standard error of the mean (SE Mean) for the actual FFA content and the predicted FFA content. For the estimation of the difference between population means ($\mu_{actual} - \mu_{predict}$), the different between sample means is 0.00203. So, the confidence interval for the difference was actually based on this estimation and the variability within the samples. Hence, the difference between the mean numbers is between −0.00514 and 0.00920 higher in actually value than the predicted value.

In addition, the t-value is 0.57 with P-value of 0.573. Therefore, the null hypothesis was accepted at $\alpha = 0.05$ level. It can be concluded that there is no statistically significant deference between the mean in actual than the predicted FFA content. Hence, the response surface model can be used to predict the FFA content in deodorizing unit.

Table 5.3 Process parameters and their optimum value (FFA < 0.05)

Process parameters	Optimal value
G760T	340.1 °C
G760P	64 bar
TE704	270.1 °C
TE705	68.6 °C
TE750	43.1 °C
STEAMB	12 bar

Optimization of FFA Content

The optimization of FFA content value was considered for all process variables. The FFA content of 0.05 was selected. Table 5.3 shows the optimal value for all the 6 process parameters in order to achieve the desired FFA content.

Conclusion

Two different models for two companies FELDA Vegetable Oil Sdn. Bhd & Kunak Refinery (Sawit Kinabalu Sdn. Bhd.) was developed. Even though both companies undergo same process and produce the same products, their data recording varies. This caused for the different models. KUNAK's data focused on the overall plant rather than Deodorizing section alone, therefore the model failed few tests, the model is not applicable to be used by the industry. Whereas, for the FELDA they provided with a sufficient amount of data regarding a specific process which is deodorizing, and their model passed all the tests. With the proposed model, FELDA can now predict their upcoming products FFA content, optimize the process conditions with the sets of data given to obtain FFA content less than or equals to 0.05 %.

Acknowledgements The author would like to acknowledge University Malaysia Sabah, FELDA Vegetable Oil Sdn. Bhd, Kunak Refinery (Sawit Kinabalu Sdn. Bhd.), Minitab Statistical Software and Kenny Then Soon Hung for providing the materials and facilities to conduct this research work.

References

Aggogeri, F., & Gentili, E. (2008). Six-Sigma methodology: An effective tool for quality management. *International Journal of Manufacturing Technology and Management, 14*, 289–298.

Brue, G. (2002). *Six Sigma for managers*. New York, NY: McGraw-Hill.

Brussee, W. (2010). *Six Sigma on a budget-achieving more with less using the principles of Six Sigma*. New York, NY: McGraw-Hill.

Evans, R. J., & Lindsay, L. M. (2014). *An introduction to Six Sigma & process improvement*. Nashville, TN: South-Western.

Hoerl, R. (2001). Six Sigma black belts: What do they need to know? *Journal of Quality Technology, 33*(4), 391–406.

Jahirul, M. (2014). Biodiesel production from non-edible beauty leaf (Calophyllum inophyllum) oil: Process optimization using response surface methodology (RSM). *Energies, 7*(8), 5317–5331.

Staudter, C., Mollenhauer, J.-P., & Roenpage, O. (2008a). *Design for Six Sigma + LeanToolset: Implementing innovations successfully*. New York, NY: Springer.

Staudter, C., Roenpage, O., & Lunau, S. (2008b). *Six Sigma + Lean Toolset: Executing improvement projects successfully*. Berlin: Springer.

Part II
Biomass and Bioenergy Technology

Chapter 6
Production of Biogas from Palm Oil Mill Effluent

Ravindra Pogaku, Kai Yieng Yong, and V. Prasarnth Raaj Veera Rao

Introduction

Energy is the key ingredient to any economic activity. Increasing demand of energy due to rapid population and industrial growth has led to the development of POME as the source of sustainable bio-energy. Bio-energy is energy resource that originates from biogenic material and is a type of biogas which can be used for fuel replacement, cooking purpose, energy generation etc. (Kristofferson and Bokalders 1991). It is produced by the biological breakdown of organic matter in the absence of oxygen (NNFCC 2011).

Malaysia is the world's largest producer and exporter of palm oil; contributing 49.5 % of world production and 64.5 % of world exports (Malaysia Palm Oil Board (MPOB) 2004). Although the expansion of palm oil industry has boosted the national economy, it also concurrently generated abundant of by-products such as palm oil mill effluent (POME), empty fruit bunch (EFB), palm kernel shell (PKS) and mesocarp fiber in palm oil mills during the processing of palm oil from fresh fruit bunch (FFB) (Yusoff 2006; Chin et al. 2013). Out of these by-products, POME still remained relatively untapped and will be a threat to the environment if directly discharged to the water course (Poh and Chong 2009). The palm oil extraction process involves use of water. Large quantity of water is required to steam and sterilize the palm fresh fruit bunches (FFB) and as well clarify the extracted oil. It is estimated that for most of the average capacity plant, 1 ton crude palm oil (CPO) produced, 5–7.5 tons of water are required, and more than 50 % of the water will end up as palm oil mill effluent (POME) (Ahmad et al. 2003). It is estimated that in Malaysia about 53 million m^3 POME is being produced (Lorestani 2006) every year based on palm oil production in 2005 (14.8 million tonnes) (Yacob et al. 2005).

R. Pogaku (✉) • K.Y. Yong • V.P.R. Veera Rao
Energy Research Unit, Faculty of Engineering, University Malaysia Sabah, Sabah, Malaysia
e-mail: ravindra@ums.edu.my; dr_ravindra@hotmail.com

P. Ravindra (ed.), *Advances in Bioprocess Technology*,
DOI 10.1007/978-3-319-17915-5_6

It is estimated that about 0.5–0.75 tonnes of POME will be discharged from mill for every tonne of fresh fruit bunch. The study in this paper will shows the ability of biogas production from POME with together with other waste supplements such as activated sludge and cow dung.

POME, when fresh is a thick brownish in colour colloidal slurry of water, oil and fine cellulosic fruit residues. It is discharged at a temperature of 80–90 °C and has a pH typically between 4 and 5 (Ma and Halim 1988; Polprasert 1989; Singh et al. 1999). POME is the liquid waste generated from the oil extraction process from FFB in palm oil mills (Borja and Banks 1995). POME is the effluent from the final stages of palm oil production in the mill. For each ton of crude palm oil (CPO) produced, about an average of 0.9–1.5 m^3 POME is generated. The biological oxygen demand (BOD), chemical oxygen demand, oil and grease, total solids and suspended solids of POME ranges from 25,000 to 35,000, 53,630, 8,370, 43,635 and 19,020 mg/L, respectively. Furthermore, its high solids concentration and acidity causes it to be unsuitable for direct discharge to water courses.

Sterilization and clarification are two main stages in palm oil mill which produces condensate and clarification sludge respectively basically form the wastewater otherwise known as palm oil mill effluent. POME is also known as the colloidal suspension of 95–96 % water, 0.6–0.7 % oil, and 4–5 % total solid (TS) including 2–4 % suspended solid (SS) originating from the mixing of sterilizer condensate, separator sludge and hydro-cyclone wastewater. In most mills, these three wastewater streams are combined together resulting in viscous brown liquid containing fine suspended solids (Ahmad et al. 2003). Table 6.1 shows the approximate composition (%) of major constituents, amino acids, fatty acids and minerals in raw POME (adapted from Habib et al. 1997).

Regulatory control over discharges from palm oil mills is instituted through Environmental Quality (Prescribed Premises) (Crude Palm Oil) Regulations, 1977 promulgated under the Environmental Quality Act, 1974 and enforced by the Department of Environmental (DOE). The palm oil mills are required to adhere to prescribed regulations, which include laws governing the discharge of mill effluent into water course sand land (Ahmad et al. 2003). On top of that, the requirement of BOD level of industrial effluent to be discharged to water course has been tightened recently by DOE where the prevailing national regulation of 100 mg/L BOD has now been reduced to 20 mg/L for mills in certain environmentally sensitive areas especially in Sabah and Sarawak (Liew et al. 2012).

Anaerobic digestion is one of the most widely used methods applied for palm oil mill effluent (POME) treatment because it digests the high-strength wastewater with lower energy consumption and generates renewable energy in the form of methane (Shanmugam and Horan 2009), thus, anaerobic digestion can be considered as one of the sources of renewable energy (Yacob et al. 2006). Anaerobic digestion is a process by which almost any organic waste can be biologically converted in the absence of oxygen (Lastella et al. 2000). In the process of degrading POME into methane, carbon dioxide and water, there is a sequence of reactions involved; hydrolysis, acidogenesis, acetogenesis and methanogenesis (Gerardi 2003).

Table 6.1 The approximate composition (%) of major constituents, amino acids, fatty acids and minerals in raw POME (adapted from Habib et al. 1997)

Major Constituents	Composition (%)	Amino acids	Composition (%)	Fatty acids	Compositions (%)	Minerals	Compositions ($\frac{\mu g}{g}$ *dry weight*)
Moisture	6.99	Aspartic acid	9.66	Caprylice acid (8:0)	2.37	Fe	11.08
Crude protein	12.75	Glutamic acid	10.88	Capric acid (10:0)	4.29	Zn	17.58
Crude lipid	10.21	Serine	6.86	Lauric acid (12:0)	3.22	P	1,4377.38
Ash	14.88	Glycine	9.43	Myristic acid (14:0)	12.66	Na	94.57
Carbohydrate	29.55	Histidine	1.43	Pentadecanoic acid (15:0)	2.21	Mg	911.95
Nitrogen-free extract	26.39	Arginine	4.25	Palmitic acid (16:0)	22.45	Mn	38.81
Total carotene	0.019	Threonine	2.58	Heptadecanoic acid (17:0)	1.39	K	8,951.55
Total	100.789	Alanine	7.70	10-Heptadecanoic acid (17:1)	1.12	Ca	1,650.09
		Proline	4.57	Stearic acid (18:0)	10.41	Co	2.40
		Tyrosine	3.16	Oleic acid (18:1n-9)	14.54	Cr	4.02
		Phenylalanine	3.20	Linoleic acid (18:2n-6)	9.53	Cu	10.76
		Valine	3.56	Linolenic acid (18:3n-3)	4.72	Ni	1.31
		Methionine	6.88	T-linolenic acid (18: 3n-6)	0	S	13.32
		Cystine	3.37	Arachidic acid (20:0)	3.56	Se	12.32
		Isoleucine	4.53	Eicosatrienoic acid (20:3n-6)	2.04	Si	10.50
		Leucine	4.86	Eicosatetraenoic acid (20:4n-6)	1.12	Sn	2.30
		Lysine	2.66	Eicosapentaenoic acid (20:5n-s)	0.36	Al	16.60
		Tryptophan	1.26	Total	95.99	B	7.60
		Total	90.84			Mo	6.45
						As	0.09
						V	0.12
						Pb	5.15
						Cd	0.44

It is noted that modeling of biogas production were generally based on the kinetic models (De Gioannis et al. 2009; Ueno et al. 2007; Rao and Singh 2004; Sosnowski et al. 2008; Derbal et al. 2009; Boubaker and Ridha 2008; Gali et al. 2009). Like the phase of bacteria growth, biogas production rate showed a rising limb and a decreasing limb which was indicated by the linear and exponential equation.

The objective of the current work was to study the production of biogas from typical palm oil mil effluent together with other waste supplements through the measure of volume and flow rate of the biogas production. This study also aims to promote environmental friendly and green engineering concept through the using of renewable and sustainable biogas from the POME. Besides this study also turns waste into wealth by generating energy. The abundance of POME (waste) can now be used to generate energy.

Feedstock Preparation

Palm oil mill effluent (POME), Activated Sludge, Palm Fiber, Palm Kernal

Eight liters fresh raw POME was collected from the first discharge point from the palm oil mill. The characteristic of the fresh POME was summarized in Table 6.2. The POME was preserved at a temperature less than 4 °C, but above the freezing point in order to prevent the wastewater from undergoing biodegradation due to microbial action (APHA 1985). POME was thawed at room temperature before ready to be used. 4 kg activated sludge was obtained from the aerobic pond and used without further purification. 10 g of palm fiber and palm kernel was obtained from the Lumadan Palm oil mill's laboratory.

Table 6.2 Characteristic of the Raw POME (from Lumadan Palm Oil Mill)

Parameter	Content
pH	4.7
BOD_3	25,000 mg L^{-1}
COD	50,000 mg L^{-1}
Suspended solids (SS)	18,400 mg L^{-1}
Total volatile solids (TVS)	34,000 mg L^{-1}
Oil and grease	3,800 mg L^{-1}
Ammonical nitrogen	36 mg L^{-1}
Total nitrogen	710 mg L^{-1}

Source: Lumadan Palm Oil Mill, 2011

Table 6.3 Typical characteristics of cow dung

Type of analysis	Cow manure
Total solids (% wet)	15.6
Volatile solids (% TS)	82.8
Total Kjeldahi nitrogen (% TS)	1.2
Total organic carbon (%TS)	–
Phosphorus	2.4 (g/kg)
Potassium (% TS)	–
pH	7.8

Source: Rene et al., 2005

Cow Dung

Four kilogram of fresh cow dung was collected from Desa Dairy Farm, Kundasang, Kota Kinabalu, Sabah, Malaysia. The samples were packed into 500 g polyethylene bags and stored at −10 °C. Frozen manure portions were thawed at room temperature before being used for experiment purpose. The characteristics of typical cow dung are shown in Table 6.3.

Experimental Design and Operation Procedure

Experimental Conditions

Temperature: 30 °C.

Pressure: 1 atm.

A 25 l plastic barrel for loading of the POME, activated sludge and cow dung were setup as shown in Fig. 6.1 below:

The weight of the empty 25 l barrel was measured. Before the experiment was started, the POME and the frozen manure portions were thawed at room temperature for around 5–6 h until it reached the room temperature. 8 l of POME was measured by using a measuring cylinder and weighed by an analytical balance before it was poured into the 25 L plastic barrel. Then, 4 kg of cow dung and 4 kg of activated sludge were weighed and they were diluted with 5 l of water to prepare slurry before pouring into the 25 L plastic barrel. Ten gram of palm kernel and palm fiber were added into the barrel.

The mixtures of POME, cow dung and activated sludge were mixed for 2 min. After mixing, 50 ml sample were taken out for pH sampling analysis. The pH of the mixture should be maintained at pH 7. 150 ml of 2 M NaOH was added to adjust the pH to neutral.

After setup, the barrel was stored in closed environment. The bulb was switched on to make sure the anaerobic process is maintained in mesophilic condition (30–35 °C). Note should be taken that the bulb condition and the surrounding

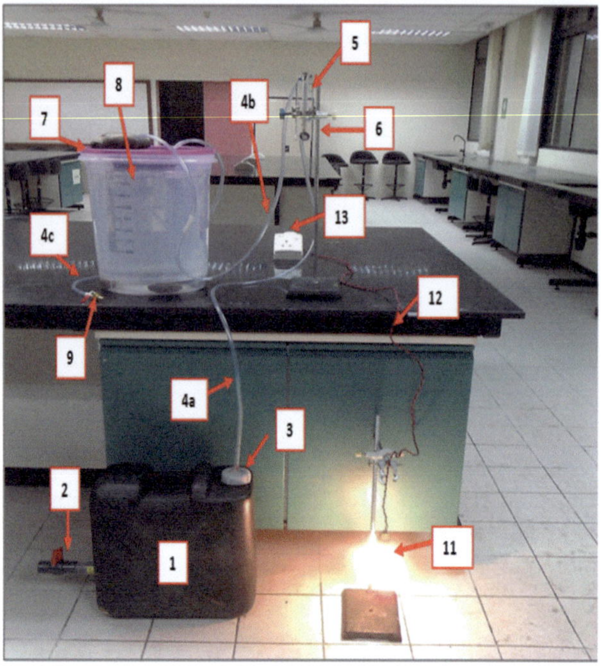

Fig. 6.1 Experimental setup. Labeling references: *1* 25 L plastic barrel; *2* PVC ball valve ¾″; *3* Lid of the 25 L plastic barrel; *4* Leveling tube 4a, 4b, 4c (total: 3 m); *5* Flow meter (lpm); *6* Retort stand with clamp; *7* 20 L plastic jar; *8* 10 L plastic jar; *9* Ball valve ½″; *10* Bunsen burner; *11* Bulb; *12* Wiring and plug; *13* Extension plug

temperature need to be checked two times daily (morning and afternoon) once the experiment was started to make sure the surrounding was always in mesophilic condition. The barrel complete loaded with the biomass was weighed again.

$$\text{Biomass used} = \text{weight of complete loaded barrel} - \text{weight of empty barrel} \tag{6.1}$$

Volume of biogas produced by the anaerobic digester and its flow rate were recorded daily for 44 days. The volume of biogas produced was estimated by water displacement method. pH was tested once in every 3 days to make sure it was maintained at pH 7. The organic waste was mixed by manually shaking the anaerobic digester slowly for 5 min once every 2 days to prevent the formation of thick scum layer on the surface that can hinder the biogas production.

Analytical Methods

A total of 23.214 kg of biomass was used in this experiment. Biogas production of the anaerobic digester was measured by using the water displacement method in the mesophilic condition of 30 ± 3 °C and atmospheric pressure of 1 atm. Parameter such as volume of biogas produced, biogas flow rate, and pH of anaerobic digester were measured according to the standard methods.

The flow rate of the biogas was read directly from the flow meter (scale: litre per minute) daily and the data was recorded. The volume of biogas produced was read from the 10 l plastic jar by taking note how much volume of water was displaced. The reading was taken daily and all the data were recorded in a table form in a logbook completed with the date of the reading was taken.

For the pH measurement, digital pH meter was used and the pH sampling analysis was done once in every 3 days to make sure the pH of the organic waste was maintained at 7. If the pH was lower than 7 or slightly acidic, then suitable amount of 2 M NaOH was added. The surrounding temperature measured once in every 3 days with digital thermometer.

Biogas production rates were simulated using linear and exponential plots. The linear equation of the two stages in ascending and descending limbs could be expressed in (6.2). Presumably biogas production rate would improve linearly with an increasing time, and after a peak, it would decrease linearly to zero as time continuously increases.

$$y = a + bt \tag{6.2}$$

Where y is the biogas production rate (L kg^{-1} day^{-1}) at time t (day), t is the time over the digestion period. a is intercept (L kg^{-1} day^{-1}) and b is slope (L kg^{-1} day^{-2}). For rising limb, b is positive whereas b is negative for failing limb.

If we assume that biogas production rate would improve exponentially with an increasing period of time and after the climax, it then decrease exponentially to zero as the time continuously increases, the exponential plot for the ascending and descending limbs could be presented in (6.3) (De Gioannis et al. 2009):

$$y = a + b\exp(ct) \tag{6.3}$$

Where y is the biogas production rate (L kg^{-1} day^{-1}) at time t (day), t is the time (day) over the digestion period, a and b are constants (L kg^{-1} day^{-1}) and c is also a constant having different unit (d^{-1}). For rising limb, c is positive whereas c is negative for falling limb.

By using the data obtained from the experiment, scale up method could be used to estimate the amount of biogas that able to be generated by Malaysia's palm oil mill industry if all the palm oil mill effluent discharged was fully utilized with the proper anaerobic treatment technology.

Results and Discussion

Biogas production and biogas accumulation were shown in Fig. 6.2a, b. Experimental result showed that the whole complete digestion period was 44 days with the retention time of 8 days. The peak biogas production rate occurred between day 29 and day 34. Biogas production started to decrease at day 36 until the completion

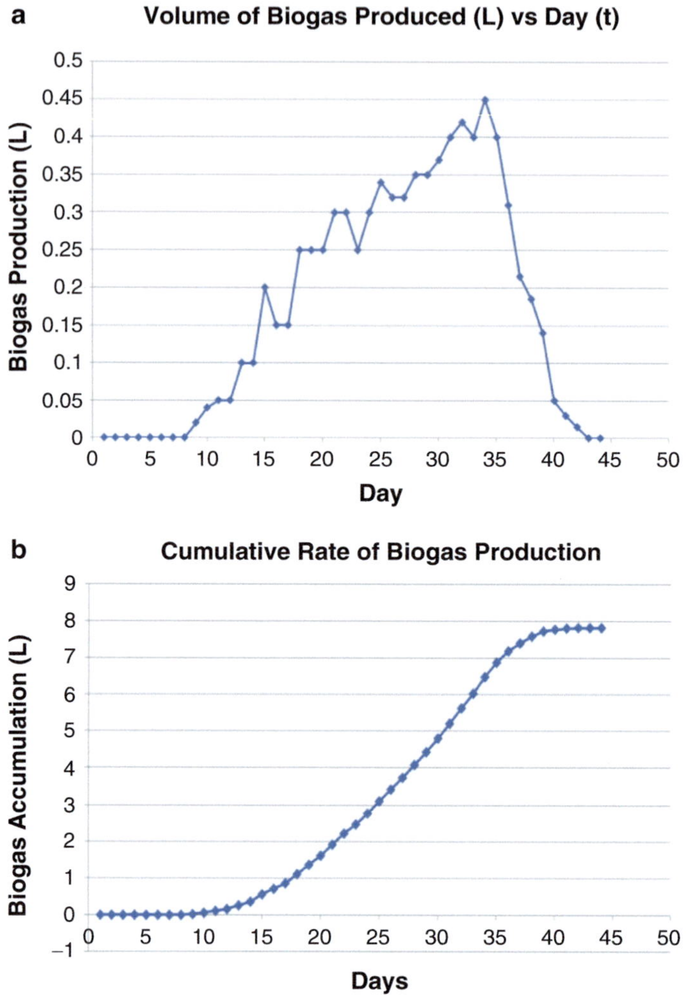

Fig. 6.2 (**a**) Biogas production versus day. (**b**) Cumulative rate of biogas production

of the experiment. The biogas accumulation curve pattern could be explained by the typical growth curve for a bacteria population. The retention time was the lag phase of the bacteria inside the cow dung. It was the period of adaption of the bacteria cells to the digester's environment. The exponential curve just after the retention time was due to the bacteria had adjusted to the new environment. The bacteria were multiply exponentially. This period of time was known as balanced growth. Short period linear curve was due to the endogenous metabolism of bacteria. Death phase of bacteria explained the stationary curve. No nutrients left in the digester caused bacteria to die.

The first 2 weeks of the digestion indicated slow production of biogas concurring with the first phase of biomass decomposition via acetogenesis process. Addition of NaOH was intensive during this time because the pH of the biomass dropped dramatically. The production of biogas increased rapidly until it reached the peak production phase where methanogenesis phase took place. Only small increment was observed after week 4 until the end of the run at day 44. The bacteria activities began to cease during this time possibly due to toxicity of high ammonia nitrogen content above 1 g L^{-1}. The nutrient provided for the bacteria also decreased when the time passed. The plots showed that the production rate was inconsistent. This could be explained by the fluctuation of temperature which caused the microorganisms' metabolism activities were varying far from the optimum condition. In addition, pH value increment gave more acidic environment which decelerated the activities of microorganism in methane production. High value of acid could cause death. The total yield of biogas was 7.825 l. Figure 6.3a, b show the temperature and pH measurement during the experiment period.

The average value of the pH during the experiment period was $6.96 \cong 7.0$ while the temperature was maintained at the average value of 30.75 °C.

From the Fig. 6.2a, the increasing and decreasing biogas production rate curve pattern was more towards either linear or exponential pattern. Hence linear and exponential equations as discussed in the analytical methods were to be used confidently. Model simulations were shown below:

Linear Plots of Biogas Production

From Fig. 6.4a, the rising limb of the biogas production rate gave the equation of $y = 0.006\,t - 0.003$. The graph intercepted at $a = -0.003$ L/kg day with the slope of $b = 0.006$ L/kg day^{-2}. Rising limb of biogas production indicated the microbial population growth was increasing.

For Fig. 6.4b, the falling limb of the biogas production rate gave the equation of $y = -0.002\,t + 0.087$. The graph intercept at $a = -0.002$ L/kg day with the slope of $b = -0.002$ L/kg day^2. R^2 of the production rate in the rising and falling limb ranged from 0.926 to 0.954. The falling limb of biogas production rate was due to the decaying of microbial kinetic growth (death phase).

Exponential Plots of Biogas Production

Figure 6.5a, b depicted the exponential plots of biogas production rates. For the rising limb of the biogas production rate, the plot gave the equation of $y = -0.001 + 0.001 \exp(0.099\,t)$. The constant value of $a = -0.001$ L/kg day, $b = 0.001$ L/kg day and $c = 0.099$ day^{-1}. Rising limb of biogas production indicated the microbial population growth was increasing.

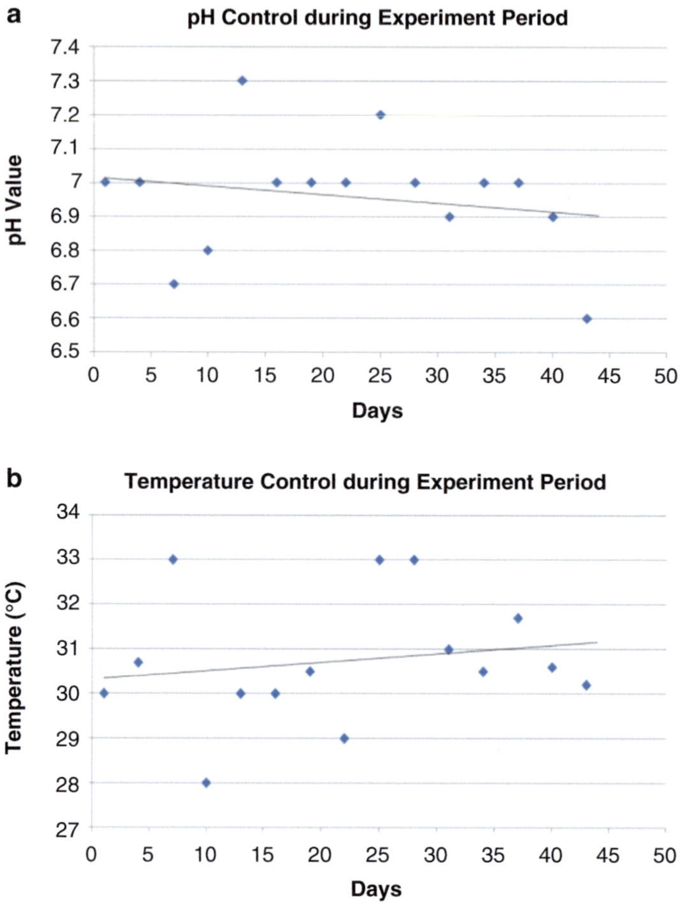

Fig. 6.3 (**a**) pH control during experiment. (**b**) Temperature control during experiment

For the falling limb, the plot of the biogas production rate gave the equation of $y = -19{,}083.881 + 19{,}084\ \exp\ (-0.46\ t)$. The constant value of $a = -19{,}083.881$ L/kg day, $b = 19{,}084$ L/kg day and $c = -0.46\ \text{day}^{-1}$ R^2 of the production rate in the rising and falling limb ranged from 0.775 to 0.940. The falling limb of biogas production rate was due to the decaying of microbial kinetic growth (death phase).

R^2 of for the rising limb of the linear plot shows better simulation than those of exponential; while for the falling limb, exponential plot shows slightly better simulation than the linear regression. Hence the biogas production rate could be best modeled with the combination of both linear and exponential models, where rising limb was best fitted by linear model ($R^2 = 0.954$) while falling limb by the exponential model ($R^2 = 0.940$).

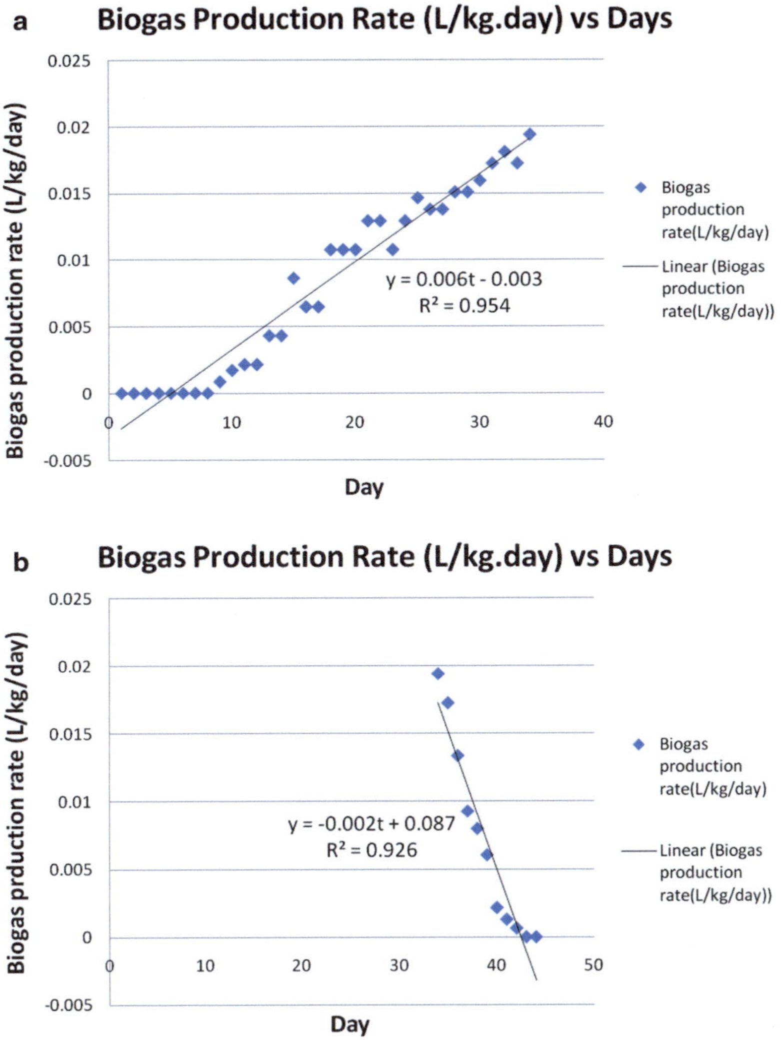

Fig. 6.4 (**a**) Biogas production linear plot (rising limb). (**b**) Biogas production linear plot (falling limb)

The experiment shows that 23.214 kg of total biomass (which contained 8 l POME) would produce 7.825 l of biogas. 62.5 % of the biogas production was estimated to be the methane gas while the rest was the carbon dioxide. Malaysia POME production was estimated to be 53 million tones/year (6.6250×10^{10} kg/year). From the experiment, 10 kg POME was used and it managed to produce 7.825 l (0.007825 m^3) of biogas. Thus 6.6250×10^{10} kg/year POME will be able to produce 6.8864×10^7 m^3/year of biogas. Table 6.4 below summarize the estimated biogas production from palm oil mill industries Malaysia.

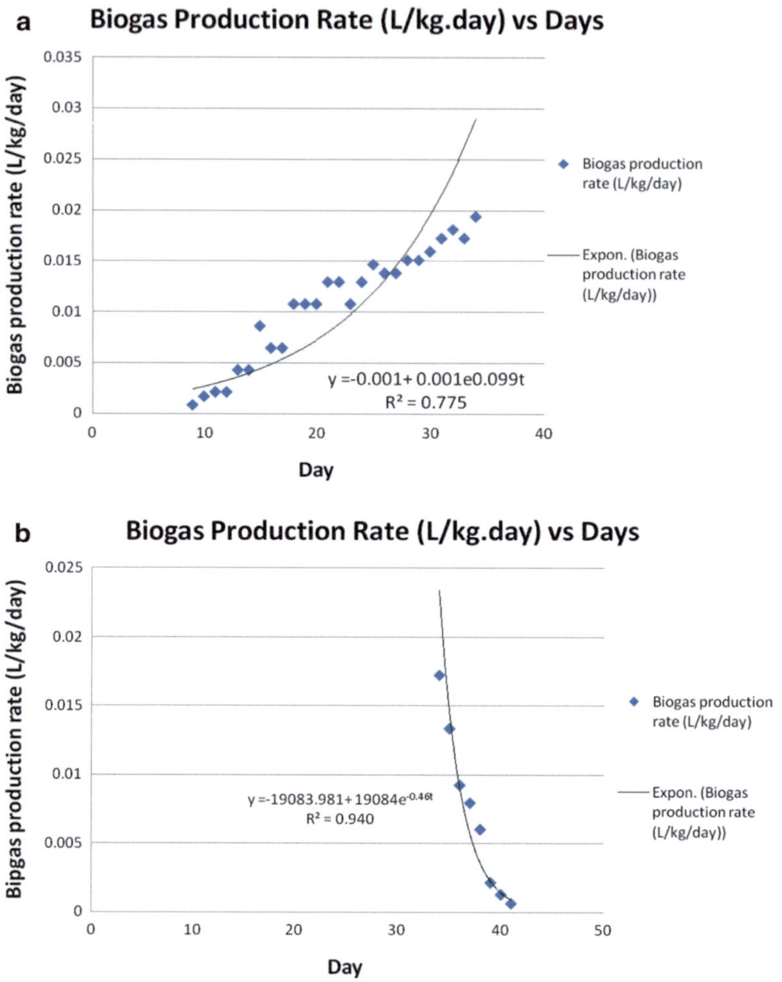

Fig. 6.5 (**a**) Biogas production exponential plot (rising limb). (**b**) Biogas production exponential plot (falling limb)

Table 6.4 Malaysia's palm oil mill effluent biogas production estimation

Malaysia palm oil mill: Total POME generated per year = 53 million m^3		
Unit	Mass (kg)	Volume (m^3)
Total biogas production	5.93×10^7	6.8864×10^7
Methane	3.71×10^7	5.7664×10^7
Carbon dioxide	2.22×10^7	1.12×10^7

Conclusions

The study proves that biogas is able to be produced from POME with an anaerobic digestion method by using cow dung and activated sludge. There is no doubt that biogas produced from POME in anaerobic digestion facilities are becoming favorably utilized to replace energy derived from fossil fuels. The overall biogas production was 0.7825 l per kg of POME mixture slurry. However, the average methane produced is only about 62.5 % with 37.5 % Carbon dioxide and traces of H_2S concentration. Biogas production by using POME and cow dung with the appropriate compositions gave a better linear plot curve fitting compared to the exponential curve fitting. From the scale up calculation, a potential of 5.93×10^7 kg of biogas could be produced each year from 5.30×10^7 m^3 POME volume production in Malaysia. Further investigation is yet to be carried out to finalize the optimization of the parameters involved in this study.

References

Ahmad, A. L., Ismail, S., & Bhatia, S. (2003). Water recycling from palm oil mill effluent (POME) using membrane technology. *Journal of Desalination, 157*(1–3), 87–95.

APHA. (1985). *Standard methods for the examination of water and wastewater* (16th ed.). Washington, DC: APHA.

Borja, R., & Banks, C. J. (1995). Comparison of an anaerobic filter and an anaerobic fluidized bed reactor treating palm oil mill effluent. *Process Biochemistry, 30*, 511–521.

Boubaker, F., & Ridha, B. C. (2008). Modelling of the mesophilic anaerobic co-digestion of olive mill wastewater with olive mill solid waste using anaerobic digestion model no. 1 (ADM 1). *Bioresource Technology, 99*, 6565–6577.

Chin, K. L., H'ng, P. S., Chai, E. W., Tey, B. T., Chin, M. J., Paridah, M. T., et al. (2013). Fuel characteristics of solid biofuel derived from oil palm biomass and fast growing timber species in Malaysia. *Bioenergy Research, 6*, 75–82.

De Gioannis, G., Muntoni, A., Cappai, G., & Milia, S. (2009). Landfill gas generation after mechanical biological treatment of municipal solid waste. Estimation of gas generation rate constants. *Waste Management, 29*, 1026–1034.

Derbal, K., Bencheikh-lehocine, M., Cecchi, F., Meniai, A. H., & Pavan, P. (2009). Application of the IWA ADM 1 model to simulate anaerobic co-digestion of organic waste with activated sludge in mesophilic condition. *Bioresource Technology, 100*, 1539–1543.

Lastella, G., Testa, C., Cornacchia, G., Notornicola, M., Voltasio, F., & Sharma, V. K. (2000). Anaerobic digestion of semi-solid organic waste: Biogas production and its purification. *Journal of Energy Conversion and Management, 43*(2002), 63–75.

Gali, A., Benabdallah, T., Astals, S., & Mata-Alvarez, J. (2009). Modified version of ADM1 model for agro-waste application. *Bioresource Technology, 100*, 2783–2790.

Gerardi, M. H. (2003). *The microbiology of anaerobic digesters* (pp. 51–57). Hoboken, NJ: Wiley.

Habib, M. A. B., Yusoff, F. M., Phang, S. M., Ang, K. J., & Mohamed, S. (1997). Nutritional values of chironomid larvae grown inpalmoil mill effluent and algal culture. *Aquaculture, 158*, 95–105.

Kristofferson, L. A., & Bokalders, V. (1991). Renewable energy technologies: Their applications in developing countries. *Journal of Agricultural System, 25*(4), 325–327. Pergamon.

Liew, W. L., Kassim, M. A., Muda, K., & Loh, S. K. (2012). Insights into efficacy of technology integration: the case of nutrient removal from palm oil mill effluent. In: Proceedings of UMT 11th international annual symposium on sustainability science and management (pp. 1203–1211). Terengganu, Malaysia.

Lorestani, A. A. Z. (2006). Biological treatment of palm oil effluent (POME) using an up-flow anaerobicsludge fixed film (UASFF) bioreactor [TD899. P4 L8692006 f rb].

Ma, A. N., & Halim, H. A. (1988). *Management of palm oil industrial wastes in Malaysia, Palm Oil Research Institute of Malaysia (PORIM)—Ministry of Primary Industries*. Kuala Lumpur: Ministry of Primary Industries.

Malaysia Palm Oil Board (MPOB). (2004). *Malaysian palm oil statistics*. Kuala Lampur: Economics and Industry Development Division.

National Non-Food Crops Centre. *NNFCC renewable fuels and energy factsheet: Anaerobic digestion*. Retrieved February 16, 2011.

Shanmugam, P., & Horan, N. J. (2009). Optimising the biogas production from leather fleshing waste by co-digestion with MSW. *Bioresource Technology, 100*(18), 4117–4120.

Poh, P. E., & Chong, M. F. (2009). Development of anaerobic digestion methods for palm oil mill effluent (POME) treatment. *Bioresource Technology, 100*, 1–9.

Polprasert, C. (1989). *Organic waste recycling*. New York, NY: Wiley.

Rao, M. S., & Singh, S. P. (2004). Bioenergy conversion studies of organic fraction of MSW: Kinetic studies and gas yield-organic loading relationships for process optimization. *Bioresource Technology, 95*, 173–185.

Yacob, S., Shirai, Y., Hassan, M. A., Wakisaka, M., & Subash, S. (2006). Start-up operation of semi-commercial closed anaerobic digester for palm oil mill effluent treatment. *Process Biochemistry, 41*, 962–964.

Singh, G., Huan, L. K., Leng, T., & Kow, D. L. (1999). *Oil palm and the environment*. Kuala Lumpur: Sp-nuda Printing, SDN, Bhd.

Sosnowski, P., Klepacz-Smolka, A., Kaczorek, K., & Lecdakowicz, S. (2008). Kinetic investigations of methane co-fermentation of sewage sludge and organic fraction of municipal solid wastes. *Bioresource Technology, 99*, 5731–5737.

Ueno, Y., Fukui, H., & Goto, M. (2007). Operation of a two-stage fermentation process producing hydrogen and methane from organic waste. *Environmental Science and Technology, 41*, 1413–1419.

Yacob, S., Hassan, M. A., Shirai, Y., Wakisaka, M., & Subash, S. (2005). Baseline study of methane emissionfrom open digesting tanks of palm oil mill effluenttreatment. *Chemosphere, 59*, 1575–1581.

Yusoff, S. (2006). Renewable energy from palm oil – Innovation on effective utilization of waste. *Journal of Cleaner Production, 14*, 87–93.

Chapter 7
Process Analysis of Microalgae Biomass Thermal Disruption for Biofuel Production

Ahmed Artan, Caleb Acquah, Michael K. Danquah, and Clarence M. Ongkudon

Introduction

Fossil fuels are the most consumed source of energy in the world today, with the transportation and electricity generation industries specifically contributing to the highest usage (Lam and Lee 2012). Numerous scientific forecasts done indicates that crude oil production is going to peak within the year range of 2030–2050 (Ivanhoe 1995; Bartlett 2000; Hallock et al. 2004). The continual reliance on conventional oil will therefore lead to economic hardships, in as much as they are also hazardous to the environment and unsustainable. Several calls have been made to look into a more sustainable and environmentally friendly approach, which must also be cost effective (Hallock et al. 2004; Harun et al. 2010; Lam and Lee 2012). According to a recent study conducted by the International Energy Agency (IEA), amongst all other renewable sources of energy, combustible renewable and waste stands out with the highest potential. Data from IEA indicates that combustible renewable energy constituted 10.0 % of the total energy, hydro energy constituting 2.2 %, and others constituting 0.7 % (geothermal, solar, wind and heat) (Lam and Lee 2012). As such, it is evident that fuels derived from combustible renewable sources are the most promising alternative to fossil fuels.

Bioethanol and Biodiesel are examples of renewable combustible fuel source and is considered as green due to its non-toxicity, biodegradability and its lower

---

A. Artan • C. Acquah • M.K. Danquah (✉)
Department of Chemical Engineering, Curtin University, Sarawak 98009, Malaysia
e-mail: mkdanquah@curtin.edu.my

C.M. Ongkudon
Institute of biotechnology, University Malaysia Sabah, Sabah, Malaysia
e-mail: clarence@ums.edu.my

P. Ravindra (ed.), *Advances in Bioprocess Technology*,
DOI 10.1007/978-3-319-17915-5_7

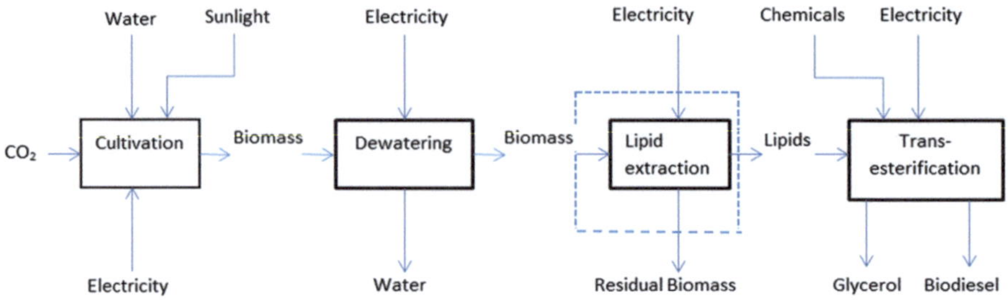

Fig. 7.1 Is a process flow sheet for the production of microalgae biodiesel, with the dashed line indicating the boundaries of this analysis

emission of GHG when burned in diesel engine (Harun et al. 2010; Ayhan and Demirbas 2011). The distinctiveness in the application of microalgae as an alternative feedstock for the production of biodiesel stems from its photosynthetic efficiency, non-existence of competition for drinking water and arable land, high growth rate, as well as, high accumulation potential for lipids and carbohydrate (Shilton et al. 2012; Chen et al. 2013). Also, they are highly economical owing to the fact that their main source of nourishment; sunlight, carbon dioxide and inorganic nutrients are found in wastewater, all of which are inexpensive and in abundance (Huntley and Redalje 2007). Chen et al. 2013, also reported that microalgae have the most potential amongst all are other renewable precursors.

However, the processes involved in; cultivation, dewatering, extraction of lipids, and trans-esterification are energy intensive. Fig. 7.1 shows the process flow diagram for the production of microalgae biodiesel. As a result, their commercial propensities are undermine. It is in the light of this that vast researches are being carried out into looking for ways of minimizing these tendencies. Griffiths and Harrison 2009, reported that the keys to determining the rate of lipids production are the appropriate choice of algae species, and an efficient method of extracting the lipids. Pursuant to this, the focal point of this research is to carry the techno-economic assessment of cell disruption with the perspective of debottlenecking. Notably, there are a number of cell disruption techniques that has been cited. These are the HPH (Samarasinghe et al. 2012), Bead mills (Spiden et al. 2013), Ultrasonic (Adam 2012) and Electroporation (Sheng et al. 2011). In general, the processes involved with these technologies are; bead beating, microwave, water bath, blender, ultrasonication, high pressure homogenizer, laser treatment and alkaline treatment (Halim et al. 2012). In the present case, the design of an industrial scale cell disruption technique was attempted. The efficiency of the cell disruption technology was compared to other existing methods on the basis of fractional energy, and subsequently illustrated with a graph in Fig. 7.4, in the energy analysis section, as a pictorial evidence for elaboration.

Methodology

According to the experiment carried out by McMillan et al. 2013, N. oculata (A species of microalgae) were grown to a concentration of about 1.8×10^8 cells/mL (7.2 g/l) before getting sampled and treated with either: microwave, ultrasonic, water bath, laser radiation, and a mechanical shear for varying times, before being microscopically analyzed for cell rupture and cell counts. The suspension was diluted to revert it to the same concentration if the counts increased within the photobioreactor (PBR) before being treated. The result of cell disruption recorded was then plotted as a function of time. The power consumed by each treatment method was monitored to provide some indication of efficacy and a figure of merit was defined by taking into account the disruption caused, energy used and volume fraction of the system utilized.

Hot water bath causes cell disruption by thermally induced pressure, leading to cells rupturing as they fail to contain their elevated internal pressure (McMillan et al. 2013). One of the reasons why the utilization of hot water bath is used for cell disruption is attributed to the production of large debris from cracked and split cells. This is an advantage which is essential in large scale operations, to ensure ease of handling and separation from soluble products. Due to this remarkable status, the industrial scale model was designed to duplicate the qualities of the hot water bath and engineered to be more adaptable for mass scale production.

The experiment was carried by taking about 30 ml sample of microalgae taken from the PBR and placed in a stainless steel vessel which was then heated by hot water that was maintained at 90 °C for 20 min. A sketch of the process scheme is as shown below (Fig. 7.2).

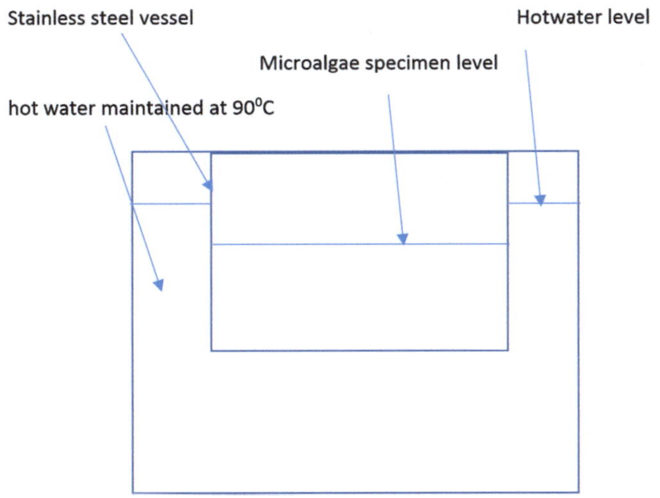

Fig. 7.2 A schematic view of the process system for hot water bath cell disruption

In order to replicate the process, a shell and tube heat exchanger was chosen due to their structural simplicity, design flexibility and low cost. However, the temperature was required to be maintained at 90 °C which prove as a challenge in the heat exchanger context. In order to achieve this set point, saturated steam at 90 °C was proposed to be the fluid that enters through the shell side, whereas, dilute microalgae culture passes through the tube side. As a result, it was expected that the steam entering will be saturated vapor at 90 °C, which will in turn leave as saturated liquid at the same temperature. The mechanism through which heat was exchanged was latent heat of vaporization, from the steam in the shell side to the diluted microalgae at the tube side. The flow rates of both the tube and the shell side was a function of the chosen design, the residence time of treatment (cell disruption) and minimum production demanded.

Heat Exchanger Design

A selected shell and tube heat exchanger needs to meet the process requirements of flow rates, composition, inlet and outlet temperatures and pressure of both tubes. The problem needs to be identified in details followed by the selection of the heat exchanger. A tentative set of exchanger design parameter is thereby selected. This is followed by rating of thermal performance and pressure drop of both streams. An evaluation of the design heat duty and pressure changes will then be analysed and validated with the requirement. If they match up with the requirement then the design is valid. However, parameters which do not meet the requirement ought to be reviewed and modified before the rating is repeated again. Finally, the cost of the heat exchanger and mechanical design is done. The basic logic structure for the process design is given below (Fig. 7.3).

Heat Transfer Coefficients

The energy balance for a shell and tube heat exchanger is as shown in the equation below:

$$Q = \left(\dot{m}c_p\right)_t (T_{t,o} - T_{t,i}) = \left(\dot{m}c_p\right)_s (T_{s,i} - T_{s,o}) \tag{7.1}$$

Where Q is the heat duty, $\dot{m}$ mass flow rate, c_p is the specific heat capacity, T is the tempearature, the subscript "t" and "s" stands for the tube side and shell side, respectively; and subscript "i" and "o" stands for the inlet and outlet of the shell or tube respectively. However, the energy transfer involves loss of the latent heat of vaporization from the shell side to the tube side. Also, the cell disruption is not just

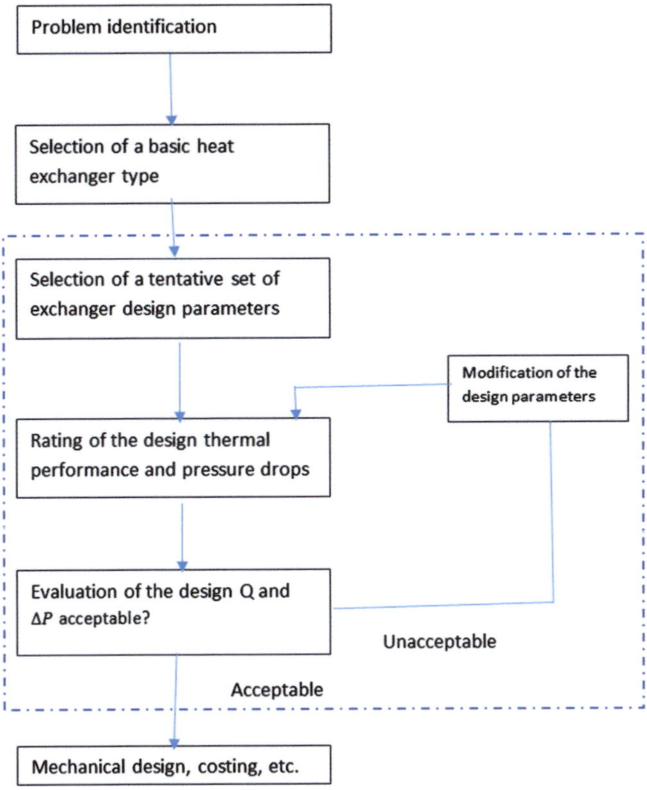

Fig. 7.3 Basic logic structure for process heat exchanger design (Bell 1980)

a function of temperature but also time, hence, the energy balance is simplified to the resultant equation below;

$$\dot{Q} = \dot{m}_s \times (h_g - h_l) \tag{7.2}$$

Where $\dot{Q}$ is the amount of heat duty required to disrupt a functional volume of microalage over a period of time (20 min), $\dot{m}_s$ stands for the rate of saturated steam in the shell side and finally h_g and h_l are the latent heat of saturated steam vapor and saturated liquid vapor respectively at 90 °C and 0.7 bar of pressure. The shell inner diameter is calculated based on the equation;

$$D_S = 0.637\sqrt{\frac{CL}{C_{tp}}} \times \left(A_O \times \frac{P_R^2 \times d_O}{L}\right)^{0.5} \tag{7.3}$$

Where C_{tp} is the tube count calculation constant that accounts for thc incomplete coverage of the shell diameter by the tubes, CL stands for the tube layout constant,

P_R is the tube pitch ratio, d_0 is the tube outer diameter, L is the length of the tube and A_0 is the area required. C_{tp}, CL and P_R are dependent on the arrangement and number of passes. It therefore becomes imperative to establish the number of tubes that can be accommodated in the reactor/exchanger so as to ascertain the limits. Number of tubes that can be accommodated can be calculated with the subsequent equation;

$$N_t = 0.875 \times \left(\frac{C_{tp}}{CL} \times \frac{D_S^2}{P_R^2 \times d^2}\right) \tag{7.4}$$

The heat transfer coefficient for the shell side can also be estimated by;

$$h_0 = \frac{k}{D_e} 0.36 Re^{0.55} Pr^{0.333} \left(\frac{\mu_b}{\mu_w}\right)^{0.14} \tag{7.5}$$

Where D_e the hydraulic diameter, k is is the thermal conductivity of steam (fluid on shell side), Re is the Reynolds number, Pr is the Prandtl's number, μ_b and μ_w are the fluid viscosity at the bulk fluid temperature and at the heat transfer surface boundary temperature respectively. The hydraulic diameter D_e is expressed as;

$$D_e = 4 \times \left(\frac{\frac{\sqrt{3} \times P_T^2}{4} - \left(\frac{\pi \times d^2}{8}\right)}{\frac{\pi \times d}{2}}\right) \tag{7.6}$$

The Reynolds number *Re* is calculated using:

$$Re = \frac{G_s D_e}{\mu} \tag{7.7}$$

Where G_s is the mass flow rate per unit cross sectional area (kg/m^{2}s) given as:

$$G_s = \frac{\dot{m}_s}{A_s} \tag{7.8}$$

And the prandtl number is as shown below;

$$Pr = \frac{C_p \mu}{k} \tag{7.9}$$

Variables that affect the velocity of the shell side fluid are the shell diameter Ds, the clearance C between adjustment tubes, the pitch size P_T and baffle spacing B. The width of the flow area of the tube located at the center of the shell is $\frac{D_S}{P_T} \times C$, and the length of the flow area is taken as the baffle spacing B. Therefore, the bundle cross flow area A_s at the center of the shell is appropriated as;

$$A_s = \frac{D_s \times C \times B}{P_T} \tag{7.10}$$

Where C and B are given as:

$$C = P_T - d_o \tag{7.11}$$

$$B = 0.4 \times D_s \tag{7.12}$$

Likewise, the heat transfer coefficient from the tube side can be calculated as shown below:

$$h_i = \frac{k}{d_i} \times 0.023 {N_{Re}}^{0.8} Pr^{0.4} \tag{7.13}$$

Where N_{Re} and Pr are the reynolds number and Prandtl number of the fluid in the tube side respectively, while d_i is the inside diameter of the tube. Finally the overall heat transfer coeffcient U can be estimated as:

$$U = \left(\frac{1}{h_o} + \left(\frac{1}{h_i} \times \frac{d_o}{d_i} \right) + \frac{d_o In\left(\frac{d_o}{d_i}\right)}{k} \right)^{-1} \tag{7.14}$$

Pressure Drop

Tube side pressure drop is calculated by knowing the number of tube passes, N_P, the length, L, of heat exchanger, the internal diameter, density of microalgae and the velocity of the microalgae in the tubes side denoted by μ_m. The pressure drop is given below;

$$\Delta P_t = \left(4f \frac{LN_P}{d_i} + 4N_P \right) \times \rho \times \frac{\mu_m^2}{2} \tag{7.15}$$

Using Gnelinski correlation;

$$f = (1.58 \, In \, Re - 3.28) \tag{7.16}$$

The shell side pressure is dependent on the number of tubes the fluid passes through in the tube bundles, thus, between the baffles and the length of each crossing. The correlation developed using the product of distance across the bundle taken as inside diameter of the shell D_s and the number of times the bundle is crossed.

The equivalent diameter, $D_{e,}$ is the same as presented earlier. The shell side pressure drop is then calculated as shown below:

$$\Delta P = \frac{f \times G_s^2 \times (N_b + 1) \times D_s}{2 \times \rho \times De \times \O_s} \tag{7.17}$$

$$\O_s = \left(\frac{\mu_b}{\mu_w}\right)^{0.14} = 1 \tag{7.18}$$

The number of baffles N_b is given by:

$$N_b = \frac{L}{B} - 1 \tag{7.19}$$

The number of times the shell fluid passes the tube bundles is expressed as:

$$N_b + 1 \tag{7.20}$$

And finally, the friction factor f is derived by:

$$f = exp(0.576 - 0, 19 \, In \, Re_s) \tag{7.21}$$

Where;

$$400 < Re_s = \frac{G_S D_e}{\mu} < 1 \times 10^6 \tag{7.22}$$

Energy Performance Analysis

The energy requirement for various extraction processes can be estimated as a function of algae concentration. This is achieved by using the moderate conditions established by Coons and colleagues. The fractional energy is the ratio between the energy used in the extraction process to the total energy of the algal water content which is the summation of lipid energy content and lipid extracted algae. The fractional energy gives an idea of how efficient cell disruption technology is. This is illustrated mathematically below.

$$F_{energy} = \frac{\overline{E}_{Extraction}}{\overline{E}_{Total}} \tag{7.23}$$

By taking the moderate condition reference (Coons et al. 2014) the total energy can be expressed as:

$$\overline{E}_{Total} = \left[\frac{6.82\ kwh}{kg\ dry\ weight\ of\ algae}\right] \times [B]_{Extraction} \tag{7.24}$$

Where $[B]_{Extraction}$ is the concentration of algae in kg dry weight per m^3. Hence, after substituting (7.21) into (7.22), the resultant fractional energy becomes:

$$F_{energy} = \frac{\overline{E}_{Extraction}}{\left[\frac{6.82\ kwh}{kg\ dry\ weight\ of\ algae}\right] \times [B]_{Extraction}} \tag{7.25}$$

This parameter is essential and will be employed for the purpose of ranking the performance, and comparing different cell disruption technologies to the designed heat exchanger technology.

Economics, Environmental and Baseline Assumptions

In order to proceed to industrial level, an assumption of the functional unit would have to be made. In this current case, this will be the amount of lipid production that is required in order to meet the company's capacity target. For the purpose of this assessment, that value was assumed at ten mega gallons of lipids per annum. The various relevant assumptions are tabulated in the table below (Table 7.1).

Table 7.1 Baseline economic assumptions

Criteria	Assumption	Reference
Working days annually	330 days (7,920 h)	–
Functional unit	10 MM gal/annum (lipids)	–
Location	Australia	–
Electricity cost	0.30 A$ per kWh	(Ergon 2014);
Steam cost	0.0094 A$ per kWh	Online technical support of Gestra Australia
Disruption	87.7 %	(McMillan et al. 2013)
Lipid yield (dry weight)	25 %	–
Algal cell density	7.2 g/L	(McMillan et al. 2013)
Emission factor	9 g of CO_2/kWh	–
Carbon taxation	A$25.40	(Asafu-Adjaye and Mahadevan 2013)

Carbon Emissions and Cost

Quantification of the total amount of GHG emitted during the process is in tonnes per annum. This is accomplished by converting the amount of energy taken to treat a functional unit of microalgae into kilowatts hour, and using the combination of the functional units as well as the emission factor.

$$\begin{aligned} Carbon\ emission = {} & Energy\ reguirement\left(\frac{kWh}{m^3}\right) \\ & \times emission\,factor\left(\frac{kg\ CO_2 - e}{kWh}\right) \end{aligned} \tag{7.26}$$

Similarly carbon costing is derived as;

$$\begin{aligned} Carbon\ cost = {} & Energy\ requirement\left(\frac{kWh}{m^3}\right) \\ & \times emission\,factor\left(\frac{kg\ CO_2 - e}{kWh}\right) \\ & \times\ Carbon\ tax\left(\frac{\$}{tone\ CO_2 - e}\right) \end{aligned} \tag{7.27}$$

Heat Exchanger Operating Cost

The total power consumption for the heat exchanger is calculated as;

$$P = \frac{1}{n}\left(\frac{\dot{m}_s}{\rho_s}\Delta P_s + \frac{\dot{m}_t}{\rho_t}\Delta P_t\right) \times A \tag{7.28}$$

Where A is the total area available for heat transfer and n is the pump efficiency taken at 0.7. The total cost is obtained through the equation below;

$$C_{tot} = C_o + EC_{steam} + carbon\ costing \tag{7.29}$$

Where EC_{steam} is the annual energy cost of steam generation, assuming the source i.e., boilers uses electricity energy as its source of power. The value for EC_{steam} is derived from the expression;

$$EC_{steam} = 2.7 \times 10^{-4} \times h_g \times \rho_s \times V_{steam} \times C_E \tag{7.30}$$

Where h_g is the saturated enthalpy of steam at inlet operating conditions, ρ_s is the density of the saturated steam, V_{steam} is the annual volume of saturated steam required to disrupt the functional unit, C_E is the cost of electricity in A\$/kWh and the value 2.7×10^{-4} is the unit conversion of kilojoules to kilowatts hour. C_0 is the annual operating cost bowed by electric charges (Taal et al. 2003) and can be calculated through the equation below;

$$C_o = P \times C_E \times H \tag{7.31}$$

Where H is the amount of working hours assumed at 7,920 h.

Operating Cost of the Alternative Cell Disruptions

Australia is the preferred location for this hypothetical case study. The essence of this is to make the experiment more relevant and practical, by synchronising the experiment into a real life scenario. For instance, the Australian government passed a legislation called the Green Energy Act 2011. This legislation is supposed to establish an emission-pricing scheme with a view of mitigating the nation's greenhouse gas emissions (Australian Parliament 2011). Also, according to Ergon energy, an electricity retailer to homes and businesses in regional Queensland, a large business with an annual consumption of more than 100 MWh per year (For which our scenario qualifies) attracts charges of 30.866 cents per kWh (Ergon Energy 2014). Considering such a scenario as presented by the government of Australia, an in depth cost analysis of the cell disruption stage was performed. To begin, with the annual energy consumed in the process was determined for each treatment method based on charges of electricity per kilowatts hour in Australia. The total amount in dollars was then evaluated. In the inflation-indexed emission reduction policy of the Australian government, the following prices (\$/t CO_2-e) were imposed: A\$23 in 2012, A\$24.15 in 2013, and A\$25.40 in 2014, with the future priced to be determined by the emission trading scheme (ETS) from 2015 onwards (Asafu-Adjaye and Mahadevan 2013). In the same vein, the carbon taxation (taking the latest which is A\$25.40 in 2014) on the amount of CO_2 emitted per annum was calculated. Finally, the summation of the two costs (cost of electricity from equation 1.32 + cost of carbon emitted from (7.24)) determined the ultimate annual cost of the cell disruption process.

$$Electricity\ cost = Extraction\ energy\left(\frac{kWh}{m^3}\right) \times\ cost\ of\ energy\left(\frac{\$}{kWh}\right) \tag{7.32}$$

This method of estimation was applied solely for the calculation of operating costs of the alternative cell disruption methods that included; bead mill, HPH, electroporation and ultrasonication. The costing for the heat exchanger is as previously elaborated.

Result and Discusions

Heat Exchanger Design

The objective function of the design is to minimize the total cost that encompasses the cost of the initial investment and power consumption. The maximum allowable pressure drop for both the shell and tube side was taken at 70,000 Pa, and all the heat conservation requirements were met. The table below shows the exchanger cell disruption technology's specification (Table 7.2).

From the table above, the steam mass flow rates is smaller compared with the mass flow rate of the algae which is a good indication of how effective this method is. The lower mass flow rate of steam signifies a drastic reduction in the overall operating cost. Of a positive note, the inlet and outlet temperatures of both tubes and shell sides were as expected. The saturated vapor steam which entered at 90 °C and lower than atmospheric pressure of 0.69 atm exited from the shell as saturated liquid at the same temperature, and in the process only losing its latent heat of vaporization. These latent heat of vaporization is the main mechanism at the heart of the exchanger. The change in pressure for both the shell and tube side were safely below the design margin of 70,000 pa.

The heat exchanger is designed as one pass both for the tube and the shell side (triangular arrangement). It should be noted however that the process is time dependent and not temperature as is usual with heat exchangers. Meaning that even after the tube side microalgae attains the desired temperature of 90 °C, it still will remain until the residence time is complete. The residence time as applied in this context means the time taken to complete a disruption percentage of 87.7 % as a mimic of the experimental properties.

Table 7.2 Design specifications for the heat exchanger

	Shell	Tube
Fluid Allocation	Saturated steam	Microalgae
Mass flow (kg/s)	0.054	6.2
T input (°C)	90	30
T output (°C)	90	90
ρ (kg/m^3)	0.450	1,010
Cp (kJ/kg K)	2.04	4.1
μ (kg/m.s)	0.315	0.0015
K (w/m K)	0.024	0.63
Pressure inlet (atm)	0.69	1
Δ *Pressure* (pa)	1,578	22.18
Passes	1	1

Energy Analysis

Lipid access is an energy intensive step that involves cell disruption and solvent extraction (Cooney et al. 2009). This section seeks to showcase the energy requirement for the heat exchanger technology in comparison with other diverse disruption technologies. The energy fraction as was earlier explained will be used as a test of the energy efficiency.

Table 7.3 below depicts different cell disruption technologies with their fractional energy at their respective algal concentration. In order to investigate on the effects of algal concentration on the fractional energies, Fig. 7.4 was plotted by inferring from Table 7.3. The presentation suggests that higher algal concentration enhances the fractional energy. However, caution should be taken while increasing this concentrations since the bulk properties of the cellular suspension is deemed to change with the concentration increment. Wileman et al. 2012, remarked that at a concentration of about 20 kg/m^3 there was an onset of non-Newtonian behaviour. Hence, any operational properties that relies on viscosity of the fluid such as, pumping, will be affected. Notably, even though comparison is achievable through the sensitivity analysis of the microalgae concentration, the results are still slightly ineffective due to the fact that microalgae cell disruption is species dependent (Halim et al. 2012)

High pressure homogenization (HPH) and ultrasonication have moderately higher extraction energy than the other methods. This could be the main reason why they registered poorly in terms of fractional energy. On the other hand, bead mill scores fairly well as shown in Fig. 7.4 which is thought to explain their popularity in mass scale production. Electroporation use simple equipment such as electrode and piezoelectric components which hypothetically translates into lower capital costs (Coons et al. 2014). That harmonises very well with the results plotted in Fig. 7.4 above, as favorable fractional energy are achieved. The heat exchanger system appears to be the most efficient technology, not only does it have an improved fractional energy but also the largest volume of scale as depicted in Fig. 7.4 and Table 7.3 above. This results however, are deemed not to be conclusive. This is attributed to the fact that, the main objective of many researchers is the maximization of yield and not minimization of energy when carrying out their experiment (Lee et al. 2012).

Economics and Environmental Analysis

Table 7.4 and Fig. 7.5 provides a summary of the environmental and economics assessment model used for the performance comparison of the different cell disruption methods. The analysis was carried by using the reported experimental

Table 7.3 Energy requirement for the different cell disruption technologies

Treatment method	$[B]_{Extraction}$ Algae concentration in kg/m^3	$\bar{E}_{Extraction}$ Energy required (kwh/m^3)	Disruption (%)	Microalgae species	$\bar{E}_{Total}$ Total energy content	System volume or flow rate	Fractional energy	Source
Heat exchanger design	7.2	16.8	87.7	*N. oculata*	51.6	2,1960 L/h	0.325	–
HPH	35	444	95–100	*N. oculata*	238	3 L/h	1.86	(Samarasinghe et al. 2012)
Bead mills	0.25	110.83	95–100	*Cholera* sp.	1.72	10 L/h	64.5	(Spiden et al. 2013)
Ultrasonic	50	316	–	*N. oculata*	343.50	0.1 L	0.92	(Adam 2012)
Electroporation	0.3	29.2	99	*Synechocytis* sp.	2.05	25 L	14.27	(Sheng et al. 2011)

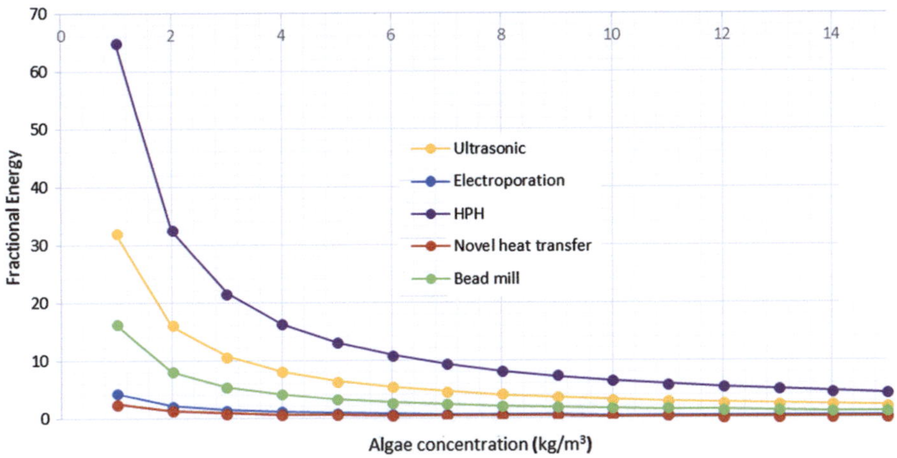

Fig. 7.4 Effects of algal concentration on fractional energy

results by other authors and expanding them to industrial scale. The result depicts annual carbon emissions and the expected total operating cost for each method. In comparison, the best disruption method in terms of economics and sustainability is the heat exchanger cell disruption technology. With the exception of electroporation, all other disruption methods are deemed uneconomical. Nevertheless, despite the potentiality of electroporation, it has only been tried on a lab scale. More so, it is worth noting that the enumerated costs especially for the alternative cell disruption methods are short of additional costs like pumping costs and cooling costs. Similarly, since this work manages to compare the operational costing only it is proposed that future work should determine the expected capital costing of the various technologies too. This is because the capital cost might greatly differ on an industrial scale and should therefore be used as a criteria for the selection of the disruption technology.

Despite the displayed figures of carbon emission, microalgae is largely considered environmentally friendly. This is because of its sequestration capability that entails the removal of CO_2 from the atmosphere. Through a multistep process of photosynthesis, plants and algae (green algae and cyanobacteria) fix CO_2 into sugar using light and water as energy and electron source, respectively (Kumar et al. 2011). The biochemistry of CO_2 fixation is given by:

$$CO_2 + H_2O + \text{Light} \rightarrow (CH_2O)_n + O_2 \qquad (7.33)$$

The use of algae for CO_2 sequestration has several advantages: mitigating CO_2, the major source of global warming as well as, producing biofuels and other interesting secondary metabolites (Kumar et al. 2011). One kilogram of algal dry cell weight utilizes around 1.83 kg of CO_2. An estimated quantity of about 54.9–67.7 Ton per annum of CO_2 can be sequestered from raceway ponds. This corresponds to an annual dry weight biomass production rate of 30–37 Ton per hectare (Brennan and

Table 7.4 An assessment of the economic and environmental impact and viability

Heat exchanger cell disruption assessment									
Treatment method	Functional unit V_{steam} $\left(\frac{m^3}{year}\right)$	Cost of energy in A $/kWh	Emission factor in $\frac{kg\ CO_2-e}{kWh}$	Carbon tax in $\frac{\$}{tone\ CO_2-e}$	Carbon emission per annum $Tone\ CO_2 - e$	C_o cost in A$	Carbon cost in A$	EC_{steam} In A$	Total cost per annum
Heat exchanger	3, 421, 440	0.03	0.009	A$25.40	28.9	63, 805	668	33, 173	97, 712
Alternative cell disruption methods assessments									
Treatment method	Functional unit (microalgae to be treated in m^3)	Cost of energy in A$/kWh	Emission factor in $\frac{kg\ CO_2-e}{kWh}$	Carbon tax in $\frac{\$}{tone\ CO_2-e}$	$\bar{E}_{Extraction}$ Energy required (kWh/m^3)	Carbon emission in $Tone\ CO_2 - e$	Carbon cost in A$	Energy cost In A$	Total cost In A$
HPH	174×10^3	0.03	0.009	A$25.40	444	68.90	1,750	2,317,680	2,319,430
Bead mills	174×10^3	0.03	0.009	A$25.40	110.83	173.56	4,408	578,532	582,940
Ultrasonic	174×10^3	0.03	0.009	A$25.40	316	494.86	12,569	1,649,520	166,2089
Electroporation	174×10^3	0.03	0.009	A$25.40	29.2	45.73	1,161	152,424	15,3585

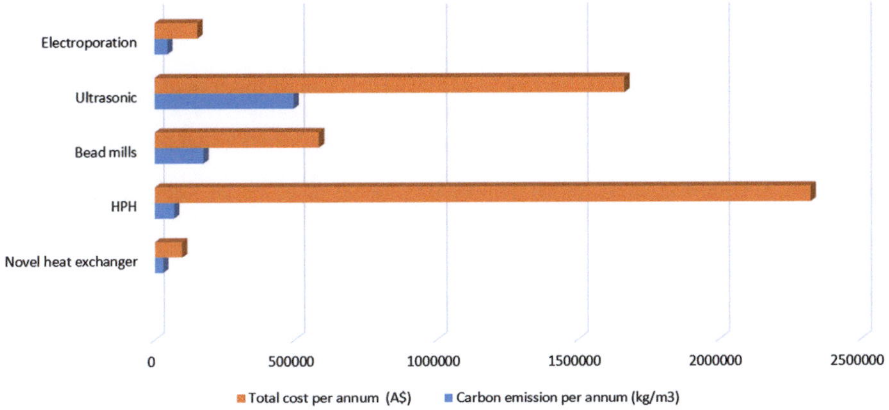

Fig. 7.5 Different cell disruption technologies with associated operating cost and carbon emissions

Owende 2010). As evidenced above, the amount of CO_2 sequestration is presumed to outdo the overall emission of the plant, hence, it is regarded as carbon green. It is yet to be seen if this kind of appraisal will hold in the Australian policy, which will lead to a waiving of the carbon taxation in this context.

Conclusion

An effective cell disruption technology was successfully designed by employing qualities of hot water bath cell disruption. As a result, a mass scale heat exchanger cell disruption technology was developed by means of thermal lysis. This new design has proven to be very efficient as it can treat a large volume of microalgae, whiles at the same time consuming minimal extraction energy. Likewise, an industrial scale analysis for the comparison of the operational cost and carbon emission further proves the potential of this design. The relationship between microalgae concentration to the energy fraction was laid out. Proposal to increase the algal concentration for maximum efficiency was given on conditions that the viscosity should be monitored. Application of this technology in conjuction with other optimised aspects in the production of biodiesel and bioethanol from microalgae will yield much fruits. From the experimental results, the prospects of it being a replacement for conventional oil are therefore bright.

References

Adam, F. (2012). "Solvent-free" ultrasound-assisted extraction of lipids from fresh microalgae cells: A green, clean and scalable process. *Bioresource Technology, 114*, 457–465.

Asafu-Adjaye, J., & Mahadevan, R. (2013). Implications of CO2 reduction policies for a high carbon emitting economy. *Energy Economics, 38*, 32–41.

Ayhan, D., & Demirbas, M. F. (2011). Importance of algae oil as a source of biodiesel. *Energy Conversion and Management, 52*(1), 163–170.

Bartlett, A. (2000). An analysis of US and world oil production patterns using Hubbert-style curves. *Mathematical Geology, 32*, 1–17.

Bell K. J. (1980). Preliminary design of shell and tube heat exchangers. In Conf. NATO Adv. Study Inst. *Heat Exchangers: Thermal-Hydraulic Fund. & Design*. pp. 559–580.

Brennan, L., & Owende, P. (2010). Biofuels from microalgae—a review of technologies for production, processing, and extractions of biofuels and co-products. *Renewable and Sustainable Energy Reviews, 14*(2), 557–577.

Chen, C., Zhao, X., Yen, H., Ho, S., Cheng, C., Lee, D., et al. (2013). Microalgae-based carbohydrates for biofuel production. *Biochemical Engineering Journal, 78*, 1–10.

Cooney, M., Young, G., & Nagle, N. (2009). Extraction of Biooils from Microalgae. *Separation & Purification Reviews, 38*(4), 291–325.

Coons, J., Kalb, D. M., Dale, T., & Marrone, B. L. (2014). Show more Getting to low-cost algal biofuels: A monograph on conventional and cutting-edge harvesting and extraction technologies. *Algal Research, 1*, 1–20.

Ergon Energy. Large business tariffs. Accessed April 2, 2014 from https://www.ergon.com.au/your-business/electricity-prices/large- business-tariffs.

Griffiths, M. J., & Harrison, S. T. L. (2009). Lipid productivity as a key characteristic for choosing algal species for biodiesel production. *J Appl Phycol, 21*, 493–507.

Halim, R., Danquah, M. K., & Webley, P. A. (2012). Extraction of oil from microalgae for biodiesel production: A review. *Biotechnology Advances, 30*(3), 709–732.

Hallock, J. L., Tharakan, P. J., Hall, C. A., Jefferson, M., & Wua, W. (2004). Forecasting the limits to the availability and diversity of global conventional oil supply. *Energy, 29*, 1673–1696.

Harun, R., Danquah, M. K., & Forde, G. M. (2010). Microalgal biomass as a fermentation feedstock for bioethanol production. *Chem. Technol, 85*, 199–203.

Huntley, M. E., & Redalje, D. G. (2007). CO2 mitigation and renewable oil from photosynthetic microbes: A new appraisal. *Bioresource Technology, 12*(4), 573.

Ivanhoe. (1995). Future world oil supplies: There is a finite limit. Accessed September 24, 2014 from http://www.oilcrash.com/articles/future.htm.

Kumar, K., Dasgupta, C. N., Nayak, B., Lindblad, P., & Das, D. (2011). Development of suitable photobioreactors for CO2 sequestration addressing global warming using green algae and cyanobacteria. *Bioresource Technology, 102*(8), 4945–4953.

Lam, M., & Lee, K. (2012). Microalgae biofuels: A critical review of issues, problems and the way forward. *Biotechnology Advances, 30*(3), 673–690.

Lee, A. K., Lewis, D. M., & Ashman, P. J. (2012). Disruption of microalgal cells for the extraction of lipids for biofuels: Processes and specific energy requirements. *Biomass and Bioenergy, 46*, 89–101.

Taal, M., Bulatov, I., Klemeš, J., & Stehlk, P. (2003). Cost estimation and energy price forecasts for economic evaluation of retrofit projects. *Applied Thermal Engineering, 23*(14), 1819–1835.

McMillan, R., Jonathan, Watson, I. A., Ali, M., & Jaafar, W. (2013). Evaluation and comparison of algal cell disruption methods: Microwave, waterbath, blender, ultrasonic and laser treatment. *Applied Energy, 103*(0306–2619), 128–134.

Samarasinghe, N., Fernando, S., Lacey, R., & Faulkner, W. B. (2012). Algal cell rupture using high pressure homogenization as a prelude to oil extraction. *Renewable Energy, 48*, 300–308.

Shilton, A. N., Powell, N., & Guieysse, B. (2012). Plant based phosphorus recovery from wastewater via algae and macrophtes. *Current opinion in biotechnology, 23*(6), 884–889.

Sheng, J., Vannela, R., & Rittmann. (2011). Evaluation of cell-disruption effects of pulsedelectric-field treatment of Synchocystis PCC 6803. *Environmental Science & Technology, 45*, 3795–3802.

Spiden, E. M., Scales, P. J., Kentish, S. E., & Martin, G. J. O. (2013). Critical analysis of quantitative indicators of cell disruption applied to *Saccharomyces cerevisiae* processed with an industrial high pressure homogenizer. *Biochemical Engineering Journal, 70*, 120–126.

Wileman, A., Ozkan, A., & Berberoglu, H. (2012). Rheological properties of algae slurries for minimizing harvesting energy requirements in biofuel production. *Bioresource technology, 104*, 432–439.

Chapter 8
Biogas from Poultry Litter: A Review on Recent Technological Advancements

Gangagni Rao Anupoju, Shruti Ahuja, Bharath Gandu, Sandhya K, Kranti Kuruti, and Venkata Swamy Yerramsetti

Introduction

Poultry industry is one of the biggest organized sectors and produces large amounts of waste that include bedding materials, excreta (manure), waste feed, feathers, broken eggs, shells, dead birds etc. waste disposal from poultry farms is one of the challenging tasks to maintain hygiene as well as improve productivity(http://www.poultryhub.org/production/husbandry-management/housing-environment/waste-management/ 2014). An adult hen produces 14 kg of egg mass and about 40 kg of excreta in a year. Poultry excreta are available as deep litter from litter-reared birds and as cage manure from cage-reared birds. The litter causes problems of fly/insect menace and offensive odour particularly at high humid areas due to the high moisture content (70–80 %) of poultry excreta (http://www.fao.org/docrep/013/al715e/al715e00.pdf 2014).Poultry litter is rich in organic and inorganic matter (http: //grist.files.wordpress.com/2008/10/a84-3-1990e.pdf 2014). Poultry litter is mainly utilized for growing crops and to a lesser extent for fish production. Excess use of poultry litter for those purposes may result in air, land and water pollution apart from spreading pathogenic organisms like *salmonella, E. coli* (Aili et al. 1990; http://www.poultryhub.org/production/husbandry-management/housing-environment/waste-management/ 2014). Poultry industry uses a great deal of inorganic source of phosphorus (P) in both layer and broiler diets that are reflected in poultry litter which may pose the problem of acceleration of eutrophication (Szogi and Vanotti 2009; http://www.poultryhub.org/production/husbandry-management/housing-environment/

G.R. Anupoju (✉) • B. Gandu • S. K • K. Kuruti • V.S. Yerramsetti
Bioenginerring and Environmental Sciences (BEES), CSIR-Indiab Institute of Chemical Technology (IICT), Tarnaka, Tarnaka, Hyderabad 500007, India
e-mail: agrao@iict.res.in; gangagnirao@gmail.com

S. Ahuja
Ahuja Engineering Services Pvt. ltd, 3017, Emerald House, S.D.Road, Secunderabad 500003, India

P. Ravindra (ed.), *Advances in Bioprocess Technology*,
DOI 10.1007/978-3-319-17915-5_8

waste-management/ 2014). Dissolved oxygen in water is depleted due to the presence of poultry litter. European countries have already restricted the amount of animal waste application to land (Bishnoi and Bajwa 1994; http://www.poultryhub.org/production/husbandry-management/housing-environment/waste-management/ 2014).

Characteristics of Poultry Waste

Poultry litter is a mixture of poultry excreta, spilled feed, feathers and material used as bedding in poultry operations. Major wastes of concern in poultry sector are bedding material used for poultry housing, poultry excreta resulting from poultry production and dead birds. The composition of poultry litter is predominantly water and carbon (C) with appreciable amounts of nitrogen (N) (1.22–1.63 %), P(0.89–1.04 %) and potassium (K) (1.34–1.7 %). In general litter composition varies with the type of feed and bird. It also contains many micronutrients such as Zinc (Zn), Copper (Cu), Iron (Fe) and selenium and trace levels of chlorine (Cl), Calcium (Ca), Magnesium (Mg), Sodium (Na), Manganese (Mn) and Arsenic (As) (Kelleher et al. 2002; Salminen and Rintal 2002a; Suleyman Sakar et al. 2009; Magbanua et al. 2001; Gangagni Rao et al. 2008a, b, 2011; Jalasutram et al. 2013; Keri B et al. 2008; Mata-Alvarez et al. 2000; Bishnoi and Bajwa 1994; Shanti et al. 2013). In deep litter system, bedding materials such as straw, saw dust, wood shavings, shredded paper, peanut, rice hulls etc., are used. During the production cycle, poultry excreta generated over a period of time is mixed with bedding material and at the end of the cycle both are removed together. N is one of the most abundant elements in the litter and exists in several forms (Kelleher et al. 2002; Salminen and Rintal 2002a, b; Suleyman Sakar et al. 2009). The composition of the litter changes constantly by microbial activity due to the changing temperature, pH, moisture and oxygen concentration. Poultry litter contains significant concentrations of organic N due to the presence of high levels of protein and amino acids in the feed (Kelleher et al. 2002; Salminen and Rintal 2002a; Suleyman Sakar et al. 2009). In fresh litter, 60–80 % of N is typically in organic form, such as urea and protein. Depending on the environmental conditions, a large percentage of organic N (40–90 %) is converted to ammonia within a year. Ammonia exists either as gas (NH_3) or in ionized state (NH_4^+), which is water-soluble. NH_3 gas is lost to the atmosphere while NH_4^+ ionis transformed by microorganisms to nitrate (a process known as nitrification) (Xiao Dong and Ernest Tollner 2003). Nitrate is highly mobile in water and may be present in run-off. Ammonium ions contribute to high alkaline pH, which is corrosive and leads to handling, storage and disposal problems. The concentration of ammoniacal nitrogen is important while considering any of the disposal techniques and minimization of ammonia is desirable for any treatment of poultry litter (Krylova et al. 1997; Gangagni Rao et al. 2008a, b, 2012; Borja et al. 1996). Poultry by-products and wastes may contain several hundred different species of microorganisms including potential pathogens such as *Salmonella sp.*, *Staphylococcus sp.*, and *Clostridum sp.*, (Salminen and Rintal 2002a). In addition the birds may accumulate various metals, drugs and other chemicals from their feed, which are added for nutritional, and pharmaceuticals purposes. Veterinary drugs and other chemical contaminants are also present in poultry litter in various concentrations (Thyagarajan et al. 2013; Xin et al. 2010; Peter Ciborowski 2001; Rajashekhar and Mohan 1994; WMFA 2014).

Waste Management in Poultry Farms

Current Practices

Poultry sector transformed into an organized sector over the years and is slowly adopting modern technologies to reduce its dependency on manual labor. However, problem of waste management in poultry farms remains the same (Thyagarajan et al. 2013; Xin et al. 2010; Peter Ciborowski 2001; Rajashekhar and Mohan 1994; WMFA 2014; http: //www.poultryhub.org/production/husbandry-management/housing-environment/waste-management/2014). In most cases, birds are kept in elevated sheds (6–7 ft high) and the waste droppings are collected below the sheds. These sheds are cleaned once a year i.e. at the end of the life cycle of the birds (deep litter system) (http: //www.poultryhub.org/production/husbandry-management/housing-environment/waste-management/2014). In some coastal areas where humidity levels are comparatively higher, poultry manure is collected once in two or three days but is not safely disposed off. It is simply spread onto nearby agricultural/farm land (land spreading). The base of every shed is filled with bedding material (straw, saw dust, wood shavings, shredded paper peanut, rice hulls, sand, coconut coir pitch viz.) to enable easy waste collection (http://www.poultryhub.org/production/husbandry-management/housing-environment/waste-management/2014). Bedding helps in absorbing moisture and reducing odor to a small extent. At the end of the life cycle of the bird (birds are kept in cages for approximately 54 weeks) the litter is cleared and is dumped/land filled. After the cages are empty, i.e. after a flock's life cycle, it takes almost a month to clean up the waste and refill the cages with a new flock. The area is sanitized before refill to kill pathogens and flies and also to reduce odour. The sanitization is very important for bio-security reasons. The waste is physically lifted by manpower and emptied into trolleys without any protective gears. Dead birds are incinerated in the farm (http: //www.poultryhub.org/production/husbandry-management/housing-environment/waste-management/2014). Exposure to methane, carbon dioxide and ammonia are highly hazardous to health (Thyagarajan et al. 2013; Xin et al. 2010; Peter Ciborowski 2001; Rajashekhar and Mohan 1994; Gert-Jan Monteny et al. 2006; WMFA 2014).

Problems Associated with Current Practices

The health of birds and people working in poultry farms is a matter of high concern since waste disposal is done only once in a year. The waste slowly degenerates naturally and produces ammonia, methane and carbon dioxide. While methane and carbon dioxide are a source of greenhouse gases, ammonia is a major noxious gas that is hazardous to birds and human health (Gert-Jan Monteny et al. 2006; WMFA 2014; http://www.poultryhub.org/production/husbandry-management/housing-environment/waste-management/2014). Increased levels of atmospheric ammonia in poultry houses reduce feed intake of birds, impedes bird growth rate and decreases egg production (Xin et al. 2010; WMFA 2014). Generally dead birds are incinerated in the farm. Though incineration of birds is biologically safe, it is expensive and may create air quality issues (WMFA 2014; Gert-Jan Monteny et al. 2006; http://www.poultryhub.org/production/husbandry-management/housing-environment/waste-management/2014). Manure spread on land without treatment may runoff into local waterways such as streams, rivers, lakes and groundwater resulting in excessive amounts of nutrients including N, P and K (http://www.fao.org/docrep/w2598e/w2598e06.htm 2014) in the eco-system. Waste can also contribute organic solids, trace heavy metals, salts, viruses, bacteria, other microorganisms and sediments (Thyagarajan et al. 2013; Xin et al. 2010; Peter Ciborowski 2001; Rajashekhar and Mohan 1994; WMFA 2014; http://www.poultryhub.org/production/husbandry-management/housing-environment/waste-management/2014).

Improvements and Good Practices

Bird droppings, bird carcasses and other wastes need to be collected and disposed on daily basis or at least once in 2–3 days. Investment needs to be made in developing more efficient systems for collection and disposal of the waste. Such systems help in foregoing the costs associated with sanitation procedures, ensure better health of birds that would increase egg production, reduce foul smell and make the environment more comfortable for farm workers (Thyagarajan et al. 2013; Xin et al. 2010; Peter Ciborowski 2001; Rajashekhar and Mohan 1994; WMFA 2014; MatiasVanotti et al. 2009). In order to reduce dependency on labor and enable daily collection of fresh manure, efficient waste collection system needs to be introduced (MatiasVanotti et al. 2009; http://www.poultryhub.org/production/husbandry-management/housing-environment/waste-management/2014). The sheds could be concreted with suitable slopes and waste could be gathered with the help of high-pressure water jets. This operation would require only one person for every five sheds.

Biomethanation of Poultry Litter

Poultry litter is treated by aerobic (Sivakumar et al. 2008; Thyagarajan et al. 2013) and anaerobic methods for degradation of organic matter (Kelleher et al. 2002; Salminen and Rintal 2002a; Suleyman Sakar et al. 2009; Benjamin et al. 2001;

Gangagni Rao et al. 2008a, b, 2011, 2012, 2013; Keri et al. 2008; Mata-Alvarez et al. 2000). It is unsuitable for compost making due to its high moisture content. Further it can lead to ammonia emission and environmental pollution (Kelleher et al. 2002; Salminen and Rintal 2002a; Suleyman Sakar et al. 2009). Poultry litter is usually treated by anaerobic methods for degradation of the organic matter. It is found to be economically very attractive option for livestock waste and poultry waste (Kelleher et al. 2002; Peter Ciborowski 2001; Salminen and Rintal 2002a, b; Suleyman Sakar et al. 2009; Benjamin et al. 2001; Gangagni Rao et al. 2008a; Keri et al. 2008; Mata-Alvarez et al. 2000; VilisDubrovskis et al. 2008). Poultry litter generates more biogas compared to piggery and cattle waste (Itodo and Awulu 1999; Callaghan et al. 1999; Rajashekhar and Mohankumar 1994). Thus, there is a great potential for generating biogas from poultry litter. There are certain limitations in using poultry litter for biogas generation (Kelleher et al. 2002; Peter Ciborowski 2001; Salminen and Rintal 2002a; Suleyman Sakar et al. 2009; Benjamin et al. 2001; Gangagni Rao et al. 2008a, b, 2011, 2013; Keri B et al. 2008; Mata-Alvarez et al. 2000; VilisDubrovskis et al. 2008; Rajashekar Reddy et al. 1996) as it is viscous in nature with high calcium content and sand/grit. Feathers pose problem and hence require pretreatment (Rajashekhar Reddy et al. 1996). Utilization of poultry litter for biogas generation not only improves sanitation in and around poultry farms but also provides rich organic manure for fishponds and crops (Thyagarajan et al. 2013; Kelleher et al. 2002).

Anaerobic digestion is a biological process in which organic matter is degraded to methane under anaerobic conditions. Methane can be used as source of energy to replace fossil fuels to reduce carbon dioxide emissions. Anaerobic digestion reduces pathogens and odours, requires little land space for treatment (Kelleher et al. 2002; Suleyman Sakar et al. 2009; Benjamin et al. 2001; Gangagni Rao et al. 2008a, b, 2011, 2012, 2013; Keri et al. 2008; Mata-Alvarez et al. 2000). In this method, most of the nutrients also remain in the treated material and can be recovered for agriculture or feed use (Salminen and Rintal 2002a; Szogi and Vanotti 2009).

Organic components of poultry litter can be classified into broad biological groups, viz; proteins, carbohydrates and lipids or fats. Carbohydrates make up the bulk of the biodegradable material (Kelleher et al. 2002; Salminen and Rintal 2002a, b; Suleyman Sakar et al. 2009; Mata-Alvarez et al. 2000).

Anaerobic treatment of poultry litter involves four distinct stages (Mata-Alvarez et al. 2000; Rajagopal et al. 2013; Costa et al. 2012) viz; hydrolysis, acid fermentation, acetogenesis and methane fermentation. Bacteria producing methane from hydrogen and carbon dioxide grow faster than those utilizing acetate (Mata-Alvarez et al. 2000; Rajagopal et al. 2013; Costa et al. 2012), so that the acetotrophic methanogens are usually rate limiting with respect to the transformation of complex macromolecules in poultry litter to biogas. The first three processes are sometimes clubbed together and denoted by acid fermentation, whereas the fourth step is referred to as methanogenic fermentation. Acid fermentation tends to cause a decrease in pH because of the production of volatile fatty acids (VFA) and other intermediates that dissociate and produce protons. As methanogens are very efficient at neutral pH values, instability may arise if, for some reason, the rate of acid removal by methane falls behind the acid production rate. Relative higher production of acid tends to decrease the pH, reducing the methanogenic activity further (Mata-Alvarez et al. 2000; Rajagopal et al. 2013; Costa et al. 2012). Anaerobic microorganisms in general show a high degree of metabolic specialization. The success of the anaerobic digestion process therefore depends upon cooperative interactions between microorganisms with different metabolic capabilities (Rajagopal et al. 2013; Costa et al. 2012).

Although anaerobic processes are often slower than aerobic processes, they are found to be economically attractive in a wide variety of applications (Mata-Alvarez et al. 2000; Rajagopal et al. 2013; Costa et al. 2012). Anaerobic digestion is a relatively efficient conversion process for poultry litter producing biogas with average methane content of 60 %. Methane produced by this process can be used as a fuel for boilers, as a replacement for natural gas or fuel oil and can also be fired in engine-generators to produce electricity for on-farm use or sale to electricity companies. The residual solid is stable and can be used as a soil fertilizer (Mata-Alvarez et al. 2000; Rajagopal et al. 2013; Costa et al. 2012).

Biomethanation of Poultry Litter in Conventional Digesters

Digester systems essentially consist of holding tanks of simple to complex design, in which a series of biological reactions occur to decompose organic materials to methane, carbon dioxide, water and a number of other simple chemicals. None of the designs can be considered as ideal, since many factors affect the design and operation. Digester types can be generally categorized into conventional batch and high rate digesters (Kelleher et al. 2002; Salminen and Rintal 2002a; Suleyman Sakar et al. 2009; Benjamin et al. 2001; Gangagni Rao et al. 2008b, 2011, 2013; Keri et al. 2008; Mata-Alvarez et al. 2000). Conventional biomethanation systems for solid waste treatment employ slow rate digesters, which are designed for very long hydraulic retention time (HRT) and consequently large reactor volumes. Many digesters of floating dome and fixed dome type biogas plants have been installed in India during the past two decades to produce methane from cow dung and agricultural wastes (http://wgbis.ces.iisc.ernet.in/energy/paper/Tr_114/chapter2.htm 2014). Most of these digesters are integral part of several cattle dung-based biogas plant programmes run in rural areas by various governmental agencies (http://www.mnre.gov.in. 2014). Conventional anaerobic digesters are also employed for digestion of vegetable waste (http: //wgbis.ces.iisc.ernet.in/energy/paper/Tr_114/chapter2.htm 2014). It is reported that although the system performed satisfactorily, a large reactor volume is required for decomposing relatively small quantity of solid waste, due to inherent disadvantage of long HRT. It is reported that a battery of several digesters is required to process large quantities of waste daily. Scum formation is another major disadvantage and this gets aggravated when reactor volume is increased. Discharge of secondary effluent in the form of digested slurry is also a disadvantage in conventional system (http: //wgbis.ces.iisc.ernet.in/energy/paper/Tr_114/chapter2.htm 2014). Studies are reported on batch type digesters for production of biogas from poultry litter where in all phases of anaerobic digestion (i.e. hydrolysis, acidification and methanogenesis) take place in one vessel (Bujoczek et al. 2000; Callaghan et al. 1999; Benjamin et al. 2001; FatmaAbouelenien et al. 2009; Hill and Bolte 2000). Hence, to maintain a feasible environment for optimal methanogenesis, conditions for various biochemical pathways need to be balanced. This can be achieved by providing high retention times leading to increased volume of digester (Mata-Alvarez et al. 2000; Peter Ciborowski 2001; Salminen and Rintal 2002b; Suleyman Sakar et al. 2009; VilisDubrovskis et al. 2008).

In India, Rajashekhar Reddy et al. (1996) have carried out detailed studies on the production of biogas and bio-manure (Shanti et al. 2013) from the poultry litter produced in poultry farms. They employed Deenabandhu and KVIC models of biogas plants in their experimental work, which are regularly used for dung digestion. Shanti et al. (2013) concluded that such batch type conventional plants without mixing are not suitable for the treatment of large quantities of poultry litter. Failure of such plants is reported within 2–3 years of operation due to scum formation at the top and choking at the bottom (VilisDubrovskis et al. 2008; Suleyman Sakar et al. 2009; Yadvika et al. 2004; Gangagni Rao et al. 2008a).

Biomethanation of Poultry Litter in High Rate Digesters

Many developed countries such as the USA, Germany, Japan, France, Spain, Belgium and Netherlands etc. have developed several variants of basic anaerobic process depending upon what is being considered as the bottleneck in the conventional digesters (Kelleher et al. 2002; Salminen and Rintal 2002a; Suleyman Sakar et al. 2009; Benjamin et al. 2001; Gangagni Rao et al. 2008a, b, 2011, 2013; Keri et al. 2008; Mata-Alvarez et al. 2000). The DRANCO (http: //www.ows.be 2014) process (Belgium), VALOGRA process (http://www.valorgainternational.fr/fr/ 2014) (France), BIMA (http: //www.entec-biogas.com/en/ 2014) process (Austria), Slurry based multistage process (Gangagni Rao et al. 2011) etc. are a few of the emerging high rate technologies for organic solid waste. These are, however, mostly proprietary in nature.

In conventional single stage anaerobic digesters, all phases (i.e. hydrolysis, acid fermentation, acetogenesis and methanogenesis) of anaerobic digestion takes place in one vessel and hence high HRT is maintained (Kelleher et al. 2002; Salminen and Rintal 2002a; Suleyman Sakar et al. 2009; Benjamin et al. 2001; Gangagni Rao et al. 2008a, b, 2011, 2013; Keri et al. 2008; Mata-Alvarez et al. 2000). This would create favorable environment for the mixed culture of organisms in a single reactor in order to balance volatile acid production and utilization rates (Vavilin et al. 2001; Banks and Wang 1999; Wang and Charles 2003). Otherwise, at low HRT, volatile acid production could exceed the utilization, leading to digester failure (Gangagni

Rao et al. 2008a; Mata-Alvarez et al. 2000; Demirel and Yenigun 2002; Banks and Wang 1999; Wang and Charles 2003).

In order to avoid digester failures and increase the overall rate of digestion, two-phase digestion systems to separate acid and methane forming phases were proposed (Gangagni Rao et al. 2008a; Mata-Alvarez et al. 2000; Demirel and Yenigun 2002; Banks and Wang 1999; Wang and Charles 2003; Raynal et al. 1998) since the metabolic characteristics and growth rates of the methanogenic and acetogenic bacteria are different. The claimed advantages of phase separation are increased stability with better control of acid phase, higher organic loading rate, increased specific activity of methanogens leading to an increase in methane production rates and increased overall chemical oxygen demand (COD) and volatile solids (VS) reduction efficiencies (Gangagni Rao et al. 2008a; Mata-Alvarez et al. 2000; Demirel and Yenigun 2002; Vavilin et al. 2001; Banks and Wang 1999; Wang and Charles 2003; Raynal et al. 1998). The speculated disadvantages of phase separation include hydrogen build-up in the first-phase reactor that could inhibit the acid producing bacteria and elimination of possible interdependent nutritional requirements of acid and methane formers (Gangagni Rao et al. 2008a; Mata-Alvarez et al. 2000; Demirel and Yenigun 2002; Vavilin et al. 2001; Banks and Wang 1999; Wang and Charles 2003; Raynal et al. 1998). In another model, hydrolysis, acidogenesis, acetogenesis and methanogenic steps are carried out in separate reactors (Kelleher et al. 2002; Salminen and Rintal 2002a; Suleyman Sakar et al. 2009; Benjamin et al. 2001; Gangagni Rao et al. 2008a, b, 2011; Keri et al. 2008; Mata-Alvarez et al. 2000; Vavilin et al. 2001; Banks and Wang 1999; Wang and Charles 2003) depending on the characteristics of organic solid waste. The organic solid waste (10–12 % slurry) is first subjected to hydrolysis and acetogenesis to produce liquid phase rich in VFA. It was reported that approximately 4,000–5,000 mg/l of VFA are produced at a HRT of 2–4 days (Vavilin et al. 2001; Banks and Wang 1999; Wang and Charles 2003). The slurry after hydrolysis and acidogenesis is separated into solid and liquid phases. The digested solids are removed from the bottom of the reactors, dried and used as soil conditioner or organic manure. The liquid portion that is rich in VFA is subjected to methanogenesis. The final step of biomethanation, viz; methanogenesis is carried out in high rate digesters such as Up-flow Anaerobic Sludge Blanket (UASB) reactor, Expanded Granular Sludge Bed reactor (EGSB), Fixed Film reactor, or Fluidized Bed reactor (Gangagni Rao et al. 2008a; Mata-Alvarez et al. 2000; Vavilin et al. 2001; Banks and Wang 1999; Wang and Charles 2003; Raynal et al. 1998). In China, a two-stage process (Aili et al. 1990) is developed for the treatment of chicken manure. However, application of the above process for higher scale operation is not known.

Studies are carried out in conventional batch digesters by installing mechanical mixers to improve the performance of batch digesters. However, it is reported that the performance of these mechanical mixers is not satisfactory (Kelleher et al. 2002; Suleyman Sakar et al. 2009). In addition, 20–30 % of the energy generated during the digestion process (Fatma Abouelenien et al. 2010; Prasad et al. 2008; Yadvikaet al. 2004) is consumed. High rate biomethanation invariably requires complete mixing to enhance the performance of the digester and accordingly digesters with novel mixing arrangements (Gangagni Rao et al. 2008a, 2011; Khursheed Karim et al. 2005a, b; Yadvika et al. 2004; Prasad et al. 2008) were developed. Self-mixed anaerobic digester (SMAD) is (Gangagni Rao et al. 2008a, b, 2011, 2012, 2013) one amongst these high rate digesters.

SMAD was developed to reduce the HRT, increase the VS loading rate, increase destruction efficiency and enhance the methane yield. Specific design features of SMAD are useful in mixing the digester contents without consuming power and de-alienate the problem of scum formation. High rate biomethanation of poultry litter is studied in single stage SMAD and its performance is also compared with conventional fixed dome anaerobic digester (CFDAD) of similar capacity (Gangagni Rao et al. 2013). The study revealed that optimized HRT, VS loading rate, VS reduction, methane yield is 24 days, 4.0 kg VS/m^3/day, 64 %, 0.15 m^3/(kg VS fed) and 40 days, 2.15 kg /m^3/day, 42 %, 0.083 m^3/(kg VS fed) for SMAD and CFDAD respectively. Better results achieved with SMAD could be attributed to specific design features and intermittent mixing of the digester contents due to self-mixing mechanism. Preliminary cost estimates revealed that installation of SMAD would be remunerative for the farmer in terms of biogas and bio-manure. Poultry litter having 10 % total solids (TS) is subjected to high rate biomethanation in multi stage configuration (SMAD-I and II in series with UASB reactor) (Gangagni Rao et al. 2008a, b, 2011, 2012, 2013). In this study, VS reduction of 58 %, methane yield of 0.16 m^3 kg^{-1} (VS reduced) and VS loading rate of 3.5 kg VS $m^{-3}day^{-1}$ at HRT of 13 days is obtained.

C to N ratio of poultry litter is in the range of 6 to 12 (Baris Calli et al. 2005, Chen et al. 2008; Gangagni Rao et al. 2008a; Hansen et al. 1998; Kayhanian 1999; Liao et al. 1995; Fatma Abouelenien et al. 2010) compared to the optimally required ratio in the range of 20–30 for biomethanation (Azeem Khalid et al. 2011; Baris et al. 2005, Chen et al. 2008; Gangagni Rao et al. 2008a; Hansen et al. 1998; Kayhanian 1999; Liao et al. 1995). Due to this reason, total ammoniacal nitrogen (TAN) generated during anaerobic degradation beyond the limit of 2,000–6,000 mg/l inhibits the process (Baris Calli et al. 2005, Chen et al. 2008; Gangagni Rao et al. 2008a; Hansen et al. 1998; Kayhanian 1999; Liao et al. 1995; Fatma Abouelenien et al. 2010; Sterling et al. 2001). In order to control TAN inhibition in the biomethanation of poultry litter leachate, novel methods are proposed (Chen et al. 2008; Gangagni Rao et al. 2008a; Hansen et al. 1998; Kayhanian 1999; Liao et al. 1995; Fatma Abouelenien et al. 2010). One of the methods is stripping of NH_3 from the anaerobic system with air (Gangagni Rao et al. 2008a).High rate biomethnation of poultry litter leachate is carried out in UASB reactor by coupling the stripper for removing ammonia (Gangagni Rao et al. 2008a). The study is aimed at the use of stripping, as ammonia inhibition control mechanism for treatment of poultry litter leachate in UASB reactor. The performance of the UASB reactor for treatment of poultry litter leachate with and without stripper is studied in detail to understand the effect of stripper as ammonia inhibition mechanism. As per this study, UASB reactor with stripper as ammonia inhibition control mechanism exhibited better performance in terms of COD reduction (96 %), methane yield

(0.26 m^3CH_4/kg COD reduced), organic loading rate (OLR) (18.5 kg COD/m^3/day) and HRT (12 h) compared to the UASB reactor without stripper (COD reduction: 92 %; methane yield: 0.21 m^3CH_4/kg COD reduced; OLR: 13.6 kg COD/m^3/day; HRT: 16 h). Improved performance is due to the reduction of TAN and free ammonia nitrogen (FAN) in the range of 75–95 % and 80–95 % respectively by the use of stripper. This method is found to be viable and feasible for implementation. But NH_3 emitted along with air from stripper needs to be disposed off safely as it could cause odour problem and consequent health hazards. Therefore in another study (Gangagni Rao et al. 2012), feasibility of coupling low cost gas phase bio-filter for the removal of NH_3 emissions from the stripper with low cost agricultural residue as a bedding material was studied and found to be economically viable to incorporate the same in the integrated system. In a similar study, Fatma Abouelenien et al. (2010) Improved the biomethanation of chicken manure by recycling the biogas for the removal of ammonia. CSIR-Indian Institute of Chemical Technology (IICT), Hyderabad, India has developed a novel high rate biomethanation technology called "ANAEROBIC GAS LIFT REACTOR (AGR)" (http://csirtech.com/khdb/viewrecord.php?recordno=20140117124455 2014) for the generation of biogas and bio-manure (Shanti et al. 2013) from poultry litter. The technology (CSIR-IICT 2014) is transferred to M/s Ahuja Engineering Services Private Limited, (AES), Hyderabad (http://www.aespl-india.com. 2014). Biogas power plants would ensure a cleaner environment and provide long-term financial benefits to farm owners by offsetting their power requirements. Safe and timely waste disposal and saving energy costs improve profitability of poultry industry. This technology is superior in terms of biogas and bio-manure production as it incorporates novel pre and post processing technologies required for biomethanation of poultry litter (CSIR-IICT 2014). The biogas produced in AGR has applications in direct heat for brooder sheds to replace liquid petroleum gas (LPG) and also to generate power by replacing diesel. Digested solid residue is being applied to the surrounding field as bio-manure (Shanti et al. 2013). Biogas can be used for brooding to replace the LPG (CSIR-IICT 2014).

Powering Poultry Farms with Biogas from Poultry Litter

Generating energy from poultry litter using biomethanation technologies for saving and even earning money is a lucrative option to tackle the problem of waste management in poultry farms (MatiasVanotti et al. 2009; Gert-Jan Monteny et al. 2006). Setting up a biogas-power plant with biomethanation technology could be a promising solution to produce energy from poultry waste (Hahn et al. 2014; Gert-Jan Monteny et al. 2006). Selection of appropriate technology that is suitable to the characteristics of the poultry litter is highly critical in deciding the success of the plant. Biomethanation produces biogas that can be used to generate energy and bio-manure (digestate) that can be used as a soil conditioner to fertilize the land. In an increasingly environment conscious world, such technologies have added benefit of providing environmentally friendly methods of disposing waste (MatiasVanotti et al. 2009; Gert-Jan Monteny et al. 2006). The advantages of using biogas technology to generate energy from poultry are; efficient in-farm usage of poultry manure to produce electricity that reduces the dependence on energy from the grid to a great extent (MatiasVanotti et al. 2009; Gert-Jan Monteny et al. 2006). Therefore setting up of such a power plant not only reduces the cost of energy consumption but also prevents business losses and disruption due to frequent power interruptions and fluctuations from the grid-based power supply.

Intangible Benefits

Setting up of poultry litter to power plant could lead to proper and regular management of waste. Another critical benefit of using such technologies is that periodic and methodical removal of poultry litter from shed to the biogas-based power plant helps in improving the health of the birds (MatiasVanotti et al. 2009; Gert-Jan Monteny et al. 2006). Periodic and methodical removal avoids the exposure of birds to NH_3 emissions emanating from uncollected waste (Xin et al. 2010; MatiasVanotti et al. 2009; Gert-Jan Monteny et al. 2006) leading to a significant increase in egg production. If necessary steps are taken up for installation of the biogas-based power plant, the egg production can be increased by 15–18 %. Utilizing poultry waste for power generation would result in methodical disposal of waste and will aid in foregoing the overhead costs of cleaning and maintaining the farm sheds filled with waste. Safe and proper disposal of waste and fulfillment of the energy needs of rural industries, using automatic/semi-manual engineering technologies will directly and indirectly contribute to the welfare of the community. Direct contribution could be in terms of better health conditions for workers in farms due to mitigation of noxious ammonia and reduction in manual cleaning activities. Indirect contribution could be towards reduction of greenhouse gases (methane and carbon dioxide) in the environment that would provide long-term benefits to the environment (Gert-Jan Monteny et al. 2006; UNFCC 2014)

Conclusions and Recommendations

Presently waste management practices are increasingly becoming important in poultry sector due to hygiene and biosecurity reasons and it is gratifying to note that industry has taken necessary initiatives. Presently, different variants of high rate biomethanation digesters for the treatment of poultry litter are available in the market and selection of appropriate technology suitable to the characteristics of the poultry litter is highly critical in deciding the success of the plant. Biogas generation for CHP applications within the farm along with production of salable by product (bio-manure) from poultry litter will improve the profitability of the poultry industry.

Acknowledgements The authors are thankful to the Department of Biotechnology, Government of India; Council for Scientific and Industrial research (CSIR), Government of India; Sri Venkateswara Veterinary University (SVVU), Hyderabad and M/s Ahuja Engineering Services Private Limited, Hyderabad for funding the projects related to biogas and bio-manure from poultry litter in different stages. The authors are also grateful to the Director-IICT, for her encouragement in carrying out this work.

References

Aili, Y., Kejun, Y., Yoaxin, Z., & Guoyuan, F. (1990). Study and design of a high performance biogas project treating chicken manure. *China Biogas, 8*(1), 30.

Khalid, A., Arshad, M., Anjum, M., Mahmood, T., & Dawson, L. (2011). The anaerobic digestion of solid organic waste. *Waste Management, 31*(8), 1737–1744.

Calli, B., Mertoglu, B., Inanc, B., & Yenigun, O. (2005). Effects of high free ammonia concentrations on the performances of anaerobic bioreactors. *Process Biochemistry, 40*, 1285–1292.

Banks, C. J., & Wang, Z. (1999). Development of a two-phase anaerobic digester for the treatment of mixed abattoir wastes. *Water Science & Technology, 40*(1), 69–76.

Magbanua, B. S., Jr., Adams, T. T., & Johnston, P. (2001). Anaerobic co-digestion of hog and poultry waste. *Bio Resource Technology, 76*, 165–168.

Bishnoi, S. R., & Bajwa, M. S. (1994). *Poultry manure for more crops. Indian poultry industry year book* (10th ed.). New Delhi: SP Gupta.

Borja, R., Sgnchez, E., & Weiland, P. (1996). Influence of ammonia concentration on thermophilic anaerobic digestion of cattle manure in Upflow Anaerobic Sludge Blanket (UASB) reactors. *Process Biochemistry, 31*(5), 477–483.

Bujoczek, G., Oleszkiewicz, J., Sparling, R., & Cenkowski, S. (2000). High solid anaerobic digestion of chicken manure. *Journal of Agricultural Engineering Research, 76*, 51–60.

Callaghan, F. J., Wase, D. A. G., Thayanith, K., & Forster, C. F. (1999). Co-digestion of waste organic solids: Batch studies. *Bioresource Technology, 67*, 117–122.

Costa, J. C., Barbosa, S. G., Alves, M. M., & Sousa, D. Z. (2012). Thermochemical pre- and biological co-treatments to improve hydrolysis and methane production from poultry litter. *Bioresource Technology, 111*, 141–147.

Chen, Y., Jay Cheng, J., & Kurt Creamer, S. (2008). Inhibition of anaerobic digestion process: A review. *Bioresource Technology, 99*(10), 4044–4064.

Demirel, B., & Yenigun, O. (2002). Two-phase anaerobic digestion processes: A review. *Journal of Chemical Technology and Biotechnology, 77*, 743–755.

Abouelenien, F., Nakashimada, Y., & Nishio, N. (2009). Dry mesophilic fermentation of chicken manure for production of methane by repeated batch culture. *Journal of Bioscience and Bioengineering, 107*(3), 293–295.

Abouelenien, F., Fujiwara, W., Namba, Y., Kosseva, M., Nishio, N., & Nakashimada, Y. (2010). Improved methane fermentation of chicken manure via ammonia removal by biogas recycle. *Bioresource Technology, 101*(16), 6368–6373.

Gangagni Rao, A., SasiKanth Reddy, T., Surya Prakash, S., Vanajakshi, J., Joseph, J., Jetty, A., et al. (2008a). Biomethanation of poultry litter leachate in UASB reactor coupled with ammonia stripper for enhancement of overall performance. *Bioresource Technology, 99*(18), 8679–8684.

Gangagni Rao, A., Surya Prakash, S., Jetty, A., Rajashekhara Reddy, A., Sarma, P. N. (2008b). Self mixed anaerobic digester. *WIPO Patent Application* WO/2008/026221.http://www.sumobrain.com/patents/wipo/Self-mixing-anaerobic-digester-useful/WO2008026221.html. Retrieved 26 Nov 2014.

Gangagni Rao, A., Surya Prakash, S., Joseph, J., Rajashekhara Reddy, A., & Sarma, P. N. (2011). Multi stage high rate biomethanation of poultry litter with self mixed anaerobic digester. *Bioresource Technology, 102*(2), 729–735.

Gangagni Rao, A., Gandu, B., & Swamy, Y. V. (2012). Mass transfer dynamics of ammonia in high rate biomethanation of poultry litter leachate. *Bioresource Technology, 109*, 234–238.

Gangagni Rao, A., Gandu, B., Sandhya, K., Kranti, K., Ahuja, S., & Swamy, Y. V. (2013). Decentralized application of anaerobic digesters in small poultry farms: Performance analysis of high rate self-mixed anaerobic digester and conventional fixed dome anaerobic digester. *Bioresource Technology, 144*, 121–127.

Monteny, G.-J., Bannink, A., & Chadwick, D. (2006). Greenhouse gas abatement strategies for animal husbandry. *Agriculture Ecosystems & Environment, 112*(2-3), 163–170.

Hahn, H., Ganagin, W., Hartmann, K., & Wachendrof, M. (2014). Cost analysis of concepts for a demand oriented biogas supply for flexible power generation. *Bioresource Technology, 170*, 211–220.

Hansen, K. H., Angelidaki, I., & Ahring, B. K. (1998). Anaerobic digestion of swine manure: inhibition by ammonia. *Water Res., 32*, 5–12.

Hill, D. T., & Bolte, J. P. (2000). Methane production from low solid concentration liquid swine waste using conventional anaerobic fermentation. *Bioresource Technology, 74*, 241–247.

Itodo, I. N., & Awulu, J. O. (1999). Effects of total solids concentrations of poultry, cattle, and piggery waste slurries on biogas yield. *Transactions of ASAE, 42*, 1853–1855.

Jalasutram, V., Kataram, S., Gandu, B., & Anupoju, G. R. (2013). Single cell protein production from digested and undigested poultry litter by Candida utilis: optimization of process parameters using response surface methodology. *Clean Technologies and Environmental Policy, 15*, 265–273.

Kayhanian, M. K. (1999). Ammonia inhibition in high-solids biogasification: An overview and practical solutions. *Environmental Technology, 20*, 355–365.

Cantrell, K. B., Ducey, T., Kyoung, S. R., & Hunt, P. G. (2008). Livestock waste-to-bioenergy generation opportunities. *Bioresource Technology, 99*(17), 7941–7953.

Kelleher, B. P., Leahy, J. J., Henihan, A. M., O'Dwyer, T. F., Sutton, D., & Leahy, M. J. (2002). Advances in poultry litter disposal technology – A review. *Bioresource Technology, 83*, 27–36.

Karim, K., Hoffmann, R., Thomas Klasson, K., & Al-Dahhan, M. H. (2005a). Anaerobic digestion of animal waste: Effect of mode of mixing. *Water Research, 39*(15), 3597–3606.

Karim, K., Hoffmann, R., Thomas Klasson, K., & Al-Dahhan, M. H. (2005b). Anaerobic digestion of animal waste: Waste strength versus impact of mixing. *Bioresource Technology, 96*, 1771–1781.

Krylova, N. I., Khabiboulline, R. E., Naumova, R. P., & Nagel, M. A. (1997). The influence of ammonium and methods for removal during the anaerobic treatment of poultry manure. *Journal of Chemical Technology and Biotechnology, 70*, 99–105.

Liao, P. H., Chen, A., & Lo, K. V. (1995). Removal of nitrogen from swine manure wastewaters by ammonia stripping. *Bioresource Technology, 54*, 17–20.

Mata-Alvarez, J., Mace, S., & Llabres, P. (2000). Anaerobic digestion of organic solid wastes; an overview of research achievements and perspectives. *Bioresource Technology, 74*, 3–16.

Vanotti, M., Szogi, A., Pilar Bernal, M., & Martinez, J. (2009). Livestock waste treatment systems of the future: A challenge to environmental quality, food safety, and sustainability. *Bioresource Technology, 100*(22), 5371–5373.

Ciborowski, P. (2001). *Anaerobic digestion of livestock manure for pollution control and energy production: A feasibility assessment*. Minnesota Pollution Control Agency. Retrieved 26 Nov 2014 from http://www.pca.state.mn.us/publications/p-gen4-02.pdf.

Prasad, K., Buendia, I., Ellegaard, L., & Angelidakia, I. (2008). Effects of mixing on methane production during thermophilicanaerobic digestion of manure: Lab-scale and pilot-scale studies. *Bioresource Technology, 99*(11), 4919–4928.

Rajagopal, R., Massé, D. I., & Singh, G. (2013). A critical review on inhibition of anaerobic digestion process by excess ammonia. *Bioresource Technology, 143*, 632–641.

Rajashekhar Reddy, A., Mohankumar, K. C. (1994). *Studies on utilization of poultry litter for the production of biogas*. Unpublished data. AICRP on Poultry Breeding, ANGRAU, Hyderabad.
Rajashekar Reddy, A., Madhukumar, K., Eswariah., Ravinder Reddy, V. (1996). Environment friendly disposal of poultry droppings by anaerobic fermentation and recycling as feed ingredient for poultry. *XX World's poultry Congress proceedings*, Vol. II, (pp. 605–611).
Raynal, J., Delgenks, J. P., & Moletta, R. (1998). Two-phase anaerobic digestion of solid wastes by a multiple liquefaction reactors process. *Bioresource Technology, 65*, 97–103.
Salminen, E., & Rintal, J. (2002). Anaerobic digestion of organic solid poultry slaughter waste – a review. *Bioresource Technology, 83*, 13–26.
Salminen, E. A., & Rintala, J. A. (2002). Semi-continuous anaerobic digestion of solid poultry slaughterhouse waste: effect of hydraulic retention time and loading. *Water Research, 36*, 3175–3182.
Shanti, M., Balaji, N. R., Suneetha, D. K. B., Rajasekhar Reddy, J., & Chiranjeevi. (2013). Studies on utilization of biogas poultry manure in crop production. *Progressive Agriculture, 13*(2), 102–109. Retrieved 26 Nov 2014 from http://www.indianjournals.com/ijor.aspx?target=ijor:pa&volume=13&issue=2&article=018.
Sivakumar, K., Saravana Kumar, V. R., Jagatheesan, P. N., Viswanathan, K., & Chandrasekaran, D. (2008). Seasonal variations in composting process of dead poultry birds. *Bioresource Technology, 99*(9), 3708–3713.
Sterling, M. C., Jr., Lacey, R. E., Engler, C. R., & Ricke, S. C. (2001). Effects of ammonia nitrogen on H and CH_E production during anaerobic digestion of dairy cattle manure. *Bioresource Technology, 77*, 9–18.
Sakar, S., Yetilmezsoy, K., & Kocak, E. (2009). Anaerobic digestion technology in poultry and livestock waste treatment — a literature review. *Waste Management & Research., 27*(1), 3–18.
Szogi, A. A., & Vanotti, M. B. (2009). Prospects for phosphorus recovery from poultry litter. *Bioresource Technology, 100*(22), 5461–5465.
Thyagarajan, D., Barathi, M., & Sakthivadivu, R. (2013). Scope of poultry waste utilization. *OSR Journal of Agriculture and Veterinary Science, 6*(5), 29–35.
Vavilin, V. A., Rytov, S. V., Lokshina, L. Y., Rintala, J. A., & Lyberatos, G. (2001). Simplified hydrolysis models for the optimal design of two-stage anaerobic digestion. *Water Research, 35*, 4247–4251.
Dubrovskis, V., Plume, I., Straume, I. (2008) Anaerobic digestion of cow and broile rmanure. http://tf.llu.lv/conference/proceedings2008/Papers/10_DubrovskisPlume.pdf. Retrieved 26 Nov 2014.
WMFA-(Waste management featured articles) http://www.thepoultrysite.com/. Retrieved 26 Nov 2014.
Xin, H., Gates, R. S., Green, A. R., Mitloehner, F. M., Moore, P. A. Jr., Wathes, C. M. (2010) Emerging issues: Social sustainability of egg, production symposium: Environmental impacts and sustainability of egg production systems, ©2011 Poultry Science Association Inc. http://www.poultryscience.org/docs/PS_877.pdf. Retrieved 26 Nov 2014.
Dong, X., & Ernest Tollner, W. (2003). Evaluation of Anammox and denitrification during anaerobic digestion of poultry manure. *Bioresource Technology, 86*, 139–145.
Santosh, Y., Sreekrishnan, T. R., Kohli, S., & Rana, V. (2004). Enhancement of biogas production from solid substrates using different techniques—a review. *Bioresource Technology, 95*(1), 1–10.
Wang, Z., & Charles, J. B. (2003). Evaluation of a two stage anaerobic digester for the treatment of mixed abattoir wastes. *Process Biochemistry, 30*(9), 1267–1273.
http://wgbis.ces.iisc.ernet.in/energy/paper/Tr_114/chapter2.htm. Retrieved 26 Nov 2014.
http://www.fao.org/docrep/013/al715e/al715e00.pdf. Retrieved 26 Nov 2014.
http://grist.files.wordpress.com/2008/10/a84-3-1990e.pdf. Retrieved 26 Nov 2014.
http://www.mnre.gov.in. Retrieved 26 Nov 2014.
http://csirtech.com/khdb/viewrecord.php?recordno=20140117124455. Retrieved 26 Nov 2014.
http://www.ows.be. Retrieved 26 Nov 2014.
http://www.valorgainternational.fr/fr/. Retrieved 26 Nov 2014.

http://www.entec-biogas.com/en/. Retrieved 26 Nov 2014.
http://www.aespl-india.com. Retrieved 26 Nov 2014.
http://www.fao.org/docrep/w2598e/w2598e06.htm. Retrieved 26 Nov 2014.
CSIR-IICT. (2014). Interactive meet and plant demonstration on biogas and bio-manure from poultry litter concluded. *Poultry Express, 10*(4), 30–36.
http://www.poultryhub.org/production/husbandry-management/housing-environment/waste-management/. Retrieved 26 Nov 2014.
http://newsroom.unfccc.int/. Retrieved 26 Nov 2014.

Chapter 9
Current Advances of Biogas Production via Anaerobic Digestion of Industrial Wastewater

Phaik Eong Poh, David Tandiono Tan, Eng-Seng Chan, and Beng Ti Tey

Introduction

In the recent years, there has been a steady increase of energy consumption in many parts of the world. In particular, the rapid development in China has caused the country's energy consumption to increase from 1,638 million tons of oil equivalents (Mtoe) in 2004 to 3,013 Mtoe in 2013 (Enerdata 2014). Though it is anticipated that the fossil fuel reserves are still able to support the current demands (Shafiee and Topal 2009), the global anthropogenic emissions of CO_2 from fossil fuel consumption has doubled from nearly 30 billion tons of CO_2 since 1970 (Hook and Tang 2013). It is therefore essential to offset the amount of anthropogenic CO_2 emissions as continuous increase of CO_2 in the environment are changing the climate patterns, leading to widespread extinction of many species and also greatly affecting mankind in different ways (McNutt 2013).

The control of anthropogenic CO_2 emission to the atmosphere can be done through the replacement of fossil fuel with renewable energy sources such as solar, wind, hydro, nuclear power, biodiesel or generation of biogas via anaerobic digestion of different feedstock etc. for power generation. AD of industrial wastewater can be a sustainable method to control global anthropogenic emission of CO_2 as it does not consume great amount of energy as opposed to aerobic digestion which requires energy for aeration (Poh and Chong 2009). As such, this minimizes the carbon footprint required to treat wastewater. Furthermore, industrial wastewaters that are highly contaminated make suitable feedstock to the anaerobic digesters for biogas production as these wastewaters were consistently generated in abundance in process plants and can be obtained at no costs.

P.E. Poh • D.T. Tan • E.-S. Chan (✉) • B.T. Tey
Chemical Engineering, School of Engineering, Monash University Malaysia, Selangor, Malaysia
e-mail: chan.eng.seng@monash.edu

P. Ravindra (ed.), *Advances in Bioprocess Technology*,
DOI 10.1007/978-3-319-17915-5_9

Therefore, this chapter provides an overview of AD process and the current practices of AD for industrial wastewater treatment. In addition, it is also aimed to discuss the challenges of AD and future direction of AD for industrial wastewater treatment.

Anaerobic Digestion (AD)

AD is a process where complex organic matters (e.g. carbohydrates, protein, oil and fats etc.) were degraded to methane, carbon dioxide and water without the presence of oxygen. The process of AD involves a sequence of reactions, namely: hydrolysis, acidogenesis, acetogenesis and methanogenesis (Gerardi 2003). A mixed population of bacteria is required for AD to be completed as each reaction involves different consortia of bacteria for degradation.

Hydrolysis

Hydrolysis is the first step to degradation of complex organic matters in AD, which involves the conversion of cellulosic materials, carbohydrate constituents and proteins in wastewater to simpler compounds such as soluble sugars or amino acids. Equations (9.1) and (9.2) are examples of hydrolysis reactions where cellulose and protein are hydrolyzed to soluble sugars and amino acids respectively with the presence of water.

$$C_6H_{10}O_5 + H_2O \rightarrow C_6H_{12}O_6 \tag{9.1}$$

$$-(H_2NCHRCOOH)_n - +H_2O \rightarrow H_2NCHRCOOH \tag{9.2}$$

Cellulosic hydrolysis (represented as 9.1) has been regarded as the rate limiting step in AD especially when the industrial wastewater contains high concentration of cellulosic materials. The rate of hydrolysis is mainly affected by pH, temperature and solid retention time (SRT) (Feng et al. 2009; Veeken et al. 2000) where higher rate can be obtained at slightly acidic condition, higher temperatures and longer SRT.

Acidogenesis

Acidogenesis is a reaction where the end product from hydrolysis (e.g. soluble sugars and amino acids) are converted to fatty acids and organic acids (e.g. acetic acid, propionate, lactate, butyrate, etc.). There are several reaction pathways to acidogenesis. Some examples of acidogenesis pathways are shown in (9.3)–(9.5) where end product from (9.3) is most desirable. This is due to the fact that acetic acid can be directly utilized by methanogens to produce methane in the biogas for power generation.

$$C_6H_{12}O_6 + 2H_2O \rightarrow 2CH_3COOH + 2CO_2 + 4H_2 \quad (9.3)$$

$$C_6H_{12}O_6 + 2H_2 \rightarrow 2CH_3CH_2COOH + 2H_2O \quad (9.4)$$

$$C_6H_{12}O_6 \rightarrow CH_3CH_2CH_2COOH + 2CO_2 + 2H_2 \quad (9.5)$$

In order to obtain higher concentration of acetic acid which is desirable for biogas production, the AD system has to be maintained under low hydrogen partial pressure. This is because β-oxidation will take place under low hydrogen partial pressure to convert long chain fatty acids to acetic or propionic acid. Zeeman and Sanders (2001) has reported on the inhibition of β-oxidation when AD system is accumulated of hydrogen. Under such conditions, (9.4 and 9.5) will be predominant in the AD system to maintain the hydrogen concentration within the system replacing (9.3) until the level of hydrogen in the system is back to normal.

Similar to hydrolysis, acidogenesis is significantly affected by the pH in the anaerobic digester where the degree of acidification increased with pH. Yu and Fang (2003) have identified that pH of 6–7 in anaerobic digester favours the production of acetate, butyrate and i-butyrate.

Acetogenesis

As mentioned in section "Acidogenesis", the presence of acetic acid in an anaerobic digester is desirable as it is an important substrate for methane conversion in AD (Gerardi 2006). Acetogenesis is a step to convert intermediates from acidogenesis to acetic acid. This step is particularly important as more than 70 % of methane from AD is produced from acetic acid. Equations (9.6) and (9.7) are examples of reactions under acetogenesis. Besides converting fatty acids and organic acids into acetic acid, acetogenic bacteria can also convert ethanol to produce acetic acid, as shown in (9.8). Homoacetogens could also consume CO_2 and H_2 to produce acetic acid (9.9)

$$CH_3CH_2COOH + 2H_2O \rightarrow CH_3COOH + CO_2 + 3H_2 \quad (9.6)$$

$$CH_3CH_2CH_2COOH + 2H_2O \rightarrow 2CH_3COOH + 2H^+ \quad (9.7)$$

$$CH_3CH_2OH + H_2O \rightarrow CH_3COOH + 2H_2 \quad (9.8)$$

$$2CO_2 + 4H_2 \rightarrow CH_3COOH + 2H_2O \quad (9.9)$$

Aside from inhibition of acidogenic reactions, high hydrogen partial pressure also inhibits acetogenesis. Hence, it is imperative to maintain an active community of homoacetogens that plays a role to maintain the hydrogen level in the AD system as the rest of the reactions produces hydrogen as by-product.

Methanogenesis

Methanogensis is the most important step in AD as methanogenesis reactions determine the amount and quality of biogas produced. It is the conversion of methane from acetate, hydrogen and carbon dioxide that were produced from previous steps as described in sections "Hydrolysis", "Acidogenesis" and "Acetogenesis". Alternative substrates such as formate, methanol and carbon monoxide can be utilized for conversion into methane. Equations (9.10) and (9.11) displays the reactions that are responsible to convert acetate, hydrogen and carbon dioxide into methane.

$$CH_3COOH \rightarrow CH_4 + CO_2 \quad (9.10)$$

$$CO_2 + 4H_2 \rightarrow CH_4 + 2H_2O \quad (9.11)$$

Methanogens that are involved in the production of methane are a group of bacteria that are strictly anaerobes. *Methanosaeta* spp. and *Methanosarcina* spp. utilize acetate as substrate while the rest of the methanogenic species are hydrogenotrophic, utilizing hydrogen and carbon dioxide as substrate for methane production.

Factors Affecting Anaerobic Digestion (AD)

Anaerobic digestion is categorised as a type biological treatment. This means that the workforce behind the treatment process comes from living organism. As most, if not all, living things are sensitive to environmental changes, the efficiency of AD system is dictated by several environmental conditions (Ling 2007).

Hydraulic Retention Time (HRT)

The quality of effluent exiting an anaerobic digester is primarily affected by the duration the particular wastewater was allowed to circulate inside bioreactor, or often termed as the hydraulic retention time (HRT). Longer HRT most likely results in higher effluent quality as contact time between pollutant as the substrates and microorganisms as the active sites is prolonged and vice versa. The HRT of an AD system is controlled by adjusting the flow rate of inlet (and thus outlet) as well as the volume of bioreactor.

pH

pH level is an important chemical factor for the solubility and dispersion of biological organic contaminants in wastewater as it influences the charge density on the particles, which serve as main raw material in the AD process. In addition, pH has also been indicated to affect the microbial activity within the population of a mixed culture (Andersson and Ingvar Nilsson 2001). Studies has shown that methanogens in anaerobic digesters are particularly sensitive towards changes in pH level and work best between pH 6.8–7.2 while methanogenic activities are completely halted at pH levels below 4 and above 9.5 (Beccari et al. 1996; Gerardi 2003).

Disturbance in the pH level of anaerobic digesters are normally experienced due to the formation of volatile fatty acids (VFAs) generated acidogenesis step of anaerobic digestion. Acids as we know are low in pH by nature and when accumulated will inevitably cause decline in the overall pH of the AD system. This in turn affects the performance of other groups of microorganisms and are normally observed by the drop in volume of biogas produced due to inhibition of the particularly pH-sensitive methanogens (Patel and Madamwar 2002).

Temperature

Biological systems such as AD system work by allowing microorganisms to digest and essentially use up organic matters in wastewater so that relatively cleaner water can be obtained. Microorganisms can generally be divided into three categories depending on the growth temperature, and hence preference towards operating temperature.

- Psychrophilic microorganisms grow and operate below 20 °C.
- Mesophilic microorganisms grow and operate between 25 and 40 °C.
- Thermophilic microorganisms grow and operate above 45 °C.

Safeguarding the AD system within the desired operational temperature is imperative to ensure anaerobic activities, growth, and reproductive continuity of the microbial population in the mixed culture sludge.

Several studies have indicated that AD at higher temperature range tends to perform better mainly in terms of treatment duration and productivity of methane (Choorit and Wisarnwan 2007; Wiegant et al. 1985). This finding could be associated with higher molecular movement of the reacting substrates and more rapid metabolic rate of microorganisms at higher temperature ranges. On the other hand, operating anaerobic digesters at temperatures above or below room temperature (thermophilic and psychrophilic) requires extra cost of heating and cooling respectively as well as difficulty in controlling process stability (Kim et al. 2002).

Agitation

Mixing or agitation generally promotes higher system efficiency by allowing better contact between substrates and microorganisms while preventing the formation of scum which could interfere with flow of gas through the surface. Localised build-up of heat, intermediates such as acids, and bases from pH regulatory system could also be avoided by implementing good agitation in anaerobic digester. Mixing can be adopted through mechanical means, biogas recirculation, or slurry recirculation (Karim et al. 2005a, b).

Investigations into the effect of mixing on the overall performance of anaerobic digesters reveals that mixing by slurry recirculation leads to better performance when compared to mechanical impeller and biogas recirculation. All three type of mixing, however, significantly improves treatment efficiency when incorporated into AD system when compared to unmixed digesters (Karim et al. 2005a, b). Additionally, slow rate of mixing is shown to lead to higher methanogenic activities in the operation of anaerobic bioreactors (Gerardi 2003).

Organic Loading Rate (OLR)

Organic loading rate is a form of measurement for the conversion capacity of a biological system. Increasing OLRs in a wastewater treatment system generally leads to decrease in efficiency of treatment, often measured in term of chemical oxygen demand (COD) removal. In contrast, rate of biogas production usually increases along with increase in OLR up to certain level where further increase in loading rate overwhelms the methanogens. As conversion of acetic acid to methane does not proceed quickly enough, the production of biogas is eventually worsened as methanogenic activity is hindered due to system upset by acid accumulation (Sánchez et al. 2005). Similarly, unconsumed hydrogen will result in high hydrogen partial pressure which inhibits acetogenesis phase, in particular the conversion of long chain fatty acids into acetates and propionates (Patel and Madamwar 2002; Zeeman and Sanders 2001).

Current AD Technology

Anaerobic technology has improved significantly in the last few decades with the development of different configuration and bioreactors to achieve high rate treatment processes, in particular to cater for industrial wastewaters. High rate anaerobic reactors enable treatment of industrial effluents at high organic loading rate with a considerably low ecological footprint (Mustafa et al. 2011). High rate anaerobic digesters are normally characterised by high concentration of mixed culture biomass which is retained through a deliberate design of the reactor (Van Lier et al. 1997).

Upflow Anaerobic Sludge Blanket (UASB)

As the name suggests, influent flows into UASB reactor from bottom of tank and then through a layer of sludge blanket as it travels upward where it is finally discharged from the top of the tank (Mustafa et al. 2011). The set-up is aimed to provide better contact between the wastewater and the mixed culture sludge through reduction in upflow velocity, compared to a downflow or horizontal flow where the organic contents in wastewater suffers greater risk of passing through the reactor without coming into contact with the microbial workforce. Similarly, solids retention is enhanced as relatively heavier solid portion tends to settle towards the bottom, far from effluent discharge at the top. This in turn contributes to greater ease in formation of granular solid which is often associated with improvement in treatment efficiency. To summarise, UASB system is favoured as it promotes good settleability, handles high biomass concentration, enables excellent solids/liquid separation, and is operable at high loading rates (Mustafa et al. 2011).

Continuous Stirred Tank Reactor (CSTR)

The basic concept of CSTR is an enclosed bioreactor where influent and equivalent effluent are continuously added and withdrawn with constant mixing which is normally provided by mechanical agitator. CSTR is often characterised as a steady state and well-mixed system that maintains effluent quality and stability to commendable extent. Relatively large area provided by mechanical impeller results in thorough mixing with little 'dead' region. CSTR, however, suffers from incapability of retaining biomass due to washouts of the microbial contents, resulting in potentially low treatment efficiency inability to handle high organic loading.

Anaerobic Fluidised Bed (AFB) Reactor

AFB reactor is characterised by the distribution of small particles such as sand or granular activated carbon to provide media on which microbes can attach. The presence of particles promotes good mass transfer as organic wastewater flows around the particles while the bed expansion prevents clogging and short-circuiting flow by creating large pore spaces. There is, however, difficulty in developing strongly attached biofilm containing the optimum blend of methanogens and other anaerobes, on top of risk of microorganism detachment from the particles and negative effects of dilution near inlet that could result from high recycle rate. Additionally, there is extra operating cost that comes from the added stream with high recycle rate.

Expanded Granular Sludge Bed (EGSB) Reactor

EGSB is a modification from AFB where the upward velocity is decreased to create partial bed fluidisation. Currently EGSB is mainly used for wastewater treatment at psychrophilic range. It was found that while COD removals of EGSB reactors are satisfactory between 70 and 93 %, biomethane production is relatively lower as compared to other high rate anaerobic digesters. Some commonly encountered problems in the operation of EGSB include scum formation, sludge floatation, and pipeline blockages. To resolve these drawbacks, pre-treatment step is normally required to remove large particles, suspended solids, oil, and grease prior to feeding into EGSB reactor (Zhang et al. 2008).

Other Technologies

Apart from those mentioned above, several other technologies that have been developed to cater for AD include the conventional ponding system, anaerobic filtration (AF), and upflow anaerobic sludge fixed film (UASFF).

Ponding system is perhaps the oldest and most common treatment method employed, owing to its simplistic setup and low operating cost. Ponding system consists of essentially an anaerobic pond with microorganisms where wastewater is continuously pumped in and extracted, usually from opposite ends. Downside of anaerobic pond is the large land requirement as it could easily takes an area equivalent to half up to a full football field. Higher methane emission has also been a major concern in implementation of anaerobic ponding system as the ponds are mostly open to the atmosphere (Yacob et al. 2006a, b).

AF is featured by the incorporation of filter within the bioreactor. It is shown to have a good stability towards fluctuation in OLR, high substrate removal efficiency within short HRT and hence small reactor volume (Borja and Banks 1994a, b). Additionally, loss of biomass is minimal as it can be well retained by the filtration system (Wang and Banks 2007). One common problem that arises from most continuous AF systems is filter clogging which limits the level of OLR to be fed despite the good system stability.

UASFF is developed by combining UASB and AF to take advantage of the benefits from both reactors and potentially eliminates the problems in each of the two. Thus, UASFF is potentially more superior in retention of biomass, ability to handle high OLR, and stability to fluctuation in loading without having problems of biomass washout and clogging.

Industrial Applications of AD Technology

The increasing global trend of industrialisation, coupled with continual effort in improving process efficiency and waste minimisation has led to the generation of large quantity of industrial wastewater with high organic content (Mustafa et al. 2011). The high strength organic effluents can be utilised as a source of energy on-site through the application of high rate anaerobic digesters. Table 9.1 lists the use of AD on the treatment of different industrial wastewater. CSTRs and UASB reactors are most commonly applied in the industrial scale while some of these technologies were only limited to investigations in a laboratory scale.

The majority of methane produced from anaerobic wastewater treatment is utilised on-site as fuel for gas engine, boiler, or burnt in flare system to be released into atmosphere as CO_2 if merely low amount is produced. When used to power gas engine for power generation, additional profit could be earned by plants adopting AD system by connecting the electricity generated to national grid. The amount of electricity supplied to the national grid will then be paid to the plant according to feed-in-tariff such as those that have been applied in several plants in Malaysia where a rate of RM0.32 (~USD0.09) is paid for every kWh sold to the national electricity grid (Chin et al. 2013). This indefinitely increases the economical attractiveness of applying AD systems in industry.

Clean Development Mechanism (CDM)

Anaerobic treatment is an energy-generating process, in contrast with aerobic treatment which typically consumes energy due to the need for constant aeration. Moreover, anaerobic treatment normally requires fewer land area as it has less height or depth restriction and produces lower amount of excess sludge which translates into lower cost in maintaining the wastewater plant (Mustafa et al. 2011).

Generation of biomethane from AD process represents the production of renewable energy with potential to earn certified emission reduction (CER) credit through

Table 9.1 Anaerobic digestion applications for different industrial wastewaters

Wastewater type	Reactor type	Feed rate (m^3/day)	OLR (kg COD/m^3 day)	COD removal (%)	Methane yield (m^3/kg COD)	Methane production (m^3)	Potential electricity generation (kWh/m^3)[a]	Reference
Brewery	AF	0.584	8	96	0.15	0.673	10.7	Leal et al. (1998)
Palm oil mill	AF	3.6	1.2–11.4	94.0	63.0	1,340	8.23	Borja and Banks (1994a, b)
Palm oil mill	UASB	17	1.27–10.63	96.7–98.4	54.2–62.0	5,730	8.06	Borja and Banks (1994a, b)
Ice-cream wastewater	UASB	80	0.5–5.0	50.0	69.6	7,650	1.91	Hawkes et al. (1995)
Dairy	UASFF	0.0025	1.8–8.4	90.1–92.0	65.3	0.758	6.45	Córdoba et al. (1995)
Olive oil purification	UASFF	1.4	2.6–17.8	75.7–90.8	69.0–75.0	856	11.8	Borja et al. (1996)
Palm oil mill	Anaerobic pond	187.5	1.4	97.8	54.4	14,000	1.90	Yacob et al. (2006a, b)
Palm oil mill	CSTR	400	3.33	80	62.5	66,600	3.70	Tong and Jaafar (2005)
Cutting-oil wastewater	Fluidised bed	15.4	11.9–51.3	67.1–95.9	20	7,910	35.8	Perez et al. (2007)

[a]Taking calorific value of CH_4 as 50 MJ/kg and gas engine operates 8000 hr/year with 40 % efficiency (Chin et al. 2013)

CDM scheme developed under Kyoto protocol (Tong and Jaafar 2005). One of the outcomes of Kyoto Protocols was the established limits to greenhouse gas emissions. When a country emits less than what is permitted, the difference between the permissible limit and actual emission is identified as excess emission capacity. This excess capacity can be sold to countries that are over their emission limit under emission trading scheme as set out in Article 17 of the Kyoto Protocol. A new commodity in term of emission reduction has since been created to be traded in 'carbon market'. On top of the trade of actual emission units, Kyoto protocol allows other units equivalent to 1 tonne of CO_2 which may be transferred under the emission trading scheme, namely removal unit (RMU), emission reduction unit (ERU), and certified emission reduction (CER). RMU is obtained through improvement in land use, land-use change and forestry (LULUCF), for instance by embarking on reforestation project. ERU is obtainable through Joint Implementation scheme by countries with emission reduction commitment who implement emission reduction project in developed countries (UNFCCC 2014). A counterpart of ERU, CER is obtainable through CDM scheme by countries with emission reduction commitment to implement emission reduction project in developing countries. RMU, ERU, and CER can all be used to meet Kyoto targets or sold to countries that are over their targets.

CDM was created in order to promote reduction in carbon emission through encouraging foreign investors to embark on renewable energy projects in developing countries (Menon et al. 2002). Carbon credits can be obtained from the projects and then sold to companies that emits excessive amount of greenhouse gases. This generates additional revenue from the renewable energy project on top of profit from the main product and the potentially saleable methane if it is produced in excess. As a result, the payback periods of anaerobic digesters can be substantially shortened through the benefits attained from CER credit and methane capture.

Improvements on Anaerobic Digestion Technology

AD technology presents dual benefits by reducing environmental pollution and at the same time produces renewable fuel. However, the technology has yet to reach maturity to completely replace fossil fuel as the main source energy globally. This section discusses technological aspects of AD that can further improve the feasibility of the system.

Robust Control Scheme

System stability is undeniably important in biological process such as anaerobic digestion to ensure consistent quality of effluent and biomethane generated. Instability of an AD system could arise from accumulation of VFAs, long fatty acid, ammonia, or sulphide that may be present in the original waste or produced during AD process itself. While the acid groups upset the reactor by significantly lowering the system pH to create unfavourable conditions, ammonia and sulphide directly harms the microorganisms due to the toxicity of the two compounds.

Process problems in AD systems often go unnoticed until they severely affect the treatment and deplete biogas production. There is little monitoring and information for the plant operator to work on in order to properly regulate the feed flow rate, composition, and operational condition (Ahring 2003).

Most anaerobic digesters are operated at much lower OLRs than what it is capable of to avoid process failure. Implementation of process control will enable anaerobic plants to operate at optimum loading and maximises the overall plant productivity. Recent development has brought about sensor that can directly measures VFAs in real time (Pind et al. 2003). This technological advancement enables continuous monitoring and consequently constant feedback through process control to ensure stability of AD operation.

Some of the types of process controls that can be implemented for anaerobic digestion include proportional-integral-derivative (PID) controller adopted from mechanistic model or more advanced models adopting meta-heuristic models such as fuzzy and neural network. The implementation of robust controllers could improve the consistency of treated effluent quality as well as biogas produced.

Enhancement on Biomethane Production

Several methods have been proposed to increase the digestibility of wastewater into biogas through biological, chemical, and mechanical means. AD systems were found to produce biogas at considerably higher rates through adopting some approaches biologically by enzyme treatment and chemically by acids and base treatment. These options, however, leads to significantly higher operational cost of the plant as enzymes and chemicals need to be replenished. Mechanically decreasing the particle size of the waste or wastewater was found to give rise the best increase in gas production relative to the capital cost invested (Ahring 2003).

Another revolutionary way to enhance the biomethane production is by Pulse Power™ technology developed by Scientific Utilisation in Alabama, USA. The equipment uses rapid-pulse high-power electrical technology to generate a disruptive shock to break large molecules into shorter fractions and has been reported to enhance the removal of volatile solid up twice as good.

Anaerobic Co-digestion

Co-digestion refers to the digestion of mixed waste from different sources, for instance municipal, agricultural, food, pharmaceutical industry, and organic fraction of municipal solid wastes. The increased diversity of organic constituents in the wastes is suggested to be the main factor that increases the overall efficiency and biogas yield of anaerobic treatment. The drawback encountered in implementing anaerobic co-digestion is fundamentally the cost and difficulty in bringing together the various types of wastes into a single waste treatment plant as well as the reduced opportunity of on-site utilisation of the generated biogas by each plant where the wastes are originated. Anaerobic co-digestion is also thought to be able to produce mixed culture population with highly adaptable microorganisms which can cater for a wider range of industrial waste when needed. In other words, it can effectively serve as stock sludge to be used as starter culture for a new AD systems.

Temperature Phase Anaerobic Digestion (TPAD)

One of the vital parameters in AD system is operating temperature that selects and determines the dominant portion in mixed culture population and the corresponding microbial growth rate (Patel and Madamwar 2002). A study conducted on the anaerobic treatment of vegetable wastes reveals that biogas production from thermophilic anaerobic digestion is higher than those from psychrophilic and mesophilic by 144 and 41 % respectively (Bouallagui et al. 2004). TPAD comprises of two-stage anaerobic digestion systems, with the first operating at thermophilic temperature range while the second operating at mesophilic temperature range. TPAD is designed to take advantage of the higher performance in thermophilic range coupled with polishing stage at mesophilic range. The arrangements has shown to be a reliable and effective means to increase stability, methane production, and contaminant removal efficiency compared to either mesophilic or thermophilic system alone. Additionally, TPAD shows higher capability to treat industrial waste at comparatively lower HRT than current conventional anaerobic systems when used to handle the same organic loading (Sung and Santha 2003). The implementation of TPAD for AD of industrial wastewater could potentially reduce subsequent polishing steps for effluent to meet regulatory standards while producing sufficient biogas for power generation.

References

Ahring, B. K. (2003). Perspectives for anaerobic digestion. In B. K. Ahring (Ed.), *Biomethanation i*. Berlin: Springer.

Andersson, S., & Ingvar Nilsson, S. (2001). Influence of pH and temperature on microbial activity, substrate availability of soil-solution bacteria and leaching of dissolved organic carbon in a Mor humus. *Soil Biology and Biochemistry, 33*(9), 1181–1191.

Beccari, M., Bonemazzi, F., Majone, M., & Riccardi, C. (1996). Interaction between acidogenesis and methanogenesis in the anaerobic treatment of olive oil mill effluents. *Water Research, 30*(1), 183–189.

Borja, R., Alba, J., & Banks, C. J. (1996). Anaerobic digestion of wash waters derived from the purification of virgin olive oil using a hybrid reactor combining a filter and a sludge blanket. *Process Biochemistry, 31*(3), 219–224.

Borja, R., & Banks, C. J. (1994a). Anaerobic digestion of palm oil mill effluent using an up-flow anaerobic sludge blanket reactor. *Biomass and Bioenergy, 6*(5), 381–389.

Borja, R., & Banks, C. J. (1994b). Treatment of palm oil mill effluent by upflow anaerobic filtration. *Journal of Chemical Technology and Biotechnology, 61*(2), 103–109.

Bouallagui, H., et al. (2004). Two-phases anaerobic digestion of fruit and vegetable wastes: bioreactors performance. *Biochemical Engineering Journal, 21*(2), 193–197.

Chin, M. J., Poh, P. E., Tey, B. T., Chan, E. S., & Chin, K. L. (2013). Biogas from palm oil mill effluent (POME): Opportunities and challenges from Malaysia's perspective. *Renewable and Sustainable Energy Reviews, 26*, 717–726.

Choorit, W., & Wisarnwan, P. (2007). Effect of temperature on the anaerobic digestion of palm oil mill effluent. *Electronic Journal of Biotechnology, 10*(3).

Córdoba, P. R., Francese, A. P., & Siñeriz, F. (1995). Improved performance of a hybrid design over an anaerobic filter for the treatment of dairy industry wastewater at laboratory scale. *Journal of Fermentation and Bioengineering, 79*(3), 270–272.

Enerdata. (2014). *Global energy statistical yearbook 2014, world energy statistics. World energy consumption & stats. Enerdata*. Retrieved November 11, 2014, from https://yearbook.enerdata.net/.

Feng, L., Wang, H., Chen, Y., & Wang, Q. (2009). Effect of solids retention time and temperature on waste activated sludge hydrolysis and short-chain fatty acids accumulation under alkaline conditions in continuous-flow reactors. *Bioresource Technology, 100*(1), 44–49.

Gerardi, M. H. (2003). *The microbiology of anaerobic digesters*. Hoboken, NJ: Wiley.

Gerardi, M.H., (2006). 'Wastewater Bacteria', New Jersey: John Wiley & Sons, Inc.

Hawkes, F. R., Donnelly, T., & Anderson, G. K. (1995). Comparative performance of anaerobic digesters operating on ice-cream wastewater. *Water Research, 29*(2), 525–533.

Hook, M., & Tang, X. (2013). Depletion of fossil fuels and anthropogenic climate change – A review. *Energy Policy, 52*, 797–809.

Karim, K., Rebecca, H., Thomas Klasson, K., & Al-Dahhan, M. H. (2005a). Anaerobic digestion of animal waste: Effect of mode of mixing. *Water Research, 39*(15), 3597–3606.

Karim, K., Thomas Klasson, K., et al. (2005b). Anaerobic digestion of animal waste: Effect of mixing. *Bioresource Technology, 96*(14), 1607–1612.

Kim, M., Young-Ho, A., & Speece, R. E. (2002). Comparative process stability and efficiency of anaerobic digestion: Mesophilic vs. thermophilic. *Water Research, 36*(17), 4369–4385.

Leal, K., et al. (1998). A mesophilic digestion of brewery wastewater in an unheated anaerobic filter. *Bioresource Technology, 65*(1–2), 51–55.

Ling, YL (2007, July) Treatability of palm oil mill effluent (POME) using black liquor in an anaerobic treatment process (*Thesis* Submitted in Fulfilment of the Requirements for the Degree of Master of Science July 2007 Acknowledgements).

McNutt, M. (2013). Climate change impacts. *Science, 341*, 435.

Menon, S., James, H., Larissa, N., & Yunfeng, L. (2002). Climate effects of black carbon aerosols in China and India. *Science, 297*(5590), 2250–2253.

Mustafa, E. E., Hale, O., Recep, K. D., & Izzet, O. (2011). Anaerobic treatment of industrial effluents: An overview of applications. In F. S. G. Einschlag (Ed.), *Waste water – Treatment and reutilization*. New York, NY: InTech.

Patel, H., & Madamwar, D. (2002). Effects of temperatures and organic loading rates on biomethanation of acidic petrochemical wastewater using an anaerobic upflow fixed-film reactor. *Bioresource Technology, 82*(1), 65–71.

Perez, M., Rodriguez-Cano, R., Romero, L. I., & Sales, D. (2007). Performance of anaerobic thermophilic fluidized bed in the treatment of cutting-oil wastewater. *Bioresource Technology, 98*(18), 3456–3463.

Pind, P. F., Angelidaki, I., & Ahring, B. K. (2003). Dynamics of the anaerobic process: Effects of volatile fatty acids. *Biotechnology and Bioengineering, 82*(7), 791–801.

Poh, P. E., & Chong, M. F. (2009). Development of anaerobic digestion methods for palm oil mill effluent (POME) treatment. *Bioresource Technology, 100*, 1–9.

Sánchez, E., Borja, R., Travieso, L., Martín, A., & Colmenarejo, M. F. (2005). Effect of organic loading rate on the stability, operational parameters and performance of a secondary upflow anaerobic sludge bed reactor treating piggery waste. *Bioresource Technology, 96*(3), 335–344.

Shafiee, S., & Topal, E. (2009). When will fossil fuel reserves be diminished? *Energy Policy, 37*, 181–189.

Sung, S., & Santha, H. (2003). Performance of temperature-phased anaerobic digestion (TPAD) system treating dairy cattle wastes. *Water Research, 37*(7), 1628–1636.

Tong SL, Jaafar AB (2005) POME biogas capture, upgrading and utilisation. In: *Proceedings of the PIPOC 2005 international palm oil congress (chemistry and technology)*. Petaling Jaya: Malaysian Palm Oil Board.

UNFCCC. (2014). *International emissions trading*. New York, NY: United Nations Framework Convention on Climate Change.

Van Lier, J. B., Rebac, S., & Lettinga, G. (1997). High-rate anaerobic wastewater treatment under psychrophilic and thermophilic conditions. *Water Science and Technology, 199–206*.

Veeken, A., Kalyuzhnyi, S., Scharff, H., & Hamelers, B. (2000). Effect of pH and VFA on hydrolysis of organic solid waste. *Journal of Environmental Engineering, 126*(12), 1076–1081.

Wang, Z., & Banks, C. J. (2007). Treatment of a high-strength sulphate-rich alkaline leachate using an anaerobic filter. *Waste Management, 27*(3), 359–366.

Wiegant, W. M., Claassen, J. A., & Lettinga, G. (1985). Thermophilic anaerobic digestion of high strength wastewaters. *Biotechnology and Bioengineering, 27*(9), 1374–1381.

Yacob, S., Hassan, M. A., Shirai, Y., Wakisaka, M., & Subash, S. (2006a). Baseline study of methane emission from anaerobic ponds of palm oil mill effluent treatment. *The Science of the Total Environment, 366*(1), 187–196.

Yacob, S., Shirai, Y., Hassan, M. A., Wakisaka, M., & Subash, S. (2006b). Start-up operation of semi-commercial closed anaerobic digester for palm oil mill effluent treatment. *Process Biochemistry, 41*(4), 962–964.

Yu, H. Q., & Fang, H. H. (2003). Acidogenesis of gelatin-rich wastewater in an upflow anaerobic reactor: Influence of pH and temperature. *Water Research, 37*(1), 55–66.

Zeeman, G., & Sanders, W. (2001). Potential of anaerobic digestion of complex waste (water). *Water Science and Technology, 44*(8), 115–122.

Zhang, Y., Yan, L., Chi, L., & Long, X. (2008). Startup and operation of anaerobic EGSB reactor treating palm oil mill effluent. *Journal of Environmental Science (China), 20*, 658–663.

Chapter 10
Bioenergy: Biofuels Process Technology

Ajay Kumar, Joginder Singh, Chinnappan Baskar, and Seeram Ramakrishna

Introduction

There is an increasing interest in the production of chemicals and fuels from renewable resources due to the continuing price increase of fossil resources, the insecurity of the availability of fossil resources in the future, and additionally environmental concerns and legislations (García et al. 2011; Baskar et al. 2012). In recent years, growing attention has been devoted to the conversion of biomass into biofuel such as ethanol, butanol, biodiesel etc. considered the cleanest liquid fuel alternative to fossil fuels (Lin and Tanaka 2006). Moreover, biomass energy can play an important role in reducing greenhouse gas emissions; since CO_2 that arises from biomass wastes would originally have been absorbed from the air, the use of biomass for energy offsets fossil fuel greenhouse gas emissions (Lynd 1996). Currently ethanol is the main bio-fuel used in the world and its use is increasingly widespread, the worldwide prospects are the expansion of the production and consumption of ethanol (Bastos 2007). Fermentation-derived butanol is a possible alternative to ethanol as a fungible biomass-based liquid transportation fuel (Pfromm et al. 2010). The transesterification of vegetable oils (VOs) with short-chain alcohols is used to produce biodiesel or by the esterification of fatty acids. During the past few years

A. Kumar (✉) • J. Singh
Department of Biotechnology, Lovely Professional University, Phagwara, Punjab, India
e-mail: kumarajaybiotech@gmail.com; joginder.15005@lpu.co.in

C. Baskar
THDC Institute of Hydropower Engineering and Technology Tehri, Uttarakhand Technical University, Dehradun, Uttarakhand, India
e-mail: baskarc@gmail.com

S. Ramakrishna
Department of Mechanical Engineering and Center for Nanofibers and Nanotechnology, National University of Singapore, Singapore
e-mail: seeram@nus.edu.sg

P. Ravindra (ed.), *Advances in Bioprocess Technology*,
DOI 10.1007/978-3-319-17915-5_10

biodiesel has attracted attention as an environmentally friendly and renewable fuel because of uncertainties concerning petroleum availability and recent increases in petroleum prices (Berchmans and Hirata 2008). Similarly Hydrogen production from biomass conversion plays a very important role in the development of hydrogen economy (Ni et al. 2006). The quality of energy crops, used for biogas production, is determined on the field. Methane production from organic substrates mainly depends on their content of substances that can be degraded to CH_4 and CO_2 (Amon et al. 2007). About 95 % of ethanol produced in the world is from agricultural products (Walter et al. 2008). Ethanol production from sugar crops such as sugarcane and sugar beet account for about 40 % of the total bioethanol produced and nearly 60 % corresponding to starch crops (Biofuels Platform, 2010a, b). Biobutanol is on the agenda of several companies and may be used in the near future as a supplement for gasoline, diesel and kerosene (Antoni et al. 2007). Assuming an oil price of US$60 per barrel, both biodiesel and bioethanol produced from wheat are not profitable in Europe. At the assumed oil price, only bioethanol and biobutanol produced on a large scale from lignocellulose-containing raw materials have the potential to be produced competitively (Festel 2008; Kumar et al. 2012). The U.S. has become the dominant ethanol producer (corn-based), although Brazil has started an ambitious program to increase production by 50 % by 2009 (sugar-based). Biodiesel production has increased at 20–100 % annual rates in recent years, particularly in Germany, France, Italy, Poland, and the United States (Renewables 2005). About half of all the hydrogen as currently produced is obtained from thermo catalytic and gasification processes using natural gas as a starting material, heavy oils and naphtha make up the next largest source, followed by coal. Currently, much research has been focused on sustainable and environmental friendly energy from biomass to replace conventional fossil fuels (Balat and Kırtay 2010). Current total annual worldwide hydrogen consumption is in the range of 400–500 billion Nm^3 (Demirbas 2009a, b). Present utilization of hydrogen is equivalent to 3 % of the energy consumption and with a growth rate estimated at 5–10 % per year (Mohan et al. 2013). Only a fraction of this hydrogen is currently used for energy purposes; the bulk serves as a chemical feedstock for petrochemical, food, electronics and metallurgical processing industries. The global market for hydrogen is already greater than US$40 billion per year (Kraus 2007); including hydrogen used in ammonia production (49 %), petroleum refining (37 %), methanol production (8 %), and miscellaneous smaller-volume uses (6 %) (Konieczny et al. 2008).

Biofuel Feedstocks

Fermentation substrate is an important factor influencing the cost of ethanol, butanol, hydrogen gas etc. production (Qureshi and Blaschek 2000). Lignocellulose is the most abundant renewable resource on the planet, and has great potential as a substrate for fermentation. Hemicelluloses are the second most abundant polysaccharides in nature, and represent about 20 to 35 % of lignocellulosic biomass

(Koukiekolo et al. 2005). Xylan or hemicellulose may contain arabinan, galactan, glucuronic, acetic, ferulic, and rcoumaric acids as well as xylose. The occurrence and quantity of these compounds depend on the sources of xylan (Olsson and Hahn-Hägerdahl 1996; Koukiekolo et al. 2005). The ethanol yield and productivity obtained during fermentation of lignocellulosic hydrolysates is decreased due to the presence of inhibiting compounds, such as weak acids, furans and phenolic compounds produced during hydrolysis (Palmqvist and Hahn-Hägerdal 2000).

Lignocellulosic Feedstocks

Lignocellulosic biomass is generally composed of hemicellulose (25–35 %), cellulose (4–50 %), and lignin (15–20 %), and these structures are illustrated in Fig. 10.1. Cellulose hydrolysis can also be achieved under harsher conditions using solutions of mineral acids (H_2SO_4) at elevated temperatures; however, the harsh conditions required for non-enzymatic deconstruction of cellulose favor the formation of degradation products such as hydroxymethylfurfural (HMF), levulinic acid, and insoluble humins (Alonso et al. 2010; Rinaldi and Schüth 2009).

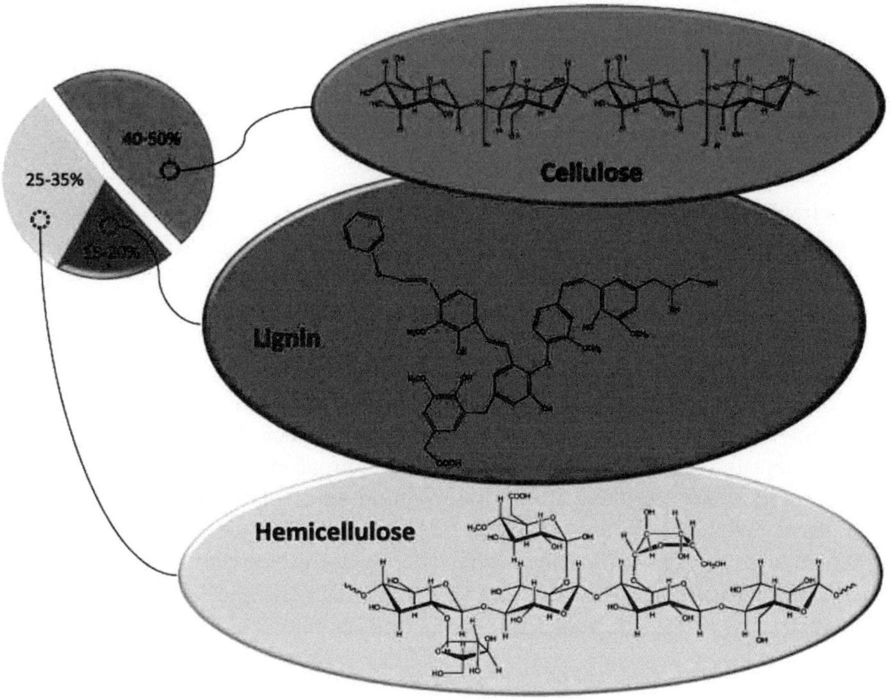

Fig. 10.1 Lignocellulose composition: cellulose, hemicellulose and lignin (Alonso et al. 2010)

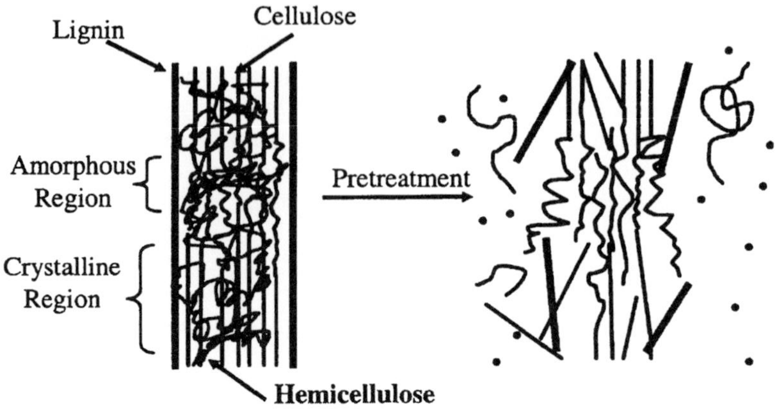

Fig. 10.2 Schematic of goals of pretreatment on lignocellulosic material (Mosier et al. 2005)

Table 10.1 Percent dry weight composition of lignocellulosic feedstock's (Mosier et al. 2005)

Feedstock	Glucan (cellulose)	Xylan (hemicellulose)	Lignin
Corn stover	37.5	22.4	17.6
Corn fiber	14.28	16.8	8.4
Pine wood	46.4	8.8	29.4
Poplar	49.9	17.4	18.1
Wheat straw	38.2	21.2	23.4
Switch grass	31.0	20.4	17.6
Office paper	68.6	12.4	11.3

Note: Because minor components are not listed, these numbers do not sum to 100 %

Pretreatment is necessary to make cellulose more prominent to be attacked by the enzymes which ultimately convert it into fermentable sugars (Fig. 10.2).

Lignocellulosic biomass normally comprises cellulose, hemicellulose and lignin i.e. near about 55–75 % carbohydrates on dry weight basis. Percent dry weight compositions of different lignocellulosic feedstocks were determined by Mosier et al. (2005) as shown in Table 10.1.

Unfortunately, neither commercial ethanol-producing cultures, nor butanol-producing cultures can hydrolyze these substrates. Hence, they need to be hydrolyzed prior to fermentation using a combination of pretreatment (acid, alkali, organosolvent, supercritical extraction or ammonia explosion) and hydrolysis (enzymes: cellulase, β-glucosidase, and xylanase) techniques (Galbe and Zacchi 2002). It should be noted that in contrast to ethanol production by yeasts, hexose and pentose sugars obtained as a result of pretreatment and hydrolysis of these residues can be used by butanol-producing cultures (Qureshi et al. 2008a, b, c).

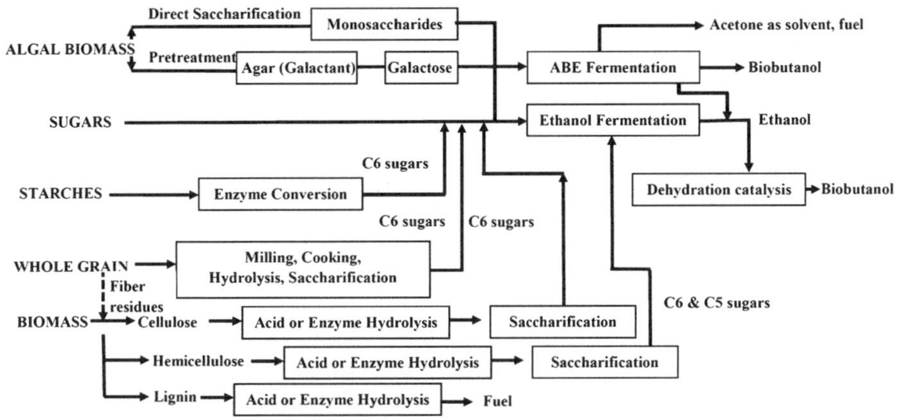

Fig. 10.3 Different feedstocks generally used for ethanol and biobutanol production

Algal Biomass Feedstocks

Algae capable of accumulating high starch/cellulose can serve as an excellent alternative to food crops for bioethanol production, a green fuel for sustainable future. Certain species of algae can produce ethanol during dark-anaerobic fermentation and thus serve as a direct source for ethanol production. Of late, oleaginous microalgae generate high starch/cellulose biomass waste after oil extraction, which can be hydrolyzed to generate sugary syrup to be used as substrate for ethanol production. Macroalgae are also harnessed as renewable source of biomass intended for ethanol production (John et al. 2011; Nguyen et al. 2009). The use of marine algal biomass with high carbohydrate contents of *Ulva lactuca* and other macroalgae like *Saccharina spp. Laminaria*, *Durvillaea*, *Ecklonia* and *Homosira* (brown algae) (Figueira et al. 2000) indicates that a more cost effective strategy might be to ferment the carbohydrates like glucose, mannitol and laminarin from these algal species to either ethanol or butane (Potts et al. 2012; Huesemann et al. 2012).

Fate of different feedstocks for ethanol and biobutanol production are shown in Fig. 10.3.

Microbial Modeling of Biofuel Production

A wide variety of biofuels can be produced through the bioconversion of substrates contained in agricultural crops and residues (Fischer et al. 2008). Bioconversion of the sugars, starches, and other organic substrates contained in agricultural residues can be converted to ethanol by a variety of yeasts, to hydrogen by a variety of fermentative bacteria and archae, to methane by a consortium of bacteria and archae, and to oils for biodiesel production by fungi and algae (Drapcho et al. 2008). All of these microbial processes can be described mathematically to simulate the bioprocess (Kumar and Murthy 2013). First generation biofuels, such as ethanol and biodiesel are already widely used, but they were selected mainly for convenience rather than their properties as fuels. In a microbial biofuel production process, bioengineered microbes are grown inside a reactor in a solution that is rich in cellulose-derived sugar (glucose and pentose) (Turner 2014).

Bioreactor modeling and design based on microbial growth and product formation kinetics may be used to optimize production of high-value biofuels or maximize utilization of feedstock nutrients. Kinetic models are normally divided into two classes: structured and unstructured one. Structured models take metabolic pathways into consideration and are generally complicated. A structured model for acetone–butanol fermentations was established by Votruba et al. (1985).

The Monod model is a widely applied model used to describe microbial growth. Suitable microbial hosts for biofuel production must tolerate process stresses such as end-product toxicity and tolerance to fermentation inhibitors in order to achieve high yields and titers (Fischer et al. 2008)

$$\mu = \frac{\mu_{max} S}{K_s + S} \tag{10.1}$$

where μ = Specific growth rate co-efficient h^{-1}
μ_{max} = Maximum Specific growth rate co-efficient h^{-1}
S = Substrate concentration mg/L
K_s = half-saturation constant, mg/L

Kinetic expressions for product formation must account for growth associated and maintenance-associated production, as in the following equation:

$$r_p = Y_{PX} r_x + m_p X. \tag{10.2}$$

Where r_x is the volumetric rate of biomass formation, Y_{PX} is the theoretical or the true yield of product from biomass, m_p is the specific rate of product formation due to maintenance, and X is biomass concentration (g/L), r_p is the volumetric rate of product formation (Doran 1995).

It is also possible that two or more substrates may simultaneously be growth-limiting, thus, a model that can describe such a system is given by:

$$\mu = \frac{\mu_{max} S_1}{K_1 + S_1} \left(\frac{S_2}{K_2 + S_2} \right) \tag{10.3}$$

Where, μ is the specific growth rate (1/h), μ_m is the maximum specific growth rate (1/h). The specific growth rate could be inhibited by medium constituents such as substrate or product. In a case of substrate inhibition, the term is given by:

$$\mu = \frac{\mu_{max}\, S}{Ks + S + \left[\frac{S^2}{K_I}\right]} \tag{10.4}$$

For a case that exhibits, product inhibition such as ethanol fermentation and ABE fermentation, the specific growth rate is written as:

$$\mu = \frac{\mu_{max}\, S}{Ks + S}\left(\frac{Kp}{Kp + P}\right) \tag{10.5}$$

Where K_p is product inhibition constant and P is the concentration of the product

Bioreactor for Biofuel Production

Basic bioreactor designs for suspended growth cultures are batch, continuous (flow) stirred tank reactor (CSTR), and CSTR with external or internal biomass recycle (Fig. 10.4). If the means of cell separation (filtration, centrifugation, settling) removes the compound with the biomass, then the product is considered particulate. Hydrogen and ethanol are examples of soluble, extracellular products, while oils produced by filamentous fungi *Pythium* are intracellular products (Drapcho et al. 2008).

Batch Bioreactor

When plotted on arithmetic paper, batch growth cure assumes a sigmoidal shape, this can be predicted by combing the Monod equation with growth equation (Shuler and Kargi 2002).

$$\frac{dX}{dt} = \frac{\mu_m\, S}{K_s + S} X \tag{10.6}$$

The relationship between microbial growth yield and substrate is

$$X - X_0 = Y_{X/S}(S_0 - S) \tag{10.7}$$

Where X_0 and S_0 are initial values and Yx/s is the cell mass yield based on limiting nutrient.

$$\frac{dX}{dt} = \frac{\mu_m\left(Y_{X/S}S_0 + X_0 - X\right)}{K_s Y_{X/S} + Y_{X/S}S_0} X \tag{10.8}$$

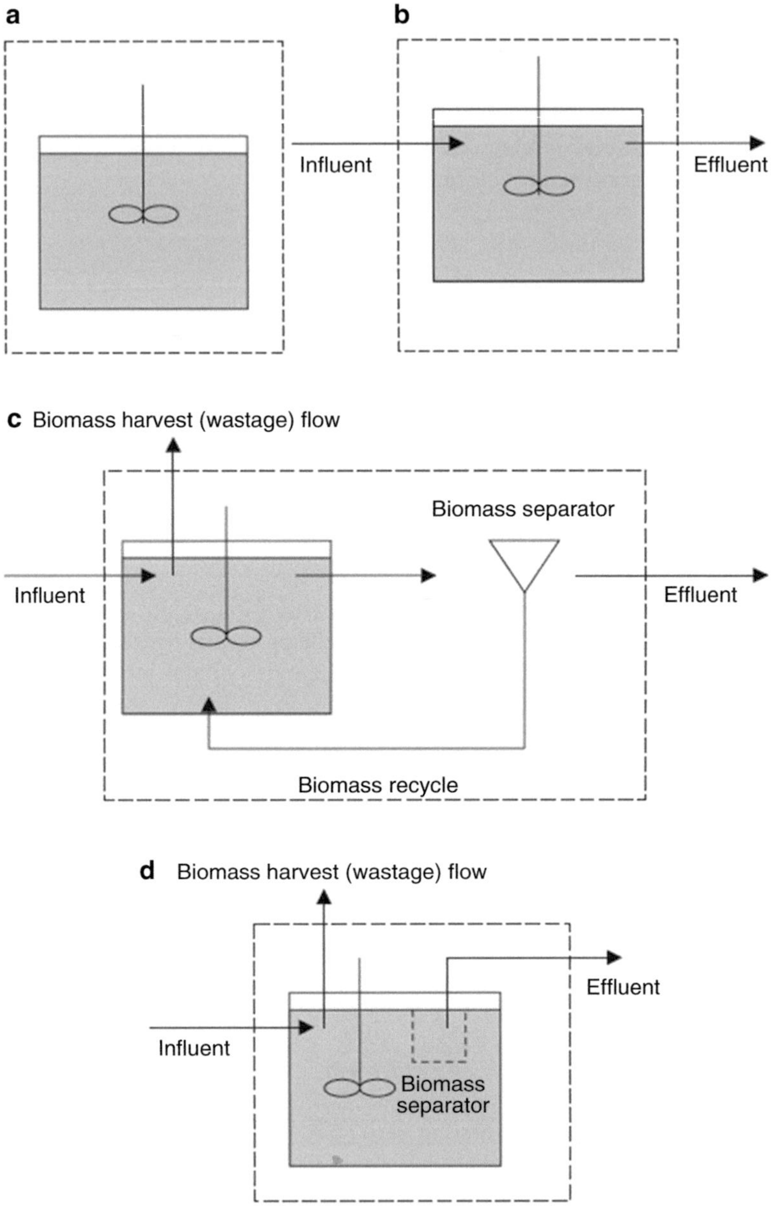

Fig. 10.4 Basic bioreactor types. (**a**) Batch. (**b**) Simple CSTR. (**c**) CSTR with external biomass recycle. (**d**) CSTR with internal biomass recycle. *Dashed lines* indicate system boundary used for developing mass balance equations

Continuous Stirred Tank Reactors

Theoretically, a continuous process can be described with the following equations:

$$\frac{dX}{dt} = (\mu - D)X \tag{10.9}$$

$$D = \frac{F}{V}, \quad \left(\frac{1}{h}\right) \tag{10.10}$$

$$T_R = \frac{V}{F}, \quad (h) \tag{10.11}$$

$$M_S = R_S - \frac{\mu}{Y_{SX}} \tag{10.12}$$

where
μ is the specific growth rate (1/h); D is dilution rate (1/h); X is biomass concentration (g/L); F is the feed flow rate (L/h); V is the volume of the bioreactor (L); T_R is the residence time (h); M_s is the maintenance value (C-mol/C-mol/h); R_s is the rate of substrate consumption (C-mol/C-mol/h); Y_{sx} is the yield of biomass per unit mass of the substrate.

Ethanol Production

Bio-ethanol is ethyl alcohol, grain alcohol, or chemically C_2H_5OH or EtOH. Bio-ethanol and bio-ethanol/gasoline blends have a long history as alternative transportation fuels. Bio-ethanol has a higher octane number (108), broader flammability limits, higher flame speeds and higher heats of vaporization. Disadvantages of bio-ethanol include its lower energy density than gasoline (bio-ethanol has 66 % of the energy that gasoline has), its corrosiveness, low flame luminosity, lower vapor pressure (making cold starts difficult), miscibility with water, toxicity to ecosystems (Spatari et al. 2005) increase in exhaust emissions of acetaldehyde, and increase in vapor pressure (and evaporative emissions) when blending with gasoline. Some properties of alcohol fuels are shown in Table 10.2.

Table 10.2 Some properties of ethanol (alcohol fuel)

S. no.	Fuel property	Ethanol
1	Octane number	108
2	Auto ignition temperature (K)	606
3	Latent heat of vaporization (MJ/Kg)	0.91
4	Lower heating value (MJ/Kg)	26.7

Source: Balat and Balat 2009

Ethanol Production from Sugar, Starch and Lignocellulosic Feedstocks

Bio-ethanol is a fuel derived from biomass sources of feedstock; typically plants such as wheat, sugar beet, corn, straw, and wood. The conversion of lignocellulosic biomass to ethanol is a three step process that involves pretreatment followed by polysaccharide hydrolysis to simple sugars followed by sugar fermentation to ethanol (Mielenz 2001). The presence of lignin in cell walls negatively impacts these conversion steps (Keating et al. 2006; Li et al. 2008).

The effect of pretreatment of lignocellulosic materials has been recognized for a long time (McMillan 1994). Pretreatments for lignocellulosic materials include mechanical comminution, alkali swelling, acid hydrolysis, steam and other fiber explosion techniques, and exposure to supercritical fluids. Mechanical comminution. Waste materials can be comminuted by a combination of chipping, grinding and milling to reduce cellulose crystallinity. Pyrolysis has also been used for pretreatment of lignocellulosic materials. When the materials are treated at temperatures greater than 300 °C, cellulose rapidly decomposes to produce gaseous products and residual char (Sun and Cheng 2002). Mild acid hydrolysis (1 N H_2SO_4, 97 °C, 2.5 h) of the residues from pyrolysis pretreatment has resulted in 80– 85 % conversion of cellulose to reducing sugars with more than 50 % glucose (Sun and Cheng 2002a, b). Steam explosion is the most commonly used method for pretreatment of lignocellulosic materials (McMillan 1994). In this method, chipped biomass is treated with high-pressure saturated steam and then the pressure is swiftly reduced, which makes the materials undergo an explosive decompression. Steam explosion is typically initiated at a temperature of 160–260 °C (corresponding pressure 0.69–4.83 MPa) for several seconds to a few minutes before the material is exposed to atmospheric pressure. The process causes hemicellulose degradation and lignin transformation due to high temperature, thus increasing the potential of cellulose hydrolysis. Ammonia fiber explosion (AFEX) is another type of physico-chemical pretreatment in which lignocellulosic materials are exposed to liquid ammonia at high temperature and pressure for a period of time, and then the pressure is swiftly reduced. Dilute acid hydrolysis such as H_2SO_4 and HCl has been successfully developed for pretreatment of lignocellulosic materials. The dilute sulfuric acid pretreatment can achieve high reaction rates and significantly improve cellulose hydrolysis (Esteghlalian et al. 1997). The mechanism of alkaline hydrolysis is believed to be saponification of intermolecular ester bonds crosslinking xylan hemicelluloses and other components, for example, lignin and other hemicellulose. The porosity of the lignocellulosic materials increases with the removal of the crosslinks (Tarkow and Feist 1969). Dilute NaOH treatment of lignocellulosic materials caused swelling, leading to an increase in internal surface area, a decrease in the degree of polymerization, a decrease in crystallinity, separation of structural linkages between lignin and carbohydrates, and disruption of the lignin structure (Fan et al. 1987).

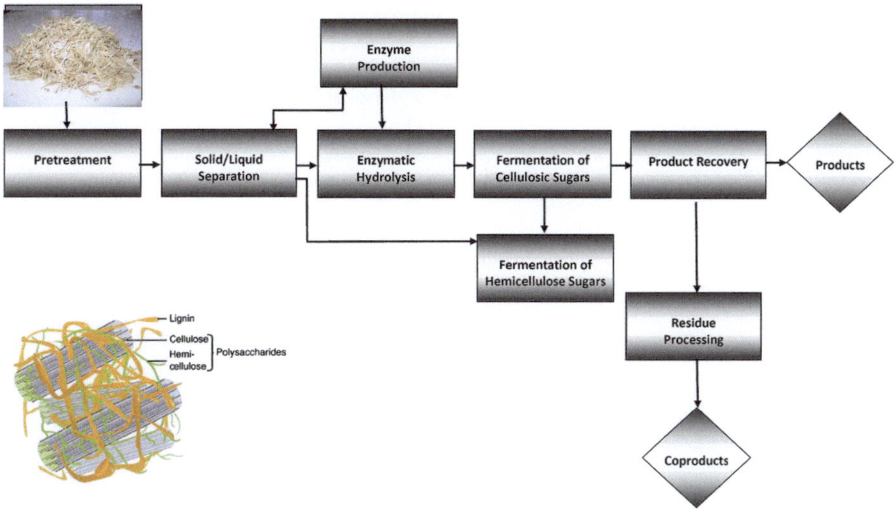

Fig. 10.5 Current process to produce biofuel from lignocellulose

Enzymatic hydrolysis of cellulose is carried out by cellulase enzymes which are highly specific (Beguin and Aubert 1994). The products of the hydrolysis are usually reducing sugars including glucose. Utility cost of enzymatic hydrolysis is low compared to acid or alkaline hydrolysis because enzyme hydrolysis is usually conducted at mild conditions (pH 4.8 and temperature 45–50 °C) and does not have a corrosion problem (Duff and Murray 1996).

Fermentation

Ethanol can be produced from lignocellulosic materials in various ways. The main features of the different ethanol processes are outlined in Fig. 10.5. All processes comprise the same main components: hydrolysis of the hemicellulose and the cellulose to monomer sugars, fermentation and product recovery and concentration by distillation (Galbe and Zacchi 2002).

The most frequently used microorganism for fermenting ethanol in industrial processes is *S. cerevisiae*, *Zymomonas mobilis* can ferment glucose to ethanol with higher yields. Since lignocellulosic hydrolysates contain pentoses, which are not readily fermented by these microorganisms, several attempts to genetically engineer *S. cerevisiae* (Walfridsson et al. 1996; Hahn-Hägerdal et al. 2007), *Z. mobilis* (Panesar et al. 2006) and the bacteria *Escherichia coli* (Decker et al. 2007) have been performed (Fig. 10.6).

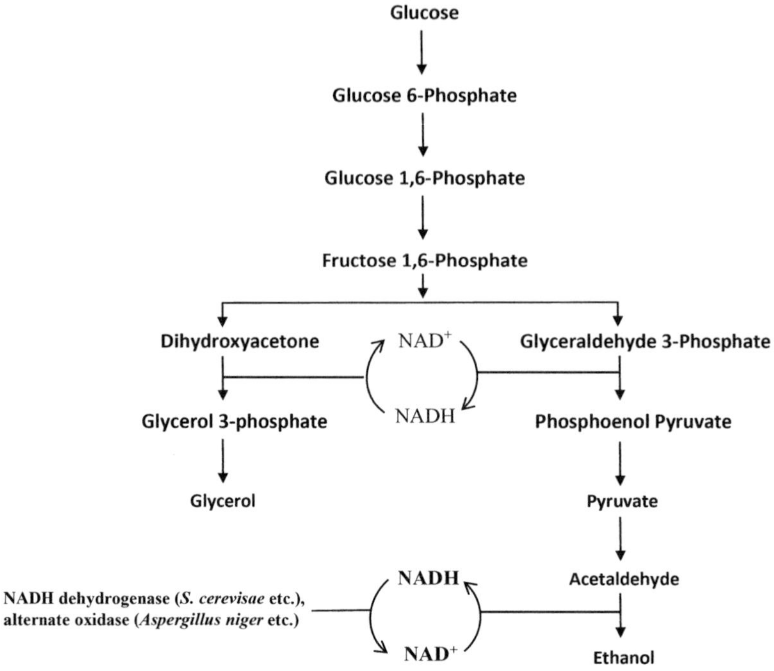

Fig. 10.6 Redox balance in biosynthetic routes to glycerol and ethanol in *S. cerevisiae* (Baskar et al. 2012a, b)

Yeast convert hexose to ethanol and carbon dioxide by glycolysis as shown by the following reaction:

$$C_6H_{12}O_6 \rightarrow 2CH_3CH_2OH + 2CO_2$$

Theoretically, 1 kg of glucose will produce 0.51 kg of bio-ethanol and 0.49 kg of carbon dioxide. However, in practice, the microorganisms use some of the glucose for growth and the actual yield is less than 100 % (Demirbas 2009a, b).

Usually by products such as glycerol, succinic acid and acetic acids are produced. Optimum temperature and pH values for yeast are 30–35 °C and 4–5 respectively (Shuler and Kargi 2002). Most notably, *C. thermohydrosulfuricum* strain 39E has the highest reported ethanol yield (1.9 mol of ethanol produced per mol of glucose fermented) of any taxonomically described thermophilic anaerobe (Ng et al. 1981). For thermophilic organisms optimum temperature may range from 50 and 60 °C. Ethanol production is triggered by anaerobic conditions. Glucose concentrations above 100 g/l are inhibitory for yeast. Ethanol and some of the other by-products are inhibitory to yeast above concentration of 5 % (v/v). Ethanol tolerance yeast strains are being developed to avoid ethanol inhibition (Balat and

Balat 2009). Simultaneous removal of ethanol from fermentation broth is another alternative for ethanol inhibitions.

Downstream Processing

Ethanol can be separated from the culture vessels during fermentation using low temperature vacuum distillation, adsorption, or membrane separation.

Butanol Production

Butanol (butyl alcohol and 1-butanol) is a four carbon primary alcohol having the molecular formula of C_4H_9OH (MW 74.12). Butanol is a colorless liquid with a distinct odor. Butanol is completely miscible with organic solvents and partly miscible with water (Lee et al. 2008a, b). Butanol represents a biofuel extender or replacement with properties clearly superior to ethanol (higher mileage, not hygroscopic, usable without engine modifications, not corrosive). In addition, it is a valuable feedstock for the chemical industry (Dürre 2011).

A sustainable bacterial fermentation route to produce biobutanol is poised for re-commercialization. Biobutanol may be produced by the acetone–butanol–ethanol (ABE) fermentation (Kumar and Gayen 2011). Today, biobutanol can compete with synthetic butanol in the chemical market (Green 2011).

Several countries have initiated new alternatives for biobutanol production from renewable feedstocks like sweet sorghum bagasse, rice bran (RB), de-oiled rice bran (DRB), corn stover, and wheat straw (Swana et al. 2011; Al-Shorgani et al. 2012; Zhang et al. 2011). By sustainable harvest based on current yields, these materials can be converted to 8.27 billion gallons of biobutanol replacing 7.55 billion gallons of gasoline annually (Swana et al. 2011). Common feedstocks used for biobutanol fermentation process are mentioned in Table 10.3.

Researchers have been re-directing their interests in biomass based fuels, which currently seem to be the only logical alternative for sustainable development in the context of economic and environmental considerations. Renewable bioresources are available globally in the form of residual agricultural biomass and wastes, which can be transformed into liquid biofuels (Nigam and Singh 2011).

Although research on genetics, fermentation, upstream processing, and downstream processing has progressed significantly, the *Clostridia* are not able to efficiently hydrolyze fiber-rich agricultural residues. For this reason, agricultural biomass must be hydrolyzed to simple sugars using economically developed methods. Dilute sulfuric acid pretreatment can be applied to agricultural residues to bring about hydrolysis. Unfortunately, during acid hydrolysis, a complex mixture of microbial inhibitors is generated. Examples of the inhibitory compounds include

Table 10.3 Feedstocks used for biobutanol fermentation process (Niemisto et al. 2013)

Feedstock source	Examples	Advantages	Disadvantages	Reference
Agricultural residues or byproducts	Bagasse, corn stover/fiber/cobs, straws (e.g. from barley, rice or wheat)	Easier upstream processing to fermentable sugars	Seasonal availability, variations in cultivation yield and quality, land-use change, transport costs (low density)	Soni et al. 1982; Qureshi et al. 2006, 2008a, b, c, 2010a, b; Marchal et al. 1984
Crop biomass	Cassava, corn	Easier upstream processing to fermentable sugars	Edible, seasonal availability, variations in yield and quality, land-use change, water need for irrigation	Thang et al. 2010; Campos et al. 2002; Ezeji et al. 2007a, b, c
Non-food crop biomass	Switchgrass, Jerusalem artichoke	Does not compete with food use	Land-use change possible if fertile land is used, potential water need	Qureshi et al. 2010a, b; Marchal et al. 1985
Wood-based biomass	Wood hydrolysates (e.g. from aspen, pine, beech or hemlock)	Non-food biomass, good availability, lower transport costs	More difficult upstream processing, indirect land-use change possible	Saddler et al. 1983; Yu et al. 1984; Sjolander et al. 1938; Maddox and Murray 1983
Industrial by-products	Apple pomace, cheese whey, distillers dry grain solids (DDGS), potato waste, brans (e.g. from rice or wheat), soy molasses, waste sulfite liquor	Better social acceptance by means of resource use efficiency and waste minimization, no land-use change	Availability and quality of the raw material may vary, additional processing may be needed to separate the feedstock from the main product	Qureshi et al. 2001; Lee et al. 2009; Nimcevic et al. 1998; Gutierrez et al. 1998; Grobben et al. 1993
Biodegradable municipal waste	Food and garden waste, starch-based packing peanuts, sludge from wastewater treatment	Better social acceptance contributes to resource efficiency and waste minimization, no land-use change	(Seasonal) and qualitative variation	Murty and Chandra 1997; Claassen et al. 2000; Jesse et al. 2002; López-Contreras et al. 2000; Kobayashi et al. 2005

Table 10.4 Characteristic properties of butanol

Characteristic	Butanol	
Formula	$CH_3(CH_2)_3OH$	 Butanol structure
Boiling point (°C)	118	
Melting point (°C)	−89.3	
Ignition temperature (°C)	35	
Flash point (°C)	365	
Density at 20 °C (g/mL)	0.8098	
Critical pressure (hPa)	48.4	
Critical temperature (°C)	287	
Heat of vaporization (MJ/kg)	0.43	
Energy density (MJ/L)	29.2	
Motor octane number	78	

furfural, hydroxymethyl furfural (HMF), and acetic, ferulic, glucuronic, rho-coumaric acids, etc. (Varga et al. 2004).

Formerly, ABE fermentation was operated as a batch process followed by distillation to recover the products. Sugars (molasses) or starch (corn, wheat, and potatoes) was used as substrates. In this process the price of the substrate accounts for up to 60 % of the cost, dramatically affecting the economic viability of ABE fermentation (Claassen et al. 1999).

The use of marine algal biomass with high carbohydrate contents of *Ulva lactuca* and other macroalgae like *Saccharina spp. Laminaria*, *Durvillaea*, *Ecklonia* and *Homosira* (brown algae) (Figueira et al. 2000) indicates that a more cost effective strategy might be to ferment the carbohydrates like glucose, mannitol and laminarin from these algal species to either ethanol or butane (Potts et al. 2012; Huesemann et al. 2012).

Important characteristics of butanol are summarized in Table 10.4. The market for biobutanol is currently worth US$5 billion and is estimated to rise to $247 billion by 2020 (Kretzers 2012).

Unfortunately, neither commercial ethanol-producing cultures, nor butanol-producing cultures can hydrolyze these substrates. Hence, they need to be hydrolyzed prior to fermentation using a combination of pretreatment (acid, alkali, organosolvent, supercritical extraction or ammonia explosion) and hydrolysis (enzymes: cellulase, β-glucosidase, and xylanase) techniques (Galbe and Zacchi 2002). It should be noted that in contrast to ethanol production by yeasts, hexose and pentose sugars obtained as a result of pretreatment and hydrolysis of these residues can be used by butanol-producing cultures (Qureshi et al. 2008a, b, c).

Feedstock's biomass pretreatment can be achieved by air dry the biomass, dry, grind, and then hydrolyze with dilute acid such as sulfuric acid with different concentrations (at 0.5, 1.0, 2.0, and 5.0 % by weight) (Potts et al. 2012). After hydrolysis the pH was adjusted to a value deemed suitable for fermentation (approximately 4.5–5). Various lignocellulosic feedstocks have been claimed for maximum solvent production by *Clostridium* (Table 10.5).

Table 10.5 Different feedstocks and strains used along with maximum solvents and productivities achieved (Jurgens et al. 2012)

Substrate	Hydrolysis method	Strain used	Yield (g/g)/ productivity (g/l h)	Total ABE (g/l)	References
Wheat straw	H_2SO_4 + enzyme	*C. beijerinckii* P260	0.60/0.42	25	Qureshi et al. (2007)
Wheat straw	H_2SO_4 + enzyme	*C. beijerinckii* P260	0.41/0.31	21.42	Qureshi et al. (2008a)
Corn fiber	H_2SO_4	*C. beijerinckii* BA101	0.39/0.10	9.3	Qureshi et al. (2008c)
Rice bran and defatted rice bran	HCl + enzyme	*C. beijerinckii* NCIMB 8052	0.31/0.26	16.42	Lee et al. (2009)
Barley straw	H_2SO_4 + enzyme	*C. beijerinckii* P260	0.43/0.39	26.64	Qureshi et al. (2010a)
Corn stover	H_2SO_4 + enzyme	*C. beijerinckii* P260	0.44/0.31	26.27	Qureshi et al. (2010b)
Wheat bran	H_2SO_4	*C. beijerinckii* ATCC 55025	0.32/0.16	11.8	Liu et al. (2010)
Rice straw	H_2SO_4 + enzyme	*C. acetobutylicum* MTCC 481	1.04a/0.017 (Only butanol yield and productivity)	3.0	Ranjan and Moholkar (2011)

In addition, during the last years the use of ionic liquids (ILs) such as [BMIM]Cl, [BMIM][PF6], [BMIM][TFSI], etc. for dissolving lignocellulosics has been examined intensively (García et al. 2011; Holm et al. 2012).

Pretreatment of lignocellulosic biomass in a microwave oven is also a feasible method which uses the higher heating efficiency of a microwave oven and it is also easy to operate (Bjerre et al. 1996). Microwave treatment utilizes thermal and non-thermal effects generated by microwaves in aqueous environments (Sun and Cheng 2002).

ABE (Acetone, Butanol and Ethanol) Fermentation

ABE hetero-fermentation produces acetate, butyrate, ethanol, and acetone, as well as butanol. The metabolism of ABE producing clostridia can be divided into the following two distinct phases: acidogenesis (acid-production) and solventogenesis (solvent-production) during the exponential and stationary phases of growth (Jones and Woods 1986).

Biobutanol is a biofuel that can be produced from renewable resources using special strains of bacteria such as *Clostridium acetobutylicum* or *Clostridium beijerinckii* (Qureshi et al. 2007).

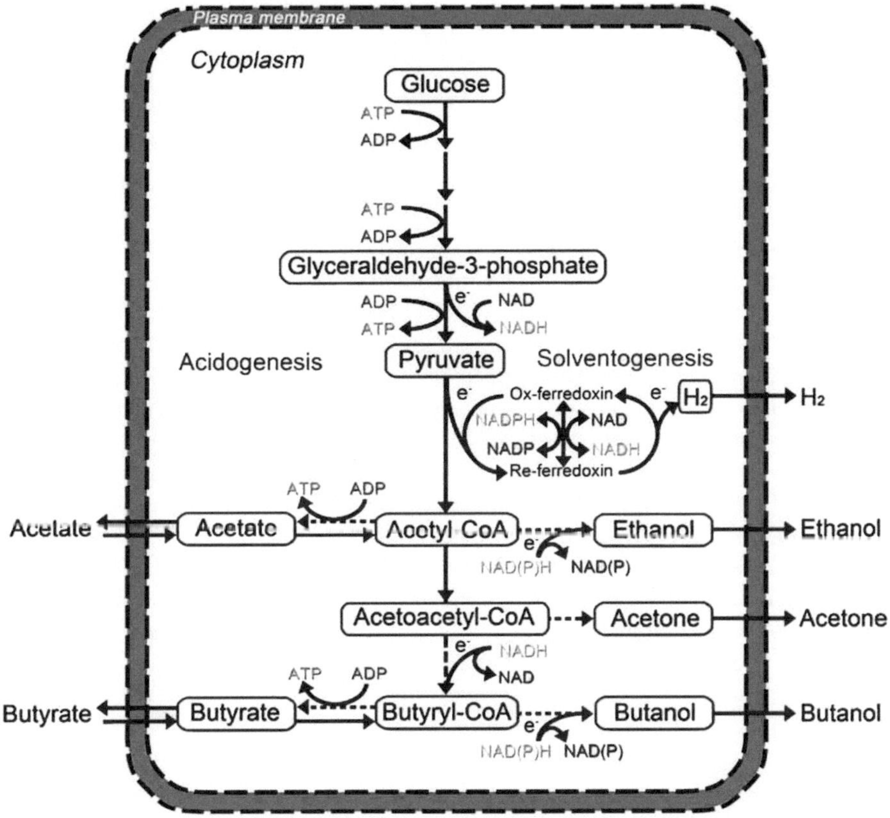

Fig. 10.7 Major redox reactions in acetone–butanol–ethanol fermentation by the bacterium *Clostridium* (Liu et al. 2013)

In a normal batch culture, solvent-producing *Clostridium* species produce hydrogen, carbon dioxide, acetate, and butyrate during the initial growth phase (acidogenic phase), which results in a decrease in the pH of the culture medium. As the culture enters the stationary growth phase, the metabolism of the cells undergoes a shift to solvent production (solventogenic phase). During the second phase of the fermentation the reassimilation of acids, this occurs concomitantly with the continued consumption of carbohydrate, normally results in an increase in the pH of the culture medium. The relationship between the breakpoint in the pH of the fermentation and the onset of solvent production, which occurs at the beginning of the second phase of the fermentation, was identified early on in the development of the industrial fermentation process (Jones and Woods 1986). Major redox reactions in acetone–butanol–ethanol fermentation by the bacterium *Clostridium* is shown in Fig. 10.7.

$$\underset{\text{Glucose}}{12\,C_6H_{12}O_6} \rightarrow \underset{\text{n}-\text{Butanol}}{6CH_3CH_2CH_2CH_2OH} + \underset{\text{Acetone}}{4CH_3COCH_3} + \underset{\text{Ethanol}}{2CH_3CH_2OH} + 18H_2 + 28CO_2 + 2H_2O$$

Acidogenesis

Bacteria grows exponentially in the first phase of fermentation (acidogenesis phase) along the formation of acids (mostly acetate and butyrate), leading to decrease of pH to 4.5 (Gheshlaghi et al. 2009). Two moles each of pyruvate, ATP and NADH are produced from one mole glucose consumed through the glycolytic pathway in the acidogenic phase. In this phase, glycolysis pathway is active to produce pyruvate consuming glucose, which is converted to Acetyl-CoA. Acetyl-CoA is the prime precursor for synthesis of acetate, butyrate, ethanol, butanol and acetone anaerobically. Acetate and butyrate are produced in acid producing phase through two analogous steps from acetyl-CoA and butyryl-CoA respectively (Kumar and Gayen 2011). An update review on key enzymes for butanol production is available (Gheshlaghi et al. 2009). When acids accumulate to sufficiently high levels, cells cannot maintain the pH gradient across membranes, and a dramatic decrease in growth occurs (Huang et al. 2010). Therefore, the shift to solvent production in Clostridia is an adaptive response to toxic effect of acidic metabolites through their re-assimilation and induced expression of genes for the stress response (Grimmler et al. 2011; Grupe and Gottschalk 1992).

Solventogenesis

As intracellular ATP is consumed by biosynthesis, solventogenesis is initiated to consume NAD(P)H accumulated during the acidogenesis (Grupe and Gottschalk 1992). The acetyl-CoA and butyryl-CoA are the key intermediates in synthesizing ethanol and butanol (Sillers et al. 2008). The reduction of acetyl-CoA and butyryl-CoA to acetylaldehyde and butyraldehyde is catalyzed by acetaldehyde dehydrogenase and butyraldehyde dehydrogenase, respectively, followed by the further reduction of acetylaldehyde and butyraldehyde to ethanol and butanol by ethanol dehydrogenase and butanol dehydrogenase (Gheshlaghi et al. 2009; Jones and Woods 1986). In both *C. acetobutylicum* and *C. beijerinckii*, the activity of butanol dehydrogenase was NADPH dependent rather than NADH dependent (Dürre 2008).

The use of excess carbon under nitrogen limitation is required to achieve high levels of solvent production (Madihah et al. 2001). Iron is one of the essential factors for the production of solvent (Kim et al. 1984). When *Clostridium acetobutylicum* was grown in batch culture under iron limitation (0.2 mg l^{-1}) at a pH of 4.8, glucose was fermented, to butanol as the major fermentation end product, and small quantities of acetic acid were produced. The final conversion yield of glucose into butanol could be increased from 20 to 30 % by iron limitation (Junelles et al. 1988). However, if the pH decreases below 4.5 before enough acids are formed, solventogenesis will be brief and unproductive. Increasing the buffering

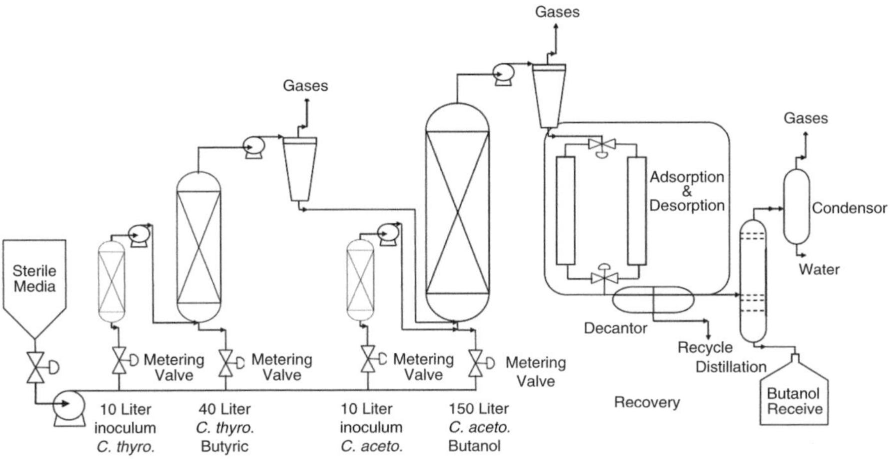

Fig. 10.8 Pilot plant for biobutanol production (*Source*: Butyl fuels, 2010 Korean Institute of Science and Technology)

capacity of the medium is a simple way to increase growth and carbohydrate utilization as well as butanol production (Bryant and Blaschek 1988).

The fermenter is inoculated with a 5 % inoculum from a 24 h culture. The batch fermentation period is usually 2–2.5 days. First rapid growth and production of acetic/butyric acids and carbon dioxide and hydrogen occur. The initial pH of the medium drops from 6.5 to nearly 4.5 during this phase. In a second phase, growth ceases and the organisms convert acetic and butyric acids to neutral acetone and butanol. The acidity of the medium decreases and gas production increases. At the end of the fermentation the pH is approximately 5 (Shuler and Kargi 2002). The final total concentration of solvents produced ranges from 12 to 20 g/L in batch fermentation, which can be separated from the fermentation broth by distillation. Classical fed-batch and continuous cultivation do not seem to be economically feasible, because of solvent toxicity and the biphasic nature of acetone–butanol fermentation, respectively. To overcome this problem, fed-batch culture has been coupled with an in situ recovery process (Ezeji et al. 2004a, b), and multistage continuous fermentation has been conducted (Godin and Engasser 1990). Pilot plant for biobutanol production has been shown in the Fig. 10.8.

Solvent Toxicity is one of the most critical problems in ABE fermentation, which ceases Clostridial cellular metabolism in the presence of 20 g/L or more solvents (Woods 1995). Moreira et al. (1981) and Jones et al. (1982) had attempted to elucidate the mechanism of butanol toxicity in *C. acetobutylicum*.

Downstream Processing

During the past two decades a significant amount of research has been performed on the use of alternative fermentation and product recovery techniques (e.g. adsorption,

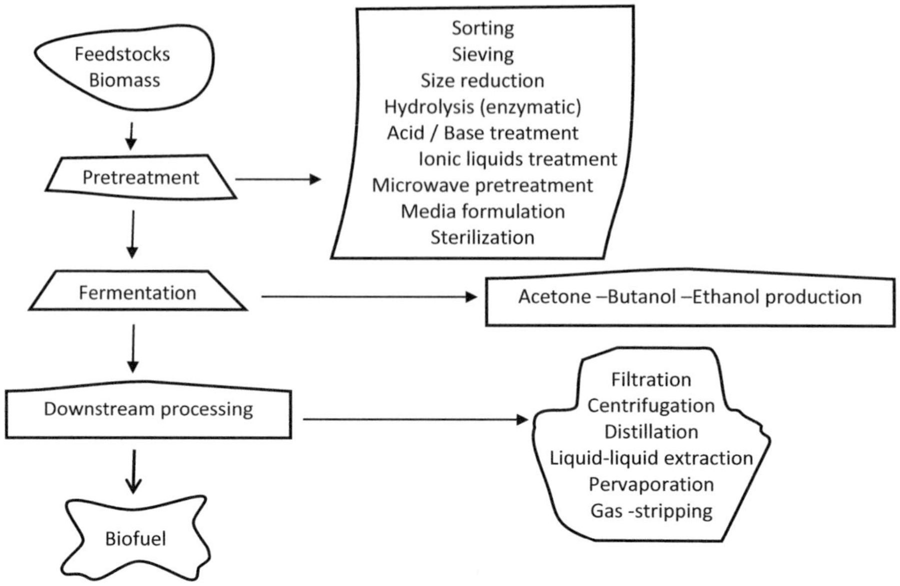

Fig. 10.9 Bioprocess stages and unit operations (Moo-Young and Chisti 1994)

gas stripping, ionic liquids, liquid–liquid extraction, pervaporation, aqueous two-phase separation, supercritical extraction, and perstraction, etc.) for biobutanol production (Ezeji et al. 2007a, b, c). The various bioprocess stages and unit operations along with pretreatment are shown in the Fig. 10.9.

The application of some of these techniques to the ABE fermentation process is described below.

1. Distillation: The cost of recovering butanol by distillation is high because its concentration in the fermentation broth is low due to product inhibition. In addition to the low product concentration, the boiling point of butanol is higher than that of water (118 °C). The usual concentration of total solvents in the fermentation broth is 18–33 g/L (using starch or glucose) of which butanol is only about 13–18 g/L. This makes butanol recovery by distillation energy intensive (Ezeji et al. 2004a, b).
2. Liquid–liquid extraction is another efficient technique to remove solvents from the fermentation broth. This approach takes advantage of the differences in the partition coefficient of the solvents. As butanol is more soluble in the extractant (organic phase) than in the fermentation broth (aqueous phase), it is selectively concentrated in the extractant. Common extractants employed include decanol and oleyl alcohol (Lee et al. 2008a, b).
3. Pervaporation is a membrane-based process that is used to remove solvents from the fermentation broth by using a selective membrane. The liquids or solvents diffuse through a solid membrane, leaving behind nutrients, sugar, and microbial

Table 10.6 Biobutanol recovery (Kumar and Gayen 2011; Heitmann et al. 2012)

Methods	Principle	Advantage	Disadvantage
Distillation	Boiling occurs when the vapor pressure of a liquid exceeds the ambient pressure	Traditional method	Expensive to perform
Gas stripping	Heating of effluent, purging with gas, condensation of solvent/water vapours	Simple to perform, low chance of clogging or fouling	Low selectivity, no complete removal of solvents, more energy required compared to membrane based processes
Liquid-liquid extraction	Contact of water—immiscible solvent with fermentation broth, recovery of acetone/butanol / isopropanol by distillation	High capacity, high selectivity, low chance of clogging or fouling	Expensive to perform, possible formation of emulsions
Pervaporation	Selective diffusion of solvents across a non-porous membrane, recovery of evaporated vapours by applying vacuum or sweep gas	High selectivity compared to membrane evaporation, simple to perform	Lower membrane flux compared to membrane evaporation, possible clogging and fouling

cells. The application of pervaporation to batch butanol fermentation has been described by several investigators (Ezeji et al. 2004a, b).

4. Gas stripping is a simple but efficient way to recover butanol from the fermentation broth. The fermentation gas is bubbled through the fermentation broth, and then passed through a condenser for solvent recovery. The stripped gas is then recycled back to the fermentor and the process continues until all the sugar in the fermentor is utilized (Lee et al. 2008a, b; Ezeji et al. 2003). Butanol recovery is based on the principle along with their advantage and disadvantage is shown in Table 10.6 and integrated systems for fermentation and in situ solvent recovery are shown in Fig. 10.10.

Biodiesel

The transesterification of vegetable oils (VOs) with short-chain alcohols is used to produce biodiesel or by the esterification of fatty acids. During the past few years biodiesel has attracted attention as an environmentally friendly and renewable fuel because of uncertainties concerning petroleum availability and recent increases in petroleum prices. Its chemical structure is that of fatty acid alkyl esters. The production of biodiesel by transesterification employing acid (H_2SO_4, HCl, etc.) or base catalyst (NaOH, KOH, $NaOCH_3$, etc.) has been industrially accepted for its

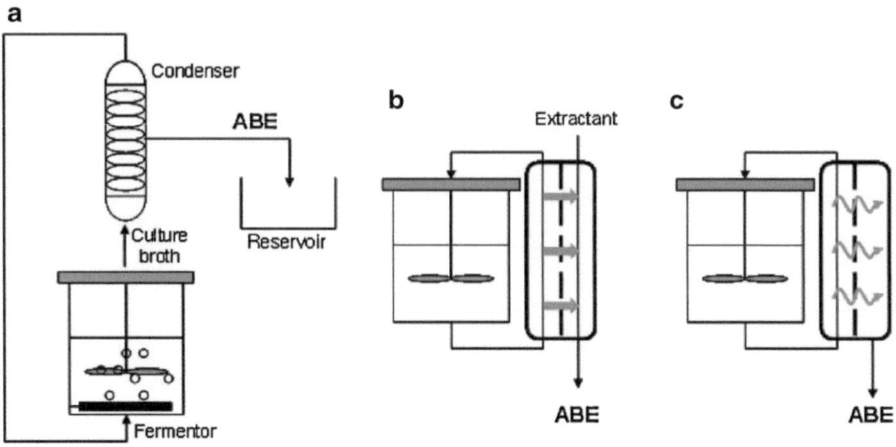

Fig. 10.10 Integrated systems for fermentation and in situ solvent recovery: fermentation coupled with (**a**) gas stripping; (**b**) liquid–liquid extraction (perstraction); (**c**) pervaporation (Lee et al. 2008a, b)

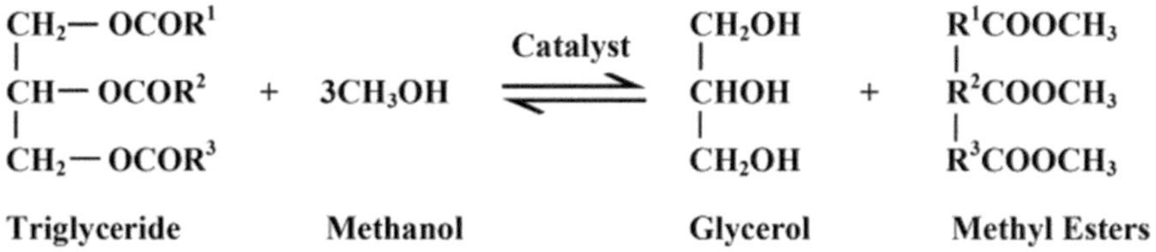

Fig. 10.11 General scheme for transesterification of triglycerides

high reaction and conversion rates. Biological catalyst (lipase) is also sufficient to carry out the reaction at lowest amount, since it is faster.

Biodiesel is defined as fatty acid methyl or ethyl esters (FAME) from vegetable oils or animal fats when they are used as fuel in diesel engines and heating systems (Marchetti and Errazu 2008). Nowadays, it is used as an alternative fuel due to depleting petroleum reserves (Sujan et al. 2009). Fatty acid methyl esters are products of the transesterification (also called methanolysis) of vegetable oils and fats with methanol in the presence of a suitable catalyst to form alkyl esters (biodiesel) and glycerin. The main chemical process to produce biodiesel is the alkaline transesterification with methanol and KOH, where the alcohol reacts in the presence of the catalyst to form alkyl esters (biodiesel) and glycerin (Neha et al. 2013). Beside these products, traces of unreacted glycerides (mono-, di- and tri-acylglycerides) can also be found. If methanol is used in this process it is called methanolysis. Methanolysis of triglyceride is represented in Fig. 10.11.

However; biodiesel has a higher cetane number, no aromatics, and contains 10 %–11 % oxygen by weight. These properties of biodiesel reduce the emissions of carbon monoxide (CO), hydrocarbons (HC), and particulate matter (PM) in the exhaust gas (Math et al. 2010). Preferred methods of production of biodiesel typically consist of reaction of oil sources with alcohols with aid of either acid or base.

Oil Sources and Methods of Biodiesel Production

Biodiesel is usually produced from food-grade vegetable oils using transestercation process. Therefore, it is said that the main obstacle for commercialization of biodiesel is its high cost. Waste cooking oils, restaurant greases, soapstocks and animal fats are potential feedstocks for biodiesel production to lower the cost of biodiesel (Canakci and Sanli 2008). The feed stock for biodiesel production is mainly soybean oil, sunflower oil, jatropha oil, canola oil, rapeseed oil, rubber seed oil and micro-algae etc. (Demirbas 2005). The cost of biodiesel is slightly higher than the petroleum based diesel mainly due to cost of edible oils which makes it more costly than the diesel fuel (Aworanti et al. 2013). Biodiesel obtained from vegetable oils has been considered a promising option but its higher viscosity is major problem which can reduce the fuel atomization (Pratas et al. 2011a, b). The petroleum based diesel fuel emits more carbon dioxide, greenhouse gases and hydrocarbon particulate matter, these are humiliation of the entire environment, regarding environmental concern, biodiesel has received more attention worldwide due to its properties such as clean, biodegradable, safe and eco-friendly (Atadashi et al. 2011). Presently many countries such as Germany, Australia and United State are already using biodiesel in replacement of traditional petroleum based diesel. In the United States, soybean oil is the most common biodiesel feedstock whereas rapeseed and palm oil are the most commonly used in Europe (Singh and Singh 2010). One of the most important disadvantages of using biodiesel is their cost. Biodiesel purification is carried out at the end of the reaction; the glycerin formed is separated from the methyl esters in a decantation funnel. The purification of methyl ester is done by washing with preheated distilled water (at 55 °C for 1 h). The pH of biodiesel should be approximately neutral (Hossain et al. 2010). The less dense phase, composed by esters, are removed and stored for further analysis and purification. The process flow diagram for biodiesel production is given in Figs. 10.12 and 10.13 represents enzymatic production process of biodiesel with immobilized lipase.

ASTM International, recognized from 2001 as the American Society for Testing and Materials, is worldwide standards organization that holds properties values of biodiesel (Table 10.7).

The combined vegetable oil and animal fat production in the United States totals about 35.3 billion pounds per year (Perlack et al. 2005). This production could

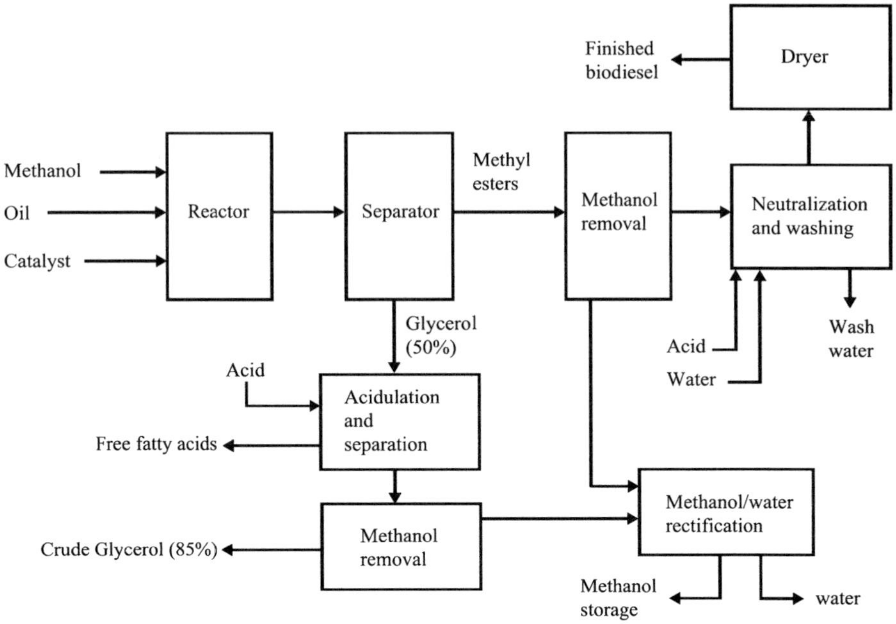

Fig. 10.12 Process flow schematic for biodiesel production

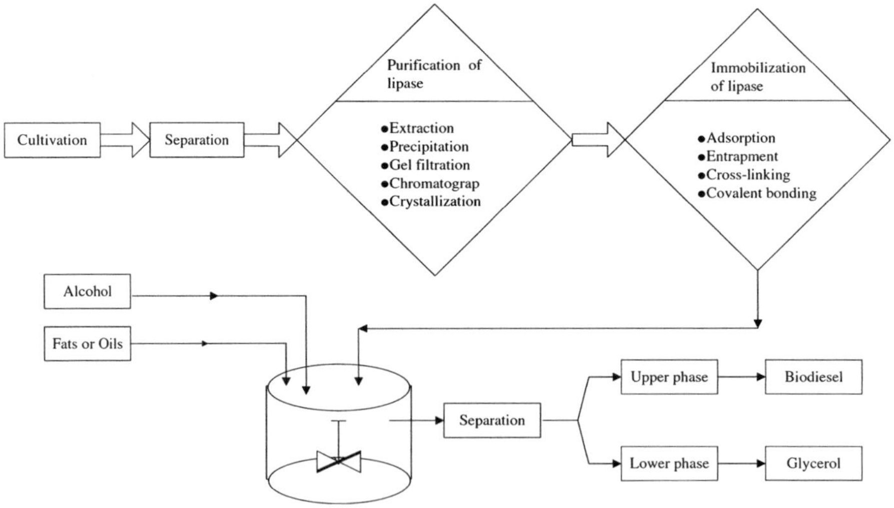

Fig. 10.13 Enzymatic production process of biodiesel with immobilized lipase (*Source*: Zhang et al. 2012)

Table 10.7 Fuel properties of biodiesel from soybean oil

S. no.	Property of biodiesel	ASTM D6751-06 standard	Soybean biodiesel
1	Density	860–890 (kg/m^3)	880 (kg/m^3)
2	Viscosity	–	90 (Redwood second)
3	Flash point	>130 (°C)	162 °C
4	Acid value	0.8 max (mg KOH/g)	0.20 (mg KOH/g)
5	Saponification value	169–280 (mg KOH/g)	137 (mg KOH/g)
6	Cloud point	−3 to 12 (°C)	10 °C

provide 4.6 billion gallons of biodiesel. Methyl ester is analysed by gas chromatography.

Production of Hydrogen

Hydrogen production plays a very important role in the development of hydrogen economy. Biomass and water can be used as renewable resources for hydrogen gas production. Biological production of hydrogen gas has significant advantages over chemical methods (Ni et al. 2006). The major biological processes utilized for hydrogen gas production are bio-photolysis of water by algae, dark and photo-fermentation of organic materials, usually carbohydrates by bacteria (Kapdan and Kargi 2006). Carbohydrate rich, nitrogen deficient solid wastes such as cellulose and starch containing agricultural and food industry wastes and some food industry wastewaters such as cheese whey, olive mill and bakers yeast industry wastewaters can be used for hydrogen production by using suitable bio-process technologies (Ghirardi et al. 2010). Conventional hydrogen gas production methods are steam reforming of methane (SRM), and other hydrocarbons (SRH), non-catalytic partial oxidation of fossil fuels (POX) and autothermal reforming which combines SRM and POX (Kapdan and Kargi 2006). Integrated biohydrogen system is shown in Fig. 10.14.

Hydrogen Production by Fermentation

Biological hydrogen production can be classified into five different groups: (1) direct biophotolysis, (2) indirect biophotolysis, (3) biological water–gas shift reaction, (4) photofermentation and (5) dark fermentation (Levin et al. 2004). Comparative biological hydrogen production process is given in Table 10.8.

Bio-hydrogen production from cellulose/starch containing agricultural wastes and food industry wastewaters is represented in Fig. 10.15.

All processes are controlled by the hydrogen-producing enzymes, such as hydrogenase and nitrogenase. The major components of nitrogenase are MoFe protein and Fe protein. Nitrogenase has the ability to use magnesium adenosine

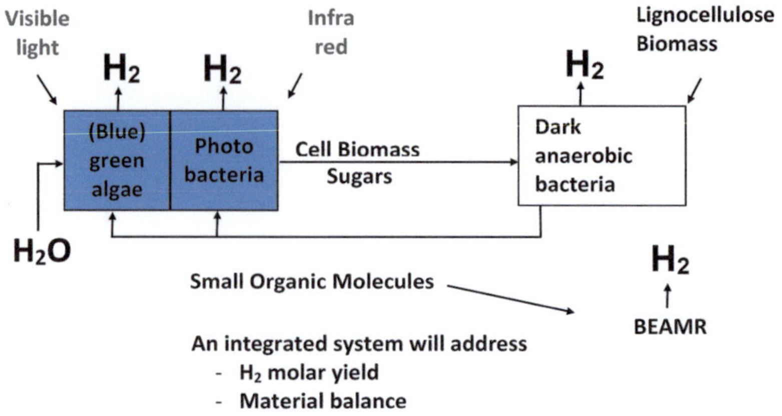

Fig. 10.14 Schematic representation of integrated biohydrogen system

Table 10.8 Major advantages and disadvantages of biological hydrogen production process

Process	Types of microorganism	Advantages	Drawback
Biophotolysis of water	Green algae or Cyanobacteria	Product: $H_2 + O_2$ Substate: $H_2O + CO_2$	Low H_2 production rate O_2 inhibition
Water-gas shift reaction	Photosynthesis or fermentative bacteria	Treatment of CO waste gas	Mass transfer limitation CO substrate limitation
Photodecomposition of organic compounds	Photosynthetic bacteria	High H_2 yield	Light requirement Low H_2 production rate
Fermentation of sugars (dark fermentation)	Fermentative bacteria	Fast rate Treatment of organic wastewater	Low H_2 yield By-product formation

triphosphate (MgATP) and electrons to reduce a variety of substrates (including protons) (Fig. 10.16). This chemical reaction yields hydrogen production by a nitrogenase-based system (Hallenbeck and Benemann 2002):

$$2e^- + 2H^+ + 4ATP \rightarrow H_2 + 4ADP + 4P_i$$

where ADP and P_i refer to adenosine diphosphate and inorganic phosphate, respectively.

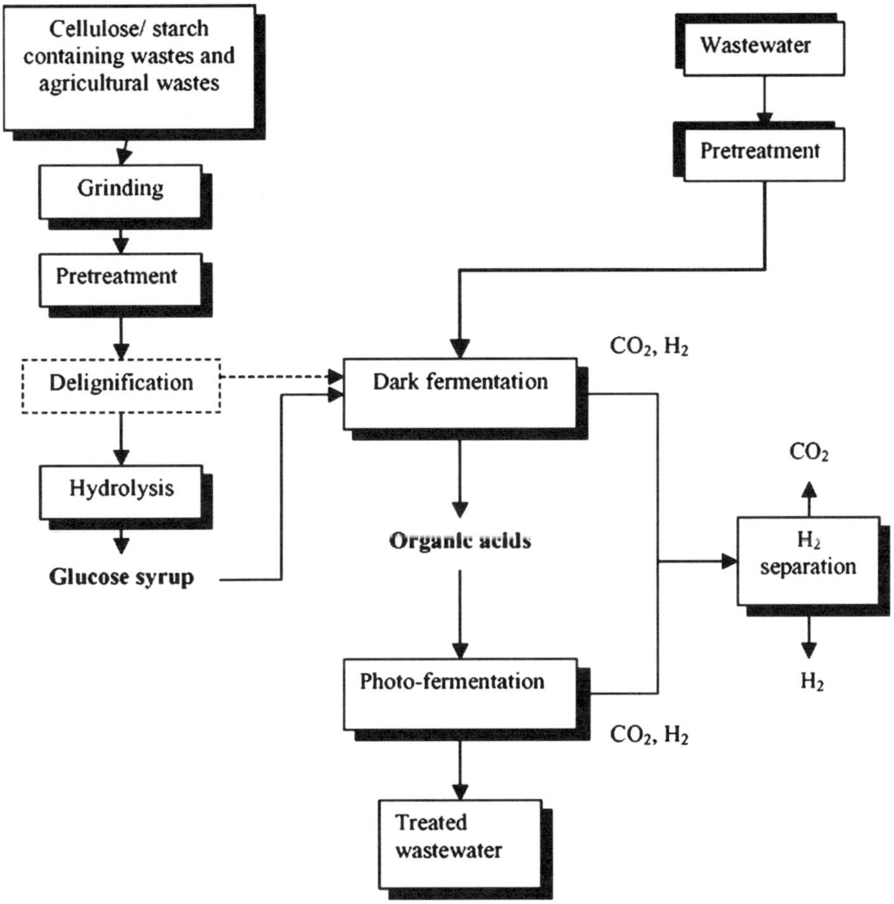

Fig. 10.15 A schematic diagram for bio-hydrogen production from cellulose/starch containing agricultural wastes and food industry wastewaters (*Source*: Kapdan dan Kargi 2006)

Hydrogen Detection and Quantification

The H_2 concentration in the gas phase is commonly measured with gas chromatography (GC) with thermal conductivity detector (TCD), using argon or nitrogen as the carrier gas. Typical GC operating conditions include temperature of 100 °C for the TCD and pressure of 151 kPa (22 psi) for the carrier gas. Silica columns (at 25 °C) or microcapillary columns may be used for separation (Drapcho et al. 2008). Fermentative hydrogen yield by different organisms is reported in Table 10.9.

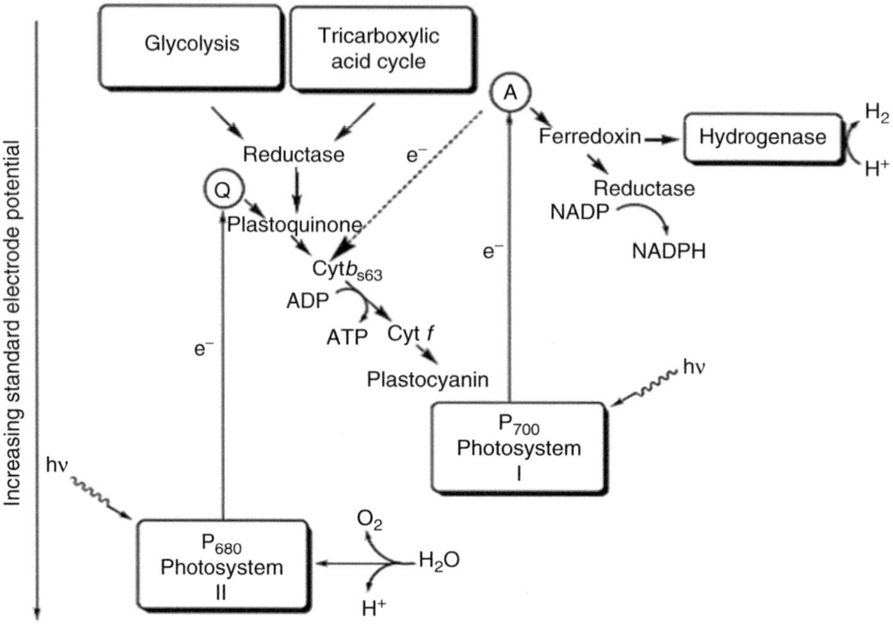

Fig. 10.16 Z scheme of photosynthetic electron flow in green plants and algae showing links to carbon metabolism and hydrogen production: *Q*, *A*, primary electron acceptors in Photosystems II and I, respectively; *dotted arrow* signifies cyclic electron flow (Melis and Happe 2001a, b)

Table 10.9 Reported fermentative hydrogen yield by different organisms

Microorganisms	Culture condition	Carbon source (g/L)	H_2 yield (mol/mol glucose)	Reference
E. coli	Batch	Glucose	0.75	Gottschalk (1986)
E. coli SR15	Batch	Glucose (10)	1.8	Yoshida et al. (2006)
Cl. butyricum strain SC-E1	Continuous	Glucose (10)	1.4	Kataoka et al. (1997)
Clostridium beijerinckii AM21B	Batch	Glucose (10)	1.3–2.0	Taguchi et al. (1992)
C. freundii	Batch	Glucose (7.7)	1.29	Kumar and Vatsala (1989)
Citrobacter intermedim	Continuous	Glucose (7.7)	0.27–1.14	Brosseau and Zajic (1982)
Citrobacter sp Y19	Batch	Glucose (5)	1.4	Oh et al. (2004)
Enterobacter cloacae IIT BT 08	Continuous	Glucose (5)	2.3	Nath and Das (2004)
Enterobacter aerogenes strain E.82005	Batch	Molasses (17 mM)	0.52–1.58	Tanisho et al. (1998)

Microbial Fuel Cells

One of the most exciting technologies for biological production of energy is the microbial fuel cell (MFC). A microbial fuel cell is a mimic of a biological system in which bacteria do not directly transfer their produced electrons to their characteristic electron acceptor. Instead, the transport process is subsequently conducted over an anode, a resistance or power user, and a cathode. Thus way, bacterial energy is directly converted to electrical energy.

Fuel Cell Design and Fabrication

Bacterial reactions can be carried out over several different temperature ranges depending on the tolerance of the bacteria, ranging from moderate or room-level temperatures (15–35 °C) to both high temperatures (50–60 °C) tolerated by thermophiles and low temperatures (<15 °C) where psychrophiles can grow. Virtually any biodegradable organic matter can be used in an MFC, including volatile acids, carbohydrates, proteins, alcohols, and even relatively recalcitrant materials like cellulose (Fig. 10.17) (Logan et al. 2006).

Hydrogen ions (protons, H^+) can accept reducing equivalents (conventionally represented as electrons, e^-) generated either photosynthetically or by the oxidation of organic and inorganic substrates inside microbial cells:

$$2e^- + 2H^+ \rightarrow H_2$$

The terminal electron donor (e.g., reduced ferredoxin) could donate electrons to the anode of a battery. Protons could then, in the presence of O_2, complete the electric circuit at the cathode by the reaction:

$$O_2 + 4e^- + 2H^+ \rightarrow 2H_2O$$

Thus, forming a highly environmentally friendly source of electric power (a battery), fueled by microbial metabolic activity (Logan and Regan 2006). That, in essence, is the definition of a microbial fuel cell (MFC).

The voltage across the external resistor or load in an MFC can be measured using a multimeter. Voltage measurements are converted to current values using Ohm's law:

$$V = IR \quad \text{where } V = \text{Voltage (V)}, \quad I = \text{Current (A)}, \quad R = \text{Resistance},$$

The power output from an MFC is calculated as

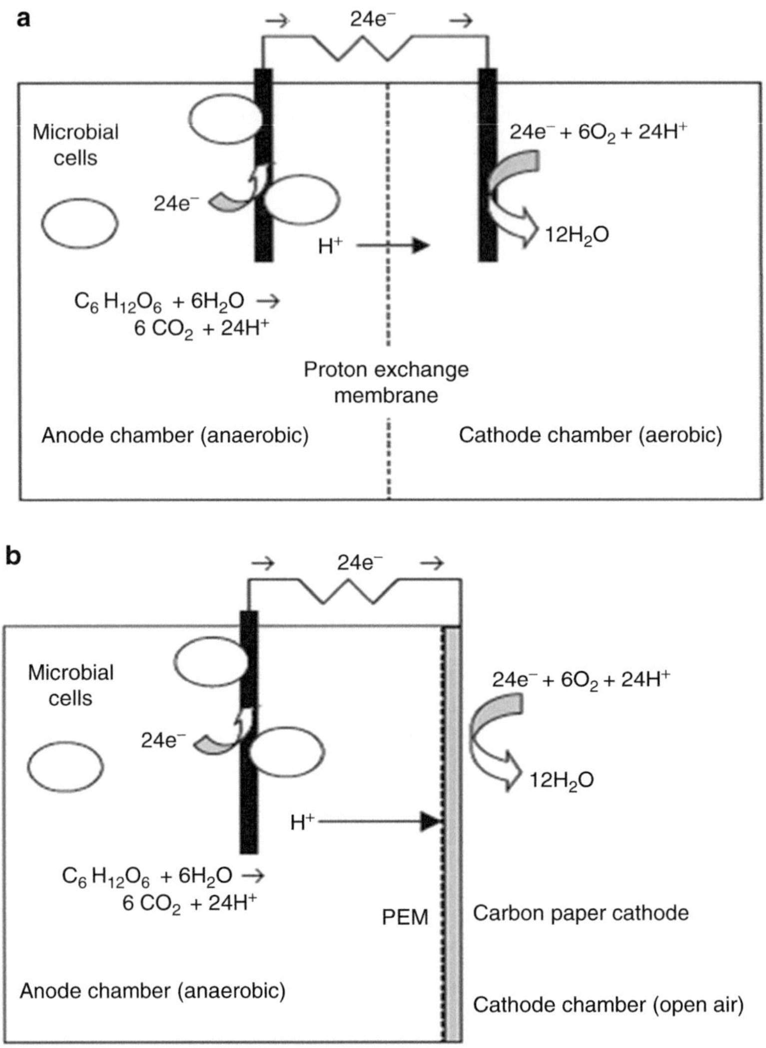

Fig. 10.17 (**a**) Diagram of two-chamber microbial fuel cell with aqueous cathode and anode chambers with solid graphite electrodes. (**b**) Diagram of single-chamber microbial fuel cell with aqueous anode chamber and air cathode chamber. The anode and cathode chambers are separated by a membrane. The bacteria grow on the anode, oxidizing organic matter and releasing electrons to the anode and protons to the solution. The cathode is sparged with air to provide dissolved oxygen for the reactions of electrons, protons and oxygen at the cathode, with a wire (and load) completing the circuit and producing power. The system is shown with a resistor used as the load for the power being generated, with the current determined based on measuring the voltage drop across the resistor using a multimeter hooked up to a data acquisition system (Drapcho et al. 2008)

$$P = IV \quad \text{where } P = \text{Power (W)}$$

Power density is used to relate power output to the anode surface area or anode chamber volume. Power density is calculated based on anode surface area as follows:

$$PD_A = \frac{IV}{A_A}$$

where PD_A = power density on area basis, W/m^2; and A_A = anode surface area, m^2.

Methane Production

During anaerobic digestion, organic matter is converted to methane and carbon dioxide by way of a series of interrelated microbial metabolisms, including hydrolysis, acetogenesis, and methanogenesis. The value and stability of the pH in an anaerobic reactor are extremely important because methanogenesis proceeds only at a high rate when the pH is maintained in the neutral range (van Haandel and Lettinga 1994; Zinder 1994). Biogas production from maize along the production process are shown in Fig. 10.18.

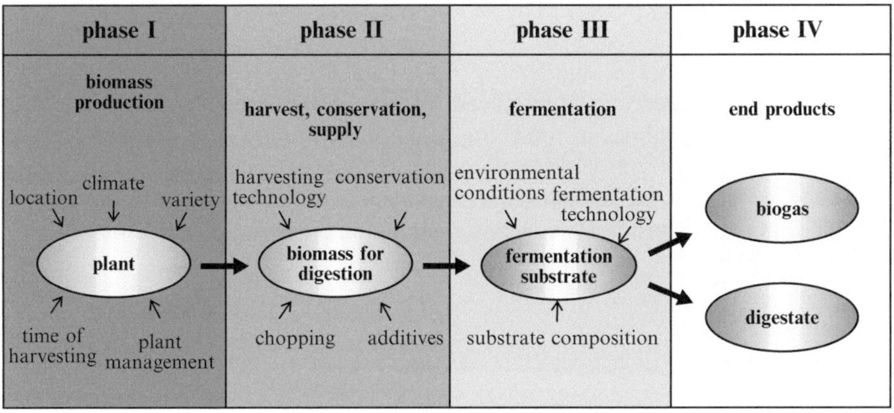

Fig. 10.18 Influences on biogas production from maize along the production process (*Source*: Amon et al. 2007)

Hydrolysis

Many of the potential biomass sources for methane production are high molecular weight, insoluble polymers such as polysaccharides, proteins, and fats that are too large to be transported across bacterial cell membranes. Polysaccharides such as cellulose and hemicellulose are hydrolyzed to glucose and xylose by cellulase and hemicellulase enzymes. Proteins and lipids are hydrolyzed to their constituent amino acids and long-chain fatty acids by proteases and lipases, respectively. The rate of hydrolysis is a function of several factors, such as pH, substrate composition, and particle size.

Fermentation (Acidogenesis)

The second phase of the overall process is fermentation that begins with the conversion of the sugar monomers to pyruvate ($C_3H_4O_3$), ATP, and the electron carrier molecule NADH by central metabolic pathways. The central metabolic pathways found within most bacteria are the Embden-Meyerhof pathway (glycolysis) and the pentose phosphate pathway. Next, these fermentative bacteria convert pyruvate and amino acids to a variety of short-chain organic acids—primarily acetate, propionate, butyrate, and succinate—and alcohols, CO_2, and H_2 through various fermentation pathways. Acid producing organisms are a mixture of facultative anaerobes, such as enteric bacteria and clostridial species which are called acid formers. The optimum temperature and pH values for this step are $T = 35$ °C and $pH = 4–6$ (Shuler and Kargi 2002).

Acetogenesis

The short-chain organic acids produced by fermentation and the fatty acids produced from the hydrolysis of lipids are fermented to acetic acid, H_2, and CO_2 by acetogenic bacteria. Syntrophic bacteria that oxidize organic acids to acetate, H_2, and CO_2 are reliant on the subsequent oxidation of H_2 by the next group, the methanogens, to lower the H_2 concentration and prevent end-product inhibition.

Methanogenesis

In the final phase, methane is produced through two distinct routes by two different microbial groups. Among methanogenic bacteria used for this purpose are

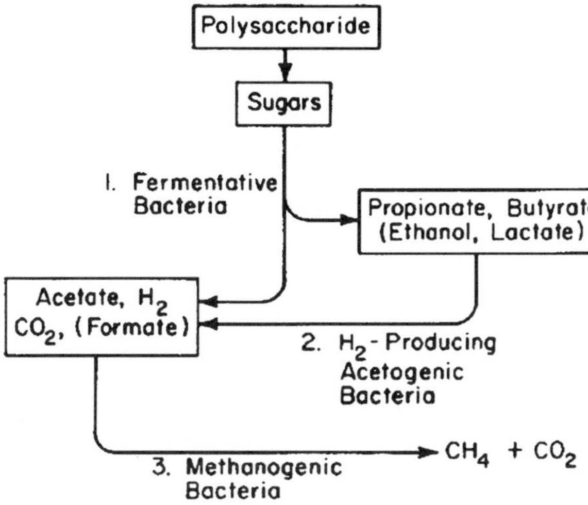

Fig. 10.19 A scheme showing the relationships of the three general metabolic groups of bacteria or stages of fermentation involved in methane production (*Source*: Bryant 1979)

Methanobacterium (nonspore-forming rods), *Methanobacillus* (spore forming rods), and *Methanococcus* and *Methanosarcina*.

The optimum temperature and pH range for methanogenic bacteria are T = 35–40 °C and pH = 7–7.8 (Shuler and Kargi 2002). The relationships of the three general metabolic groups of bacteria or stages of fermentation involved in methane production are shown in Fig. 10.19. One route is by the action of the lithotrophic H_2-oxidizing methanogens that use H_2 as electron donor and reduce CO_2 to produce methane. In the second route, the organotrophic acetoclastic methanogens ferment acetic acid to methane and carbon dioxide.

$$4H_2 + CO_2 \rightarrow CH_4 + 2H_2O$$

$$CH_3COOH \rightarrow CH_4 + CO_2$$

Conclusions and Future Prospects

Bio-fuels are being promoted in the transportation sector. More recently, ethanol produced from sugar and starch-based feedstocks has become another important biofuel. Other biofuels such as lignocellulosic ethanol, biodiesel, biohydrogen, and bioelectricity have been the focus of vigorous research, and the technologies for their production are being developed, although most of these are not quite ready for commercialization. Currently, a large amount of studies regarding the utilization of lignocellulosic biomass as a feedstock for producing fuel ethanol is being carried out worldwide (Balat and Balat 2009). Bioconversion of lignocellulosic biomass to ethanol is significantly hindered by the structural and chemical complexity of biomass, which makes these materials a challenge to be used as feedstocks for cellulosic ethanol production (Zheng et al. 2009). But in addition to that, the technology of recombinant DNA will provide important advances for the development of fuel ethanol industry. The development of genetically modified microorganisms capable of converting starch or biomass directly into ethanol and with a proven stability under industrial conditions will allow the implementation of the consolidated bioprocessing of the feedstocks (Cardona and Sánchez 2007). The willingness of mankind to pay high prices for energy in the future is a great uncertainty. Hence, the biobutanol production can aid in extending the life of petroleum oil reserves and diminish environmental concerns (Kumar and Gayen 2011). The acetone-butanol fermentation was the first large-scale fermentation process developed which is sensitive to contamination. Therefore this fermentation contributed much to the knowledge of how to run sterile processes on an industrial scale. As an alternative plan, many research projects have been initiated for the efficient use of lignocellulosic biomass, algal biomass, etc. which should be accomplished in the future (Mosier et al. 2005; Kumar et al. 2009). Sucrose from sugar cane is also an excellent substrate in certain regions of the world. An optimal Bioprocess for butanol production can be developed by integrating the fermentation and downstream processes with strain development (Lee et al. 2008a, b). Researchers are also attempting an aerobic production of biobutanol using genetically engineered organisms like *E. coli, S. cerevisiae* etc. (Atsumi and Liao 2008; Steen et al. 2008). It will be the milestone to attract the attention of government, commercial, and research organizations for further support in implementing the innovative fermentation and extraction technology. The study on biodiesel synthesis showed that the quantity of catalyst, the temperature and reaction time are the main factors affecting the production of methyl esters both for short chain methyl esters and long chain methyl esters (Riadi et al. 2014). Biological methods offer distinct advantages for hydrogen production such as operation under mild conditions and specific conversions. However, raw material cost is one of the major limitations for bio-hydrogen production. Utilization of some carbohydrate rich, starch or cellulose containing solid wastes and/or some food industry wastewaters is an attractive approach for bio-hydrogen production (Kapdan and Kargi 2006). Microbial production of electricity may become an important form of bioenergy in future because MFCs offer the possibility of extracting electric current from a wide range of soluble or dissolved complex organic wastes and renewable biomass. A large number of substrates have been explored as feed. The major substrates that have been tried include various kinds of artificial and real wastewaters and lignocellulosic biomass (Pant et al. 2010). Currently, biogas production from energy crops is mainly based on the anaerobic digestion of maize. Among them, methane produced by anaerobic digestion has been used by the human race for hundreds, if not thousands, of years. In the near future, biogas production from energy crops will increase and it has to be considered that energy crops are grown in versatile, sustainable crop rotations (Strauß et al. 2012).

References

Alonso, D. M., Bond, J. Q., & Dumesic, J. A. (2010). Catalytic conversion of biomass to biofuels. *Green Chemistry, 12*(9), 1493–1513.

Al-Shorgani, N. K. N., Kalil, M. S., & Yusoff, W. M. W. (2012). Biobutanol production from rice bran and de-oiled rice bran by *Clostridium saccharoperbutylacetonicum* N1-4. *Bioprocess and Biosystems Engineering, 35*(5), 817–826.

Amon, T., Amon, B., Kryvoruchko, V., Zollitsch, W., Mayer, K., & Gruber, L. (2007). Biogas production from maize and dairy cattle manure—Influence of biomass composition on the methane yield. *Agriculture, Ecosystems & Environment, 118*(1), 173–182.

Antoni, D., Zverlov, V. V., & Schwarz, W. H. (2007). Biofuels from microbes. *Applied Microbiology and Biotechnology, 77*(1), 23–35.

Atadashi, I. M., Aroua, M. K., & Aziz, A. A. (2011). Biodiesel separation and purification: A review. *Renewable Energy, 36*(2), 437–443.

Atsumi, S., & Liao, J. C. (2008). Metabolic engineering for advanced biofuels production from *Escherichia coli. Current Opinion in Biotechnology, 19*(5), 414–419.

Aworanti, O. A., Agarry, S. E., & Ajani, A. O. (2013). Statistical optimization of process variables for biodiesel production from waste cooking oil using heterogeneous base catalyst. *British Biotechnology Journal, 3*(2), 116–132.

Balat, M., & Balat, H. (2009). Recent trends in global production and utilization of bio-ethanol fuel. *Applied Energy, 86*(11), 2273–2282.

Balat, H., & Kırtay, E. (2010). Hydrogen from biomass – Present scenario and future prospects. *International Journal of Hydrogen Energy, 35*(14), 7416–7426.

Baskar, C., Baskar, S., & Dhillon, R. S. (2012a). *Biomass conversion: The interface of biotechnology, chemistry and materials sceince*. London: Springer. ISBN 978-3-642-28417-5.

Baskar, C., Baskar, S., & Dhillon, R. S. (2012b). *Biomass conversion: The interface of biotechnology, chemistry and materials science*. New York, NY: Springer.

Bastos, V. D. (2007). Etanol, alcoolquímica e biorrefinarias. *BNDES Setorial, 25*, 5–38.

BEguin, P., & Aubert, J. P. (1994). The biological degradation of cellulose. *FEMS Microbiology Reviews, 13*(1), 25–58.

Berchmans, H. J., & Hirata, S. (2008). Biodiesel production from crude *Jatropha curcus* L. Seed oil with a high content of free fatty acids. *Bioresource Technology, 99*, 1716–1721.

Bjerre, A. B., Olesen, A. B., & Fernqvist, T. (1996). Pretreatment of wheat straw using combined wet oxidation and alkaline hydrolysis resulting in convertible cellulose and hemicellulose. *Biotechnology and Bioengineering, 49*(5), 568–577.

Bramucci, M. G., Flint, D., Miller, E. S., Nagarajan, V., Sedkova, N., Singh, M., et al. (2008). *US20080274526*.

Brosseau, J. D., & Zajic, J. E. (1982). Hydrogen-gas production with Citrobacter intermedim and Clostridium pasteurianum. *Journal of Chemical Technology and Biotechnology, 32*(3), 496–502.

Bryant, M. P. (1979). Microbial methane production—Theoretical aspects. *Journal of Animal Science, 48*(1), 193–201.

Bryant, D. L., & Blaschek, H. P. (1988). Buffering as a means for increasing growth and butanol production by *Clostridium acetobutylicum. Journal of Industrial Microbiology, 3*(1), 49–55.

Campos, E. J., Qureshi, N., & Blaschek, H. P. (2002). Production of acetone butanol ethanol from degermed corn using *Clostridium beijerinckii* BA101. *Applied Biochemistry and Biotechnology, 98–100*, 553–561.

Canakci, M., & Sanli, H. (2008). Biodiesel production from various feedstocks and their effects on the fuel properties. *Journal of Industrial Microbiology & Biotechnology, 35*(5), 431–441.

Cardona, C. A., & Sánchez, Ó. J. (2007). Fuel ethanol production: Process design trends and integration opportunities. *Bioresource Technology, 98*(12), 2415–2457.

Claassen, P. A. M., Budde, M. A. W., & López-Contreras, A. M. (2000). Acetone, butanol and ethanol production from domestic organic waste by solventogenic clostridia. *Journal of Molecular Microbiology and Biotechnology, 2*, 39–44.

Claassen, P. A. M., Van Lier, J. B., Contreras, A. L., Van Niel, E. W. J., Sijtsma, L., Stams, A. J. M., et al. (1999). Utilisation of biomass for the supply of energy carriers. *Applied Microbiology and Biotechnology, 52*(6), 741–755.

Decker, S. R., Sheehan, J., Dayton, D. C., Bozell, J. J., Adney, W. S., Hames, B., et al. (2007). Biomass conversion. In J. A. Kent (Ed.), *Kent and Riegel's handbook of industrial chemistry and biotechnology* (pp. 1449–1548). New York, NY: Springer.

Demirbas, A. (2005). Biodiesel production from vegetable oils via catalytic and non-catalytic supercritical methanol transesterification methods. *Progress in Energy and Combustion Science, 31*(5), 466–487.

Demirbas, A. (2009a). *Biohydrogen for future engine fuel demands*. London: Springer.

Demirbas, A. (2009b). *Biorenewable liquid fuels. Biofuels: Securing the planet's future energy needs* (pp. 103–230). New York, NY: Springer.

Doran, P. M. (1995). *Bioprocess engineering principles*. Waltham, MA: Academic.

Drapcho, C. M., Nhuan, N. P., & Walker, T. H. (2008). *Biofuels engineering process technology* (pp. 334–335). New York, NY: McGraw-Hill.

Duff, S. J., & Murray, W. D. (1996). Bioconversion of forest products industry waste cellulosics to fuel ethanol: A review. *Bioresource Technology, 55*(1), 1–33.

Dürre, P. (2008). Fermentative butanol production. *Annals of the New York Academy of Sciences, 1125*(1), 353–362.

Dürre, P. (2011). Fermentative production of butanol – The academic perspective. *Current Opinion in Biotechnology, 22*(3), 331–336.

Esteghlalian, A., Hashimoto, A. G., Fenske, J. J., & Penner, M. H. (1997). Modeling and optimization of the dilute-sulfuric-acid pretreatment of corn stover, poplar and switchgrass. *Bioresource Technology, 59*(2), 129–136.

Ezeji, T. C., Qureshi, N., & Blaschek, H. P. (2003). Production of acetone, butanol and ethanol by *Clostridium beijerinckii* BA101 and in situ recovery by gas stripping. *World Journal of Microbiology and Biotechnology, 19*(6), 595–603.

Ezeji, T. C., Qureshi, N., & Blaschek, H. P. (2004a). Acetone butanol ethanol (ABE) production from concentrated substrate: Reduction in substrate inhibition by fed-batch technique and product inhibition by gas stripping. *Applied Microbiology and Biotechnology, 63*(6), 653–658.

Ezeji, T. C., Qureshi, N., & Blaschek, H. P. (2004b). Butanol fermentation research: Upstream and downstream manipulations. *The Chemical Record, 4*(5), 305–314.

Ezeji, T., Qureshi, N., & Blaschek, H. P. (2007a). Butanol production from agricultural residues: Impact of degradation products on *Clostridium beijerinckii* growth and butanol fermentation. *Biotechnology and Bioengineering, 97*(6), 1460–1469.

Ezeji, T., Qureshi, N., & Blaschek, H. P. (2007b). Production of acetone-butanol-ethanol (ABE) in a continuous flow bioreactor using degermed corn and *Clostridium beijerinckii*. *Process Biochemistry, 42*, 34–39.

Ezeji, T. C., Qureshi, N., & Blaschek, H. P. (2007c). Bioproduction of butanol from biomass: From genes to bioreactors. *Current Opinion in Biotechnology, 18*(3), 220–227.

Fan, L. T., Gharpuray, M. M., & Lee, Y. H. (1987). Enzymatic hydrolysis. In L. T. Fan, M. M. Gharpuray, & Y. H. Lee (Eds.), *Cellulose hydrolysis* (pp. 21–119). Berlin: Springer.

Festel, G. W. (2008). Biofuels – Economic aspects. *Chemical Engineering & Technology, 31*(5), 715–720.

Figueira, M. M., Volesky, B., Ciminelli, V. S. T., & Roddick, F. A. (2000). Biosorption of metals in brown seaweed biomass. *Water Research, 34*(1), 196–204.

Fischer, C. R., Klein-Marcuschamer, D., & Stephanopoulos, G. (2008). Selection and optimization of microbial hosts for biofuels production. *Metabolic Engineering, 10*(6), 295–304.

Galbe, M., & Zacchi, G. (2002). A review of the production of ethanol from softwood. *Applied Microbiology and Biotechnology, 59*(6), 618–628.

García, V., Päkkilä, J., Ojamo, H., Muurinen, E., & Keiski, R. L. (2011). Challenges in biobutanol production: How to improve the efficiency? *Renewable and Sustainable Energy Reviews, 15*(2), 964–980.

GHeshlaghi, R. E. Z. A., Scharer, J. M., Moo-Young, M., & Chou, C. P. (2009). Metabolic pathways of clostridia for producing butanol. *Biotechnology Advances, 27*(6), 764–781.

Ghirardi, M. L., Kosourov, S., Maness, P., Smolinski, S., & Seibert, M. (2010). *Algal hydrogen production. Encyclopedia of Industrial Biotechnology*. New York: Wiley.

Godin, C., & Engasser, J. M. (1990). Two-stage continuous fermentation of *Clostridium acetobutylicum*: Effects of pH and dilution rate. *Applied Microbiology and Biotechnology, 33*(3), 269–273.

Gottschalk, G. (1986). Bacterial fermentations. In G. Gottschalk (Ed.), *Bacterial metabolism* (pp. 208–282). New York, NY: Springer.

Green, E. M. (2011). Fermentative production of butanol – The industrial perspective. *Current Opinion in Biotechnology, 22*(3), 337–343.

Grimmler, C., Janssen, H., Krauβe, D., Fischer, R. J., Bahl, H., Dürre, P., et al. (2011). Genome-wide gene expression analysis of the switch between acidogenesis and solventogenesis in continuous cultures of Clostridium acetobutylicum. *Journal of Molecular Microbiology and Biotechnology, 20*(1), 1.

Grobben, N. C., Eggink, G., Cuperus, F. P., & Huizing, H. J. (1993). Production of acetone, butanol and ethanol (ABE) from potato wastes; fermentation with integrated membrane extraction. *Applied Microbiology and Biotechnology, 39*, 494–498.

Grupe, H., & Gottschalk, G. (1992). Physiological events in *Clostridium acetobutylicum* during the shift from acidogenesis to solventogenesis in continuous culture and presentation of a model for shift induction. *Applied and Environmental Microbiologyl, 58*(12), 3896 3902.

Gutierrez, N. A., Maddox, I. S., Schuster, K. C., Swoboda, H., & Gapes, J. R. (1998). Strain comparison and medium preparation for the acetone-butanol-ethanol (ABE) fermentation process using a substrate of potato. *Bioresource Technology, 66*, 263–265.

Hahn-Hägerdal, B., Karhumaa, K., Fonseca, C., Spencer-Martins, I., & Gorwa-Grauslund, M. F. (2007). Towards industrial pentose-fermenting yeast strains. *Applied Microbiology and Biotechnology, 74*(5), 937–953.

Hallenbeck, P. C., & Benemann, J. R. (2002). Biological hydrogen production; fundamentals and limiting processes. *International Journal of Hydrogen Energy, 27*(11), 1185–1193.

Heitmann, S., Krings, J., Kreis, P., Lennert, A., Pitner, W. R., Górak, A., et al. (2012). Recovery of n-butanol using ionic liquid-based pervaporation membranes. *Separation and Purification Technology, 97*, 108–114.

Holm, J., Lassi, U., Romar, H., Lahti, R., Kärkkäinen, J., & Lajunen, M. (2012). Pretreatment of fibre sludge in ionic liquids followed by enzyme and acid catalysed hydrolysis. *Catalysis Today, 196*(1), 11–15.

Hossain, A. B. M. S., Nasrulhaq, B. A., Salleh, A., & Chandran, S. (2010). Biodiesel production from waste soybean oil biomass as renewable energy and environmental recycled process. *African Journal of Biotechnology, 9*(27), 4233–4240.

Huang, H., Liu, H., & Gan, Y. R. (2010). Genetic modification of critical enzymes and involved genes in butanol biosynthesis from biomass. *Biotechnology Advances, 28*(5), 651–657.

Huesemann, M. H., Kuo, L. J., Urquhart, L., Gill, G. A., & Roesijadi, G. (2012). Acetone-butanol fermentation of marine macroalgae. *Bioresource Technology, 108*, 305–309.

Jesse, T. W., Ezeji, T. C., Qureshi, N., & Blaschek, H. P. (2002). Production of butanol from starch-based waste packing peanuts and agricultural waste. *Journal of Industrial Microbiology and Biotechnology, 29*(3), 117–123.

John, R. P., Anisha, G. S., Nampoothiri, K. M., & Pandey, A. (2011). Micro and macroalgal biomass: A renewable source for bioethanol. *Bioresource Technology, 102*(1), 186–193.

Jones, D. T., Van der Westhuizen, A., Long, S., Allcock, E. R., Reid, S. J., & Woods, D. R. (1982). Solvent production and morphological changes in Clostridium acetobutylicum. *Applied and Environmental Microbiology, 43*(6), 1434–1439.

Jones, D. T., & Woods, D. R. (1986). Acetone-butanol fermentation revisited. *Microbiological Reviews, 50*(4), 484.

Junelles, A. M., Janati-Idrissi, R., Petitdemange, H., & Gay, R. (1988). Iron effect on acetone-butanol fermentation. *Current Microbiology, 17*(5), 299–303.

Jurgens, G., Survase, S., Berezina, O., et al. (2012). Butanol production from lignocellulosics. *Biotechnology Letters, 34*(8), 1415–1434.

Kapdan, I. K., & Kargi, F. (2006). Bio-hydrogen production from waste materials. *Enzyme and Microbial Technology, 38*(5), 569–582.

Kataoka, N., Miya, A., & Kiriyama, K. (1997). Studies on hydrogen production by continuous culture system of hydrogen-producing anaerobic bacteria. *Water Science and Technology, 36*(6), 41–47.

Keating, J. D., Panganiban, C., & Mansfield, S. D. (2006). Tolerance and adaptation of ethanologenic yeasts to lignocellulosic inhibitory compounds. *Biotechnology and Bioengineering, 93*(6), 1196–1206.

Kim, B. H., Bellows, P., Datta, R., & Zeikus, J. G. (1984). Control of carbon and electron flow in Clostridium acetobutylicum fermentations: Utilization of carbon monoxide to inhibit hydrogen production and to enhance butanol yields. *Applied and Environmental Microbiology, 48*(4), 764–770.

Kobayashi, G., Eto, K., Tashiro, Y., Okubo, K., Sonomoto, K., & Ishizaki, A. (2005). Utilization of excess sludge by acetone–butanol–ethanol fermentation employing Clostridium saccharoperbutylacetonicum N1-4 (ATCC 13564). *Journal of Bioscience and Bioengineering, 99*(5), 517–519.

Konieczny, A., Mondal, K., Wiltowski, T., & Dydo, P. (2008). Catalyst development for thermocatalytic decomposition of methane to hydrogen. *International Journal of Hydrogen Energy, 33*, 264e72.

Koukiekolo, R., Cho, H. Y., Kosugi, A., Inui, M., Yukawa, H., & Doi, R. H. (2005). Degradation of corn fiber by Clostridium cellulovorans cellulases and hemicellulases and contribution of scaffolding protein CbpA. *Applied and Environmental Microbiology, 71*(7), 3504–3511.

Kraus, T. (2007). Hydrogen fuel – An economically viable future for the transportation industry. *Duke Journal of Economics, 19*, 39.

Kretzers, I. (2012). Conference report: Euroscicon conference on innovations in renewable energies. *Biofuels, 3*(4), 375–376.

Kumar, P., Barrett, D. M., Delwiche, M. J., & Stroeve, P. (2009). Methods for pretreatment of lignocellulosic biomass for efficient hydrolysis and biofuel production. *Industrial & Engineering Chemistry Research, 48*(8), 3713–3729.

Kumar, M., & Gayen, K. (2011). Developments in biobutanol production: New insights. *Applied Energy, 88*(6), 1999–2012.

Kumar, M., Goyal, Y., Sarkar, A., & Gayen, K. (2012). Comparative economic assessment of ABE fermentation based on cellulosic and non-cellulosic feedstocks. *Applied Energy, 93*, 193–204.

Kumar, D., & Murthy, G. S. (2013). Stochastic molecular model of enzymatic hydrolysis of cellulose for ethanol production. *Biotechnology for Biofuels, 6*(1), 1–20.

Kumar, G. R., & Vatsala, T. M. (1989). Hydrogen production from glucose by Citrobacter freundii. *Indian Journal of Experimental Biology, 27*(9), 824–825.

Lee, S. Y., Park, J. H., Jang, S. H., Nielsen, L. K., Kim, J., & Jung, K. S. (2008). Fermentative butanol production by Clostridia. *Biotechnology and Bioengineering, 101*(2), 209–228.

Lee, S. Y., Park, J. H., & Papoutsakis, E. (2008). *WIPO Patent 2008072921.*

Lee, J., Seo, E., Kweon, D. H., Park, K., & Jin, Y. S. (2009). Fermentation of rice bran and defatted rice bran for butanol production using *Clostridium beijerinckii* NCIMB 8052. *Journal of Microbiology and Biotechnology, 19*, 482–490.

Levin, D. B., Pitt, L., & Love, M. (2004). Biohydrogen production: Prospects and limitations to practical application. *International Journal of Hydrogen Energy, 29*(2), 173–185.

Li, X., Weng, J. K., & Chapple, C. (2008). Improvement of biomass through lignin modification. *The Plant Journal, 54*(4), 569–581.

Lin, Y., & Tanaka, S. (2006). Ethanol fermentation from biomass resources: Current state and prospects. *Applied Microbiology and Biotechnology, 69*(6), 627–642.

Liu, C. G., Xue, C., Lin, Y. H., & Bai, F. W. (2013). Redox potential control and applications in microaerobic and anaerobic fermentations. *Biotechnology Advances, 31*(2), 257–265.

Liu, Z., Ying, Y., Li, F., Ma, C., & Xu, P. (2010). Butanol production by Clostridium beijerinckii ATCC 55025 from wheat bran. *Journal of Industrial Microbiology & Biotechnology, 37*(5), 495–501.

Logan, B. E., Hamelers, B., Rozendal, R., Schröder, U., Keller, J., Freguia, S., et al. (2006). Microbial fuel cells: Methodology and technology. *Environmental Science & Technology, 40* (17), 5181–5192.

Logan, B. E., & Regan, J. M. (2006). Electricity-producing bacterial communities in microbial fuel cells. *Trends in Microbiology, 14*(12), 512–518.

López-Contreras, A. M., Claassen, P. A. M., Mooibrock, H., & De Vos, W. M. (2000). Utilisation of saccharides in extruded domestic organic waste by *Clostridium acetobutylicum* ATCC 824 for production of acetone, butanol and ethanol. *Applied Microbiology and Biotechnology, 54*, 162–167.

Lynd, L. R. (1996). Overview and evaluation of fuel ethanol from cellulosic biomass: Technology, economics, the environment, and policy. *Annual Review of Energy and the Environment, 21* (1), 403–465.

Maddox, I. S., & Murray, A. E. (1983). Production of n-butanol by fermentation of wood hydrolysate. *Biotechnology Letters, 5*, 175–178.

Madihah, M. S., Ariff, A. B., Sahaid, K. M., Suraini, A. A., & Karim, M. I. A. (2001). Direct fermentation of gelatinized sago starch to acetone–butanol–ethanol by Clostridium acetobutylicum. *World Journal of Microbiology and Biotechnology, 17*(6), 567–576.

Marchal, R., Blanchet, D., & Vandecasteele, J. P. (1985). Industrial optimization of acetone-butanol fermentation: A study of the utilization of Jerusalem artichokes. *Applied Microbiology and Biotechnology, 23*, 92–98.

Marchal, R., Rebeller, M., & Vandecasteele, J. P. (1984). Direct bioconversion of alkali-pretreated straw using simultaneous enzymatic hydrolysis and acetone butanol production. *Biotechnology Letters, 6*, 523–528.

Marchetti, J. M., & Errazu, A. F. (2008). Comparison of different heterogeneous catalysts and different alcohols for the esterification reaction of oleic acid. *Fuel, 87*(15), 3477–3480.

Math, M. C., Kumar, S. P., & Chetty, S. V. (2010). Technologies for biodiesel production from used cooking oil. A review. *Energy for Sustainable Development, 14*(4), 339–345.

McMillan, J. D. (1994). Pretreatment of lignocellulosic biomass. In M. E. Himmel, J. O. Baker, & R. P. Overend (Eds.), *Enzymatic conversion of biomass for fuels production, ACS symposium series (USA)*. Washington, DC: American Chemical Society.

Melis, A., & Happe, T. (2001). Hydrogen production. Green algae as a source of energy. *Plant Physiology, 127*(3), 740–748.

Mielenz, J. R. (2001). Ethanol production from biomass: Technology and commercialization. *Current Opinion in Microbiology, 4*(3), 324–329.

Mohan, S. V., Srikanth, S., Velvizhi, G., & Babu, M. L. (2013). Microbial fuel cells for sustainable bioenergy generation: Principles and perspective applications. In V. K. Gupta & M. G. Tuohy (Eds.), *Biofuel technologies: Recent developments* (pp. 335–368). Berlin: Springer.

Moo-Young, M., & Chisti, Y. (1994). Biochemical engineering in biotechnology. *Pure and Applied Chemistry, 66*, 117.

Moreira, A. R., Ulmer, D. C., & Linden, J. C. (1981). Butanol toxicity in the butylic fermentation. *Biotechnology & Bioengineering Symposium, 11*, 567–579.

Mosier, N., Wyman, C., Dale, B., et al. (2005). Features of promising technologies for pretreatment of lignocellulosic biomass. *Bioresource Technology, 96*(6), 673–686.

Murty, M. V. S., & Chandra, T. S. (1997). Fermentability of hemicelluloses extracted from municipal waste and commercial xylans by *Clostridium* sp. *Applied Microbiology and Biotechnology, 47*, 212–217.

Nath, K., & Das, D. (2004). Improvement of fermentative hydrogen production: Various approaches. *Applied Microbiology and Biotechnology, 65*(5), 520–529.

Neha, P., Chintan, B., Pallavi, D., & Neha, T. (2013). Use of sunflower and cottonseed oil to prepare biodiesel by catalyst assisted transesterification. *Research Journal of Chemical Sciences, 2231*, 606X.

Ng, T. K., Ben-Bassat, A., & Zeikus, J. G. (1981). Ethanol production by thermophilic bacteria: fermentation of cellulosic substrates by cocultures of Clostridium thermocellum and Clostridium thermohydrosulfuricum. *Applied and Environmental Microbiology, 41*(6), 1337–1343.

Nguyen, M. T., Choi, S. P., Lee, J., Lee, J. H., & Sim, S. J. (2009). Hydrothermal acid pretreatment of Chlamydomonas reinhardtii biomass for ethanol production. *Journal of Microbiology and Biotechnology, 19*(2), 161–166.

Ni, M., Leung, D. Y., Leung, M. K., & Sumathy, K. (2006). An overview of hydrogen production from biomass. *Fuel Processing Technology, 87*(5), 461–472.

Niemisto, J., Saavalainen, P., Pongrácz, E., & Keiski, R. L. (2013). Biobutanol as a potential sustainable biofuel-assessment of lignocellulosic and waste-based feedstocks. *Journal of Sustainable Development of Energy, Water and Environment Systems, 1*(2), 58–77.

Nigam, P. S., & Singh, A. (2011). Production of liquid biofuels from renewable resources. *Progress in Energy and Combustion Science, 37*(1), 52–68.

Nimcevic, D., Schuster, M., & Gapes, J. R. (1998). Solvent production by *Clostridium beijerinckii* NRRL B592 growing on different potato media. *Applied Microbiology and Biotechnology, 50*, 426–428.

Oh, Y. K., Kim, S. H., Kim, M. S., & Park, S. (2004). Thermophilic biohydrogen production from glucose with trickling biofilter. *Biotechnology and Bioengineering, 88*(6), 690–698.

Olsson, L., & Hahn-Hägerdal, B. (1996). Fermentation of lignocellulosic hydrolysates for ethanol production. *Enzyme and Microbial Technology, 18*(5), 312–331.

Palmqvist, E., & Hahn-Hägerdal, B. (2000). Fermentation of lignocellulosic hydrolysates. I: inhibition and detoxification. *Bioresource Technology, 74*(1), 17–24.

Panesar, P. S., Marwaha, S. S., & Kennedy, J. F. (2006). Zymomonas mobilis: An alternative ethanol producer. *Journal of Chemical Technology and Biotechnology, 81*(4), 623–635.

Pant, D., Van Bogaert, G., Diels, L., & Vanbroekhoven, K. (2010). A review of the substrates used in microbial fuel cells (MFCs) for sustainable energy production. *Bioresource Technology, 101* (6), 1533–1543.

Perlack, R. D., Wright, L. L., Turhollow, A. F., Graham, R. L., Stokes, B. J., & Erbach, D. C. (2005). *Biomass as feedstock for a bioenergy and bioproducts industry: The technical feasibility of a billion-ton annual supply*. Oak Ridge, TN: Oak Ridge National Lab.

Pfromm, P. H., Amanor-Boadu, V., Nelson, R., Vadlani, P., & Madl, R. (2010). Bio-butanol vs. bio-ethanol: A technical and economic assessment for corn and switchgrass fermented by yeast or Clostridium acetobutylicum. *Biomass and Bioenergy, 34*(4), 515–524.

Potts, T., Du, J., Paul, M., May, P., Beitle, R., & Hestekin, J. (2012). The production of butanol from Jamaica bay macro algae. *Environmental Progress & Sustainable Energy, 31*(1), 29–36.

Pratas, M. J., Freitas, S. V., Oliveira, M. B., Monteiro, S. C., Lima, Á. S., & Coutinho, J. A. (2011a). Biodiesel density: Experimental measurements and prediction models. *Energy & Fuels, 25*(5), 2333–2340.

Pratas, M. J., Freitas, S., Oliveira, M. B., Monteiro, S. C., Lima, Á. S., & Coutinho, J. A. (2011b). Densities and viscosities of minority fatty acid methyl and ethyl esters present in biodiesel. *Journal of Chemical & Engineering Data, 56*(5), 2175–2180.

Qureshi, N., & Blaschek, H. P. (2000). Economics of butanol fermentation using hyper-butanol producing *Clostridium Beijerinckii* BA101. *Food and Bioproducts Processing, 78*(3), 139–144.

Qureshi, N., Ezeji, T. C., Ebener, J., Dien, B. S., Cotta, M. A., & Blaschek, H. P. (2008a). Butanol production by Clostridium beijerinckii. Part I: Use of acid and enzyme hydrolyzed corn fiber. *Bioresource Technology, 99*(13), 5915–5922.

Qureshi, N., Li, X. L., Hughes, S., Saha, B. C., & Cotta, M. A. (2006). Butanol production from corn fiber xylan using Clostridium acetobutylicum. *Biotechnology Progress, 22*(3), 673–680.

Qureshi, N., Lolas, A., & Blaschek, H. P. (2001). Soy molasses as fermentation substrate for production of butanol using *Clostridium beijerinckii* BA101. *Journal of Industrial Microbiology and Biotechnology, 26*, 290–295.

Qureshi, N., Saha, B. C., Dien, B., Hector, R. E., & Cotta, M. A. (2010a). Production of butanol (a biofuel) from agricultural residues: Part I—Use of barley straw hydrolysate. *Biomass and Bioenergy, 34*(4), 559–565.

Qureshi, N., Saha, B. C., Hector, R. E., Hughes, S. R., & Cotta, M. A. (2008b). Butanol production from wheat straw by simultaneous saccharification and fermentation using Clostridium beijerinckii: Part I—Batch fermentation. *Biomass and Bioenergy, 32*(2), 168–175.

Qureshi, N., Saha, B. C., Hector, R. E., Dien, B., Hughes, S., Liu, S., et al. (2010b). Production of butanol (a biofuel) from agricultural residues: Part II—Use of corn stover and switchgrass hydrolysates. *Biomass and Bioenergy, 34*(4), 566–571.

Qureshi, N., Saha, B. C., & Cotta, M. A. (2007). Butanol production from wheat straw hydrolysate using *Clostridium beijerinckii. Bioprocess and Biosystems Engineering, 30*, 419–427.

Qureshi, N., Saha, B. C., & Cotta, M. A. (2008c). Butanol production from wheat straw by simultaneous saccharification and fermentation using *Clostridium beijerinckii*: part II—Fedbatch fermentation. *Biomass and Bioenergy, 32*, 176–183.

Ranjan, A., & Moholkar, V. S. (2011). Comparative study of various pretreatment techniques for rice straw saccharification for the production of alcoholic biofuels. *Fuel*. doi:10.1016/j.fuel.2011.03.030.

Riadi, L., Purwanto, E., Kurniawan, H., & Oktaviana, R. (2014). Effect of bio-based catalyst in biodiesel synthesis. *Procedia Chemistry, 9*, 172–181.

Rinaldi, R., & Schüth, F. (2009). Acid hydrolysis of cellulose as the entry point into biorefinery schemes. *ChemSusChem, 2*(12), 1096–1107.

Saddler, J. N., Ernest, K. C., Mes-Hartree, M., Levitin, N., & Brownell, H. H. (1983). Utilization of enzymatically hydrolyzed wood hemicelluloses by microorganisms for production of liquid fuels. *Applied and Environmental Microbiology, 45*(1), 153–160.

Shuler, M. L., & Kargi, F. (2002). *Bioprocess engineering*. Upper Saddle River, NJ: Prentice Hall.

SillersR, C. A., Tracy, B., et al. (2008). Metabolic engineering of the non-sporulating, non-solventogenic *Clostridium acetobutylicum* strain M5 to produce butanol without acetone demonstrate the robustness of the acid-formation pathways and the importance of the electron balance. *Metabolic Engineering, 10*(6), 321–332.

Singh, S. P., & Singh, D. (2010). Biodiesel production through the use of different sources and characterization of oils and their esters as the substitute of diesel: A review. *Renewable and Sustainable Energy Reviews, 14*(1), 200–216.

Sjolander, N. O., Langlykke, A. F., & Peterson, W. H. (1938). Butyl alcohol fermentation of wood sugar. *Industrial and Engineering Chemistry, 30*, 1251–1255.

Sonderegger, M., Jeppsson, M., Larsson, C., Gorwa-Grauslund, M. F., Boles, E., Olsson, L., et al. (2004). Fermentation performance of engineered and evolved xylose-fermenting Saccharomyces cerevisiae strains. *Biotechnology and Bioengineering, 87*(1), 90–98.

Soni, B. K., Das, K., & Ghose, T. K. (1982). Bioconversion of agro-wastes into acetone butanol. *Biotechnology Letters, 4*, 19–22.

Spatari, S., Zhang, Y., & MacLean, H. L. (2005). Life cycle assessment of switchgrass-and corn stover-derived ethanol-fueled automobiles. *Environmental Science & Technology, 39*(24), 9750–9758.

Steen, E. J., Chan, R., Prasad, N., Myers, S., Petzold, C. J., Redding, A., et al. (2008). Metabolic engineering of Saccharomyces cerevisiae for the production of n-butanol. *Microbial Cell Factories, 7*(1), 36.

Strauß, C., Vetter, A., & Von Felde, A. (2012). Biogas production and energy crops biogas energy crops. In R. A. Meyers (Ed.), *Encyclopedia of sustainability science and technology* (pp. 1097–1145). New York, NY: Springer.

Sujan, S. A., Hirata, S., & Minowa, T. (2009). An approach to produce biodiesel from non-edible Jatropha curcus oil through dual step process: Preesterification and transesterification. *Bangladesh Journal of Scientific and Industrial Research, 44*(3), 347–352.

Sun, Y., & Cheng, J. (2002). Hydrolysis of lignocellulosic materials for ethanol production: A review. *Bioresource Technology, 83*(1), 1–11.

Swana, J., Yang, Y., Behnam, M., & Thompson, R. (2011). An analysis of net energy production and feedstock availability for biobutanol and bioethanol. *Bioresource Technology, 102*(2), 2112–2117.

Taguchi, F., Takiguchi, S., & Morimoto, M. (1992). Efficient hydrogen production from starch by a bacterium isolated from termites. *Journal of Fermentation and Bioengineering, 73*(3), 244–245.

Tanisho, S., Kuromoto, M., & Kadokura, N. (1998). Effect of CO_2 removal on hydrogen production by fermentation. *International Journal of Hydrogen Energy, 23*(7), 559–563.

Tarkow, H., & FEIST, W. C. (1969). *A mechanism for improving the digestibility of lignocellulosic materials with dilute alkali and liquid ammonia.*

Thang, V. H., Kanda, K., & Kobayashi, G. (2010). Production of acetone–butanol–ethanol (ABE) in direct fermentation of cassava by *Clostridium saccharoperbutylacetonicum* N1-4. *Applied Biochemistry and Biotechnology, 161*, 157–170.

Turner, W. J. (2014). *Understanding and improving microbial biofuel tolerance as a result of efflux pump expression through genetic engineering and mathematical modeling* (Doctoral dissertation). The University of Vermont

Van Haandel, A. C., & Lettinga, G. (1994). *Anaerobic sewage treatment: A practical guide for regions with a hot climate*. New York, NY: John Wiley & Sons.

Varga, E., Réczey, K., & Zacchi, G. (2004). Optimization of steam pretreatment of corn stover to enhance enzymatic digestibility. *Applied Biochemistry and Biotechnology, 114*(1–3), 509–523.

Votruba, J., Volesky, B., & Yerushalmi, L. (1985). Mathematical model of a batch acetone-butanol fermentation. *Biotechnology and Bioengineering, 28*, 247.

Walfridsson, M., Bao, X., Anderlund, M., Lilius, G., Bülow, L., & Hahn-Hägerdal, B. (1996). Ethanolic fermentation of xylose with Saccharomyces cerevisiae harboring the Thermus thermophilus xylA gene, which expresses an active xylose (glucose) isomerase. *Applied and Environmental Microbiology, 62*(12), 4648–4651.

Walter, A., Rosillo-Calle, F., Dolzan, P., Piacente, E., & Borges da Cunha, K. (2008). Perspectives on fuel ethanol consumption and trade. *Biomass and Bioenergy, 32*(8), 730–748.

Woods, D. R. (1995). The genetic engineering of microbial solvent production. *Trends in Biotechnology, 13*, 259–264.

Worldwatch Institute. (2005). *Renewables, 2005: Global status report*. Washington, DC: Worldwatch Institute.

Yoshida, N., Nakasato, M., Ohmura, N., Ando, A., Saiki, H., Ishii, M., & Igarashi, Y. (2006). Acidianus manzaensis sp. nov., a novel thermoacidophilic archaeon growing autotrophically by the oxidation of H_2 with the reduction of Fe^{3+}. *Current Microbiology, 53*(5), 406–411.

Yu, E. K. C., Deschatelets, L., & Saddler, J. N. (1984). The bioconversion of wood hydrolyzates to butanol and butanediol. *Biotechnology Letters, 6*, 327–332.

Zhang, T., Du, N., & Tan, T. (2011). Biobutanol production from sweet sorghum bagasse. *Journal of Biobased Materials and Bioenergy, 5*(3), 331–336.

Zhang, B., Weng, Y., Xu, H., & Mao, Z. (2012). Enzyme immobilization for biodiesel production. *Applied Microbiology and Biotechnology, 93*(1), 61–70.

Zheng, Y., Pan, Z., & Zhang, R. (2009). Overview of biomass pretreatment for cellulosic ethanol production. *International Journal of Agricultural and Biological Engineering, 2*(3), 51–68.
Zinder, S. H. (1994). Syntrophic acetate oxidation and "reversible acetogenesis". In H. L. Drake (Ed.), *Acetogenesis* (pp. 386–415). New York, NY: Springer.

Chapter 11
Optimization Study of Catalytic Co-gasification of Rubber Seed Shell and High Density Polyethylene Waste for Hydrogen Production Using Response Surface Methodology

Bridgid Lai Fui Chin, Suzana Yusup, Ahmed Al Shoaibi, Pravin Kannan, Chandrasekar Srinivasakannan, and Shaharin Anwar Sulaiman

Introduction

Co-gasification of plastic and biomass waste mixtures have received much interest and attention in recent years as an alternative means for the replacement of the depleting amount of fossil fuels, overcoming the greenhouse gas emissions, and as well as the increasing development of waste processing technology to reduce the negative environmental impact of plastic waste (Chin et al. 2014; Pinto et al. 2002; Van Kasteren 2006). It is said that gasification is able to provide more benefits compared to other thermochemical methods such as pyrolysis and combustion (Corella et al. 1998).

Although most studies are carried out in the area of gasification, however, limited studies are conducted related to co-gasification of plastic and biomass mixtures. Moghadam et al. (2013) studied on the binary mixtures of palm kernel shell (PKS) with polyethylene (PE) waste using catalytic steam gasification in pilot scale

B.L.F. Chin
Department of Chemical Engineering, Faculty of Engineering & Science,
Curtin University, Sarawak campus, CDT 250, 98009 Miri, Sarawak, Malaysia

S. Yusup (✉)
Biomass Processing Lab, Department of Chemical Engineering,
Centre for Biofuel and Biochemical Research, Green Technology MOR,
Universiti Teknologi PETRONAS, Bandar Seri Iskandar, Tronoh 31750,
Perak Darul Ridzuan, Malaysia
e-mail: drsuzana_yusuf@petronas.com.my

A. Al Shoaibi • P. Kannan • C. Srinivasakannan
Department of Chemical Engineering, The Petroleum Institute, P.O. Box 2533,
Abu Dhabi, United Arab Emirates

S.A. Sulaiman
Department of Mechanical Engineering, Universiti Teknologi PETRONAS,
Bandar Seri Iskandar, Tronoh 31750, Perak Darul Ridzuan, Malaysia

P. Ravindra (ed.), *Advances in Bioprocess Technology*,
DOI 10.1007/978-3-319-17915-5_11

fluidized bed gasifier with a height of 2,500 mm and internal diameter of 150 and 200 mm in gasification and free board zone respectively at a feeding rate of 1.2 kg/h. It is found that the temperature of the gasifier has significant influence on the gas composition and yield in the process. High H_2 and gas yield can be achieved when operating at high temperature. The highest H_2 yield can be achieved when operating at temperature of 800 °C, steam to feedstock ratio of 3 kg/kg and polyethylene/PKS ratio of 0.25 kg/kg. Van Kasteren (2006) investigated the reaction between the mixtures of wood and PE using steam gasification in a bench scale bubbling fluidized bed gasifier with a height of 500 mm and inner diameter of 65 mm. An empirical modeling approach using MODDE software (Umetrics AB, Malmö, Sweden) is used to predict the optimum operating conditions of CO and H_2 efficiency. It is found that the optimum conditions to achieve a maximum CO and H_2 efficiency of 42 % is at temperature of 900 °C, equivalence ratio (ER) of 0.15, amount of PE in the feed of 0.11 g/g and amount of steam added of 0.42 g/g feed. The optimum CO_2 and H_2 is achievable at high temperature, low PE content and low ER. It is found that ER ≤ 0.2 controls the formation of CO_2. Furthermore, low supply of steam is preferable in the process for energy reduction consumption.

In the present work, the catalytic co-gasification of plastic high density polyethylene (HDPE) waste and biomass rubber seed shell (RSS) with an argon gas and steam supplied into the thermogravimetric analysis (TGA) equipment coupled with mass spectrometer (MS) system to assess the effect of temperature, HDPE particle size, RSS particle size and percentage of HDPE in the mixture on the response variable of H_2 production whilst the response surface methodology (RSM) is employed to find the optimum conditions to achieve maximum H_2 production within the range of process parameters studied. The optimum gasification condition in this study is the basis for the syngas production in a larger scale gasification process using these feedstock.

Experimental

Materials and Sample Preparation

The raw materials used in this work are RSS from Vegpro Trading, Malaysia and HDPE plastic froom Shen Foong Plastic Industries Sdn Bhd, Malaysia. These materials are ground and sieved to a particle size ranging from 0 to 710 μm. Appropriate homogenized HDPE/RSS mixtures are prepared in respective weight ratios. The characteristic of the materials used in this study are presented in Table 11.1. Both

Table 11.1 Characteristics of rubber seed shell (RSS) and high density polyethylene (HDPE)

	Ultimate analysis (wt%, dry ash basis)					Proximate analysis (wt%, dry basis)			
Sample	C	H	N	S	O[a]	MC	VM	FC[a]	Ash
RSS	44.31	4.38	0.51	0.13	50.67	8.59	80.98	6.62	3.81
HDPE	81.45	12.06	0.34	0.79	5.36	0.00	99.46	0.00	0.34

[a]By difference

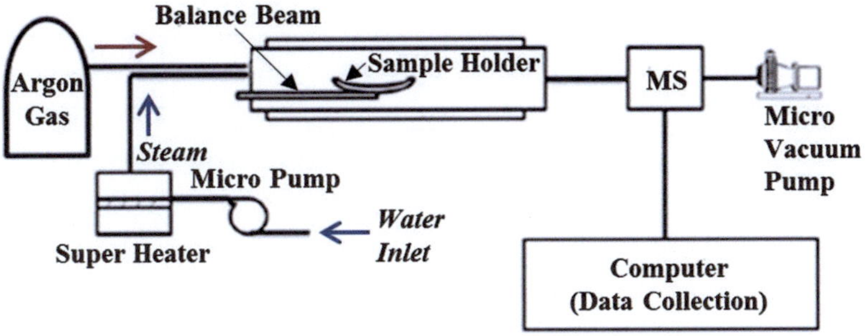

Fig. 11.1 Experimental setup of thermogravimetric analysis (TGA) equipment coupled with mass spectrometer (MS)

ultimate and proximate analyses of these feedstock are conducted in LECO CHNS-932 elemental analyzer and thermogravimetry analyzer EXSTAR TG/DTA 6300 (Seiko instrument Inc.) respectively. A commercial nickel powder (index number 028-002-01-4, Merck) is selected as the catalyst in the gasification process of HDPE/RSS mixtures. The particle size of the nickel powder is in the range of ~10 μm.

Experimental Apparatus and Procedure

The experiments are performed using thermogravimetry analyzer EXSTAR TG/DTA 6300 (Seiko Instrument Inc.) coupled with the mass spectrometer (Pfeiffer Vacuum Thermostar) as illustrated in Fig. 11.1. Approximately 5 mg of sample is placed on a ceramic crucible in a TGA equipment under an inert atmosphere of argon. A flow rate of 100 mL min^{-1} of argon gas is fed into the system for 20 min at a temperature of 110 °C. Subsequently, all samples are heated from 110 to 900 °C at heating rate of 20 °C min^{-1}. During heating, the TGA equipment is used to measure mass variation of the materials and furnace temperature meanwhile the mass spectrometer (MS) is used to measure the gaseous products such as H_2, CO, CO_2 and CH_4. Steam is generated by a superheater at 110 °C and is injected into the TGA at a flowrate of 0.005 mL/min when the temperature inside the TGA equipment reached 110 °C to avoid any occurrence of condensation within the system. The equivalence ratio (ER) for gasification is 0.25 respectively. All experiments are repeated minimum of two times.

Design of Experiments Using Response Surface Methodology

An empirical modeling approach known as Response Surface Methodology (RSM) coupled with Central Composite Design (CCD) using the Design-Expert version 8.0.7.1 (Stat-Ease, Inc.) software is selected for the design of experiments and to

Table 11.2 Process parameters in Central Composite Design (CCD): coded and natural values

Variable	Parameter code	Unit	Level				
			$-\alpha$	-1	0	+1	$+\alpha$
Temperature	*A*	°C	500	600	700	800	900
RSS particle size	*B*	mm	0.125	0.250	0.375	0.500	0.625
HDPE particle size	*C*	mm	0.125	0.250	0.375	0.500	0.625
Percentage of plastics in mixture	*D*	wt%	0	10	20	30	40

determine the relationship between the independent variables and possible interactions of the dependent variables. The independent variables also known as process parameters for this study are temperature, RSS particle size, HDPE particle size, and percentage of plastics in the mixture meanwhile the dependent variable are H_2 production. With the involvement of CCD, a model equation is produced for the purpose of prediction and verification. Furthermore, the optimization of each individual dependent variables as the function of independent can be obtained (Isa et al. 2011).

The prediction of the optimal point in the quadratic equation model is based on (11.1) (Isa et al. 2011):

$$y = \beta_0 + \sum_{i=1}^{k} \beta_i x_i + \sum_{i=1}^{k} \beta_{ii} x_i^2 + \sum_{i=1} \sum_{j=i+1} \beta_{ij} x_i x_j + \varepsilon \quad (11.1)$$

where $\beta_0, \beta_i, \beta_{ii}$ are regression coefficients for the intercept, linear, quadratic, and interaction coefficients respectively and x_i and x_j are the coded independent variables.

Table 11.2 displays the coded and actual values of the independent variables used in the design of experiments. The independent variables are coded to two levels: low (-1) and high (+1) and the axial points are coded as $-2(-\alpha)$ and +2 ($+\alpha$). In this study, α is the distance of the axial point from center and makes the design rotatable and the value is 2, which is determined using (11.2) (Brown and Brown 2012). The experiments are conducted with a single response in CCD, which is the H_2 production. The four factors (F) which are tested are temperature, RSS particle size, HDPE particle size, and percentage of plastic in the mixtures are considered at five levels specifically -1,+1, 0, $-\alpha$, and $+\alpha$.

$$\alpha = \left(2^F\right)^{1/4} = \left(2^4\right)^{1/4} = 2 \quad (11.2)$$

where F is the number of factors.

The number of experiments, N is determined based on (11.3) (Brown and Brown 2012). A total of 30 experiments are required to support the RSM data in CCD which comprises of 16 factorial points from 25 full factorial CCD for the four variables, eight axial points and six replicates at the center points are employed in the study. In order to determine the experimental error and the reproducibility of the data, the center points are normally repeated around 4–6 times (Abnisa et al. 2011).

$$N = 2^n + 2n + m = 2^4 + (2 \times 4) + 6 = 30 \tag{11.3}$$

where n is number variables and m is the replicate number of experiments.

The experiments are conducted based on a full design matrix generated corresponding to the CCD design as shown in Table 11.3. Upon completion of all the experimental runs, the response of H_2 production is fitted into a suitable model using regression analysis. Three sets of experiments are carried out at the optimum conditions for the purpose of validating the accuracy of the model proposed.

Table 11.3 Experimental design matrix and results

	Parameters				Response
Run	Temperature (°C)	RSS particle size (mm)	HDPE particle size (mm)	Percentage of plastics in mixture (wt%)	H_2 (vol%)
1	800	0.250	0.250	10	46.676
2	700	0.125	0.375	20	50.123
3	600	0.500	0.250	30	47.751
4	800	0.500	0.250	10	45.952
5	500	0.375	0.375	20	44.781
6	700	0.375	0.625	20	43.031
7	600	0.500	0.250	10	45.324
8	900	0.375	0.375	20	49.230
9	800	0.500	0.500	30	44.355
10	600	0.500	0.500	30	44.208
11	700	0.375	0.375	0	44.466
12	700	0.375	0.375	40	46.603
13	700	0.625	0.375	20	43.072
14	800	0.250	0.500	30	47.396
15	700	0.375	0.375	20	39.980
16	800	0.250	0.500	10	46.338
17	700	0.375	0.375	20	38.569
18	700	0.375	0.125	20	49.868
19	800	0.250	0.250	30	46.545
20	700	0.375	0.375	20	38.612
21	600	0.500	0.500	10	41.032
22	700	0.375	0.375	20	38.625
23	600	0.250	0.500	30	47.123
24	700	0.375	0.375	20	38.621
25	600	0.250	0.250	10	48.634
26	800	0.500	0.250	30	48.475
27	600	0.250	0.500	10	48.132
28	700	0.375	0.375	20	39.262
29	600	0.250	0.250	30	46.502
30	800	0.500	0.500	10	41.930

Results and Discussion

Analysis of Variance (ANOVA)

The aim of using analysis of variance (ANOVA) is to study the effect of operating parameters such as temperature, HDPE particle size, RSS particle size, percentage of plastics in the mixture and their possible interactions on the dependent response variable i.e. H_2 production. Moreover, this method can deduce the results of the system where several factors are effective and can be varied simultaneously (Luo et al. 2010).

The summary of the ANOVA results obtained from Design-Expert version 8.0.7.1 (Stat-Ease, Inc.) software are presented in Table 11.4. The test of significant terms, the R-squared test, and the lack-of-fit are the pre-requisite to determine the reliability of the model (Luo et al. 2010). The confidence level of 95 % is used to determine the significant effect of the factors and their interaction effects. The R-squared (R^2) test is a statistical measure of the overall predictive capabilities of the model produced. Few authors (Cui et al. 2006; Luo et al. 2010) stated when new terms are added into the model, the R^2 value increases. Therefore, a large

Table 11.4 Analysis of variance (ANOVA) for H_2 production

Term	Sum of squares	Degree of freedom	Mean of squares	F value	p-Value (Prob > F)
Model	370.26	14	26.45	24.06	<0.0001[a]
A	2.57	1	2.57	2.34	0.1468[b]
B	43.80	1	43.80	39.84	<0.0001[a]
C	35.09	1	35.09	31.91	<0.0001[a]
D	6.63	1	6.63	6.03	0.0268[a]
AB	2.13	1	2.13	1.93	0.1846[b]
AC	0.0004731	1	0.0004731	0.0004303	0.9837[b]
AD	0.73	1	0.73	0.66	0.4285[b]
BC	17.24	1	17.24	15.68	0.0013[a]
BD	10.18	1	10.18	9.26	0.0082[a]
CD	0.55	1	0.55	0.50	0.4907[b]
A^2	104.61	1	104.61	95.15	<0.0001[a]
B^2	93.97	1	93.97	85.47	<0.0001[a]
C^2	90.25	1	90.25	82.09	<0.0001[a]
D^2	68.92	1	68.92	62.69	<0.0001[a]
Residual	16.49	15	1.10		
Lack-of-fit	14.89	10	1.49	4.55	0.0540[b]
Pure error	1.63	5			

$R^2 = 0.9574$, $adj\text{-}R^2 = 0.9176$, Predicted $R^2 = 0.7726$, Adeq Precision = 13.673, Mean = 44.71
A temperature, B RSS particle size, C HDPE particle size, D percentage of plastics in mixture
[a]Significant term
[b]Insignificant term

model with weak predictability could produce higher R^2 value. This is because the calculation of adjusted-R^2 (*adj-*R^2) is included which penalize the statistics of R^2 as additional terms are included in the model. The acceptability of the model can be checked using the lack-of-fit test which is used to determine whether there are any discrepancies between measured and expected values which could contribute to random or systematic error (Luo et al. 2010). With the confidence level of 95 %, if the *p*-value is greater than 0.05, the individual terms in the model showed the insignificant effect on the response. After analyzing the three tests, the adequacy of the model is analyzed from the internally studentized plot and the predicted versus actual response plot for H_2 production. Furthermore, surface plot is also presented to show the interaction of the response variable with two different variables based on a model equation generated from the software.

Table 11.4 summarizes the ANOVA result on the H_2 production. The ANOVA result shows the overall model for H_2 production is statistically significant. This is evidence from the F-value of 24.06 and *p*-value of <0.0001. The significance of the individual input parameters and their interactions for H_2 production are also investigated. It is found that the RSS particle size (*B*) shows the highest F-value of 39.84 and *p*-value of <0.0001. The significance of the individual input parameters and their interactions for H_2 production are also investigated. It is found that the RSS particle size (*B*) shows the highest F-value of 39.84 and *p*-value of <0.0001 which indicates the most effecting parameter to influence the H_2 production among the individual input parameters. Furthermore, the other individual input parameters that influence the H_2 production are HDPE particle size (*C*) and percentage of plastics in mixture (*D*) with *p*-values of <0.0001 and 0.0268 respectively, which are indicated by the value of *p*-value lower than 0.05. Likewise, the significant model interactive terms are RSS particle size and HDPE particle size (*BC*) and RSS particle size and percentage of plastics in the mixture (*BD*) with *p*-values of 0.0013 and 0.082 respectively. Meanwhile, the insignificant model terms are temperature and RSS particle size (*AB*), temperature and HDPE particle size (*AC*), temperature and percentage of plastics in the mixture (*AD*) and HDPE particle size and percentage of plastics in the mixture (*CD*) which are seen by the *p*-values of 0.1846, 0.9837, 0.4285 and 0.4905 respectively.

The p-value for the lack-of-fit test of the model is 0.0540 which indicates that the lack-of-fit is not significant relative to the pure error. The R^2 value of the model is 0.9574 refers that 95 % of the observed variability. The predicted R^2 value of the model is 0.7726 which is in a reasonable agreement with a *adj-*R^2 with the value of 0.9176. The value of Adeq Precision is 13.673 > 4 indicates an adequate signal that can be used to navigate the design space. The quadratic model for H_2 production in terms of coded factors is represented in (11.4).

$$\begin{aligned} H_2 \text{ production (vol\%)} = {} & 38.94 + 0.33A - 1.35B - 1.21C + 0.53D + 0.36AB \\ & - 0.0005438AC + 0.21AD - 1.04BC + 0.80BD \\ & + 0.19CD + 1.95A^2 + 1.85B^2 + 1.81C^2 + 1.59D^2 \end{aligned} \tag{11.4}$$

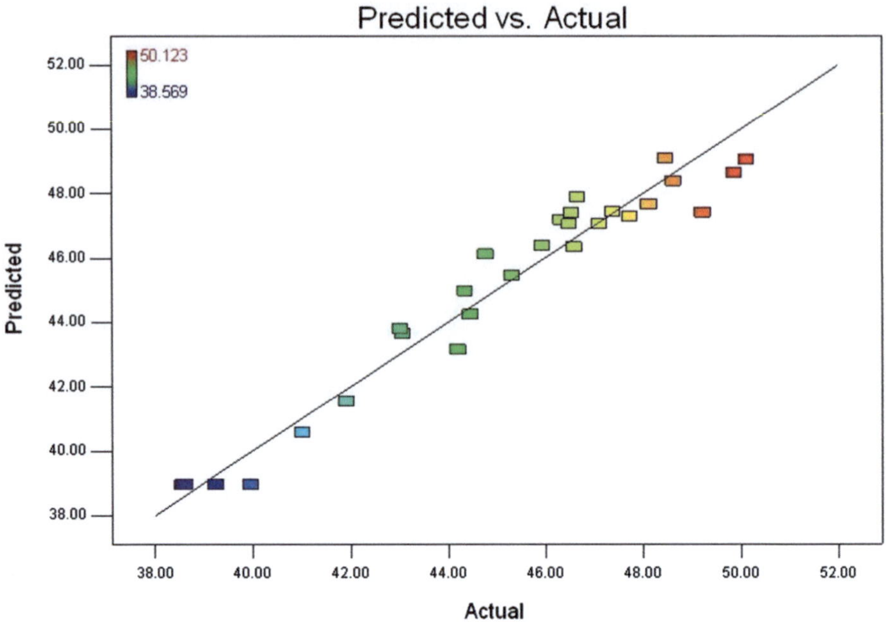

Fig. 11.2 Predicted versus actual values for H_2 production

Predicted Versus Actual Response

Figure 11.2 shows the predicted versus actual response values for H_2 production. The minimum and maximum values for H_2 production are 50.123 and 38.569 vol% respectively.

Figure 11.3 shows the internally studentized plot for H_2 production. It is observed that the figure shows no distinct patterns which indicate that there are no biases exist.

Effect of Operating Variables on the H_2 Production

Figure 11.4 presents the 3-dimensional response surface plot for the H_2 production on the effect of temperature and RSS particle size. The results show that there is no significant change in the H_2 production when temperature varies from 600 to 800 °C at lower RSS particle size of 0.250 mm. However, a slight increment by 1.6 vol% of H_2 production when temperature changes from 600 to 800 °C at higher RSS particle size of 0.500 mm. At lower temperature of 600 °C, an increment by 2.1 vol% in the H_2 production when RSS particle size reduces from 0.500 to 0.250 mm is observed. Similarly, at higher temperature of 800 °C, the H_2 production increases by 1.7 vol%

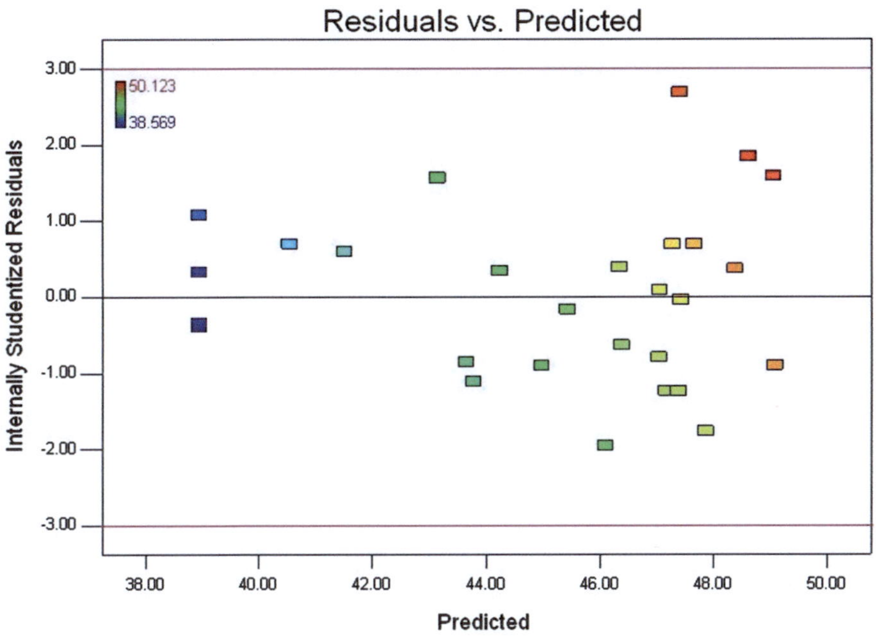

Fig. 11.3 Internally studentized residuals plot for H_2 production

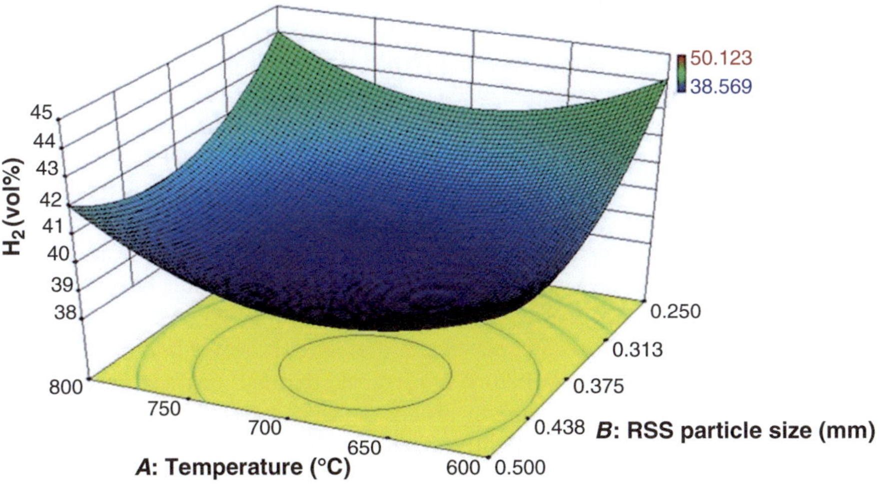

Fig. 11.4 3-Dimensional response surface plot representing combined effects of temperature (600–800 °C) and RSS particle size (0.250–0.500 mm) at constant HDPE particle size of 0.375 mm, and 20 wt% plastics in the mixture on H_2 production

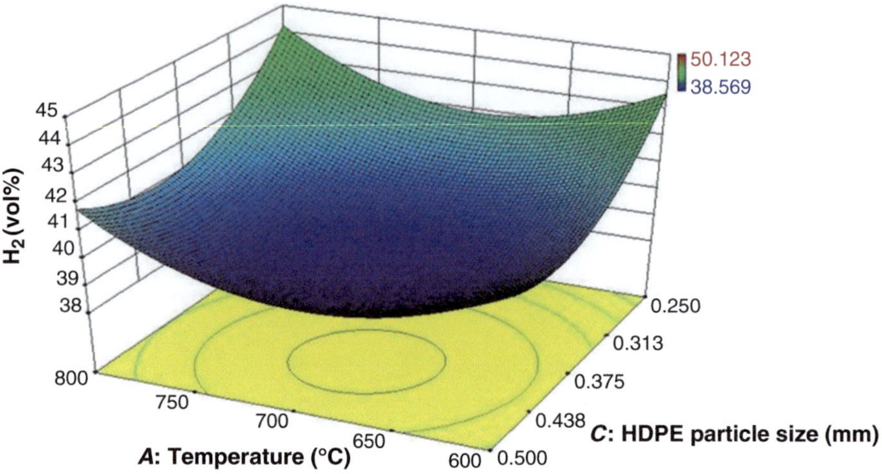

Fig. 11.5 3-Dimensional response surface plot representing combined effects of temperature (600–800 °C) and HDPE particle size (0.250–0.500 mm) at constant RSS particle size of 0.375 mm, and 20 wt% plastics in the mixture on H_2 production

when RSS particle size decreases from 0.500 to 0.250 mm. This could be explained that smaller particle size has larger surface area which directly enhanced both heat and mass transfer that allows faster heating rate in producing more light gases but less tar and char. The effect of the biomass particle size on reactor temperature is in good agreement with published data on gasification as reported in literature (Lv et al. 2004; Turn et al. 1998).

Figure 11.5 illustrates the 3-dimensional response surface plot for the H_2 production on the effect of temperature and HDPE particle size. The result shows no significant change in H_2 production when temperature increases from 600 to 800 °C for both HDPE particle size of 0.250 and 0.500 mm. However, at lower temperature of 600 °C, the H_2 production increases by 2.2 vol% when HDPE particle size reduces from 0.500 to 0.250 mm. Similarly, at higher temperature of 800 °C, H_2 production increases by 2.4 vol% when HDPE particle size reduces from 0.500 to 0.250 mm. As mentioned earlier, smaller particle size has larger surface area which directly enhances both heat and mass transfer that permits faster heating rate in producing more light gas (Lv et al. 2004; Turn et al. 1998). Although smaller particle size for both biomass and plastic are preferred, however both biomass and plastic responds differently when it undergoes thermal degradation process. Plastic particle melt and vaporize promptly with negligible solid residues meanwhile biomass decreases slightly. This clearly shows that plastic is more reactive compared to biomass as the conversion of plastic to gaseous products is faster. By incorporating higher temperature with low plastic particle size, the conversion of plastic to gaseous products can be achieved to the desired temperature in a shorter time.

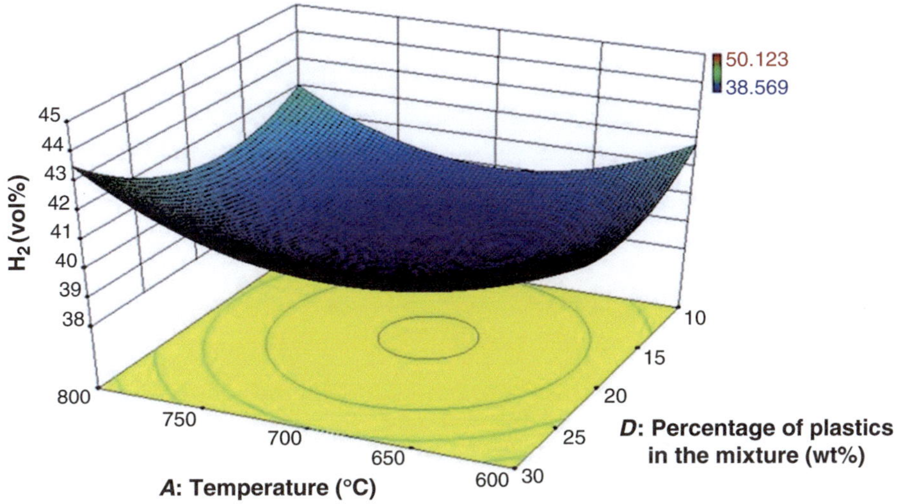

Fig. 11.6 3-Dimensional response surface plot representing combined effects of temperature (600–800 °C) and percentage of plastics in the mixture (10–30 wt%) at constant RSS particle size of 0.375 mm, and HDPE particle size of 0.375 mm on H_2 production

Figure 11.6 depicts the 3-dimensional response surface plot for the H_2 production on the effect of temperature and percentage of plastics in the mixture. The results illustrate that there is no significant change in the H_2 production when temperature increases from 600 to 800 °C at lower percentage of plastics in the mixture of 10 wt %. However, at higher percentage of plastics in the mixture of 30 wt%, there is a slight increased in the H_2 production by 2.5 vol%. At higher temperature of 800 °C, the H_2 production increased from 41.771 to 42.906 vol% when percentage of plastics in the mixture increases from 10 to 30 wt%. The highest H_2 production of 42.906 vol % at 800 °C was found in the range of percentage of plastics in the mixture studied. A similar observation is observed by Moghadam et al. (2013) on the H_2 production when the percentage of plastics in the mixture increased. The increased of H_2 production is likely attributed to the cracking reactions occurring.

Figure 11.7 demonstrates the 3-dimensional response surface plot for the H_2 production on the effect of RSS particle size and HDPE particle size. The analysis shows no significant change in the H_2 production when RSS particle size increases from 0.250 to 0.500 mm at lower HDPE particle size of 0.250 mm. Conversely, at higher HDPE particle size of 0.500 mm, H_2 production decreases from 41.691 to 38.818 vol% when RSS particle size increases from 0.250 to 0.500 mm. It is observed that the H_2 production decreases slightly from 43.688 to 41.6912 vol% when HDPE particle size increases from 0.250 to 0.500 mm at low level particle size of RSS. It is concluded that the effect of increasing particle size of HDPE is more dominant compared to RSS particle size for H_2 production.

Figure 11.8 shows the 3-dimensional response surface plot for the H_2 production on the effect of RSS particle size and percentage of plastics in the mixture.

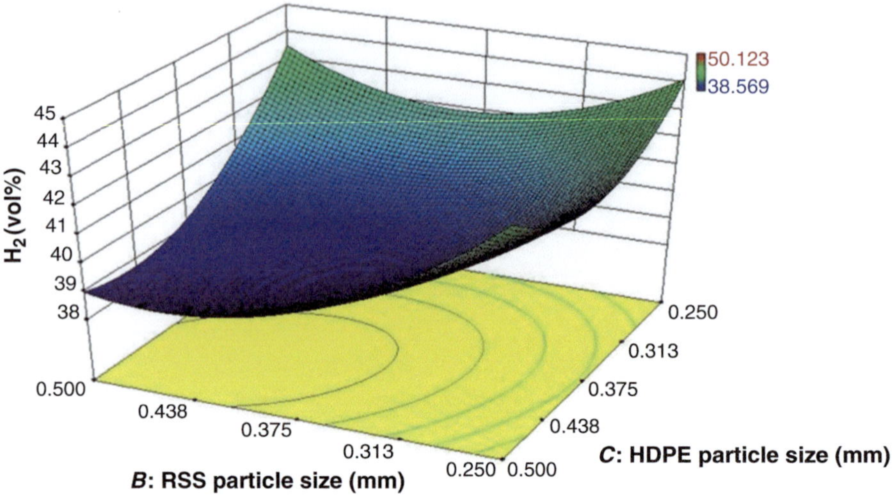

Fig. 11.7 3-Dimensional response surface plot representing combined effects of RSS particle size (0.250–0.500 mm) and HDPE particle size (0.250–0.500 mm) at constant temperature of 700 °C and percentage of plastics in the mixture (20 wt%) on H_2 production

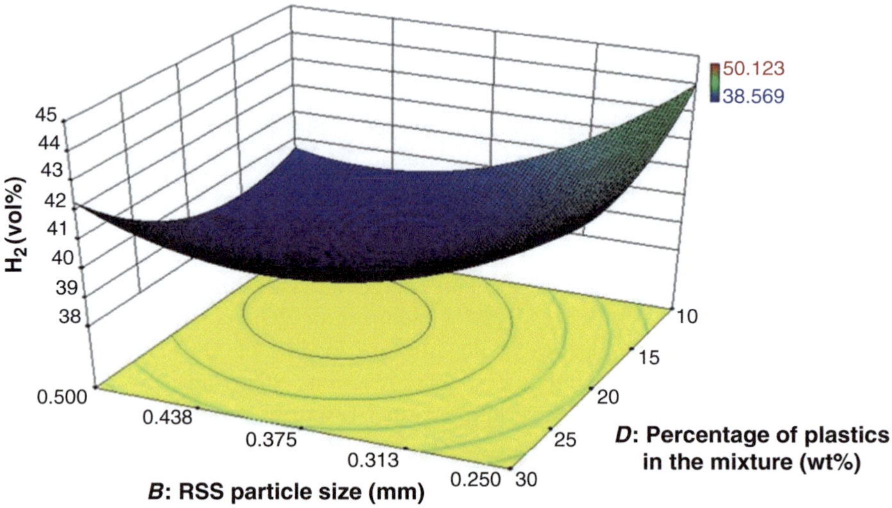

Fig. 11.8 3-Dimensional response surface plot representing combined effects of RSS particle size (0.250–0.500 mm) and percentage of plastics in the mixture (10–30 wt%) at constant temperature of 700 °C and HDPE particle size of 0.375 mm on H_2 production

The graphical analysis presents that the H_2 production decreases by 3.8 vol% when RSS particle size increases from 0.250 to 0.500 mm at lower percentage of plastics in the mixture of 10 wt%. Meanwhile, the H_2 production decreases by 0.9 vol% when RSS particle size increases from 0.250 to 0.500 mm at higher percentage of

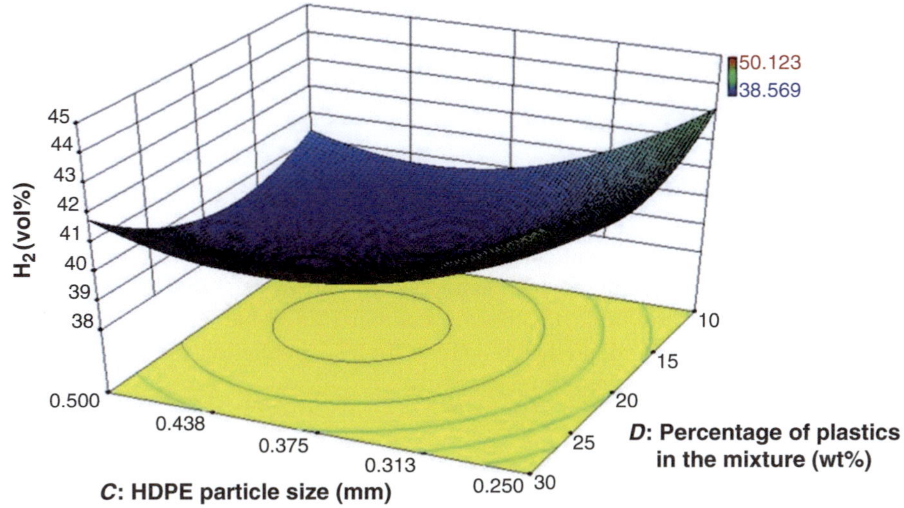

Fig. 11.9 3-Dimensional response surface plot representing combined effects of HDPE particle size (0.250–0.500 mm) and percentage of plastics in the mixture (10–30 wt%) at constant temperature of 700 °C and RSS particle size of 0.37 mm on H_2 production

plastics in the mixture of 30 wt%. It is observed that at higher RSS particle size of 0.500 mm, the H_2 production reduces by 2.4 vol% when percentage of plastics in the mixture varies from 10 to 30 wt%. The highest H_2 production of 43.246 vol% is achieved when using these input parameters of which temperature is 700 °C, 0.250 mm of RSS particle size, 0.375 mm of HDPE particle size, and 10 wt% of plastics in the mixture.

Figure 11.9 presents the 3-dimensional response surface plot for the H_2 production on the effect of HDPE particle size and percentage of plastics in the mixture. The result elucidates no significant change in the H_2 production when HDPE particle size increases from 0.250 to 0.500 mm at higher percentage of plastics in the mixture of 30 wt%. Conversely, the H_2 production decreases from 42.790 to 40.177 vol% at lower percentage of plastics in the mixture of 10 wt%. In addition, the H_2 production reduced slightly from 42.800 to 41.726 vol% when percentage of plastics in the mixture increases from 10 to 30 wt% using RSS particle size of 0.250 mm.

Optimization

After assessing the combined effects of the operating parameters such as temperature, HDPE particle size, RSS particle size, and percentage of plastics in the mixture in the TGA-MS system, the optimum input parameters which could provide the maximum values for H_2 production from catalytic co-gasification of RSS and HDPE mixtures are temperature of 700 °C, RSS particle size of 0.125 mm, HDPE particle size of 0.375 mm, and percentage of plastics in the mixture of 20 wt%.

Table 11.5 Validation of optimum conditions for H_2 production

	Experimental run			
Input parameter	First run	1	2	3
H_2 production	50.123	48.876	53.590	52.340

Reproducibility of Experimental Results

The reproducible results are validated and compared with the optimized value of H_2 production which is predicted by the model as shown in Table 11.5. For H_2 production, the first run indicates the run number 2 in Table 11.3. It is reported that the experimental and predicted values for H_2 production are found to be 50.123 ± 2.127 and 38.945 vol% respectively.

Conclusions

In the present work, an extensive experimental statistical analysis of the combined effects such as temperature, high density polyethylene (HDPE) particle size, rubber seed shell (RSS) particle size, percentage of plastic in the mixtures on the H_2 production from the catalytic co-gasification of HDPE/RSS mixtures using a central composite design (CCD) and response surface methodology are conducted. The experiments are carried out in a non-isothermal thermogravimetric analyzer (TGA) coupled with mass spectrometer (MS). Thus, this research reveals that the optimum conditions to produce a maximum amount of H_2 production in this system are at temperature of 700 °C, RSS particle size of 0.125 mm, HDPE particle size of 0.375 mm and percentage of plastics in the mixture of 20 wt%. The experimental and predicted values for H_2 production are 50.123 ± 2.127 and 38.945 vol% respectively.

References

Abnisa, F., Wan Daud, W. M. A., & Sahu, J. N. (2011). Optimization and characterization studies on bio-oil production from palm shell by pyrolysis using response surface methodology. *Biomass and Bioenergy, 35*, 3604–3616.

Brown, J. N., & Brown, R. C. (2012). Process optimization of an auger pyrolyzer with heat carrier using response surface methodology. *Bioresource Technology, 103*, 405–414.

Chin, B. L. F., Yusup, S., Shoaibi, A., et al. (2014). Comparative studies on catalytic and non-catalytic co-gasification of rubber seed shell and high density polyethylene mixtures. *Journal of Cleaner Production, 70*, 303–314.

Corella, J., Orío, A., & Aznar, P. (1998). Biomass gasification with air in fluidized bed: reforming of the gas composition with commercial steam reforming catalysts. *Industrial and Engineering Chemistry Research, 37*, 4617–4624.

Cui, L.J., Lin, W.G., & Yao, J.Z. (2006). Influences of temperature and coal particle size on the flash pyrolysis of coal in a fast-entrained bed. *Chemical Research in Chinese Universities, 22*, 103–110.

Isa, K. M., Daud, S., Hamidin, N., et al. (2011). Thermogravimetric analysis and the optimisation of bio-oil yield from fixed-based pyrolysis of rice husk using response surface methodology (RSM). *Industrial Crops and Products, 33*, 481–487.

Luo, S., Xiao, B., Hu, Z., et al. (2010). Effect of particle size on pyrolysis of single-component municipal solid waste in fixed bed reactor. *International Journal of Hydrogen Energy, 35*, 93–97.

Lv, P. M., Xiong, Z. H., Chang, J., et al. (2004). An experimental study on biomass air-steam gasification in a fluidized bed. *Bioresource Technology, 95*, 95–101.

Moghadam, R. A., Yusup, S., Lam, H. L., et al. (2013). Hydrogen production from mixture of biomass and polyethylene waste in fluidized bed catalytic steam co-gasification process. *Chemical Engineering Transactions, 35*, 565–570.

Pinto, F., Franco, C., André, M., et al. (2002). Co gasification study of biomass mixed with plastic wastes. *Fuel, 81*, 291–297.

Turn, S., Kinoshita, C., Zhang, Z., et al. (1998). An experimental investigation of hydrogen production from biomass gasification. *Journal of Hydrogen Energy, 23*, 641–648.

Van Kasteren, J. M. N. (2006). Co gasification of wood and polyethylene with the aim of CO and H_2 production. *Journal of Material Cycles and Waste Management, 8*, 95–98.

Chapter 12
Studies on Effect of Process Parameters Variation on Bio-oil Yield in Subcritical and Supercritical Hydrothermal Liquefaction of Malaysian Oil Palm Biomass

Yi Herng Chan, Suzana Yusup, Armando T. Quitain, Raymond R. Tan, Yoshimitsu Uemura, Mitsuru Sasaki, and Hon Loong Lam

Introduction

The global energy crisis associated with the depletion of non-renewable energy sources such as fossil fuels and the environmental impacts due to their utilization have raised worldwide concerns on the needs to resort to renewable energy sources. The outlook for various renewable energy sources such as wind, solar, hydropower, geothermal, nuclear and biomass has been intensively researched worldwide in order to reduce dependence on non-renewable energy sources as well as mitigate environmental problems.

Bio-oil or biocrude is a dark brown liquid product derived from biomass through pyrolysis or liquefaction reaction. Bio-oil is comprised of oxygenated compounds, various organic acids and many other organic compounds such as aldehydes,

Y.H. Chan • S. Yusup (✉)
Biomass Processing Lab, Department of Chemical Engineering, Centre for Biofuel and Biochemical Research, Green Technology MOR, Universiti Teknologi PETRONAS, Perak, Malaysia
e-mail: drsuzana_yusuf@petronas.com.my

A.T. Quitain • M. Sasaki
Graduate School of Science and Technology, Department of Applied Chemistry and Biochemistry, Faculty of Engineering, Kumamoto University, Kumamoto, Japan

R.R. Tan
Chemical Engineering Department, Centre for Engineering and Sustainable Development Research, De La Salle University, Manila, Philippines

Y. Uemura
Department of Chemical Engineering, Centre for Biofuel and Biochemical Research, Green Technology MOR, Universiti Teknologi PETRONAS, Perak, Malaysia

H.L. Lam
Department of Chemical and Environmental Engineering, The University of Nottingham, Malaysia Campus, Serangor, Malaysia

P. Ravindra (ed.), *Advances in Bioprocess Technology*,
DOI 10.1007/978-3-319-17915-5_12

ketones, phenols, alcohols and polyaromatic hydrocarbons (PAHs). As a result, raw bio-oil needs to be further processed or upgraded to usable liquid transportation fuels and value-added products, such as food flavorings, resins, agro-chemicals, fertilizers and fine chemicals (Hew et al. 2010).

Generally, there are two thermochemical conversion methods that can be used to produce bio-oil from biomass; pyrolysis and liquefaction. Pyrolysis of biomass is the thermal decomposition and degradation of organic substances in the absence of oxygen near atmospheric pressure and high temperatures of up to 400 °C and above to generate a mixture of condensable vapor, gases and char as products (Abdullah and Gerhauser 2008). The vapor produced in the reaction will then be rapidly quenched and condensed to form bio-oil. Liquefaction of biomass is a process whereby biomass is treated with a solvent at high temperature and pressure. The heated and pressurized solvent acts as a good medium and reactant to break down the complex structure of biomass, thereby producing liquid products (Savage et al. 2010). Bio-oil produced is then extracted from the liquid products using organic solvents. Recently, it is found that liquefaction of biomass is a promising method to convert biomass to bio-oil as this process is attractive in terms of energy consumption and cost reduction compared to pyrolysis process because drying of biomass is not necessary. This gives an advantage to liquefaction process as most biomass usually contain high moisture content and drying them will require substantial energy consumption due to the large latent heat of water vaporization (Qian et al. 2007). In addition, liquefaction processes usually take place at a lower temperature (about 250–400 °C) and higher pressure (5–20 MPa) compared to that of pyrolysis. Apart from that, systematically controlled liquefaction operation is able to produce better quality bio-oil with yield comparable to that of fast pyrolysis process and low amounts of gas or char byproducts (Akhtar et al. 2010).

In contrast to pyrolysis process in which biomass decomposes thermolytically via de-polymerization and cracking in an inert condition, liquefaction produces bio-oil from the solvolytic break-down of biomass structure through hydrolysis and hydrogenation reactions, thus leading to different product properties (such as yield and composition of products, energy density, thermal and storage stabilities) (Jena and Das 2011). In addition, the use of solvent in liquefaction process dilutes the concentration of products. This prevents the cross-linking and reverse reactions which usually occur in the case of slow pyrolysis and gasification at lower temperature and lead to the formation of tar and char (Aysu and Kucuk 2013; Liu and Zhang 2008).

The properties of solvent change drastically from ambient condition to subcritical and supercritical conditions. Due to these changes in the physical-chemical properties of the solvents, it has enabled subcritical and supercritical fluids to be utilized and employed in various applications, such as in food and flavoring industries, petrochemicals and pharmaceuticals sectors, extraction of useful compounds from plant materials and environmental protection (Sahu 2003), in addition to conversion of solid biomass to biofuels. Water becomes an excellent reaction medium when it is under elevated temperatures and pressures (Kruse et al. 2013). Good properties such as low dielectric constant, weak hydrogen bond, high

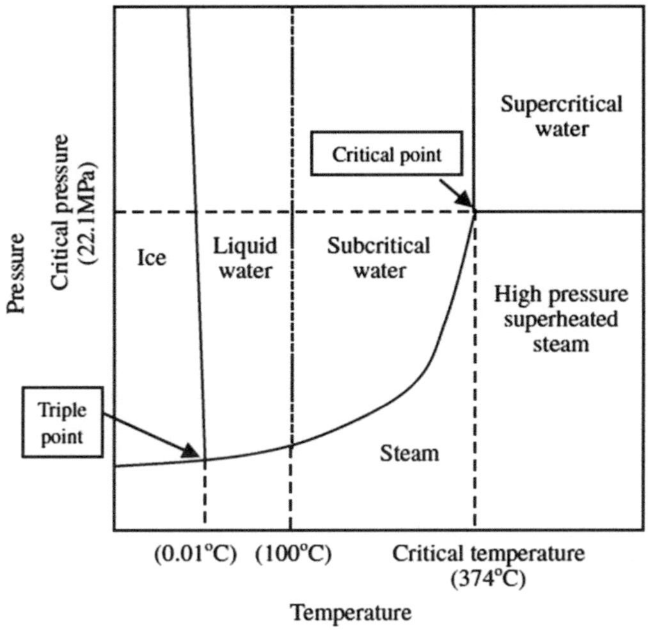

Fig. 12.1 Phase diagram of water (Okajima and Sako 2014)

isothermal compressibility and reactivity as compared to ambient liquid water has enabled subcritical and supercritical water to be utilized in various synthesis and degradation reactions, which include biomass degradation and decomposition to form gas, liquid and solid products. In this context, water not only acts as a solvent and medium where liquefaction of biomass occurs, it also serves as a reactant and catalyst in the process. As water is changed from its ambient conditions to heated and pressurized state, the ionic product of water increases, promoting the partial dissociation of water molecules into H_3O^+ and OH^- ions, enhancing reactions which are usually catalyzed by acids or bases (Kruse et al. 2013; Brunner 2009). Besides, the lower dielectric constant and weak hydrogen bond due to the decrease in dipole moment as water changes from ambient condition to subcritical and supercritical conditions makes water an excellent solvent to absorb and extract organic compounds and non-ionic species which are normally insoluble (Brunner 2009). The phase diagram of water is shown in Fig. 12.1.

In the present work, the effects of three process parameters that include temperature, pressure and reaction time on the bio-oil yield are investigated. The temperature, pressure and reaction time are varied between 330–390 °C, 25–35 MPa and 30–240 min, respectively. Then, a basic life cycle assessment (LCA) of a conceptual liquefaction process is constructed and its impacts to the environment are discussed.

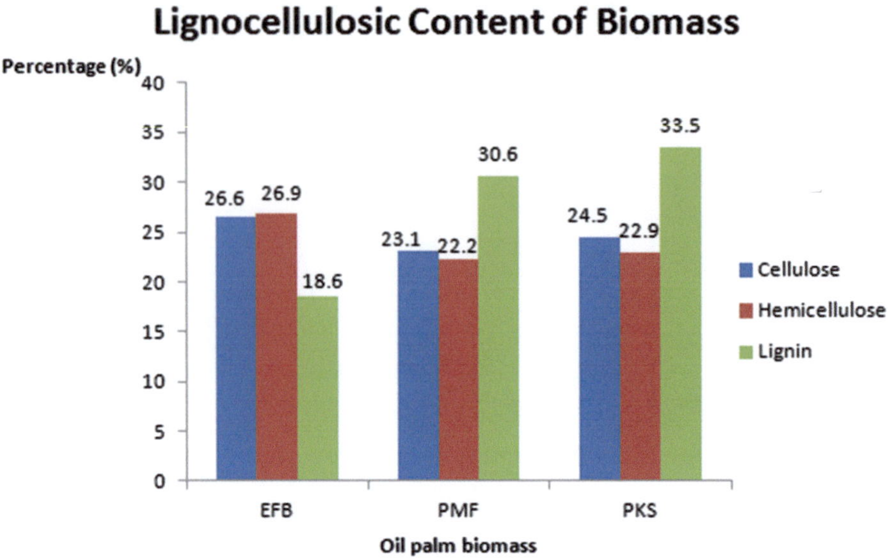

Fig. 12.2 Lignocellulosic contents of EFB, PMF and PKS

Methodology

Biomass Feedstock Pre-treatment and Characterization

Oil palm biomass feedstocks (EFB, PMF, PKS) used in this study were collected from FELCRA Nasaruddin Oil Palm Mill, Bota, Perak, Malaysia. These biomass feedstocks were treated before being characterized and utilized in all the experimental runs. EFB, PMF and PKS were washed thoroughly to remove dust, sand and other impurities. Then, they were dried in an oven at 80 °C for 48 h to remove moisture. Lastly, they were grinded with FRITSCH Cutting Mill and sieved to particle size of <710 μm which will be utilized in the experiments. Biomass samples were sent to Forest Research Institute Malaysia (FRIM) for analysis on the structural content which consists of hemicellulose, cellulose and lignin, while the remaining components were assumed to be extractives and ash, as reported in the literature pertaining to characterization of lignocellulosic biomass (Kelly-Yong et al. 2007). The lignocellulosic content of EFB, PMF and PKS are shown in Fig. 12.2.

Experimental Apparatus and Procedures

Effect of Temperature and Pressure

In this study, the liquefaction condition is categorized into two conditions, namely subcritical condition and supercritical condition. The conditions of hydrothermal

Table 12.1 Reaction conditions (temperature and pressure) for liquefaction using water

Liquefaction conditions	Temperature (°C)	Pressure (MPa)	Density (kg/m^3)	Average weight of biomass feedstock (g)	Average weight of distilled water (g)
Subcritical water	330	25	680.56	0.5989	5.9889
	330	30	692.60	0.6095	6.0949
	330	35	703.19	0.6188	6.1880
	360	25	588.90	0.5182	5.1823
	360	30	614.08	0.5404	5.4039
	360	35	632.95	0.5570	5.5700
Supercritical water	390	25	216.95	0.1909	1.9092
	390	30	466.45	0.4105	4.1048
	390	35	528.40	0.4650	4.6499

liquefaction were set by selecting appropriate temperatures and pressures corresponding to the respective conditions, with the selection of liquefaction temperature at 330, 360 or 390 °C and pressure at 25, 30 or 35 MPa. As the liquefaction experiments involved subcritical and supercritical water, suitable temperature range of 330–390 °C was chosen as high temperatures (above 400 °C) would increase the dominance of gasification reaction (Meryemoğlu et al. 2014). The pressure range of 25–35 MPa was chosen since it is above the critical pressure of water (22.1 MPa) to ensure that water is in the subcritical and supercritical state (Okajima and Sako 2014). EFB, PMF and PKS were subjected to the reaction conditions as shown in Table 12.1.

The liquefaction of oil palm biomass was carried out in an Inconel batch reactor with internal volume of 8.8 ml. Prior to the experiment, the densities of water at the experimental conditions (Table 12.1) were determined using Water V3.3 software developed by Summit Research Corporation (Santa Fe, USA). Knowing the density of water at a particular temperature and the volume of the reactor, suitable amount of distilled water was loaded into the reactor such that it would produce the estimated equilibrium pressure when the system reaches the set reaction temperature and does not fluctuate throughout the course of liquefaction. Biomass was loaded into the reactor with biomass to water ratio of 1:10. The selection of biomass to water ratio of 1:10 in this study was based on the works reported in the literature, in which liquefaction experiments were optimized at this condition (Bach et al. 2014; Chen et al. 2012; Anastasakis and Ross 2011; Mazaheri et al. 2010a, b). The average weights of biomass feedstock and distilled water loaded into the reactor are shown in Table 12.1. Once the biomass sample and distilled water were loaded into the reactor, the mixture was then inserted to the furnace and heated from atmospheric condition to the desired reaction temperature. In this study, the reaction time of 1 h was selected as a basis because this time frame was based on the literature studies (Barreiro et al. 2013) in which completion of the reaction were observed. Furthermore, within this time frame, most of the solid biomass sample was able to be liquefied in the reactor used. After 1 h, the reactor was removed from the furnace and quenched in water to room temperature. The reactor was washed and its content was extracted thoroughly with toluene as an organic extracting solvent, which has good solubility towards degradation products of lignocellulosic materials and has been used for the extraction of degradation products of lignin-model compounds (Radoykova et al. 2013). The product mixture was then allowed to separate into three levels of distinct phases based on the difference in density, with toluene-soluble extract which contained bio-oil on the top, followed by toluene insoluble phase (solid residues) and aqueous phase in the middle and bottom levels, respectively. The toluene-soluble phase was then filtered to remove any suspended solid particles. After that, the toluene was separated from the filtrate using rotary evaporator operating under vacuum pressure (88 hPA) at 57 °C. Bio-oil was obtained after toluene was evaporated. The bio-oil yield was calculated using (12.1) as follows. The experiment at each reaction condition was repeated twice to ensure the repeatability of the data obtained.

$$\text{Bio-oil yield (wt\%)} = \frac{\text{Weight of bio-oil obtained (g)}}{\text{Weight of biomass used (g)}} \times 100 \qquad (12.1)$$

Effect of Reaction Time

The effect of liquefaction time on the bio-oil yields was also studied. In this study, experiments were conducted for temperatures 330, 360 and 390 °C at 30 MPa and at different reaction times of 30 min, 1 h, 2 h and 4 h. The procedures and steps in conducting the liquefaction experiments and collecting bio-oil were similar to that as described in section "Effect of Temperature and Pressure". The kinetic data of bio-oil yields obtained was discussed and analyzed.

Life Cycle Assessment (LCA)

The LCA conducted was based on a conceptual industrial biomass liquefaction process for bio-oil production, with some of the stages mimicking the experimental steps employed in this study, such as the milling of biomass feedstock and recovery of bio-oil using toluene as organic solvent. To systematically conduct the life cycle assessment, the International Organization for Standardization developed the standard ISO 14040 series LCA standards (ISO 14040, 1997). Based on the standards defined, there are four major phases in a complete and comprehensive LCA. The LCA reported in this study is coherent with the mentioned standards: (1) goal and scope definition, (2) life cycle inventory (LCI), (3) life cycle impact assessment (LCIA), and (4) interpretation.

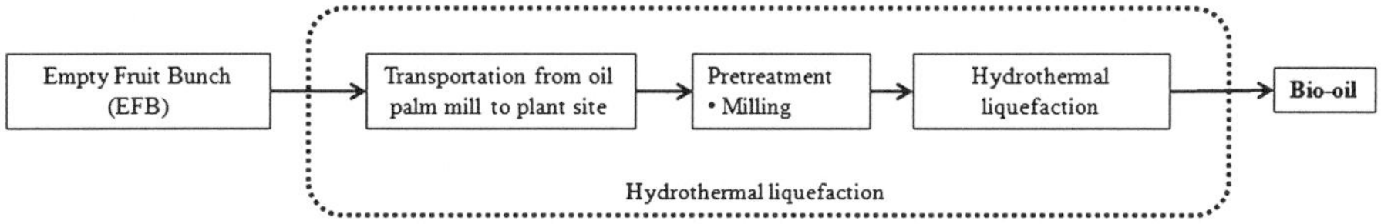

Fig. 12.3 System boundary defined by dashed-line box for bio-oil production from hydrothermal liquefaction of EFB

Goal and Scope Definition

The goal of this LCA study is to identify and compare the environmental impacts of thermochemical conversion process of converting oil palm empty fruit bunch (EFB) to bio-oil via hydrothermal liquefaction. Figure 12.3 shows the system boundary of this study. In this study, the functional unit selected was one kilogram (1 kg) of bio-oil produced.

Life Cycle Inventory (LCI)

In this study, empty fruit brunch is assumed to be collected and transported from oil palm mill to the processing plant site using diesel truck. The transportation distance is assumed to be 100 km. In this study, transportation distance will not be a factor influencing the outcome of the assessment. The feedstock (EFB) will undergo pre-treatment stage before being subjected to hydrothermal liquefaction for bio-oil production. In this context, it is assumed that in the industrial hydrothermal liquefaction process, there is no drying of EFB feedstock in the pre-treatment process. The electricity consumption for milling and other auxiliary services at this stage was determined and adjusted based on the data reported in the literature (Iribarren et al. 2012). After being milled at the pre-treatment stage, the EFB feedstock is subjected to hydrothermal liquefaction. There are several sub-processes that are involved in this stage, which include hydrothermal liquefaction of EFB to bio-oil and other by-products, extraction of bio-oil using toluene as solvent, and bio-oil recovery. In this context, the bio-oil yield was based on the data obtained from experimental work conducted in this study whereas the yields for by-products such as ash, char and gaseous products were obtained from the literature (Akhtar and Amin 2011; Butler et al. 2011; Sulaiman and Abdullah 2011; Abdullah and Gerhauser 2008) and aqueous product yield was computed based on mass balance. The electricity consumption of this stage was adjusted and modified based on the physical property of typical bio-oils (Chan et al. 2014; Chen et al. 2012) and a similar hydrothermal liquefaction process reported in the literature (Liu et al. 2013). Due to the lack of information reported in the literature in the context of solvent loss in the extraction of bio-oil, it is estimated that the solvent lost in the hydrothermal liquefaction process is similar to that of soybean oil extraction using hexane (Sangaletti-Gerhard et al. 2014).

Life Cycle Impact Assessment (LCIA)

In this study, the emissions related to the process stages studied in the system boundary (Fig. 12.3) were used to evaluate five environmental impact categories. These impact categories are climate change, acidification, eutrophication, photo-oxidant formation, human toxicity. The impacts were calculated from various emissions based on the conversion factors reported in the Handbook of Life Cycle Assessment (Guinée 2002).

Interpretation of LCA Results

The end results of this LCA study are the impacts of the conceptual industrial process of biomass liquefaction for bio-oil production to the environment based on the impact categories listed in Section "Life Cycle Impact Assessment (LCIA)". The results were interpreted and discussed.

Results and Discussion

Effect of Temperature and Pressure

The bio-oil yields from EFB, PMF and PKS at different liquefaction conditions are shown in Table 12.2. The optimum hydrothermal liquefaction condition for bio-oil production was found to be at the supercritical condition of water (390 °C and 25 MPa) for EFB, PMF and PKS.

In this study, it is observed that the bio-oil yield increased with decreasing density and dielectric constant of water. The optimum hydrothermal liquefaction conditions for bio-oil production was at 390 °C and 25 MPa for EFB, PMF and PKS, which corresponds to the lowest density and dielectric constant of water

Table 12.2 Bio-oil yields from hydrothermal liquefaction of EFB, PMF and PKS

			Bio-oil yield (wt%)		
Reaction condition	T (°C)	P (MPa)	EFB	PMF	PKS
Subcritical water	330	25	15.72 ± 1.05	16.40 ± 0.21	22.83 ± 0.56
	330	30	16.96 ± 1.18	15.72 ± 0.81	25.62 ± 1.87
	330	35	15.68 ± 0.68	14.79 ± 0.21	23.67 ± 0.27
	360	25	25.31 ± 0.44	22.71 ± 0.83	26.55 ± 1.29
	360	30	26.02 ± 0.44	23.22 ± 0.54	27.54 ± 0.70
	360	35	22.76 ± 0.28	21.75 ± 0.01	23.44 ± 1.24
Supercritical water	390	25	**37.39 ± 0.67**	**34.32 ± 1.87**	**38.53 ± 1.46**
	390	30	28.17 ± 0.35	27.57 ± 1.03	31.16 ± 0.81
	390	35	30.16 ± 0.98	24.07 ± 1.04	29.35 ± 0.71

Table 12.3 Densities and dielectric constants of water

Liquefaction conditions	Temperature (°C)	Pressure (MPa)	Density of water (kg/m^3)	Dielectric constant
Subcritical water	330	25	680.56	17.74
	330	30	692.60	18.23
	330	35	703.19	18.66
	360	25	588.90	13.41
	360	30	614.08	14.31
	360	35	632.95	14.99
Supercritical water	390	25	216.95	3.18
	390	30	466.45	8.91
	390	35	528.40	10.78

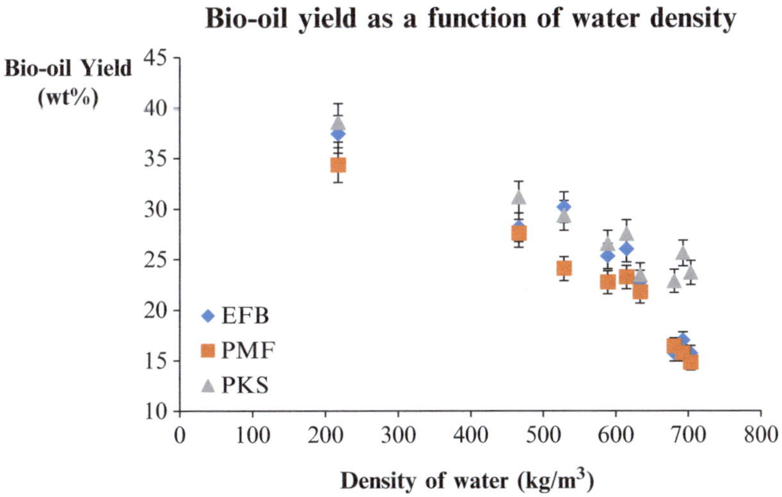

Fig. 12.4 Bio-oil yield in relation with water density

explored in this study. The densities and dielectric constants of water in all experimental runs were determined using Water V3.3 software developed by Summit Research Corporation (Santa Fe, USA) and are shown in Table 12.3.

Low density of supercritical water at 390 °C and 25 MPa leads to high diffusivity and compressibility of water, and hence water is able to penetrate more efficiently into the matrix structure of biomass, achieving higher degrees of liquefaction (Wen et al. 2009; Kruse and Dinjus 2007). In addition, low dielectric constant of supercritical water at 390 °C and 25 MPa increases the solvation power of water to dissolve and extract organic materials which are normally water insoluble (Brunner 2009; Kruse and Dinjus 2007;). Hence, supercritical fluid has enhanced solubility for organic compounds compared to a conventional liquid or gas solvent, making supercritical fluids ideal for separation and extraction of useful products (Wen et al. 2009). Figures 12.4 and 12.5 show the trends of bio-oil yield with respect to densities and dielectric constants of water, respectively.

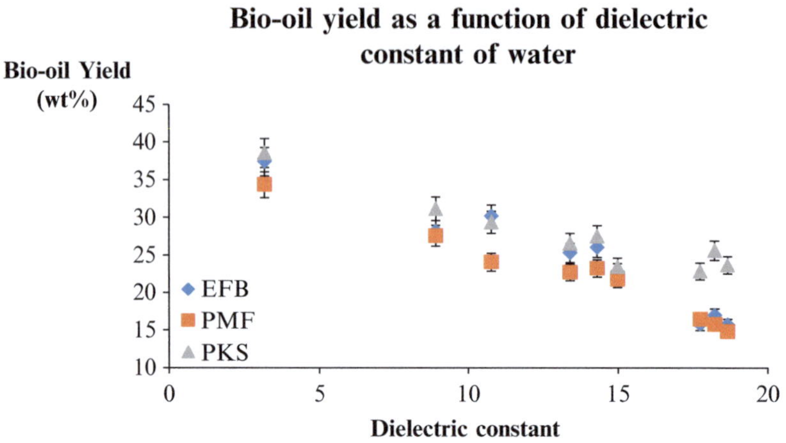

Fig. 12.5 Bio-oil yield in relation to dielectric constant of water

The difference in the bio-oil yields from EFB, PMF and PKS is most probably due to the influence of chemical composition of the feedstocks (hemicellulose, cellulose and lignin). At subcritical conditions, major lignocellulosic contents of biomass being degraded are hemicellulose and cellulose. Hemicellulose composed of various side-chains and hence has a less uniform structure and lower degree of crystallinity (Toor et al. 2011). It can be easily decomposed at lower temperatures between 210 and 330 °C (Zhou et al. 2013). Cellulose consists of long polymers of glucose units without branches and thus has a higher degree of crystallinity. This requires higher temperatures of 300–375 °C to break down and degrade the cellulose from the matrix structure of biomass (Zhou et al. 2013). At supercritical conditions (above $T_c = 374.3$ °C; $P_c = 22.1$ MPa), lignin is the major lignocellulosic component being degraded. Lignin has the highest thermal stability compared to hemicellulose and cellulose due to its highly crossed-linked polyphenolic aromatic structure with no ordered repeating units. As such, lignin can be decomposed at a wider range of temperatures between 150 and 1,000 °C (Zhou et al. 2013). The maximum bio-oil yield obtained in this study was 38.53 wt% from PKS due to its highest lignin content compared to that of EFB and PMF. Hence, the difference in the bio-oil yields from different types of biomass was attributed by the chemical composition of the biomass, as was also indicated by previous work (Demirbas 2000a, b).

Generally, optimizing bio-oil yield is primarily finding a balance between reactions that lead to the formation and destruction of bio-oil components. In this context, temperature and pressure are the primary parameters to be controlled. From Table 12.2, higher temperature increased the bio-oil yield as it increased the rate of decomposition and cracking of lignocellulosic components from the matrix structure of biomass to bio-oil components. As the thermal stability of lignocellulosic materials increases in the order of hemicellulose, cellulose and lignin, higher temperature enhanced the extent of decomposition of these materials, thus giving rise to increase in bio-oil yield. At higher temperatures, hydrolysis of biomass structure proceeds dominantly via radical mechanism, leading to higher yield of low-polarity compounds, which are captured as bio-oil using toluene (Kus 2012). Similar trend was also observed from the literature (Aysu and Kucuk 2013). However, it is important to note that beyond the temperature range explored in this study (above 390 °C), further increase in temperature from a particular optimum temperature may cause the bio-oil yield to decrease, as commonly observed in several works (Chen et al. 2012; Akhtar and Amin 2011; Qian et al. 2007). This is due to the occurrence of competing reactions such as gasification, condensation, cyclization and repolymerization at higher temperatures that convert bio-oil components formed in the initial stages of the liquefaction process to char, water-soluble and gaseous products which cannot be extracted as bio-oil (Valdez et al. 2012; Savage et al. 2010).

The dependency of bio-oil yields on the pressure of the system did not show a clear trend in this study. Higher pressure is known to increase the solvent density and solubility of the target biomass components, allowing solvent to penetrate more efficiently into the biomass molecular structure, enhancing degradation of biomass structure and extraction of bio-oil components. However, increased pressure may also lead to increase in local solvent density, causing a caging effect for the C-C bonds in biomass, which inhibits C-C bonds cleavage and fragmentation (Akhtar and Amin 2011). In subcritical conditions, the properties of water (density and dielectric constant) do not change as much as they do as in the case of supercritical conditions. In supercritical conditions, decreased bio-oil yield at higher pressures may be attributed to the gasification of bio-oil components into permanent gases due to increasing free radical reactions (Kruse et al. 2013), hence lead to lower amounts of toluene-extractable compounds.

Effect of Reaction Time

The bio-oil yield at different reaction time at various conversion temperatures (330, 360 and 390 °C) and 30 MPa for hydrothermal liquefaction of EFB, PMF and PKS are reported in this section. Figures 12.6, 12.7 and 12.8 presents the kinetic data

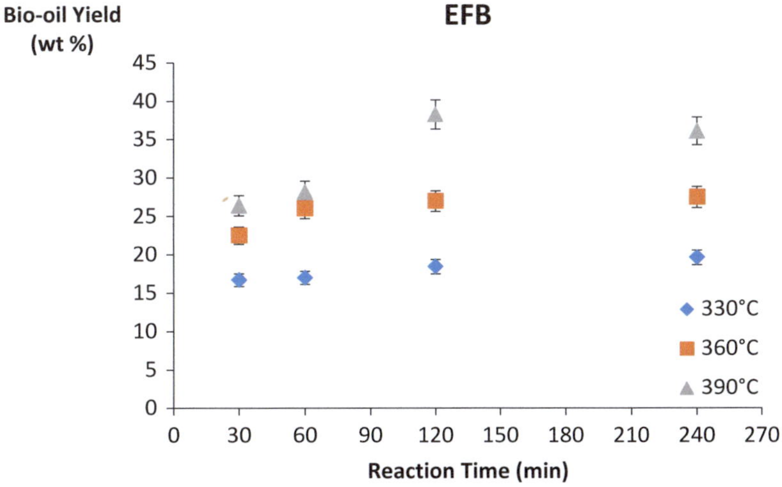

Fig. 12.6 Bio-oil yield for liquefaction of EFB at different reaction time duration

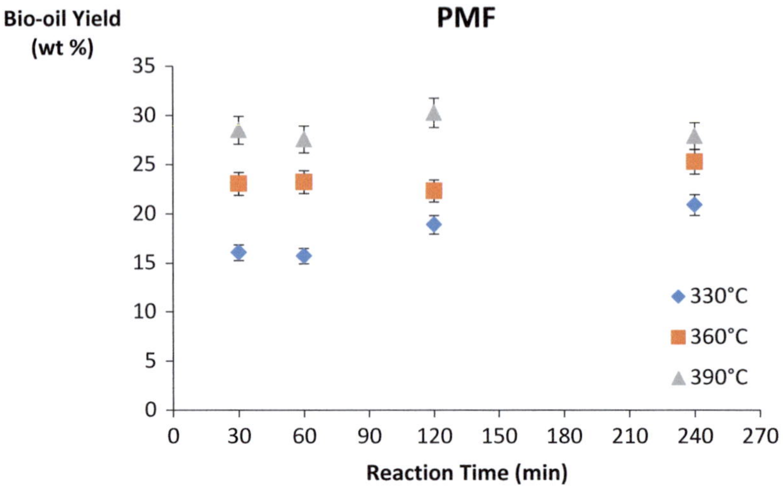

Fig. 12.7 Bio-oil yield for liquefaction of PMF at different reaction time duration

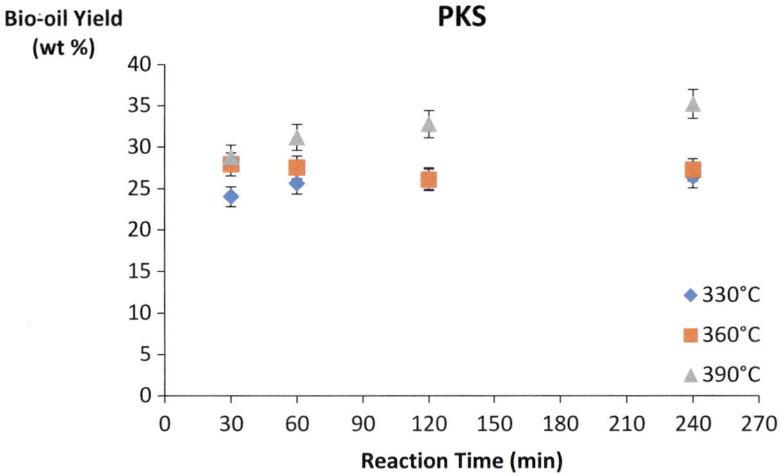

Fig. 12.8 Bio-oil yield for liquefaction of PKS at different reaction time duration

for bio-oil yield with respect to varying reaction time for EFB, PMF and PKS, respectively.

Based on the data obtained, it is observed that the reaction time had less significant effect on the bio-oil yield compared to that of reaction temperature. In this context, it is seen that the largest variation in bio-oil yield due to the variation in reaction temperature across the entire range explored in this study was higher compared to the entire range of reaction time explored in this study. The largest variation in bio-oil yield from liquefaction of EFB, PMF and PKS due to the increase in temperature from 330 to 390 °C, is 19.83, 12.42 and 8.8 wt%, respectively, at the same reaction time. However, the increase in reaction time from 30 to 240 min caused lower impact to the bio-oil yield from EFB, PMF and PKS, with the largest difference of 11.85, 5.17 and 6.4 wt%, respectively. This shows that bio-oil yield was more sensitive to reaction temperature compared to reaction time, indicating that hydrothermal liquefaction was a thermal controlled process and temperature is a critical parameter for organic conversion via hydrothermal liquefaction, as reported by Wang et al. (2013) and Jena et al. (2011).

In all cases, at a fixed reaction time, the bio-oil yield from liquefaction of EFB, PMF and PKS generally increased with increasing reaction temperature. This is expected because as temperature increases, the rate of decomposition, fragmentation and hydrolysis of lignocellulosic content increases, hence giving rise to an increase in bio-oil yield (Aysu and Kucuk 2013; Demirbas 2000a, b). At high temperature of 390 °C, all the three components (cellulose, hemicelluloses and lignin) in the biomass was decomposed more effectively via hydrolysis compared to liquefaction at lower temperature of 330 and 360 °C. Besides, lower density and dielectric constant of water at 390 °C compared to that of 330 and 360 °C was able to liquefy EFB, PMF and PKS and produce bio-oil more effectively.

In this study, the effect of reaction time on the bio-oil yield was insignificant, as shown in Fig. 12.6, 12.7 and 12.8. Across the entire range of reaction time explored in this study, the bio-oil yield for EFB, PMF and PKS does not change for more than 5.17 wt%, for all cases, except bio-oil yield from EFB and PKS at 390 °C, which caused a difference of 11.85 and 6.4 wt% in bio-oil yield, respectively. This may be attributed to the properties of water at supercritical condition, as the density and dielectric constant of water dropped drastically beyond the critical point of water (Table 12.3).

The optimum reaction time for bio-oil production from EFB and PMF was 120 min, with bio-oil yield of 38.24 and 30.26 wt%, respectively. At prolonged reaction time of 240 min, the bio-oil yield decreased slightly, probably due to the secondary decomposition of bio-oil to solid and gaseous products, via multiple reactions such as recondensation, repolymerization and gasification. This trend is also reported by several previous studies (Bach et al. 2014; Valdez et al. 2012; Anastasakis and Ross 2011; Jena et al. 2011). For PKS, the optimum condition for bio-oil production was 240 min, as shown in Fig. 12.8. At 390 °C, the bio-oil yield from liquefaction of PKS continued to increase when the reaction time was further prolonged from 120 to 240 min. This may be attributed to the high lignin content of the PKS, which requires longer time to decompose and form bio-oil completely. At lower temperatures of 330 and 360 °C, the bio-oil yield reached plateau from 120 min onwards. This may be due to the decomposition rate of the biomass reaching an equilibrium rate, hence prolonged reaction time did not change the bio-oil yield significantly (Miao et al. 2012).

Life Cycle Assessment (LCA)

Life Cycle Inventory (LCI) Analysis

The life cycle inventory data for conceptual industrial hydrothermal liquefaction processes are presented in Table 12.4. These data were based on relevant literature

Table 12.4 LCI of bio-oil production via hydrothermal liquefaction

Parameters/item	Amount	Unit	Remark	Assumption	References
Transportation					
Fresh empty fruit bunch (wet basis)	6.87	kg	–	–	–
Travel distance	100	km	–	Assumption	–
Transportation energy: diesel truck	0.63	MJ	0.91 MJ diesel/ ton-km	–	Joelsson and Gustavsson (2010)
Pretreatment (milling)					
Electricity	0.49	kWh	Calculated based on weight of biomass for pretreatment	–	Iribarren et al. (2012)
Hydrothermal liquefaction					
Pretreated EFB	2.67	kg	–	Dried from moisture content of 65 wt % to 10 wt%	–
Bio-oil	1.00	kg	Based on yield obtained from experimental work; 37.39 wt%	–	–
Ash	0.13	kg	Based on typical ash content of EFB; 5 %		Butler et al. (2011); Sulaiman and Abdullah (2011); Abdullah et al. (2010); Abdullah and Gerhauser (2008)
Gas	0.56	kg	Based on typical yield of 20.8 wt%	–	Mazaheri et al. (2010a, b)
Char	0.35	kg	Calculated based on correlation of char yield and lignin content of EFB	–	Akhtar and Amin (2011)
Aqueous product	0.63	kg	Calculated from mass balance	–	–
Make up solvent (toluene)	0.01	kg	–	Assume solvent loss is similar to that of soybean extraction using hexane	Sangaletti-Gerhard et al. (2014)
Heat (liquefaction and solvent recovery)	1.29	MJ	Adjusted from data provided in the literature based on density of typical bio-oils	Natural gas fired boiler with 80 % efficiency	Chan et al. (2014); Liu et al. (2013); Chen et al. (2012)
Electricity	0.04	kWh	Adjusted from data provided in the literature based on density of typical bio-oils	–	Chan et al. (2014); Liu et al. (2013); Chen et al. (2012)

as mentioned in section "Life Cycle Assessment". The emissions associated with each activity are summarized in Table 12.5. These emissions were based on the data published and reported in the literature (Economic Input–Output Life Cycle Assessment 2014; Sathre 2014; Iribarren et al. 2012; Shekarchian et al. 2011). The emissions considered in this study include carbon dioxide (CO_2), carbon monoxide (CO), volatile organic compounds (VOC), nitrogen oxides (NO_x), particulate matters (PM), sulfur dioxide (SO_2), methane (CH_4) and nitrous oxide (N_2O). In this study, as char is the by-product of the hydrothermal liquefaction process and it can be intended for soil amendment, the carbon abatement associated with the generation of char was also considered (Hammond et al. 2011).

Life Cycle Impact Assessment (LCIA)

The environmental impacts due to the emissions associated with the activities considered in this study are global warming potential (GWP), acidification, eutrophication, toxicity and photo-oxidant formation. The emissions determined in section "Life Cycle Inventory (LCI) Analysis" were normalized to the environmental impacts based on the conversion factors reported in the Handbook of Life Cycle Assessment (Guinée 2002). The conversion factors are shown in Table 12.6. Figure 12.9 shows the plot of GWP indicating the emission and abatement of CO_2 in kg equivalent whereas Figure 12.10 shows the plot of other environmental impacts.

Interpretation of LCA Results

Based on Figure 12.9, it is estimated that for every kilogram (kg) of bio-oil produced, 3.55 kg of CO_2 equivalent is produced throughout its life cycle. In this context the contribution to the global warming potential (GWP) is due to the release of greenhouse gases, particularly CO_2, CH_4 and N_2O (Table 12.6). The emission of CO_2 estimated in this LCA study contributes 99.78 % of the GWP, with 98.49 % of CO_2 emitted due to the electricity consumption associated with equipment at the pretreatment stage and bio-oil production stage of the process. Similar situation is reported in the literatures (Fortier et al. 2014; Ning et al. 2013; Khoo 2009). As bio-char is produced as by-product in the process, its prospective application to the land is able to reduce atmospheric greenhouse gas levels (Hammond et al. 2011). Hence, the carbon abatement (carbon credit) due to the production of bio-char was also considered and it is shown in Figure 12.9, where 1.26 kg of CO_2 equivalent is abated per 1 kg of bio-oil produced. This has led to 35.50 % reduction of the overall emission of CO_2 equivalent, giving a net GWP of 2.29 kg CO_2 equivalent per kg of bio-oil produced.

Table 12.5 Environmental emissions associated with the hydrothermal liquefaction process

	Emissions (kg)							
Parameters/item	CO_2	VOC	CO	NO_x	PM	SO_2	CH_4	N_2O
Transportation								
Transportation energy: diesel truck	0.054	0.0050	0.067	0.037	0.0075	0.0030	0.00006	0.0000025
Pretreatment (milling)								
Electricity	3.13	0.0000056	0.0016	0.0099	0.000019	0.024	0.0000052	0.0000029
Hydrothermal liquefaction								
Char	−1.26	–	–	–	–	–	–	–
Make up solvent (toluene)	0.048	0	0.00002	0.00002	0	0.00002	0	0
Heat (liquefaction and solvent recovery)	0.075	0.0000033	0.00005	0.00011	0.0000045	0.0000004	0.00023	0.0000002
Electricity	0.23	0.0000004	0.00012	0.00073	0.0000014	0.0018	0.0000004	0.0000002

Table 12.6 Conversion factor for normalization to environmental impacts

Emissions	Environmental impacts				
	Global warming potential, kg CO_2 eq	Acidification, kg SO_2 eq	Eutrophication, kg PO_4^{3-} eq	Toxicity, kg 1,4-DCB eq	Photo-oxidant formation, kg ethylene
CO_2	1	–	–	–	–
CO	–	–	–	–	0.0027
VOC	–	–	–	–	–
NO_x	–	0.7	0.13	–	–
PM	–	–	–	0.82	–
SO_2	–	1	–	0.096	0.048
CH_4	21	–	–	–	0.006
N_2O	310	–	–	–	–

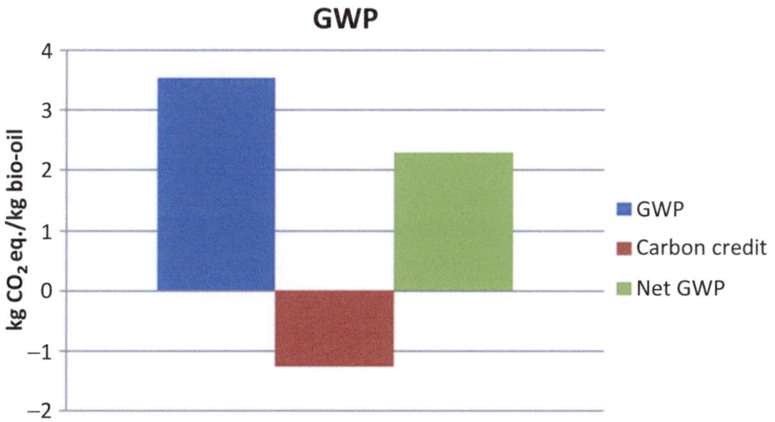

Fig. 12.9 GWP of hydrothermal liquefaction process case study

Compared to GWP, the emission leading to other environmental impacts such as acidification, eutrophication, toxicity and photo-oxidant formation are less severe, as shown in Figure 12.10. For every kilogram of bio-oil produced, 0.062 kg SO_2 equivalent (acidification), 0.006 kg PO_4^{3-} equivalent (eutrophication), 0.009 kg 1,4-dichlorobenzene (DCB) equivalent (toxicity) and 0.002 kg ethylene equivalent (photo-oxidant formation) is produced, respectively. The emission of NO_x contributes to about 54 % of the calculated acidification impact, with 77.60 % of NO_x emitted at the biomass transportation stage due to the use of diesel trucks, similar to the case study reported in the literature (Ning et al. 2013). The remaining 46 % of acidification is contributed by the emission of SO_2, particularly in the biomass pretreatment stage, which makes up 83.30 % of total SO_2 released due to electricity consumption. Similar situation is also reported in the literature (Iribarren et al. 2012). The impact of eutrophication associated with the process is less severe compared to GWP and acidification. In this context, the biomass

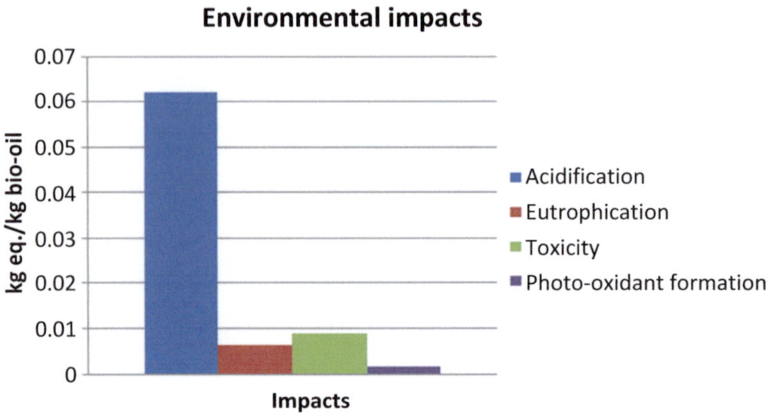

Fig. 12.10 Environmental impacts of hydrothermal liquefaction process case study

transportation stage contributes 77.60 % to the overall eutrophication impact due to the release of NO_x as a result of diesel truck use. The impact of toxicity is mostly contributed by the release of particulate matter (PM) to the air, contributing 69.37 % of the overall toxicity impact. In this study, 99.67 % of particulate matter is estimated from the biomass transportation stage due to the combustion of diesel in the trucks. Last but not least, photo-oxidant formation has the least impact to the environment, with SO_2 emission dominates 88.00 % of the overall impact mainly due to the electricity consumption in the biomass pretreatment stage.

Limitations of This LCA Study

This LCA study conducted is based on a general conceptual industrial process of biomass liquefaction to bio-oils. As such, many technical data, such as the breakdown of utility and electrical consumption at each stage of the actual liquefaction process are not available. The data used in this LCA study were based on the reported data in the literature, which in reality, may deviate from the actual process due to the lack of primary data. Besides, the system boundary in this LCA study excluded the utilization of bio-oil. Hence, the overall cradle-to-grave life cycle analysis of bio-oil is beyond the scope of this LCA study.

Conclusions

In the present work, supercritical water is found to be an effective green solvent in liquefying EFB, PMF and PKS to bio-oil, generating 37.39, 34.32 and 38.53 wt% of bio-oil, respectively, for a reaction time of 1 h. Low density and dielectric constant of supercritical water enables high diffusivity of water into the biomass matrix structure as well as high solubility of organic (bio-oil) compounds in water. In this study, the effects of pressure and reaction time on bio-oil yield are less significant compared to that of temperature. The life cycle assessment of a conceptual industrial biomass liquefaction process for bio-oil production reveals that global warming potential (GWP) is the most affected environmental impact, with a net 2.29 kg CO_2 equivalent per kg of bio-oil produced. The other environmental impacts that include acidification, eutrophication, toxicity and photo-oxidant formation are less significant.

References

Abdullah, N., & Gerhauser, H. (2008). Bio-oil derived from empty fruit bunches. *Fuel, 87*, 2606–2613.

Abdullah, N., Gerhauser, H., & Sulaiman, F. (2010). Fast pyrolysis of empty fruit bunches. *Fuel, 89*, 2166–2169.

Akhtar, J., & Amin, N. A. S. (2011). A review on process conditions for optimum bio-oil yield in hydrothermal liquefaction of biomass. *Renewable and Sustainable Energy Reviews, 15*, 1615–1624.

Akhtar, J., Kuang, S. K., & Amin, N. A. S. (2010). Liquefaction of empty palm fruit bunch (EPFB) in alkaline hot compressed water. *Renewable Energy, 35*, 1220–1227.

Anastasakis, K., & Ross, A. B. (2011). Hydrothermal liquefaction of the brown macro-alga Laminaria Saccharina: Effect of reaction conditions on product distribution and composition. *Bioresource Technology, 102*, 4876–4883.

Aysu, T., & Kucuk, M. M. (2013). Liquefaction of giant fennel (Ferula orientalis L.) in supercritical organic solvents: Effects of liquefaction parameters on product yields and character. *Journal of Supercritical Fluids, 83*, 104–123.

Bach, Q.-V., Sillero, M. V., Tran, K.-Q., Skjermo, J. (2014). Fast hydrothermal liquefaction of a Norwegian macro-alga: Screening tests. *Algal Research*. 10.1016/j.algal.2014.05.009

Barreiro, D. L., Prins, W., Ronsse, F., & Brilman, W. (2013). Hydrothermal liquefaction (HTL) of microalgae for biofuel production: State of the art review and future prospects. *Biomass and Bioenergy, 53*, 113–127.

Brunner, G. (2009). Near critical and supercritical water. Part I. Hydrolytic and hydrothermal processes. *Journal of Supercritical Fluids, 47*, 373–381.

Butler, E., Devlin, G., Meier, D., & McDonnell, K. (2011). A review of recent laboratory research and commercial developments in fast pyrolysis and upgrading. *Renewable and Sustainable Energy Reviews, 15*, 4171–4186.

Carnegie Mellon University Green Design Institute. (2014). *Economic Input-Output Life Cycle Assessment (EIO-LCA) US 2002 (428 sectors) Purchaser model [Internet]*. Retrieved June 22, 2014, from http://www.eiolca.net/

Chan, Y. H., Vi, D. K., Yusup, S., Lim, M. T., Zain, A. M., & Uemura, Y. (2014). Studies on catalytic pyrolysis of empty fruit bunch (EFB) using Taguchi's L9 orthogonal array. *Journal of the Energy Institute, 87*, 227–234.

Chen, Y., Wu, Y., Zhang, P., Hua, D., Yang, M., Li, C., et al. (2012). Direct liquefaction of Dunaliella tertiolecta for bio-oil in sub/supercritical ethanol-water. *Bioresource Technology, 124*, 190–198.

Demirbas, A. (2000a). Effect of lignin content on aqueous liquefaction products of biomass. *Energy Conversion & Management, 41*, 1601–1607.

Demirbas, A. (2000b). Liquefaction of olive husk by supercritical fluid extraction. *Energy Conversion & Management, 41*, 1875–1883.

Fortier, M.-O. P., Roberts, G. W., Stagg-Williams, S. M., & Sturm, B. S. M. (2014). Life cycle assessment of bio-jet fuel from hydrothermal liquefaction of microalgae. *Applied Energy, 122*, 73–82.

Guinée, J. B. (Ed.). (2002). *Handbook on Life Cycle Assessment: Operational guide to the ISO standards*. Dordrecht: Kluwer.

Hammond, J., Shackley, S., Sohi, S., & Brownsort, P. (2011). Prospective life cycle carbon abatement for pyrolysis biochar systems in the UK. *Energy Policy, 39*, 2646–2655.

Hew, K. L., Tamidi, A. M., Yusup, S., Lee, K. T., & Ahmad, M. M. (2010). Catalytic cracking of bio-oil to organic liquid product (OLP). *Bioresource Technology, 101*, 8855–8858.

Iribarren, D., Peters, J. F., & Dufour, J. (2012). Life cycle assessment of transportation fuels from biomass pyrolysis. *Fuel, 97*, 812–821.

ISO 14040. (1997). *Environmental management – Life cycle assessment – Principles and framework*. Geneva: International Organization for Standardisation.

Jena, U., & Das, K. C. (2011). Comparative evaluation of thermochemical liquefaction and pyrolysis for bio-oil production from microalgae. *Energy and Fuels, 25*, 5472–5482.

Jena, U., Das, K. C., & Kastner, J. R. (2011). Effect of operating conditions of thermochemical liquefaction on biocrude production from Spirulina platensis. *Bioresource Technology, 102*, 6221–6229.

Joelsson, J. M., & Gustavsson, L. (2010). Reduction of CO_2 emission and oil dependency with biomass-based polygeneration. *Biomass and Bioenergy, 34*, 967–984.

Kelly-Yong, T. L., Lee, K. T., Mohamed, A. R., & Bhatia, S. (2007). Potential of hydrogen from oil palm biomass as a source of renewable energy worldwide. *Energy Policy, 35*, 5692–5701.

Khoo, H. H. (2009). Life cycle impact assessment of various waste conversion technologies. *Waste Management, 29*, 1892–1900.

Kruse, A., & Dinjus, E. (2007). Hot compressed water as reaction medium and reactant: Properties and synthesis reactions. *Journal of Supercritical Fluids, 39*, 362–380.

Kruse, A., Funke, A., & Titirici, M. M. (2013). Hydrothermal conversion of biomass to fuels and energetic materials. *Current Opinion in Chemical Biology, 17*, 515–521.

Kus, N. S. (2012). Organic reactions in subcritical and supercritical water. *Tetrahedron, 68*, 949–958.

Liu, X., Saydah, B., Eranki, P., Colosi, L. M., Mitchell, B. G., James, R., et al. (2013). Pilot-scale data provide enhanced estimates of the life cycle energy and emissions profile of algae biofuels produced via hydrothermal liquefaction. *Bioresource Technology, 148*, 163–171.

Liu, Z., & Zhang, F. S. (2008). Effects of various solvents on the liquefaction of biomass to produce fuels and chemical feedstocks. *Energy Conversion and Management, 49*, 3498–3504.

Mazaheri, H., Lee, K. T., Bhatia, S., & Mohamed, A. R. (2010a). Subcritical water liquefaction of oil palm fruit press fiber for the production of bio-oil: Effect of catalysts. *Bioresource Technology, 101*, 745–751.

Mazaheri, H., Lee, K. T., Bhatia, S., & Mohamed, A. R. (2010b). Sub/supercritical liquefaction of oil palm fruit press fiber for the production of bio-oil: Effects of solvents. *Bioresource Technology, 101*, 7641–7647.

Meryemoğlu, B., Hasanoğlu, A., Irmak, S., & Erbatur, O. (2014). Biofuel production by liquefaction of kenaf (Hibiscus cannabinus L.) biomass. *Bioresource Technology, 151*, 278–283.

Miao, C., Chakraborty, M., & Chen, S. (2012). Impact of reaction conditions on the simultaneous production of polysaccharides and bio-oil from heterotrophically grown Chlorella sorokiniana by a unique sequential hydrothermal liquefaction process. *Bioresource Technology, 110*, 617–627.

Ning, S. K., Hung, M. C., Chang, Y. H., Wan, H. P., Lee, H. T., & Shih, R. F. (2013). Benefit assessment of cost, energy, and environment for biomass pyrolysis oil. *Journal of Cleaner Production, 59*, 141–149.

Okajima, I., & Sako, T. (2014). Energy conversion of biomass with supercritical and subcritical water using large-scale plants. *Journal of Bioscience and Bioengineering, 117*, 1–9.

Qian, Y., Zuo, C., Tan, J., & He, J. (2007). Structural analysis of bio-oils from sub- and supercritical water liquefaction of woody biomass. *Energy, 32*, 196–202.

Radoykova, T., Nenkova, S., & Valchev, I. (2013). Black liquor lignin products, isolation and characterization. *Journal of Chemical Technology and Metallurgy, 48*, 524–529.

Sahu, S. (2003). Supercritical Fluid Extraction: A Cleaner Technology Option For The Industry. In R. Sanghi & M. M. Srivastava (Eds.), *Green chemistry: Environment friendly alternatives* (pp. 123–145). New Delhi: Narosa Publishing House.

Sangaletti-Gerhard, N., Romanelli, T. L., Vieira, T. M., Ferreirade, S., Navia, R., & Regitano-d'Arce, M. A. B. (2014). Energy flow in the soybean biodiesel production chain using ethanol as solvent extraction of oil from soybeans. *Biomass and Bioenergy, 66*, 39–48.

Sathre, R. (2014). Comparing the heat of combustion of fossil fuels to the heat accumulated by their lifecycle greenhouse gases. *Fuel, 115*, 674–677.

Savage, P. E., Levine, R. B., & Huelsman, C. M. (2010). Hydrothermal Processing of Biomass. In M. Crocker (Ed.), *Thermochemical conversion of biomass to liquid fuels and chemicals* (pp. 192–221). Cambridge: The Royal Society of Chemistry.

Shekarchian, M., Moghavvemi, M., Mahlia, T. M. I., & Mazandarani, A. (2011). A review on the pattern of electricity generation and emission in Malaysia from 1976 to 2008. *Renewable and Sustainable Energy Reviews, 15*, 2629–2642.

Sulaiman, F., & Abdullah, N. (2011). Optimum conditions for maximizing pyrolysis liquids of oil palm empty fruit bunches. *Energy, 36*, 2352–2359.

Toor, S. S., Rosendahl, L., & Rudolf, A. (2011). Hydrothermal liquefaction of biomass: A review of subcritical water technologies. *Energy, 36*, 2328–2342.

Valdez, P. J., Nelson, M. C., Wang, H. Y., Lin, X. N., & Savage, P. E. (2012). Hydrothermal liquefaction of Nannochloropsis sp.: Systematic study of process variables and analysis of the product fractions. *Biomass and Bioenergy, 46*, 317–331.

Wang, F., Chang, Z., Duan, P., Yan, W., Xu, Y., Zhang, L., et al. (2013). Hydrothermal liquefaction of Litsea cubeba seed to produce bio-oils. *Bioresource Technology, 149*, 509–515.

Wen, D., Jiang, H., & Zhang, K. (2009). Supercritical fluids technology for clean biofuel production. *Progress in Natural Science, 19*, 273–284.

Zhou, H., Long, Y. Q., Meng, A. H., Li, Q. H., & Zhang, Y. G. (2013). The pyrolysis simulation of five biomass species by hemi-cellulose, cellulose and lignin based on thermogravimetric curves. *Thermochimica Acta, 566*, 36–43.

Chapter 13
Agro-Residues as Fuel and as a Feedstock for Other Products

Anupama R, Niju S, Meera Sheriffa Begum K.M, and Anantharaman N

Introduction

The life of conventional fossil fuels has become limited in the present era, where the use of energy and their sources has been growing faster than the world population. Along with increase in population, the energy needs is also increasing. The world energy consumption would increase by 53 % by 2035. On the global scale, increase in the emission rates of greenhouse gases produced from the use of these conventional fossil fuels presents a threat to the world climate. With declining fossil fuel resources and increased demand for their products, there is an urgent need to develop economical, eco friendly, energy-efficient processes and resources for the production of fuels and chemicals. Biomass, which contributes to one-seventh of the world-wide energy consumption and for as much as 43 % of the energy consumption in some developing countries, has a great potential to be a renewable source and can be replaced with the conventional fossil fuels. In the present investigation, four types of raw materials namely, rice straw, cassava, cotton seeds and red grams outer cover were pyrolyzed and the physical properties of the bio-oil obtained were determined. The composition and functional groups of cotton seed oil were characterized by Gas Chromatography equipped with Mass Spectrometry (GC-MS) and Fourier Transform Infrared spectroscopy (FTIR) analyses.

Anupama R • Niju S • M.S. Begum K.M • Anantharaman N (✉)
Department of Chemical Engineering, National Institute of Technology, Tiruchirappalli, Tamilnadu 620015, India
e-mail: nanant@gmail.com

P. Ravindra (ed.), *Advances in Bioprocess Technology*,
DOI 10.1007/978-3-319-17915-5_13

Biomass

Biomass is a name given to anything that is living on earth's surface. These are substances that use solar energy for their survival as one of the essential components. Plants use the process of photosynthesis for the production of biomass. It is a general term used to describe all biologically produced matter. Biomass is one of the most important potential and the only carbon-containing renewable energy resource. Biomass mainly consists of cellulose, hemicelluloses, lignin, lipids, proteins, simple sugars and starches and among these compounds, cellulose, hemicelluloses, and lignin are the three main constituents.

Advantages of Biomass

1. Biomass is renewable and available essentially in all countries in the world and at anytime.
2. High availability of the resource.
3. Great variety of biomass types and possible energy uses.
4. Biomass contains negligible sulfur, nitrogen and metal contents. Hence, it helps in minimizing the acid rain.
5. It can help mitigate climate change, water pollution and soil erosion.
6. Valuation of wastes.
7. Fixation of rural population and depletion of fossil fuels.
8. Usage of biomass does not result in a net increase in the CO_2 concentration in the atmosphere because the CO_2 released during combustion is taken by the biomass from the atmosphere by photosynthesis during their growth.

Sources of Biomass

The major sources of biomass include wood and wood wastes; agricultural crops and their waste by-products, municipal solid waste, animal wastes, waste from food processing and aquatic plants and algae. Biomass resources are categorized into three types. They are

1. **Wastes**: Agricultural production wastes, agricultural processing wastes, crop residues, mill woodwastes, urban wood wastes, and urban organic wastes.
2. **Forest Products**: Wood, logging residues, trees, shrubs, sawdust, bark, etc., from forest clearings.
3. **Energy Crops**: Short rotation woody crops, herbaceous woody crops, grasses, starch crops (corn, wheat and barley), sugar crops (cane and beet), oil seed crops (soya bean, sunflower, safflower).

Biomass-Conversion Processes

Energy from biomass can be converted to useful forms by different processes. They are thermo chemical processes and bio-chemical processes. Thermo-chemical conversion is classified into four processes. (1) Combustion, (2) Gasification, (3) Liquefaction, and (4) Pyrolysis.

Amongst the thermo-chemical processes, pyrolysis has received increasing attention because they produce high energy pyrolytic oils in addition to char and gas. Pyrolysis is a thermal decomposition process that takes place in the absence of oxygen to convert biomass into solid charcoal, liquid (bio-oil), and gases at higher temperatures. Pyrolysis processes are of three types. They are (1) slow or conventional pyrolysis, (2) fast pyrolysis, and (3) flash pyrolysis.

Slow or Conventional Pyrolysis

Conventional or slow pyrolysis process, is done at relatively long vapor residence time (minutes to days), at low heating rate (3–5 °C/min) and moderate temperatures of around 300–400 °C. This process is mostly used to maximize char yield which increases up to 30 % when compared to other processes.

Fast Pyrolysis

Fast pyrolysis is a pyrolysis process occurring at high heating rate (as high as 50–100 °C/min), short residence time (minutes to seconds) and higher temperatures of around 450–600 °C. It particularly favors the formation of liquid products, but inhibits the formation of solid chars. Bio-oil, the major product from fast pyrolysis, is a potential liquid fuel that bio-oil derived from wood pyrolysis.

Flash Pyrolysis

Flash pyrolysis is a pyrolysis process occurring at relatively high temperatures of above 700 °C; at very high heating rates and similarly short residence times. It particularly maximizes gas yields with minimum liquid and char production.

Case Study: Preparation of Bio-oil from Agroresidues Using Pyrolysis

Materials

The raw materials rice straw, cassava, cotton seeds and red gram outer cover have been collected in the form of solid.

Experimental Procedure

Pyrolysis experiment was carried out at two different temperatures (400 and 500 °C) for rice straw, cassava, cotton seeds and red gram. Pyrolysis is done in a reactor (made of SS—316) material with a diameter 5.5 cm and a height 18.5 cm and is placed in an electrically heated furnace. The temperature of the furnace is maintained by a highly sensitive PID controller. Pyrolysis was done by taking some amount of raw material in the reactor and the temperature is raised. The condensable products were condensed by cooling with water. The temperature is measured by a Cr-Al: K type thermocouple fixed inside the reactor and in the furnace. Figure 13.1 shows the schematic diagram of the pyrolysis experimental set up and Fig. 13.2 represents the photographic view of the experimental setup.

Physio-chemical Properties of the Bio-oil

The samples obtained from pyrolysis of agro residues like rice straw, cassava, cotton seeds and red gram were analyzed for their physical and chemical properties and the values obtained are presented in Table 13.1.

It is evident from Table 13.1, that the oil from agro residues has densities and viscosities higher than the conventional fuels. This makes it difficult for one to use the oil from agro residues in IC engines as their properties will demand a re-design of pumping system. The ash content is also higher for bio oil. This could be a major handicap especially when one prefers to use it in IC engines. However, their flash

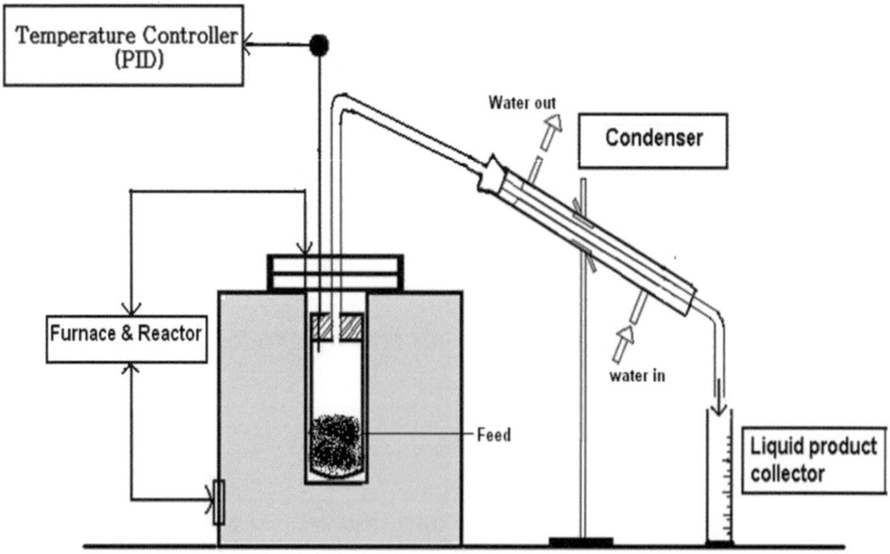

Fig. 13.1 Schematic diagram of pyrolysis experimental set-up

Fig. 13.2 Photographic view of pyrolysis experimental set-up

Table 13.1 Physiochemical properties of the bio-oil obtained from different agroresidues

Raw materials	Temperature (°C)	% Yield	Density (g/cc)	Viscosity (MPas)	Calorific value (Cal/g)	Flash point (°C)	Fire point (°C)
Rice straw	400	14.8	1.005	1.18 at 28.3 °C	–	–	–
	500	22	1.0258	1.32 at 30.3 °C	–	–	–
Cassava	400	27.7	1.037	1.44 at 26.8 °C	–	–	–
	500	31.7	1.035	1.56 at 26.9 °C	–	–	–
Cotton seeds	400	34.86	0.9978	2 at 26.9 °C	1,104	135	145
	500	34.13	1.0089	2.04 at 28.8 °C			
Red grams	400	30.01	0.9998	1.2 at 30 °C	–	–	–
	500	32.07	1.024	1.24 at 30.3 °C	–	–	–

and fire points are high which make them safe for storage. The calorific value is also low compared to conventional fuels. All this proves that though the oil has certain properties to be used as a fuel, their direct use is not feasible in IC engines. However, one can use them in furnaces and boilers. They can also be used in the manufacture of soaps, cosmetics, additives in organic products, surfactant, lubricant, paints, varnishes, emulsifying agent etc.

FTIR Analysis of Cotton Seed Oil

To have an idea on the functional groups present in the oil, bio oil obtained from cotton seeds was chosen. The functional groups present in the cotton seed oil were identified using Fourier Transform Infrared spectroscopy (FTIR). Figure 13.3 shows the FTIR analysis of cotton seed oil and Table 13.2 shows the functional groups present in cotton seed oil.

GC-MS Analysis of Cotton Seed Oil

The cotton seed oil obtained was characterized by using gas chromatography equipped with mass spectrometry (GC-MS). Figure 13.4 shows the GC spectrum of cotton seed oil and it is found that the produced cotton seed oil contains palmitic acid, linoleic acid, oleic acid and stearic acid (Table 13.3).

Figures 13.5, 13.6, 13.7 and 13.8 represent the mass spectra of palmitic acid, linoleic acid, oleic acid and stearic acid present in the cotton seed oil.

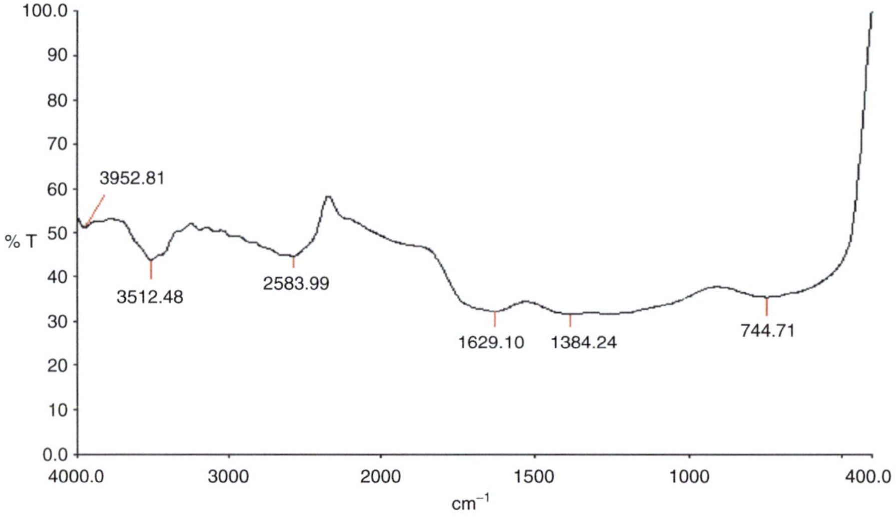

Fig. 13.3 FTIR analysis of cotton seed oil

Table 13.2 Functional groups present in cotton seed oil

Wave number (cm^{-1})	Type of vibration	Bond
3,512.48	Stretching	OH
2,583.99	Overtone	COOH
1,629.10	Bending	NH_2
1,384.29	Bending	CH_3
744.71	Rocking	CH_2

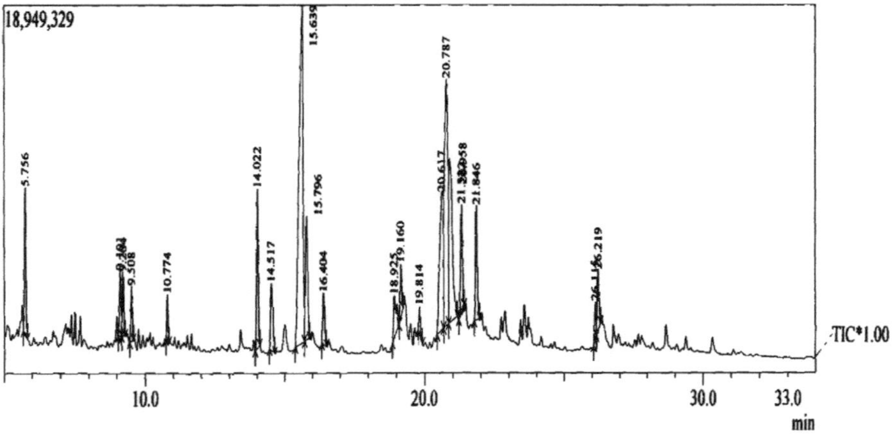

Fig. 13.4 GC spectrum of cotton seed oil

Table 13.3 Fatty acids found in cotton seed oil

Retention time	Area%	Name of compound	Molecular formula
15.639	12.45	Palmitic acid	$C_{16}H_{32}O_2$
20.787	5.04	Linoleic acid	$C_{18}H_{32}O_2$
20.958	7.96	Oleic acid	$C_{18}H_{34}O_2$
21.322	1.89	Stearic acid	$C_{18}H_{36}O_2$

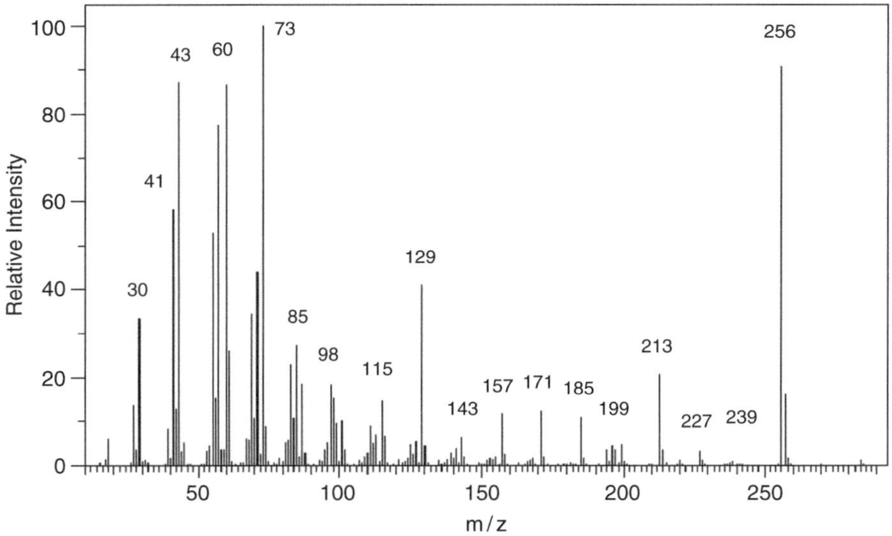

Fig. 13.5 Mass spectrum of palmitic acid

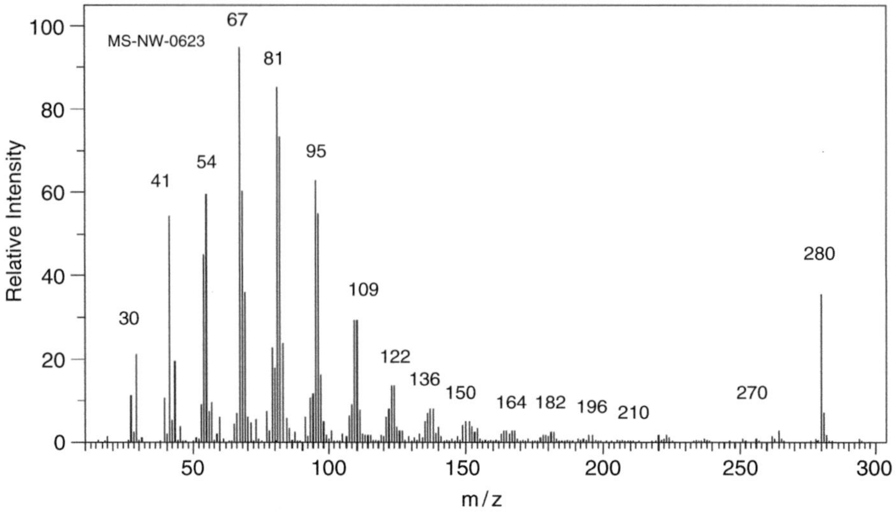

Fig. 13.6 Mass spectrum of linoleic acid

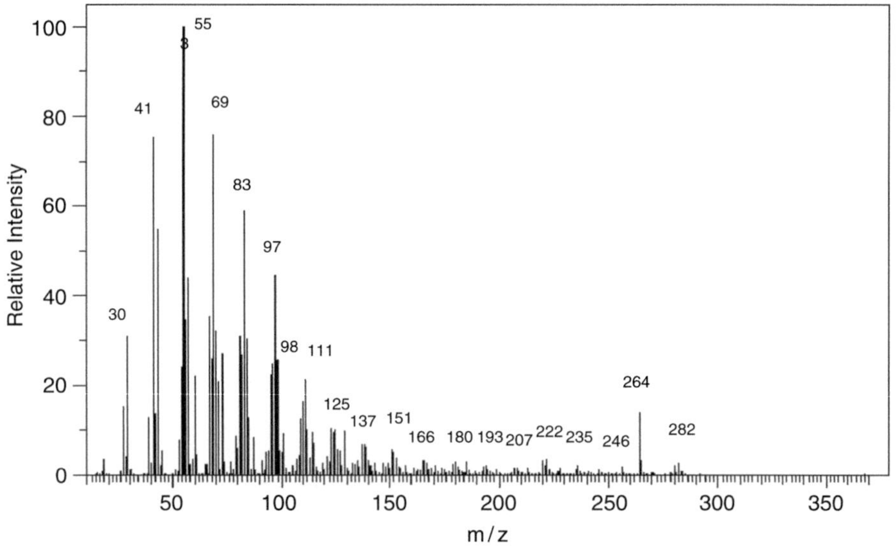

Fig. 13.7 Mass spectrum of oleic acid

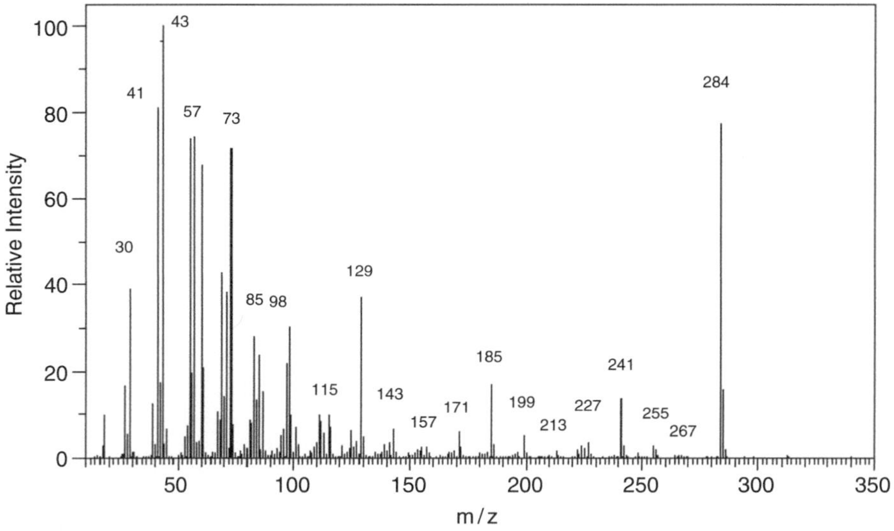

Fig. 13.8 Mass spectrum of stearic acid

Conclusions

From the experimental investigation done in this research work, the following conclusions can be drawn.

- The oil obtained has certain properties to be used as a fuel; their direct use is not feasible in IC engines. Further post treatment is necessary for application of this fuel in diesel engines. However it can be use in furnaces or boilers.
- Conventional diesel can also be blended with these oils (at lesser percentage) and can be used in IC engines without any modification of engine.
- The maximum bio oil yield obtained for rice straw, cassava and red gram are 22, 31.7 and 32.07 and 34.86 % for cotton seeds.
- From the FTIR analysis, it is found that the produced cotton seed oil contains OH group, COOH group, NH_2 group, CH_3 group and aromatic groups.
- From the GCMS analysis, it is found that the produced cotton seed oil contains palmitic acid, linoleic acid, oleic acid and stearic acid.
- All this prove that the oil has rich value not only as a fuel but also in the manufacture of host of other products which will have applications as lubricants, surface coatings, soaps, cosmetics, etc.

References

Onay, O., & Kockar, O. M. (2003). Slow, fast and flash pyrolysis of rapeseed. *Renewable Energy, 28*, 2417–2433.

Chapter 14
Biogas as Clean Fuel for Cooking and Transportation Needs in India

Virendra Kumar Vijay, Rimika Kapoor, Abhinav Trivedi, and Vandit Vijay

Introduction

Energy is an essential ingredient for economic growth, social development, human welfare and improving the standards of living. This growing consumption of energy has resulted in increased dependence on fossil fuels, which is causing environmental problems. Concerns about depletion of fossil fuels, energy security, and emission of greenhouse gas (GHG) have prompted renewable energy studies. There are several feasible renewable energy technologies in the areas of solar, wind, ocean, geothermal and biomass, but adoption of these technologies faces economic constrains along with lack of user friendly technical know-how.

Biomass plays a key role in transformation to a low carbon economy. Among biomass sources, turning waste to biogas is an option with a large potential, having possibilities to replace and therefore reduce our dependence on fossil fuels. Biogas is an energy source which is produced from biodegradable/organic wastes, and hence contributes simultaneously to waste management and to building a sustainable environment. Biogas is produced by anaerobic digestion of biodegradable wastes and is a mixture of combustible gas. Production of biogas involves complex physiochemical and biological processes depending upon different factors like the type of substrate, temperature, pH etc. Main products of the anaerobic digestion are biogas and slurry. Biogas constitutes of different component gases the majority of them being methane (CH4) and carbon dioxide (CO2) with traces of hydrogen sulphide (H2S), moisture and hydrogen (H2) gas. It is a renewable fuel, an energy source that can be applied in many different applications.

V.K. Vijay (✉) • R. Kapoor • A. Trivedi • V. Vijay
Centre for Rural Development and Technology, Indian Institute of Technology Delhi, India, New Delhi 110016, India
e-mail: vkvijay@rdat.iitd.ac.in

P. Ravindra (ed.), *Advances in Bioprocess Technology*,
DOI 10.1007/978-3-319-17915-5_14

Biogas produced can be directly used for thermal applications or for power generation. Alternatively biogas can be upgraded to biomethane. Biomethane is defined as methane produced from upgrading biogas with properties close to natural gas, is an interesting fuel to support the transition from fossil fuels to renewables and to achieve the greenhouse gas emission reduction targets in different ways. In principal, biomethane can be used for exactly the same applications as natural gas, if the final composition is in line with the different natural gas qualities on the market. Therefore, it can be used as a substitute for transport fuels, to produce combined heat and power (CHP), heat alone or serve as feedstock for the chemical sector. It can be transported and stored in the facilities and infrastructure available for natural gas. Biomethane can be produced by upgrading biogas. Biogas upgrading includes increasing the energy density by separating carbon dioxide from methane. Furthermore, water, hydrogen sulphide and other contaminants are removed, sometimes before the upgrading process to avoid corrosion or other problems in downstream applications. Today, a range of technologies for CO2-separation are on the market. It is difficult to specify the exact characteristics for an upgrading technology, since the design and operating conditions vary between the different manufacturers, sizes and applications. The key quality criteria for the upgrading technologies are the energy demand and the methane loss during upgrading.

Status and scenario of biogas production in India

In India, feedstock for biogas production is present in variable quantities at the site. According to the quantity of the feedstock, biogas plants are available in different ranges. The types of biogas plants are summarised below.

- *Household—Domestic/family size/small scale biogas plants*—These plants are mainly for households having small number of animals. These plants are in the capacity range of 1–10 cubic meter biogas per day.
- *Community size/Cooperative biogas plants*—These biogas plants are for the feed stocks from small dairies, vegetable and fruit markets, poultry farms, hostels, restaurants, etc. These plants are in the capacity range of 100–1,000 cubic meter biogas per days. Examples of this sized biogas plants are community and institutional biogas plants. If the waste is collected by a number of people/households of a local group such as village or residents of a particular housing community and fed in a centralised digester for the biogas production are categorised as community biogas plants whereas institutional biogas plants are those in which people contributing the waste belong to an institution such as hospitals, hostels/mess.
- *Industrial Scale biogas plants*—These plants are mainly for feed stocks from waste water treatment plants, industries such as distilleries, pulp and paper, sugar, food processing, etc producing waste in large quantities. These are commercial level plants. The capacity range of these plants is above 5,000 cubic meter biogas per day.

Organic wastes are widely available in India, biogas can be produced at different scales. With India being second largest country in the world in biogas technology implementation, about 4.75 million family size (1–10 m^3/day) biogas plants has already been installed against potential of 12 million such biogas plants (Bamboriya 2014). There are numerous cattle farms, dairies, and village cattle communities with a large number of cattle which have the potential for producing biogas at a community scale. Large-scale biogas can be produced in industries like distilleries, food processing, pulp and paper etc., and at sewage treatment plants and landfills in urban areas. For biogas upgrading, family-size biogas plants are not considered here as the amount of biogas produced is only a few cubic meters and is not viable for bottling. Resources like industrial wastes from distilleries, food processing, pulp and paper, waste water treatment plants and landfills, etc. are also not considered as these produce large volumes of biogas for which large-scale upgrading and bottling becomes viable. Hence, for small-scale biogas upgrading and bottling medium and community level resources are considered like cattle manure, dairy farms, canteens, hostels, community toilets, institutional and community biogas plants, etc.

Resource Availability and Potential for Biogas Production

Animal manure—India has a large livestock population with around 300 cattles. The quantity of dung will grow from an estimated current total output of two million tonnes day^{-1} to over three million tonnes day^{-1} in 2022. Since livestock are common in rural areas, their dung serve as a basic energy fuel for cooking needs. Rural households prepare dung cake and use it along with firewood to fuel cook stoves. Rural dispersed small farmers with 1–8 grazing animals versus large commercial farms of 50–1,000 animals clustered in pre-urban areas is the structure of farming in India (Harsdoraff 2012). In India cattle population exceeds 304 million figure till 2012, majority of which is with rural families in a average group of 1–8 cattle's/family. Based on the dung available from this many cattle's around 18,240 million m^3 of biogas can be generated annually (Bamboria, n.d.). Upgrading and bottling of small-scale biogas produced from domestic-size biogas plants is neither energy-efficient nor economically viable, hence the biogas potential from these is not considered in this report.

Poultry waste—The poultry population in India is 649 million till 2012 and is fifth largest in egg production as reported by National Dairy Development Board. With increasing consumption of poultry products, number of poultry farms are increasing. Poultry meat consumption increased upto 350 % from 0.8 kg/capita/annum to 2.8 kg/capita/annum from 2000 to 2012 as per ICRA statistics. Considering this growth Indian poultry industry has grown 8–10 %/annum in last decade. With this huge number poultry waste is becoming a cause of concern. Converting poultry waste into biogas seems to be best option for waste mitigation. Biogas production potential from poultry waste across all the poultry farms in the country is estimated around 438,227 m^3 day^{-1} (Rao et al. 2010). There need to put collaborative efforts from industries and government agencies to tap this enormous potential.

Crop residue—India being an agrarian country produces a record 257 million tonnes of food grain for year 2011–2012 and produces more than 500 million tonnes of agricultural waste. These agricultural waste in the form of crop residue is valuable for rural population. These residues are being used as animal feed, thatched roofs, fuel for cooking and huge chunk is burned in open field, this leads to emission of green house gases. Biomethanation of these crop residues provides a clean alternative against direct burning of crop residues in cook stoves. Biogas as a clean fuel for kitchen and for other purposes such as lighting and its usage in rural industries will be widely accepted. The biogas potential of crop residues and agricultural waste is estimated as 45.8 million m^3 day^{-1} or 16,717 million m^3 $year^{-1}$ (Rao et al. 2010).

Community Biogas Plants/Dairies/Clusters of Small Dairies

Dairy and dairy clusters—Indian dairying is characterised by very small individual producers owning 1–3 milk animals. Today, around 75 million smallholders are engaged in low productive dairying activities especially in semi urban and rural areas. Dairy operations in India can be classified on the basis of the number of cattle: large, medium, semi-medium, small, marginal, and landless (Global Methane Initiative 2011). Semi-medium, medium, and large operations account for only 40 % of the Indian dairy herd, thus a large amount of cattle dung can be collected from dairies and dairy clusters for biogas productions. Dung from dairies is easily to collect for biogas digester feeding. Therefore, medium to community-size production of biogas becomes a viable option in dairies. The biogas generation potential from small, medium and large dairies is 14,792 million m^3 per annum.

Community biogas plants—the majority of the cattle dung biogas plants in India are small-scale household level plants. Since only prosperous villagers have an adequate number of cattle, however, most small farmers and landless labour and artisans in the villages cannot have biogas plants. The common needs of the villagers such as organic fertilisers in large quantities, lighting and water supply cannot be met from privately owned individual biogas plants, which are used mostly for cooking and the sludge for fertilising the fields. At present, the number of community biogas plants established in villages to cater to the cooking fuel, organic fertiliser or electricity requirements of the village is very small. These plants are placed in urban communities where different households contribute waste cattle dung or food waste/MSW, benefit from the gas produced for cooking and lighting purposes, and share the system. Larger biogas plants can also be found in institutional buildings like schools, hospitals, jails and monasteries. Many of the community and institutional biogas plants are in remote areas and scattered around the country.

Vegetable market waste—Another substrate for biomethanation in medium-size biogas plants is vegetable market yard waste (mandi waste). At present, collection, transportation and disposal of vegetable/fruit market waste is a problem for most cities and towns. If this waste could be segregated and digested in a biogas digester, both biogas and fertiliser could be produced. In addition to this, the management of public health problems arising out of such waste could be dealt with effectively. All cities in India, districts and tehsils have vegetable markets which produce plenty of vegetable waste irrespective of the size of the market. In large cities and towns it could be to the extent of a few hundred tonnes. An estimated production of fruits and vegetables in India is 150 million tones and the total waste generated is 50 million tones per annum. That is, 30 % of the estimated production of Fruits and Vegetables (Velmurugan and Ramanujam 2011). One tonne of vegetable waste yields around 80 m^3 of biogas per day; hence from the total vegetable market waste generated the biogas production potential is about 4,000 million m^3 $year^{-1}$.

Food waste/canteen waste—Food waste is an untapped energy source that mostly ends up rotting in landfills, thereby releasing greenhouse gases into the atmosphere. Food waste is difficult to treat or recycle since it has high moisture content and is mixed with other wastes during collection. The growing number of hotels, canteens, restaurants, and townships in India are facing problems related to waste disposal: currently this waste is disposed of into sewers, dumped in low-lying areas/dump sites/landfill sites. A lot of food waste is generated during weddings in India. Wedding halls/community halls generate waste which can be utilised for biogas production in the medium scale. A leading newspaper in India quoted "A survey shows that annually, Bangalore alone wastes 943 tonnes of high calorie, high quality food during weddings. About 84,960 marriages are held at 531 kalyana mantapas (marriage halls) in Bangalore every year". Community/Institutional biogas plants can be built to harness the potential of the high quality calorie rich food waste (The Economic Times 2012) Considering that 30 % of population resides in urban areas where the probability of waste collection is at a maximum, and that 1 m^3 of biogas is generated from 6 kg of food waste, the total biogas generation potential can be estimated as 5,780 million m^3 $year^{-1}$.

Human waste/community toilets—According to estimates by Energy Alternatives India, about 0.12 million tonnes of faecal sludge is generated in India per day. Anaerobic digestion of human excreta is best way for managing this kind of waste. Biogas produced from digesters connected to public toilets can be utilised for cooking or localised lighting. Around 200 toilet linked biogas plants of sizes 35–60 m^3 have been designed by Sulabh International in different states of country (Pathak 2006). Biogas produced from human excreta is used for different purposes e.g. cooking, lighting, warming oneself during winter, heating water and electricity generation. The engine to covert biogas to electricity is run 100 % on biogas (Tambwe 2012).

Industrial Waste—Agro-industries, distilleries, dairy, slaughter houses and pulp and paper industry are group of major industries among others which are generating biological waste. This waste is either drained out in rivers or routed towards dumping sites. Biomethanation of Industrial waste give a way towards waste management and energy generation as an advantage.

Distilleries: Currently, about 60 million m^3 $year^{-1}$ of spent wash is generated from distilleries in India. The 325 distilleries in India produce 4.02 million m^3 $year^{-1}$

of alcohol. It is estimated that from the spent wash available in the country approximately 1,500 million m^3 of biogas can be generated per year (Global Methane Initiative 2011).

Dairy Industry Waste—The dairy industry is the most promising industry for biogas production. In dairy cattle manure can be directly utilised for biogas production. Waste generated during dairy product processing can also contributes in biogas production. India's total dairy industry waste can generate around 80 million m^3 biogas $year^{-1}$ (Rao et al. 2010).

Pulp and Paper industry—The pulp and paper industry is one of the key industrial sectors contributing to the Indian economy. There are 759 paper mills in India. It is estimated that a paper mill having a capacity of 17 metric tonne per day paper production, generates an approximate 1.02 metric tonne per day of pulp waste (Dasgupta and Das 2002). Around 412,278 m^3 day^{-1} or 153 million m^3 $year^{-1}$ biogas generation can be achieved from anaerobic digestion of black liquor (Rao et al. 2010).

Sugar Industry—India being the second largest producer of sugarcane after Brazil, sugar industries in India produces around 18.5 million tonnes of sugar every year. For every tonne of sugar cane crushed nearly 1,000 l of wastewater is pumped out of the industry. Since bagasse is used as an alternative fuel for as a solid waste and molasses is being already utilised by distilleries, utilisation of press mud and waste water for biomethanation is a suitable alternative. The biogas generation potential of utilising press mud is 2.9 million m^3 day^{-1} and from waste water obtained from sugar factories is 0.6 million m^3 day^{-1} (Global Methane Initiative 2011).

Municipal waste—Total sewage generation from urban centres in India is around 38 billion litres day^{-1}. Sewage sludge generation in India is increasing at a faster rate as more and more sewage treatment plants (STP) are developed. In India, wastewater disposal systems are usually managed by local bodies. Estimates suggest that there is a potential of about 226 MW from sewage sludge—from treated and untreated sewage together. The total municipal solid waste produced in India is 97,173 tonnes day^{-1} with a biogas generation potential of 9.23 million m^3 day^{-1} (Rao et al. 2010) or 3,369 million m^3 $year^{-1}$. Hence, instead of using large-scale wastewater or sewage treatment systems for a city or a town for biogas generation, small to medium-scale treatment systems can be installed for housing societies and housing clusters, and biogas can be generated at a medium scale to serve the housing society or cluster.

As discussed above there is a huge potential for biogas production and hence biogas upgrading and bottling. There is also a need to treat these wastes to improve environmental conditions and reduce methane emissions that can cause climate change. In addition to gaseous fuel, biogas plants can provide high quality organic manure with nutrients which improves soil fertility for sustainable production and improved productivity. Thus, there is a huge potential for the installation of medium size biogas plants in the country. The potential can be translated to an aggregated estimated capacity of approximately 48,382.5 million m^3 of biogas generation annually.

Upgradation and Bottling of Biogas for Cooking and Vehicular Applications in India

The major application areas for the scope of utilisation of bottled biogas in rural and urban parts of India are:

(a) **Cooking**: Domestic (replacement of household LPG cylinders, Kerosene Oil) Commercial (replacement of commercial LPG cylinders in for use in hotels, bakeries, canteens, etc.)
(b) **Industrial**: For use in production and manufacturing process, captive power
(c) **Automotive Sector**: For use as fuel in automobiles, cars, two/three wheelers, commercial vehicles (as a replacement of motor spirit—petrol and high speed diesel)

The major petroleum fuels used in India for the above applications are Liquefied Petroleum Gas (LPG), Compressed Natural Gas (CNG), Kerosene, Diesel and Petrol. While usage of kerosene is prominent among kitchens of small towns and villages, people in metropolitans have access to LPG for cooking needs. LPG is alternatively used in industries and in some cases as vehicular fuel too. Compressed natural gas is being widely used for vehicular purpose and is being supplied in major cities as Piped Natural Gas (PNG) for cooking purposes. Diesel and Petrol are major transport fuels in India and these can be easily replaced by upgraded and bottled biogas. Upgraded and bottled biogas can replace CNG, Petrol and Diesel by installing gas kits in the vehicle. Upgraded biogas has higher methane content than that of CNG. Upgraded biogas when compressed is termed as Compressed Biogas (CBG) and it contains 93 % methane, while base CNG contains 89.14 % of methane (Subramanian et al. 2013). The consumption pattern of the above mentioned petroleum fuels from 2005 to 2013 in India are shown in following Tables and figures. Table 14.1 and Fig. 14.1 shows the consumption of the major petroleum

Table 14.1 Consumption pattern of major petroleum products in India

	(In million metric tonnes)				
	Natural gas	LPG	Petrol	Diesel (high speed diesel oil)	Kerosene
2005–2006	22.35	10.46	8.65	40.19	9.54
2006–2007	21.96	10.85	9.29	42.90	9.51
2007–2008	22.46	12.17	10.33	47.67	9.37
2008–2009	22.64	12.34	11.26	51.71	9.30
2009–2010	33.18	13.14	12.82	56.24	9.30
2010–2011	36.55	14.33	14.19	60.07	8.93
2011–2012	33.14	15.35	14.99	64.75	8.23
2012–2013 (P)	28.37	15.61	15.75	69.17	7.50
Growth rate of 2012–2013 over 2011–2012 (%)	–14.39	+1.66	+5.02	+6.83	–8.83

Source: Jeyalakshmi (2014)

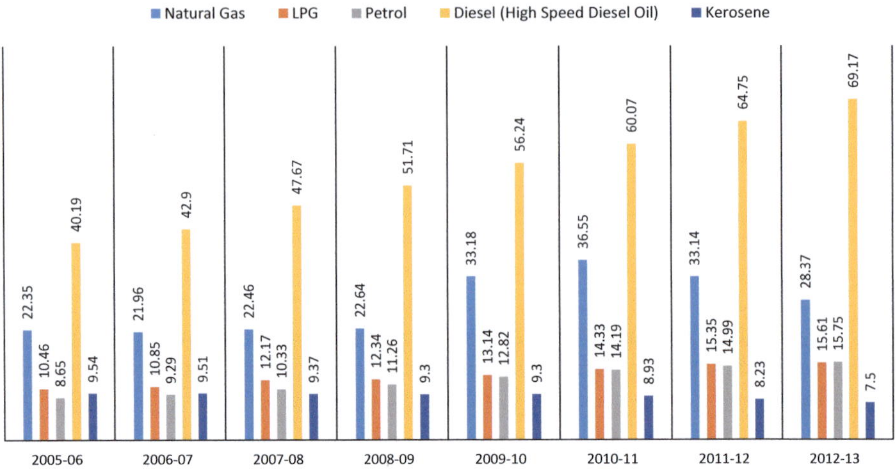

Fig. 14.1 Consumption of major petroleum products in India during 2005–2013. (All figures in MMT.) (*Source*: Jeyalakshmi 2014)

Table 14.2 Conversion of raw biogas to upgraded biogas on mass basis and energy basis

Biogas form	Amount	Unit	Remarks
Total raw biogas potential	48,382.5	Million m^3 $year^{-1}$	Considering the total raw biogas potential from the available organic feed stocks in India as evaluated in Table
Total upgraded biogas potential	18,946.59	Million kg $year^{-1}$	Considering 55 % methane in raw biogas and density of upgraded biogas as 0.712 kg m^{-3}
Total energy potential of biogas	807,503.5	Million MJ $year^{-1}$	Considering calorific value of upgraded biogas (biomethane) as 42.6 MJ/kg

Source: Subramanian et al. (2013)

products during the during year 2005–2013 in terms of energy in two major sectors, i.e. cooking and transportation.

The total biogas generation potential from all the sources mentioned in preceding pages excluding wastewater is approximately 48,383 million m^3 biogas annually. The raw biogas if upgraded and bottled can be utilised as a vehicle fuel or as a cooking fuel for commercial purposes. The contribution of upgraded biogas in terms of energy in the cooking and transportation sector as a percentage of the total petroleum fuels consumption can be evaluated and is presented in Table 14.2.

As calculated above by referencing the source Indian Petroleum and Natural Gas Statistics—2011–2012 data and the references quoted in the text above, the contributions of upgraded biogas in the transportation and cooking sector as a percentage of total petroleum fuels consumption for the year 2011–2012 are approximately 86.8 % (If used to replace transport fuel) and 83.4 % (If used to replace cooking fuel) respectively as shown in Fig. 14.2.

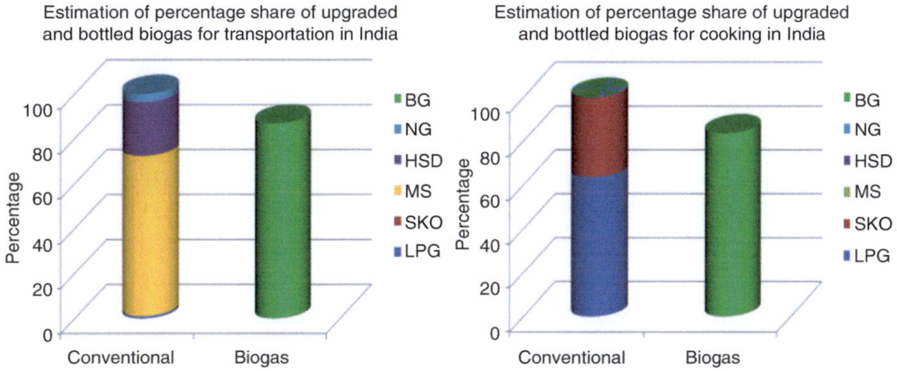

Fig. 14.2 Contribution of upgraded biogas in the transportation and cooking sector as a percentage of total petroleum fuels consumption for the year 2011–2012 (*BG* upgraded biogas, *NG* natural gas, *HSD* high speed diesel fuel, *MS* motor spirit or petrol, *SKO* superior kerosene oil, *LPG* liquefied petroleum gas)

Adaptability of Upgraded Biogas in Natural Gas Grid and Vehicles in India

With a large number of Indian cities implementing the natural gas vehicle programme, consumption has increased rapidly within the past decade. As per the Ministry of Petroleum, Government of India, the estimated reserves of Natural gas in country as on March, 2013 was 1,354.76 billion cubic meter (BCM). There is an uprising trend in production of natural gas in India due to discovery of new gas reserves, which increased from 31.33 BCM in 2005–2006 to 39.78 BCM in 2012–2013. Of the total produced natural gas 33.46 % is utilised in power generation, 27.87 % in fertilizer industry and 5.20 % as domestic fuel (Jeyalakshmi 2014). India's future demand for gas could reach 113.61 billion m^3 $year^{-1}$ by 2015 and 135 billion m^3 $year^{-1}$ by 2025, depending on how the gas market develops (Roychowdhury 2010). India currently has around 12,000 km of natural gas pipeline. Most of these gas pipelines are in the northern and western regions and much development is needed in southern, eastern and central regions. The network density is low when compared with some of the more developed natural gas markets.

Presently, the transportation sector in India, with around 1.1 million natural gas vehicles, consumes less than 2 % of the total natural gas consumption (Mathur 2012). It is expected that within the next decade the number of natural gas vehicles will increase to over 5.8 million. It is also expected that the natural gas pipeline network will increase to 15,000 km and implementation of city gas distribution networks will cover around 150–200 cities by 2014. This would further increase the share of natural gas imports to India (Roychowdhury 2010). There are many CBG (Compressed Biogas) filling stations in Sweden, such development is yet to begin in India, as no commercial upgraded biogas facilities are currently in operation.

Hence, the potential of organic waste that can be translated to an optimistic aggregated estimated capacity of raw biogas production is 54 million m^3 $year^{-1}$ or 24 million kg $year^{-1}$ of upgraded biogas (NB: This is just an approximation of the total organic waste available in India based on the waste data and rough calculations). Bottled biogas can be easily dispensed through the present available natural gas infrastructure in the country. Hence, small-scale biogas bottling systems would help in contributing to the natural gas demand in India for the local transportation market.

Assessment of Possible Scenarios for the Adoption of Upgraded Biogas in India

Bottled biogas can be utilised in two different scenarios: (1) Captive/in-house use and (2) Selling of bottled biogas as a fuel either for cooking or for vehicles. The following section depicts some potential scenarios for harnessing this technology. Upgraded and bottled biogas can be utilised as a transport fuel or can be used as a replacement of cooking fuel—LPG. Bottling facilitates easy transportation of biogas to remote places. If the barriers in the promotion of this technology are overcome and support is provided to entrepreneurs in the form of tax incentives, policies and subsidies, then this technology has all the benefits of fulfilling energy demands, reducing dependence on fossil fuels and waste management in rural, urban as well as in remote areas.

Scenario I: *Captive/In-house use*
This scenario represents the various models where biogas production, upgrading and bottling is done in the same place. This type of scenario is possible in locations like cattle sheds, fruit and vegetable markets, sewage treatment plants, community toilets and housing societies etc.

Model A: Biogas upgrading and bottling in rural areas (Cattle sheds).

The scope of implementation of this technology in rural areas is immense as apart from the small-scale—individual household biogas plants, rural areas have cattle sheds (dairies) which generate wastes at different scales which in turn have a potential of producing medium to large scale biogas. Biogas upgrading and bottling systems can become a feasible option where there is a potential of producing above 500 m^3 day^{-1} of biogas. There is a large scope for biogas upgrading and bottling in dairies as waste can be converted into bottled biogas there. Small-scale bottling systems can be installed in cattle sheds. This bottled biogas can be used as a transportation fuel in the captive vehicles like vans, tractors, etc. of the cattle sheds. Apart from this, if a suitable market is found outside the farm, then the bottled biogas can be transported in a cascade of cylinders to nearby dispensing stations, where dedicated systems are available for metering and monitoring the quality of the gas for selling it as a vehicle fuel instead of CNG cylinders. Bottled biogas can also be sold as a vehicle fuel in villages by deploying a compressing, dispensing and metering system. Hence it can be sold as a vehicle fuel for the private vehicles or tractors of villagers.

Model B: Biogas upgrading and bottling in communities like restaurants, hostels, fruit and vegetable markets, community toilets.

Biogas upgrading and bottling can become a feasible option in urban or rural areas where community biogas plants can be employed, like community toilets, fruit and vegetable markets and marriage halls, etc. Biogas can be upgraded and bottled at the site of production and can be used as vehicle fuel for the captive vehicles of the communities like trucks, lorries and auto rickshaws. Bottled biogas can also be used as a replacement for LPG in cooking, for example in marriage halls. As in scenario I(a), if a suitable market is found outside the community, then the bottled biogas can be transported and sold as a vehicle fuel to replace CNG cylinders or by direct dispensing and metering.

Model C: Biogas upgrading and bottling in urban areas serving housing societies/ housing clusters, industries producing organic effluents (medium-scale biogas production).

In urban areas organic wastes are available in abundance in sewers and landfills. Apart from this, wastes from fruit and vegetable markets, restaurants and hotels are also available, which are presently sold as piggery feed or dumped in landfills. These municipal wastes can be used for medium-scale biogas generation and small-scale biogas upgrading and bottling systems can be installed at these sites. This biogas after upgrading and bottling can be dispensed into private vehicles or those of the housing societies/housing clusters or filled in a cascade of cylinders and transported to centralised dispensing stations.

Scenario II: *Selling of upgraded biogas as a fuel*
Another option for implementing biogas upgrading and bottling system as an entrepreneurship model in rural areas is in dairy clusters, village clusters and clusters of housing societies in urban areas. Wastes from different dairies can be collected and transported to a centralised site for biogas production, upgrading and bottling then bottled biogas can be used for vehicle fuel. Another option for offsite upgrading and bottling is a mobile biogas upgrading unit where upgradation and bottling can be done on enrichment plant stationed over a vehicle.

Model A: Biogas upgrading and bottling at a location away from the site of production of waste (collection of waste from different locations and transportation to a centralised site for biogas production and upgrading).

Another option for waste collection is from village clusters/dairy clusters in rural areas. In urban areas, waste can be collected from dedicated waste production sites for example housing societies, restaurants, hostel messes, sewers etc. Centralised collection systems can be installed in which people dispose of waste in a centralised place. From here the waste is mixed and shredded and fed into digesters known as community biogas plants or medium/large-scale biogas plants, depending upon the scale of production of biogas. Upgraded bottled biogas can be produced at the biogas production site using small-scale upgrading technologies, and can be sold to local people at nominal prices. This model is shown in Fig. 14.3. The upgraded bottled biogas can be filled in a cascade of biogas cylinders for transportation to

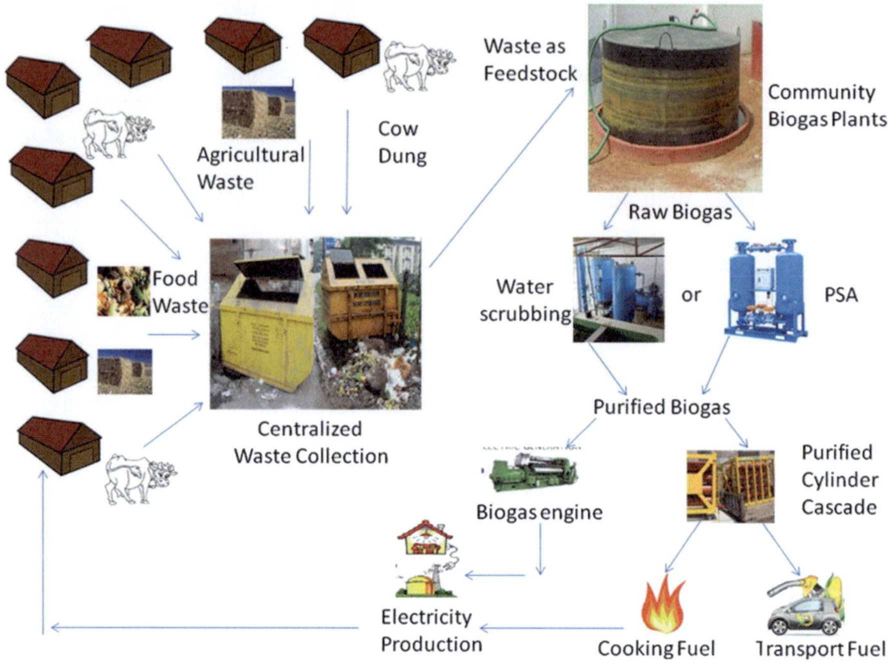

Fig. 14.3 Schematic diagram of biogas entrepreneurship (*Source*: Author)

remote rural areas and or in cylinders dispensing into a vehicle. This bottled biogas can be used as a transportation fuel for the captive vehicles of the community. Apart from this, if a suitable market is found outside the community, then the bottled biogas can be transported and used as in Scenario I (Model A).

Model B: Onsite upgrading & bottling of biogas from various plants using mobile unit.

Another option for upgrading and bottling is a mobile biogas upgrading unit serving a cluster of biogas plants in villages or in urban areas a cluster of housing where there is no biogas upgrading and bottling system. The biogas producer can hire the mobile upgrading unit and hence biogas can be upgraded at the production site without implanting an upgrading and bottling unit. The cost of the upgrading and bottling plant is saved by the biogas producer. In this option the mobile upgrading unit is mounted on a trolley attached to a vehicle. The unit can service more than one biogas plant within a cluster. The trolley-mounted machine as shown in Fig. 14.4 can be transported to digesters in different locations where raw biogas is produced. The raw biogas can be upgraded and used to fill CNG cylinders for storage at high pressure which are then transported to the nearby dispensing stations, giving an uninterrupted supply of upgraded biogas (Leonard and Massie 2006). The scenario can be site-specific depending on the local situation. After bottling, biogas can be used as a transportation fuel for the captive vehicles of the village or community. Apart from this, if a suitable market is found outside the community, then the bottled biogas can be transported and used as in Scenario I (Model A).

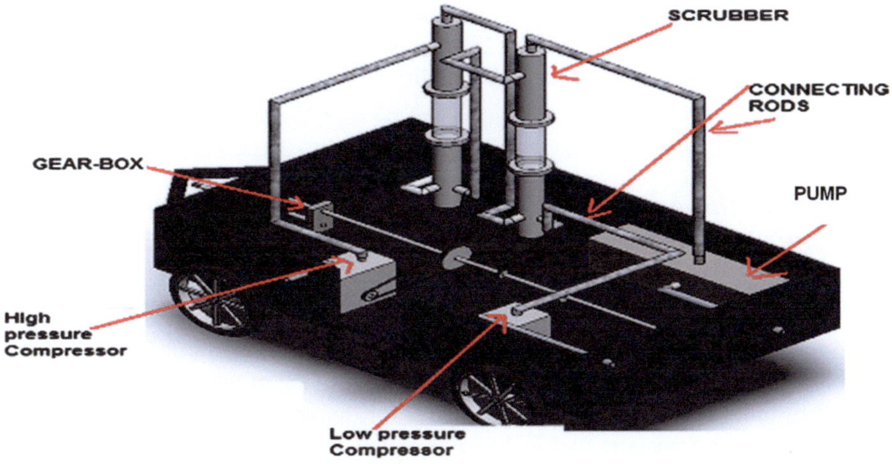

Fig. 14.4 View of the mobile unit (*Source*: Author)

Model C: Setting up of upgradation and bottling plant at the site of waste generation (for example, large cattle sheds, dairies, large vegetable and market yards, residential colonies) and selling of upgraded biogas to the local people for cooking and transportation purpose.

Another scope of implementation of biogas upgrading and bottling plants for commercial purposes is at sites where huge quantity of organic waste is generated having a potential to produce medium to large scale of biogas for example, large cattle sheds, dairies, large vegetable and market yards, industries such as sugar mills, pulp and paper, distilleries, residential colonies in urban areas. Biogas upgrading and bottling systems can become a feasible option where there is a potential of producing above 500 m^3 day^{-1} of biogas. The upgraded and bottled biogas can be sold as a fuel to local vendors for cooking in nearby restaurants, hotels, hostels etc. If a small dispensing system along with metering device is installed then the upgraded biogas can be sold as a transport fuel for local vehicles. Apart from this, if a suitable market is found at far of places, then the bottled biogas can be transported in a cascade of cylinders to the site of utilisation.

Urban areas are densely populated and have large scope for small-scale biogas upgrading and bottling systems. If residential colonies societies etc can be grouped in 20,000–25,000 households, then waste from these households can be collected at a centralised place. Biogas can be produced, upgraded and bottled for consumers for applications like cooking, as a replacement for LPG, or as a vehicle fuel for private vehicles of the residents of the societies. For example, if on average one household gives 250 g day^{-1} of biodegradable waste, then a small community of 25,000 households will be able to generate approximately 6.5 tonnes day^{-1}. This quantity of waste can generate approximately 500 Nm^3 day^{-1} of biogas, which is sufficient for running a small-scale biogas upgrading system in entrepreneurial mode.

Bottled biogas can be used either in community kitchens in the housing society or as a vehicle fuel in community vehicles like laundry vans etc. Such systems are a need of the time and in future may become a reality everywhere.

Barriers to the Adoption of Small-Scale Biogas Upgrading and Bottling Industry

Despite the feedstock potential, easy adoption feasibility and the various benefits of using bottled biogas as a transport fuel in India, the development of this industry is quite slow. There are many reasons for this debility in the growth of this industry. It is evident that there is a conducive situation for the adaption of biogas upgrading and bottling as a transport fuel in India as stated above. The main barriers in the promotion of bottling can be attributed to many factors as listed below. Stringent rules, regulations and policies, lack of awareness, high initial investment cost and scale of the technology, etc. contribute to the slow adoption of this technology sector, but the main factor is the non-existence of standards, policies and tariffs particularly for bottled biogas adoption for vehicular use. Unfortunately there is insufficient promotion of the available technology for biogas utilisation. The main barriers to the promotion of biogas bottling as a transport fuel in India are listed and discussed below.

Policies and Regulations

The high investment required to start an industry requires financial support; this may be in form of capital grants or low interest loans. The major issue for this industry is that at the current level of high initial investment and 50 % subsidy, the payback period for biogas bottling plants is too high when the bottled biogas is for vehicle use. The profit margin is also small at the present level of central financial assistance and subsidies.

During 2008–2009 the Ministry of New and Renewable Energy (MNRE), Government of India undertook an initiative to demonstrate an integrated technology package on medium-size (200–1,000 Nm^3 day^{-1}) Biogas Generation and Fertiliser Program (BGFP) for generation, upgrading, bottling and piped distribution of biogas. A 50 % subsidy of the total project cost for compressed biogas plant (biogas bottling) was granted by the Ministry of New and Renewable Energy. The available term loan was up to 30 % from financing institutions with a 20 % entrepreneur's share. Installation of such plants aimed at production of Compressed Biogas (CBG) of the quality of Compressed Natural Gas (CNG) to be used as vehicle fuel in addition to meeting other needs. The upgraded biogas, compressed at 200 bar pressure and filled into CNG cylinders for various applications, should be in accordance with the approval given by Petroleum Explosive and Safety Organisation (PESO) India. The Compressed Biogas project is eligible to obtain Carbon Emissions Reduction certificates for methane avoidance and replacing fossil fuels (compression and utilisation of methane gas vehicle fuel).

Biogas Standards in India

In most of the developed countries, standards and policies for use of upgraded biogas in vehicles as a transport fuel exist or are in the formulation stage. In India, the Bureau of Indian Standards has formulated standards for upgraded biogas use in stationary engines and in vehicles. But these standards can only be implemented if the other legal authorisations are fulfilled for utilisation of biogas in vehicles or as a transport fuel. Standards for biomethane composition have been developed through the Bureau of Indian Standards (BIS) and published in the year 2013 (IS 16087: 2013; for further details please refer to the Bureau of Indian Standards). This standard prescribes the requirements and the methods of sampling and test for the biogas (biomethane) applications in stationary engines, automotive and thermal applications and supply through piped network. The purpose of these standards and policies is to provide general guidelines for upgraded biogas composition and its filling into CNG cylinders. In the biomethane standards, the composition of upgraded biogas suitable for filling in CNG cylinders (at 200 bar) is as shown in Table 14.3.

Legal Authorisation

Presently there are no existing norms for the use of high pressure compressed biogas for bottling and use in vehicles. Legal authorisations from Petroleum Explosives Safety Organisation (PESO), Ministry of Environment and Forest, the Ministry of Industry and the Central Pollution Control Board (CPCB) must be fulfilled for biogas bottling and its use in vehicles. As there is no single window clearance, therefore, the procedure is time consuming.

Table 14.3 Standards for biogas composition in India

Biogas component	Percentage
CH_4, %, Min	90
Moisture, mg m^{-3}, Max	16
H_2S, mg m^{-3}, Max	30.3
$CO_2 + N_2 + O_2$, %	10
CO_2, %, Max (v/v) (when intended for filling in cylinders)	4.0
O_2, %, Max (v/v)	0.5

Source: BIS Standard 16087: 2013

Feedstock Availability

As mentioned above, large quantities of organic wastes are available in India. For effective utilisation, systems should be developed in such a way so as to harness the complete potential of the available wastes. But due to the lack of awareness, unorganised collection and transportation of wastes, limited financial support and low tariffs these are not utilised to the maximum of the available potential.

Cost of Small-Scale Biogas Upgrading and Bottling

The operation and maintenance costs of biogas upgrading and bottling project are quite significant. These consist of labour, feedstock, and fixed cost etc:

- Purchase, collection and transportation of the feedstock.
- Water supply for cleaning and mixing the feedstock.
- Supervision and maintenance of the plant.
- Drying, processing, storage and disposal of the slurry.
- Production of bio fertiliser.
- Biogas upgrading and bottling unit.
- Gas distribution and utilisation, supply chain management.

The running cost of a biogas upgrading plant include professional management is important. Currently, limited experience exists in reliable cost calculation for biogas upgrading plants in India.

Barriers in the Collection, Segregation and Transportation of the Waste

There is widespread availability of organic wastes in India, but one barrier in the promotion of this technology is the lack of proper technologies and strategies for the collection, segregation and transportation of biodegradable waste. If proper systems are developed for waste management and feeding it into digesters then the project can become viable for implementation of biogas upgrading and bottling plants.

Lack of Awareness

There is a need of spreading awareness about the role of this technology for waste management, energy security and biofertilisers. The main bottlenecks in financing of biogas bottling plants in India are the lack of knowledge about biogas projects in general from the side of the financing decision-makers. There is lack of human resource development programmes for project developers on the project financing, technical knowledge and economics of small-scale gas bottling system in India.

Conclusion

Bottled biogas is a renewable energy source that can be produced from biodegradable/organic wastes, and hence can help both in waste management and in building a clean and sustainable environment. Centralised biogas upgrading and bottling is an economically viable option for biogas produced at medium to large scales.

There is a huge potential for the installation of biogas plants in this country. With 4.75 million family size (1–10 m^3/day biogas) plants, focus has been widened towards installing cooperative (100–1,000 m^3/day) to industrial size (>1,000 m^3/day) biogas plants in the country. If all the sources mentioned in the above sections of chapter are to be utilised for biogas production an estimated capacity of approximately 48,383 million m^3 of biogas can be generated annually. Based on the Indian Petroleum and Natural Gas Statistics—2011–2012 data and the references quoted in the text above the contribution of upgraded biogas in the transportation and cooking sector as a percentage of total petroleum fuels consumption for the year 2011–2012 is approximately 86.8 and 83.4 % respectively. In the present report the above mentioned evaluation was made to evaluate the amount of upgraded biogas that can be obtained in urban and rural areas which can substitute for fossil fuels in vehicles as well as for cooking. In this report, various real life situations are analysed where the available organic wastes can be harnessed for biogas production and hence by upgrading and bottling the gas it can be utilised in local vehicles and cooking. If the above mentioned examples are disseminated in urban and rural areas in large numbers, then bottled biogas can help to substitute fossil fuel. Small-scale biogas upgrading and bottling technology is thus a step towards helping in finding a replacement of transport and cooking fuel. However, central financial assistance and other incentives are required to reach the biogas utilisation levels close to the full technical biogas potential of India. There is an urgent need for developing commercially viable systems for different situations. Hence, it is concluded that the present systems available for bottling which are being demonstrated in some developing countries like India are low cost and economically viable. In India IIT Delhi has developed a biogas upgrading and bottling unit using high pressure water scrubbing technology, and based on this development and pilot-scale set ups in the field, the government has undertaken initiatives to provide financial support to any such commercial projects in India. In 2013, Bureau of Indian Standards has brought out national standards for biogas (biomethane) which specifies biogas (biomethane) applications in stationary engines, automotive and thermal applications and supply through piped network. The purpose of these standards and policies is to provide general guidelines for upgraded biogas composition and its filling into CNG cylinders. Therefore, it can be concluded that:

(a) The technologies for biogas upgrading should be standardised for minimum methane loss and less energy requirement.
(b) Water scrubbing and PSA technologies are the suitable technologies for small scale biogas upgrading and bottling in Indian context.
(c) Indigenous high pressure compressors with low gas flow rates (5–50 m^3/h) should be developed which could be affordable by the small entrepreneurs.
(d) CO_2 recovery systems should also be developed to make the whole system more economically viable.
(e) Training, workshops and dissemination activities for users, manufacturers and entrepreneurs.
(f) Policies and financial support from the government are necessary for promotion of biogas upgrading in India and other developing countries. Appropriate grants should be made available to support the initiation of this industry. Special allowances and incentives for the promotion of biogas bottling projects should be made.
(g) Bank loans and central subsidies should be provided for the promotion of biogas upgrading and bottling plants.
(h) Biogas upgrading and bottling plants less than 25 Nm^3/h are generally not economically viable due to the large capital investment required for plant and machinery.
(i) Involvement of more industries/companies is required in this sector to bring down the capital investment in the technology.
(j) Mobile biogas upgrading unit may be useful in rural as well as urban areas for collection of biogas from the existing plant and its centralised upgrading and bottling.

References

Bamboriya, M. L. (2014). Biogas production mechanism. *Akshay Urja, 8*(2), 18–21.

Bamboriya, M.L. (n.d.), *Biogas Bottling in India – a case study*. Retrieved December 21, 2014, from http://mnre.gov.in/filemanager/UserFiles/casestudy_biogas_bottling_in_India_mlbamboria.pdf.

Bureau of Indian Standards, BIS. (2013). Indian standard for biogas (biomethane) – specification, IS 16087, pp 1–6.

Dasgupta, S., & Das, S. K. (2002). Paper pulp waste—A new source of raw material for the synthesis of a porous ceramic composite. *Bulletin of Materials Science, 25*(5), 381–385.

Global Methane Initiative. (2011). *Resource assessment for livestock and agro-industrial wastes – India*. Retrieved January 4, 2015, from https://www.globalmethane.org/documents/ag_india_res_assessment.pdf.

Harsdoraff, M. (2012). *The economics of cow dung creating green jobs in the dairy industry in India, case study – A green value chain analysis of the cow dung and dairy industry in Jabalpur*. Green Jobs in Asia Regional Conference, International Labor Organization.

ICRA. (n.d.). Indian poultry industry – Broiler meat and table egg. *Investment Information and Credit Rating Agency of India Limited*. Retrieved December 24, 2014, from www.icra.in/Files/ticker/SH-2013-1-ICRA-Poultry.pdf.

Jeyalakshmi, S. (2014). Energy statistics 2014 (twenty first Issue). Central Statistics Office, Government of India, pp 1–107

Leonard, T., & Massie, C. (2006). *Mobile biogas processing system and method*. US Patent Application 2006/0213370. Minneapolis, MN (US).

Mathur, A. S. (2012). *Basic statistics on Indian petroleum and natural gas, 2011–12. Report* (pp. 1–51). New Delhi: Ministry of Petroleum and Natural Gas (Economic Division), Government of India.

Pathak, B. (2006). Operation, impact and financing of sulabh, human development report, occasional paper. *Human Development Report Office Annual Report*.

Rao, V. P., Baral, S. S., Dey, R., & Mutnuri, S. (2010). Biogas generation potential by anaerobic digestion for sustainable energy development in India. *Renewable and Sustainable Energy Reviews, 14*, 2086–2094.

Roychowdhury, A. (2010). *CNG programme in India: The future challenges*. Fact sheet series (pp 1–37). Centre for Science and Environment. Retrieved December 11, 2014, from http://www.cseindia.org/userfiles/cngfuture_pdf.pdf.

Subramanian, K. A., Mathad, V. C., Vijay, V. K., & Subbarao, P. M. V. (2013). Comparative evaluation of emission and fuel economy of an automotive spark ignition vehicle fuelled with methane enriched biogas and CNG using chassis dynamometer. *Applied Energy, 105*, 17–29.

Tambwe, M. (2012, May 21). Making cents from Loos, the sulabh story. *Africa News Services*.

The Economic Times (2012) Wedding food worth Rs 339 crore goes waste. Retrieved December 15, 2014, from http://m.economictimes.com/news/news-by-industry/cons-products/food/wedding-food-worth-rs-339-crore-goes-waste/articleshow/16775621.cms.

Velmurugan, B., & Ramanujam, R. A. (2011). Anaerobic digestion of vegetable wastes for biogas production in a fed-batch reactor. *International Journal of Emerging Sciences, 1*(3), 478–486.

Chapter 15
Thermochemical Processing of Biomass

Sarma V. Pisupati and Aime H. Tchapda

Torrefaction

Introduction

Torrefaction was initially developed for coffee processing (Thiel 1897; Offrion 1900) for drying and roasting coffee beans, making them brittle and giving them their distinctive flavor. With the growing interest of biomass as a feedstock for producing energy, fuel, chemicals and materials, and given the need for improving the properties and uniformity of biomass feedstock, torrefaction has become very attractive in the recent years. Although researchers in this area are yet to come up with a generally accepted definition for torrefaction, a common definition can be devised by examining various features of the process and attributes of the output product. Therefore torrefaction can be defined as "a thermochemical process in an inert or limited oxygen environment where biomass is slowly heated to within a specified temperature range and retained there for a stipulated time such that it results in near complete degradation of its hemicellulose content while maximizing mass and energy yield of solid product" (Basu 2013). It is therefore a low-temperature (200–300 °C) and slow heating rate thermal conversion process.

Heterogeneity is one of the most obvious drawbacks with biomass. The physical, chemical and morphological characteristics vary significantly among various biomass feedstock. The supply of a single type of biomass feedstock is intermittent and is highly seasonal. Therefore merging various sources of biomass having different characteristics is needed in order to satisfy an uninterrupted supply of feedstock.

S.V. Pisupati (✉) • A.H. Tchapda
John and Willie Leone Family Department of Energy and Mineral Engineering, Pennsylvania State University, University Park, PA 16802, USA
e-mail: sxp17@psu.edu

P. Ravindra (ed.), *Advances in Bioprocess Technology*,
DOI 10.1007/978-3-319-17915-5_15

Torrefaction as a pretreatment operation for biomass feedstock has the ability to increase the uniformity of the final product. The seasonal influences which is obvious in the initial biomass is considerably reduced (Girard and Shah 1989; Bergman et al. 2005). Due to its high oxygen content, biomass is highly hygroscopic. This hygroscopicity, coupled with the fibrous nature of biomass, its tenacity and versatility result in a very poor grindability for most biomass feedstock. Fluidization and flowability are also adversely affected. Moreover, biomass fuels have a low calorific value, hampering their long distance transportation. Torrefaction is a technology to take care of all these issues. This mild thermal treatment not only destroys the fibrous structure and tenacity of the biomass, but also increases the calorific value and inverts its hydrophilic nature (Bergman et al. 2003; Tumuluru et al. 2012). Torrefaction therefore increases its handling characteristics for storage. Torrefaction can significantly decrease the power need for the processing operation (up to 85 %) and at the same time the capacity of the milling plant can increase considerably (by a factor of 2–6) while nearly 85–90 % of the original biomass energy is retained in the solid product (Bergman et al. 2003); issues such as bridging and blockage of feeding systems as well as poor pulverization characteristics are resolved (van der Drift 2004; Zulfiqar et al. 2006; Tchapda and Pisupati 2014). As a slow heating and low temperature process, torrefaction cannot be well understood without a thorough look at the structural composition and arrangement of biomass components.

Biomass Constitution and Relevance to Torrefaction

Cellulose, hemicellulose and lignin are the major constituents of lignocellulosic biomass. Cellulose and hemicelluloses are polysaccharides while lignin is an oxygenated polymer of phenylpropane units. Beside these three major constituents, a variable quantity of extraneous chemicals known collectively as extractives and a small amount of inorganic elements are also present. Cellulose is a polymer of glucose. Its specific structure promotes the ordering of the polymer chains, forming a tightly packed and highly crystalline structure which makes it water insoluble and resistant to depolymerization and conferring rigidity to plant cells. Hemicelluloses are low-molecular weight polysaccharides having a random, amorphous structure with little strength and surrounding cellulose fibers. They act as a matrix for the cellulose and increase the packing density of the cell wall (Jerrold and Roger 2005). The structure of lignin can be summarized as a three-dimensional amorphous macro-molecule of three types of monolignols or phenylpropane units consisting of a phenolic unit with an aliphatic side chain located in the para-position. The three types of monolignols characterized are: Guaiacyl lignin which consists almost exclusively of coniferyl alcohol (softwood lignin), guaiacyl-syringyl lignin which consists of sinapyl alcohol and coniferyl alcohol (hardwood lignin), and Graminaceous lignin which contains both sinapyl alcohol and coniferyl alcohol, but predominantly p-coumaryl alcohol (Lignin in grasses and other

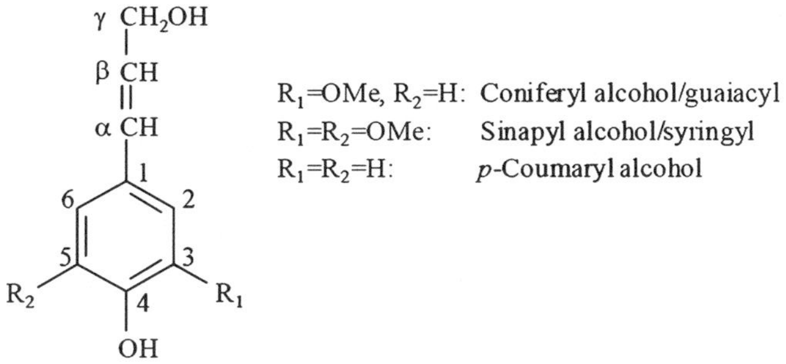

Fig. 15.1 Illustrated structure of lignin

monocotyledons) (Henriksson et al. 2010; Brunow and Lundquist 2011). Lignin is intimately mixed with the carbohydrate components (Fig. 15.1). It interpenetrates hemicelluloses and encrusts the cellulose microfibrils acting as a cementing agent. It confers to lignocellulosic materials their hardness and rigidity. Cellulose, hemicelluloses, and lignin can be classified from a morphological point of view as framework, matrix, and encrusting substances, respectively (Wardrop 1964).

Hemicelluloses can co-crystallize with cellulosic glucan chains at the surface of the cellulose microfibrils. The cocrystallization is believed to involve formation of hydrogen bonds from the $-CH_2$ OH groups present in cellulose chains to the glucosidic oxygens in the adjacent hemicellulose chains. This association would form a tightly bound monolayer of the hemicellulose on the surface of the cellulose microfibril, and would function as part of the 'glue' which holds the microfibrils together in the cell wall (Smith 1977). There is also chemical bonding between lignin and hemicellulose but not between lignin and cellulose. The highly branched structure of hemicelluloses prevents it from forming a crystalline arrangement.

Lignocellulosic materials are hydrophilic and take up moisture when exposed to high relative humidity or immersed in water. The adsorbed water in cell wall clusters around polar groups (mainly –OH and –COO– in biomass). The cell wall of cellulosic biomass materials contains hydroxyl and other oxygen containing groups, which attract moisture through hydrogen bonding. The hydroxyl group in cellulose is less accessible to water molecules because of the crystalline structure of cellulose whereas it is easily accessible in hemicelluloses due to their branched arrangement. Although lignin also contains some hydroxyl groups, it is less polar than hemicellulose and cellulose, therefore its equilibrium moisture content is lower (Feldman 1985; Pettersen et al. 2005; Rowell 2005).

Two main types of linkages dominate the structure of cellulose: the 1-4 β D-glucosidic bond connecting glucose units together and the hydrogen bond. The beta linkages confer to cellulose its "straightness" because the alternating glucose molecules are at 180° turns to each other. This conformation favors very easy hydrogen bonding, not only between monomers in the same fibril, but also with neighboring fibrils, making the entire structure much more stable. Fructosic and glucosidic bonds are dominant types of bonds found in hemicelluloses.

Unlike cellulose, the conformation of hemicellulose does not reveal hydrogen bonds and carboxyl groups are present as carboxyl or as esters or even as salts in the molecule. The various lignin monolignols are connected by different ethers and carbon–carbon bonds, while the aliphatic double bond is often absent (Ralph et al. 1999; Boerjan et al. 2003). The ether bond between the β-carbon and the 4-carbon, the β-O-4′ bond, is the most common bond in all lignins and is also the main bond that is broken during biomass processing (Henriksson et al. 2010).

The structural arrangement of lignocellulosic materials described in the previous paragraphs has a noticeable implication on its thermal degradation at low temperature. The initial heating of biomass is marked by the removal of the unbound and bound moisture, mostly from hemicelluloses. Continued heating results in the softening of lignin and depolarization/condensation of lignin, then hemicelluloses. It is well established that the thermal degradation of biomass starts by the decomposition lignin (130–250 °C), followed by hemicelluloses (150–230 °C) while minor changes occur in cellulose in this temperature range. From 200 to 300 °C, severe thermal degradation is observed in hemicelluloses while lignin gradually devolatilized and thermal degradation is initiated for cellulose (Shafizadeh 1982;

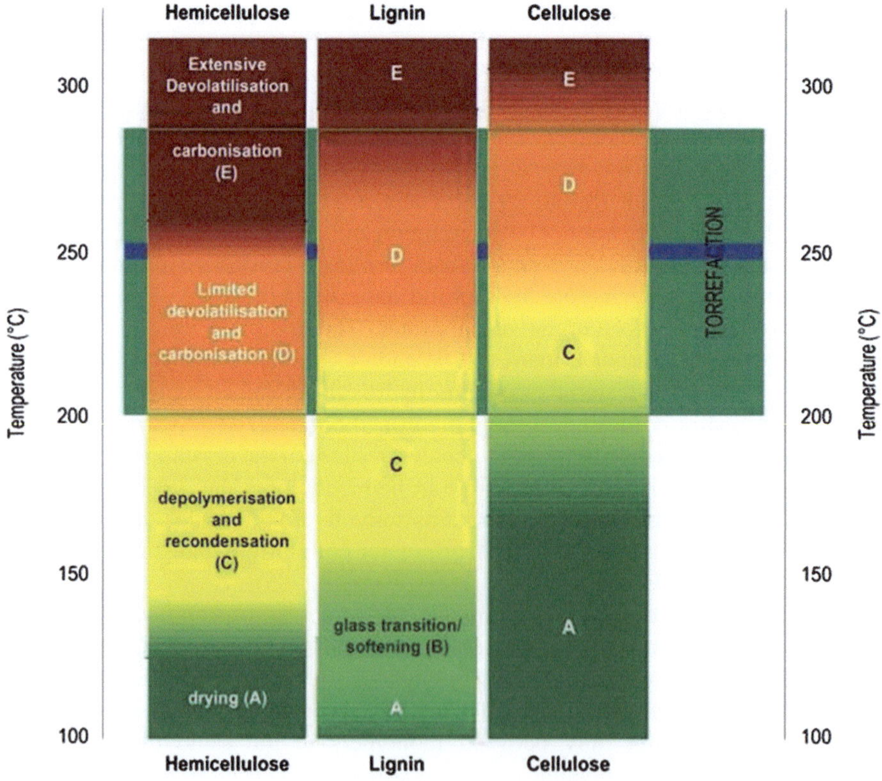

Fig. 15.2 Thermal degradation regimes of biomass components (Stelte 2014)

Koukios 1993; Sule 2012). This initial thermal degradation of biomass is illustrated in Fig. 15.2.

Interestingly, the thermal degradation of cellulose and hemicelluloses begins later than that of lignin, but are completed earlier; hence lignin has a wider degradation temperature range than cellulose and hemicellulose. The thermal degradation of lignin extends up to 900 °C while that of cellulose is limited to 400 °C and hemicelluloses has been reported to extend up to 350 °C (Yang et al. 2007; Lv et al. 2010). A plausible justification of this behavior comes from the chemical structure of cellulose, hemicellulose and lignin. Cellulose has a pretty uniform structure; although its thermal degradation starts at higher temperature, its complete degradation is confined in a narrow temperature range because of the consistency in its chemical arrangement. Unlike cellulose, hemicelluloses is a group of structurally diverse polysaccharides with branched, amorphous structure and lower degree of polymerization making it amenable to lower thermal stability. However, the diverse chemical complex found in this group of polysaccharides widens their thermal degradation range (Werner et al. 2014). The thermal decomposition range of lignin is even broader given the presence of weak bonds in the aliphatic side chain as well as very strong carbon to carbon bonds in the aromatic ring (Brebu and Vasile 2010; Li et al. 2015).

Biomass Torrefaction Process Fundamentals

Biomass torrefaction is considered as a pre-treatment technology in order to solve some of the shortcomings mentioned earlier. This pre-treatment process is aimed at improving the following properties:

- Grindability.
- Hydrophobicity.
- Energy density.
- Rheological properties (flowability, fluidization).
- Storability and durability.
- Homogeneity.

Grindability

Biomass grindability is one of the fuel characteristics of concern for power plants because it dictates whether or not existing coal mills could be used for reducing biomass particles to appropriate size in order to achieve complete conversion. The fibrous structure of biomass materials makes biomass grinding a highly energy intensive process. Torrefaction can significantly reduce the energy requirement for grinding biomass, ease the grinding operation and allow the use of existing coal grinding equipments for biomass. The energy required for grinding torrefied biomass is reduced even when its moisture content is higher than raw biomass as

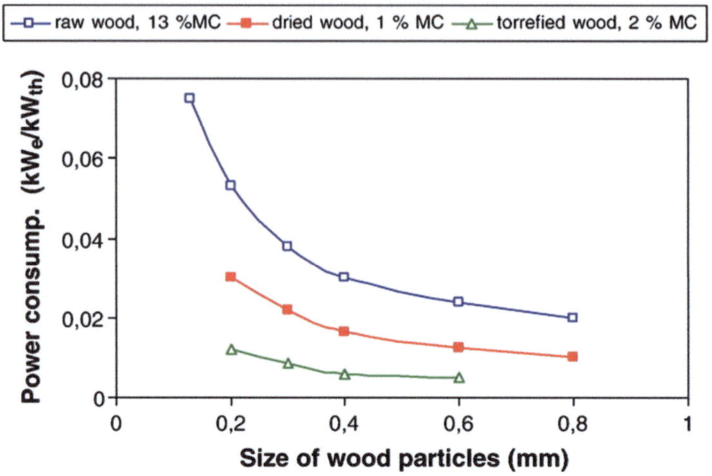

Fig. 15.3 Energy consumption for grinding raw and torrefied biomass (Svoboda et al. 2009)

illustrated in Fig. 15.3 (Svoboda et al. 2009). Phanphanich and Mani have found that the specific energy consumption for grinding torrefied biomass can be reduced by up to ten times for torrefied wood chips and up to six times for torrefied logging residues (Phanphanich and Mani 2011). Arias et al. compared the grindability of raw and torrefied Eucalyptus chips and observed that grinding raw Eucalyptus chips produces a mixture of spherical and cylindrical particles whereas the grinding of torrefied Eucalyptus chip produced essentially only spherical particles, depicting the improvement of the of the grinding process after torrefaction. They concluded that the benefits acquired from torrefaction is far worthy than the 10 % loss of energy from the original biomass (Arias et al. 2008).

Ibrahim et al. found that increasing the severity of the thermal treatment improves the grindability of both softwoods and hardwoods (Ibrahim et al. 2013) while Bridgeman et al. found that the higher temperature severity caused torrefied Miscanthus and Willow to exhibit similar grindability as coal (Bridgeman et al. 2010). Various authors have found a similar trend with woody as well as herbaceous biomass (Deng et al. 2009; Chen et al. 2011a, b; Chew and Doshi 2011; van der Stelt et al. 2011; Wang et al. 2011; Agar and Wihersaari 2012; Kim et al. 2012; Shang et al. 2012; Lu and Chen 2013; Ohliger et al. 2013; Saleh et al. 2013; Satpathy et al. 2014; Mei et al. 2015). However increasing the severity (temperature and time) also translates to more energy use for pretreatment, therefore an optimum level of severity should be developed and this is likely to depend on the type of biomass. Chen et al. found that the grindability of the torrefied wood could be improved in a significant way if the torrefaction temperature was as high as 250 °C and the torrefaction time longer than 1 h for woody biomass (Lauan). Abdullah and Wu (2009) studied the grindability of Mallee wood torrefied in the temperature range of 300–500 °C for 30 min and observed that torrefaction carried out above 330 °C had negligible effect on grindability.

Although many authors have observed an improved grindability of biomass upon torrefaction, very few of them have been able to investigate the mechanism justifying this increased grindability. According to Sohi et al. (2013), the lignin content of biomass may be a key indicator of the grindability of torrefied biomass. Repellin et al. (2010) give a rough outline of the reasons behind the improved grindability of heat treated lignocellulosic materials. In fact the softening of biomass under heating is mainly attributed to lignin as it reaches the glass transition temperature. Above the glass transition temperature, the resulting plastic behavior of lignin, a consequence of cross-links in its structure, allows it to spread, interpenetrate and cover the cellulose and hemicelluloses. Upon cooling, the heat treated material becomes brittle with low impact resistance favorable to grinding.

Hydrophobicity

The equilibrium moisture content (EMC) and the immersion test are two tests commonly used to measure the hydrophobicity of torrefied biomass. EMC can be defined as the moisture content in biomass that is in thermodynamic equilibrium with the moisture in the surrounding atmosphere at a given relative humidity, temperature and pressure. Beside temperature, relative humidity and pressure, factors inherent to biomass can also affect the EMC. Biomass composition, mechanical handling and previous moisture history strongly affect the EMC (Silakul and Jindal 2002).

EMC applies static desiccator technique where torrefied biomass samples are exposed to constant relative humidity levels maintained by saturated salt solutions (Bellur et al. 2009). Various authors have used this technique to prove that torrefied biomass is more reluctant to absorb/adsorb moisture than raw biomass (Acharjee et al. 2011; Medic 2012; Chen et al. 2014).

In the immersion test, raw and torrefied biomass are submerged in water for a fixed duration and the absorbed moisture is estimated by measuring the weight change of the immersed sample. Felfli et al. (2005) found through this method that torrefied wood briquette is unaffected by immersion in water and that the resistance to absorb water was increased by the severity of torrefaction. Pimchuai et al. also used this method on various biomasses (rice husks, sawdust, peanut husks, bagasse, and water hyacinth) torrefied at 250–300 °C in 1–2 h and confirmed that the hydrophobicity of torrefied biomass was improved compared to raw biomass.

As mentioned earlier, the destruction of polar groups in the original biomass (mainly –OH and –COO–) during torrefaction reduces the ability of the resulting product to attract water molecules by hydrogen bonding, making it hydrophobic. Another plausible theory is that during torrefaction, condensable volatiles (mainly tar) which are apolar by nature, condensed on the pore surface, acting as a barrier to moisture.

Energy Density

Torrefaction induces a partial devolatilization which results in the loss of 10–20 % of the energy contained in the original biomass. However, there is also a decrease of the biomass mass ranging from 30 to 50 %, depending of the torrefaction severity (temperature and time). Therefore the energy density of biomass can potentially be increased 1.2- to 3-fold. The increase of energy density of biomass is one of the main reasons making torrefaction an attractive pre-treatment technology. In fact, the lower energy density of biomass restricts its long distance transportation. Through torrefaction, the viability of biomass transportation over long distance can be achieved. The justification of this observation is what is appealing, but unfortunately has been flimsily clarified. First of all, torrefaction induces dehydration and decarboxylation reactions, the former releasing moisture and the later carbon dioxide, both of which have no useable energy. In terms of elemental modification, a sizeable amount of oxygen is released, decreasing the atomic oxygen-to-carbon ratio (O/C) of the solid fraction substantially, meanwhile the decrease of the hydrogen-to-carbon ratio (H/C) is little. This behavior is reminiscent of the progression of higher plant materials to type III kerogen in the coalification process (Schobert 2013) where a progression toward a carbon rich material is observed. The data from various torrefaction experiments have been used to plot the van Krevelen diagram presented in Fig. 15.4. It is apparent from the graph that torrefaction alters biomass properties towards peat, lignite, subbituminous coal and bituminous coal depending on the degree of severity.

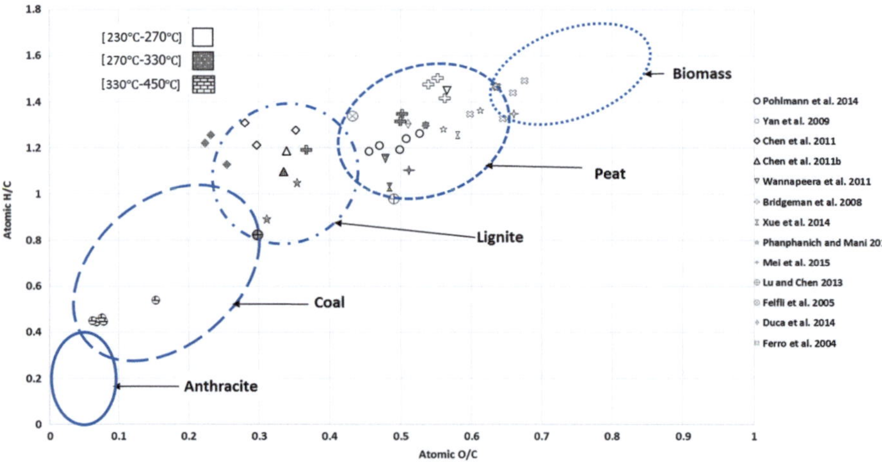

Fig. 15.4 van Krevelen diagram of torrefied biomass. Computed with data from the following authors: Ferro et al. (2004), Felfli et al. (2005), Bridgeman et al. (2008), Yan et al. (2009), Chen et al. (2011a, b), Phanphanich and Mani (2011), Wannapeera et al. (2011), Chen et al. (2011b), Lu and Chen (2013), Duca et al. (2014), Pohlmann et al. (2014), Xue et al. (2014), Mei et al. (2015)

Rheological Properties

Material handling is essential for the successful operation of biomass thermal conversion systems, given the poor rheological properties of biomass. Biomass materials are inherently difficult to handle as a consequence of its fibrous and tenacious nature. The intrinsic cohesion of these materials is negligible because the adhesion of individual particles is insignificant. However, due to fiber interlocking and elastic wind-up effects, their strength values are large, resulting in arching and hang-up tendencies. However the arching and hang-up tendencies are inversely proportional to the bulk density. Torrefaction once again is convenient as it takes away the fibrous and tenacious tendency of biomass. The resulting ground particles exhibits a better sphericity compared to untreated biomass (Phanphanich and Mani 2011) conferring a higher rheological properties.

Storability, Durability and Homogeneity

Torrefied biomass is low in moisture and is highly hydrophobic. These characteristics reduces the biological activity in torrefied biomass therefore improving its storability. It is well known that hemicelluloses are easily decomposed by various microorganisms including fungi, bacteria and actinomycetes both aerobic and anaerobic whereas lignin is one of the most resistant organic substances for microorganisms to degrade. It is also noticed that cellulose undergoes rapid degradation when it is associated to hemicelluloses whereas its degradation becomes very slow when it is associated with lignin (Husain and Kelman 1959). Since torrefaction gets rid of hemicelluloses and keeps most of the lignin and cellulose intact, the resistance to bio-degradation is improved and further reinforced by the resulting low moisture content and high hydrophobicity.

The gradual decrease of the O/C and H/C ratio with torrefaction severity increases the carbon content of various torrefied biomass materials, contributing to the convergence of their properties. It has been observed that physical and chemical variations observed in various biomass can be eliminated through torrefaction (Girard and Shah 1989) allowing the use of biomass from various sources.

Pyrolysis

Introduction

Pyrolysis is a thermal decomposition of organic materials in the absence of oxygen, producing a solid residue rich in carbon, condensable volatiles (bio-oil) and non-condensable gases (producer gas). The technology can be traced back to ancient Egypt where tar for caulking boats and certain embalming agents was made by pyrolysis of wood. The resulting solid residue has been given various appellations: char, biochar, charcoal, coke, biocoal, and has drawn many attention in the Bronze Age, since only its combustion could provide the required temperature to melt tin with copper in order to produce bronze. Nowadays, pyrolysis of biomass is still intensively used in the chemical industry to produce activated carbons for various applications, charcoal for metallurgical application, and even methanol. In developing countries, pyrolysis of wood sometimes referred to as carbonization is carried out to produce charcoal for cooking and heating; the process is very inefficient and environmentally unfriendly since it is done in an artisanal way, releasing all the volatiles in the atmosphere. The products of pyrolysis (char, bio-oil and producer gas) are commonly used as fuel with or without upgrading, as feedstock for other industries. Recent interests have grown on using biochar for soil amendment because of the potential of biochar to increase the cation exchange capacity (CEC) and the moisture retention of the soil, as well as a way to sequester carbon (Liang et al. 2006; Lee et al. 2013; Mukherjee and Lal 2013; Ouyang et al. 2013). Pyrolysis has also been considered as a pretreatment process, just like torrefaction, for biomass. In this regard, the brittle char produced is easily ground into a fine powder and mixed with the bio-oil produced during pyrolysis to form what has been called a bio-slurry. This energy dense (compared to the initial biomass) and less bulky product can now be transported to relatively long distance for gasification/combustion. This form is particularly suitable for high pressure slurry fed entrained flow gasifiers or pulverized fuel combustion (Henrich and Dinjus 2002; Henrich and Weirich 2004; Raffelt et al. 2004; van der Drift and Boerrigter 2006; Henrich et al. 2007, 2009).

The thorough study of pyrolysis is capital for a good understanding of other thermal conversion processes of biomass as gasification and combustion. This is because pyrolysis is actually the first mechanism taking place during biomass gasification or combustion and significantly impacts the conversion behavior of the resulting char. The impact of pyrolysis on gasification and combustion is in turn dictated by the heating rate of the fuel, the final pyrolysis temperature, the residence time and the pressure. The first three factors (heating rate, temperature and residence time) serve as the basis for defining the pyrolysis modes. Three pyrolysis modes have been outlined: slow pyrolysis, flash pyrolysis and fast pyrolysis. There is no common agreement in the literature regarding the delineation of these three pyrolysis modes.

Pyrolysis Modes

Slow Pyrolysis

This mode of pyrolysis is described by a slow heating rate (0.1–1 °C/s), operating temperature around 500 °C and very long residence time, using large size of the fuel (Naik et al. 2010). Slow pyrolysis is often carried out for the production of charcoal and is also referred to as carbonization. In practice, the term residence is not appropriate when referring to slow pyrolysis or carbonization, instead the reaction time is more applicable since slow pyrolysis is generally operated in fixed bed settings.

A more comprehensive definition considers pyrolysis to be slow if the time, $t_{heating}$, required to heat the fuel to the pyrolysis temperature is much longer than the characteristic pyrolysis reaction time, t_r (Basu 2010). The characteristic pyrolysis reaction time, t_r is estimated as the inverse of the pyrolysis reaction rate constant (Probstein and Hicks 2006). The low heating rate and large size of the fuel in slow pyrolysis allows the re-polymerization of the heavy volatiles.

The Biot number of the heating conditions is also a parameter used to delineate the boundary between slow and flash/fast pyrolysis. The Biot number is the ratio of the external to internal heat transfer. In slow heating, the heat is transferred to the interior faster than it can be supplied and the whole particle is essentially isothermal for Biot number less than 0.1. When the Biot number is greater than 1, the particle surface is heated faster than the inside of the particle creating temperature gradient within the particle; as a result, volatiles are quickly released from inside out (Reed and Gaur 1997).

Flash Pyrolysis

Flash pyrolysis is carried out at heating rates greater than 2 °C/s and operating temperature around 700 °C. The residence time of the fuel in this mode of operation is in the range of 10–30 s (Bridgwater 2012). This pyrolysis is meant to optimize liquid products (tar and bio-oil). The size of the fuel in this mode is smaller than in slow pyrolysis in order to allow complete heating of the fuel particle, given the lower residence time compared to slow pyrolysis.

Fast Pyrolysis

Fast pyrolysis is characterized by very high heating rates (200–10^5 C/s), short residence times and high temperature reaching 1,200 °C and higher. Gaseous products are optimized in these conditions as a result of the high operation temperature suitable for cracking heavy volatiles into light gases. The particle size in this pyrolysis mode is even smaller than in flash pyrolysis.

Fundamentals of Pyrolysis

Reaction Mechanisms

Pyrolysis involves multiple series and parallel competing reactions including dehydration, cracking, isomerization, dehydrogenation, aromatization, coking and condensation. Pyrolysis is a very challenging process to describe accurately because of multiple physical and chemical events happening simultaneously and

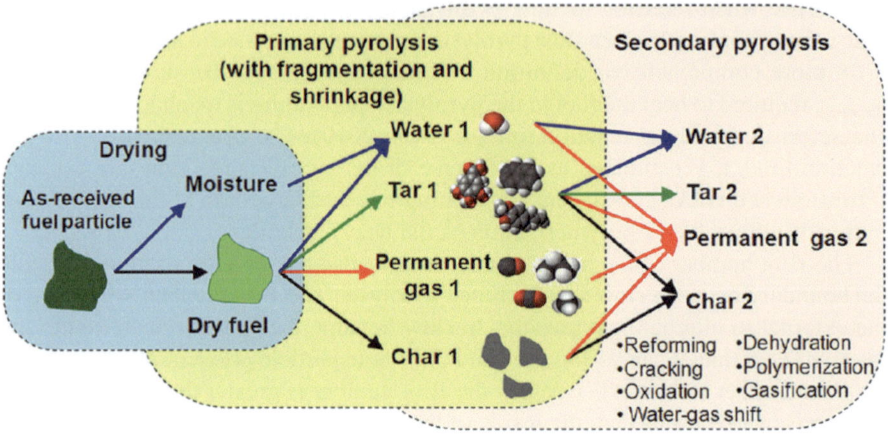

Fig. 15.5 Basic illustration of pyrolysis (Neves et al. 2011)

varying depending on the fuel composition, and operation condition. Therefore different fuels may behave differently in the same process condition while the same fuel may exhibit different behaviors when subjected to different pyrolysis conditions.

Pyrolysis can be viewed as an increased in temperature severity of the torrefaction process discussed in section "Torrefaction". Figure 15.5 presents a basic illustration of the pyrolysis mechanism. Pyrolysis begins with heat transfer to the biomass particle which triggers chemical reactions in the particle with the release of liquid and gaseous products; thus a mass transfer mechanism is also involved. The pyrolysis process can be segregated into primary pyrolysis and secondary pyrolysis. The primary pyrolysis entails the decomposition of the initial fuel particle into intermediate products while secondary pyrolysis consists of the degradation of the intermediates. The primary pyrolysis can be recapped into three main mechanisms: fragmentation, depolymerization and char formation (Fig. 15.6).

Char Formation

The heaviest fraction of the biomass, usually containing the inorganic elements is not volatilized, it turns into a solid char. Some heavy fraction of tar is also returned to char as depicted in Fig. 15.5. Lower heating rates, lower temperatures, and larger particle sizes favor this pathway as they promote intra and inter molecular rearrangement reactions resulting in a high degree of reticulation and in a higher thermal stability of the char residue (Scheirs et al. 2001; McGrath et al. 2003) for which the highly polycyclic aromatic structure is just a consequence (Pastorova et al. 1994; McGrath et al. 2003). The formation and consolidation of benzene rings in a polycyclic structure is the fundamental mechanism in this pathway. One important feature in this repolymerization process is the release of water vapor or non-condensable gases as it proceeds (Banyasz et al. 2001; Scheirs et al. 2001; Van de Velden et al. 2010; Collard et al. 2012).

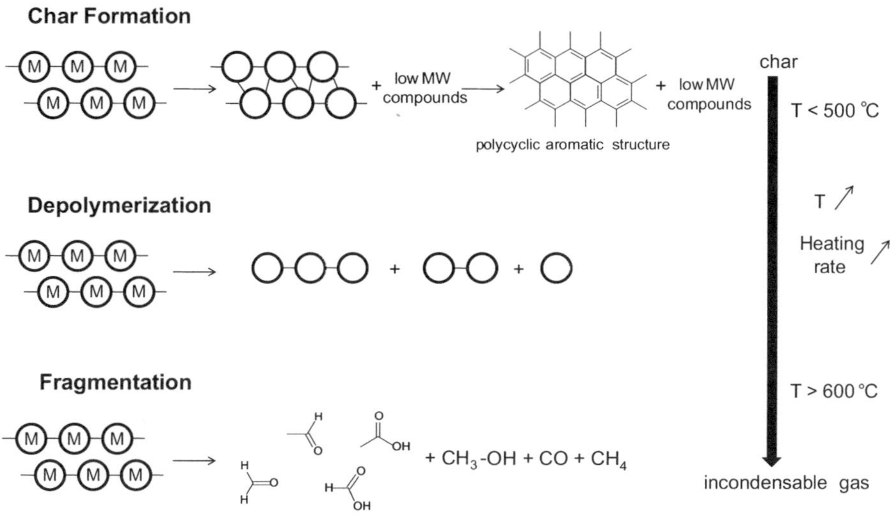

Fig. 15.6 Pyrolysis pathways as influenced by temperature (Collard and Blin 2014)

Depolymerization

Depolymerization reactions kick in when biomass is heated to higher temperature (>300 °C). The various polymers present in biomass depolymerize to short chains of some 200 sugar units (Lomax et al. 1991). Lignin and hemicelluloses and the amorphous segments of cellulose decompose preferentially. The crystalline segments of cellulose begins its decomposition later at much higher temperature. The initial depolymerization of cellulose occurs at the boundaries between the crystalline and amorphous regions by random cleavage of glucosidic linkages (Broido et al. 1973). Lignin decomposition is spread over a wide range of temperature, therefore lignin is the main contributor to the formation of char.

Depolymerization results in the breaking of the bonds between the monomer units of the polymers. The rupture is followed by stabilization reactions of the new fragments (Scheirs et al. 2001; Mamleev et al. 2009). It is theorized that intramolecular ketal formation and/or retro aldol decomposition are the main mechanisms governing the depolymerization of biomass components during depolymerization. Two fragments emerge from the intramolecular ketal rupture, one of which is terminated by an anhydro-sugar. Two fragments also emerge from retro-adol reactions, each of which carries a glycol-aldehyde entity (Lomax et al. 1991). These mechanisms are influenced by residence time and pressure which alter the various aldehydes and the unsaturated fragments formed. Depolymerization continuously decreases the degree of polymerization of the chains/fragments until formation of volatiles (Lédé et al. 2002; Mamleev et al. 2009; Azeez et al. 2011).

Fragmentation

Fragmentation, also referred to as open ring reactions, is the scission of the principal intermediates such as levoglucosan and cellobiosan to form lower molecular weight products. The ring scission generally occurs at the C-O bonds known to be less stable than the C-O bonds (Madorsky et al. 1956). Fragmentation contributes to the linkage of many covalent bonds of the intermediates, even within the monomer units (Van de Velden et al. 2010; Lu et al. 2011) with formation of small chain organic compounds, mostly intermediate to small molecular weight volatiles (Jakab et al. 1995; Van de Velden et al. 2010; Mullen and Boateng 2011).

Secondary Reactions of Pyrolysis

Secondary pyrolysis reactions take place in the gas phase and are characterized by the further conversion of intermediates fragments (C_{2-4} oxygenates) and the cracking of C_{2-4} hydrocarbons to a mixed gas of moderate heating value composed mainly of CO, CO_2, H_2, H_2O and CH_4. Secondary reactions are enhanced at higher temperatures and shift the pyrolysis products towards gases. Reforming of the hydrocarbons contained in the gas mixture occurs at temperatures beyond 1,000 °C resulting in a mixture of mainly CO and H_2. The pyrolysis reaction mechanisms described here are very general and may face some alteration depending on the heating rate and operating temperature with direct impact on product (char, bio-oil and gases) quality and quantity as well the composition of gas species.

Peculiarity of Biomass Slow Pyrolysis

As biomass is slowly heated to 100 °C, initial mass loss is observed due to evaporation of water. Towards 160 °C mainly bound water is released and the calorific value of the pyrolysis vapor is still negligible. At 180 °C and higher, the decomposition of lignin and hemicelluloses kicks in and condensable vapors are the main products released along with some non-condensable gases. The composition of the condensed vapor at this stage, sometimes called pyroligneous acid, is made of water, acetone, methanol, acetic acid, phenols and heavier tars (Nachenius et al. 2013). When the temperature reaches 250 °C, the degradation of all biomass components including cellulose is fully initiated. Prolonged heating (which fundamentally differentiates slow pyrolysis from torrefaction) at this temperature region results in a reduction of the degree of polymerization of cellulose which can be linked to the generation of free radicals, elimination of water, formation of carbonyl, carboxyl and hydroperoxide groups and the evolution of carbon dioxide (Bhuiyan et al. 2000; Bhuiyan et al. 2001). These reactions result in a charred residue formed by cross-linking reactions after long heating periods and in the liberation of water (Garcia-Perez et al. 2009). Beyond 300 °C various reactions, essentially depolymerizations, are triggered.

Owing to the various pyrolysis modes that are driven by the heating rate and temperature, heat and mass transfer during the respective processes will not be similar. Therefore the respective representation of their kinetics will be different. However, given the slight similarity between flash and fast pyrolysis, a common representation is acceptable.

Kinetics of Biomass Pyrolysis

The design and optimization of biomass pyrolysis reactors requires analytical description of the process. A thorough understanding of the kinetic law governing the pyrolysis process is necessary for system optimization. The complexity of the biomass pyrolysis process involving numerous complex reactions yields a large number of gaseous and liquids products as well as a char residue, making the derivation of exact reaction mechanisms describing any single feature of the process extremely difficult. Simplifications have led to the development of lumped models containing conceptual or pseudo-reactions for modeling pyrolysis. Available models can be arranged into three main groups: one step models, model with competing reactions and models with secondary reactions.

One Step Models

The one step global model considers pyrolysis happening in a single decomposition step producing volatiles and char. Arrhenius type first order kinetic is usually applied to calculate kinetic parameters based on experimental results. The kinetic parameters obtained in this method are very specific to operating conditions where the data used to generate the parameters were generated. Here, only the reduction in the mass of the solid material is quantified and secondary reactions are ignored, which limits its applicability to situations where in-depth knowledge of the volatile formation is not necessary (such as fast pyrolysis). This could be suitable for use when dealing with a simplified solid substrate such as cellulose, but such a reaction model does not have the flexibility to account for variation of the constituent content within more complex materials such as biomass (Nachenius et al. 2013).

Models with Competing Parallel Reactions

One of the first competing models for biomass pyrolysis is the Broido-Nelson model, developed on pure cellulose samples. This model suggests that the thermal decomposition of cellulose takes two different routes: (1) an intermolecular dehydration leading to char formation and non-condensable gases and (2) a char-free depolymerization leading to tars (Broido and Nelson 1975). Secondary pyrolysis reactions are not taken into account in this model. Another model very popular in the coal conversion community and sometimes applied to biomass was developed by Kobayashi (1976). This model considers two competing reactions each producing volatiles and char but one prevails at lower temperatures while the other dominates at higher temperatures. This model also lumps secondary pyrolysis reactions into each of the parallel paths. However, the effect of decreasing char yields for increased temperatures is carefully adjusted by this model. Studies on pyrolysis of biomass particles of various sizes have been modeled using this model (Nunn et al. 1985a, b; Zaror and Pyle 1986; Samolada and Vasalos 1991).

Models with Multi-step Reactions

Successive and parallel competing reactions seem to express a realistic picture of the thermal decomposition of biomass. The stepwise mechanism take into account the presence of primary and secondary reactions whereas the parallel feature accounts for the competing reactions leading to the char, tar and light gases. Bradbury et al. (1979) were one of the first authors to theorize that biomass (cellulose), upon heating, is mobilized into an active form ("active cellulose") then subsequently proceeds with two competing reactions: one producing condensable volatiles and the other producing char and non-condensable gases. This model is similar to what was proposed by Broido and Nelson (1975) couple of years before with the exception of the initiation stage where it is first converted to an activated form. The two models have given the notorious appellation Broido-Shafizadeh model in the biomass pyrolysis community and have been extensively used and commented (Wichman and Melaaen 1993; Antal and Varhegyi 1995; Di Blasi 1998; Wooten et al. 2003) and sometimes questioned (Varhegyi et al. 1994). This model has undergone various modifications/improvements; Chan et al. (1985) included a tar cracking reaction as well as dehydration reactions while Di Blasi and Russo (1993) added tar cracking and repolymerization reactions of intermediate fractions. Figure 15.7 shows a modification of the Broido-Shafizadeh model according to Thurner and Mann (1981).

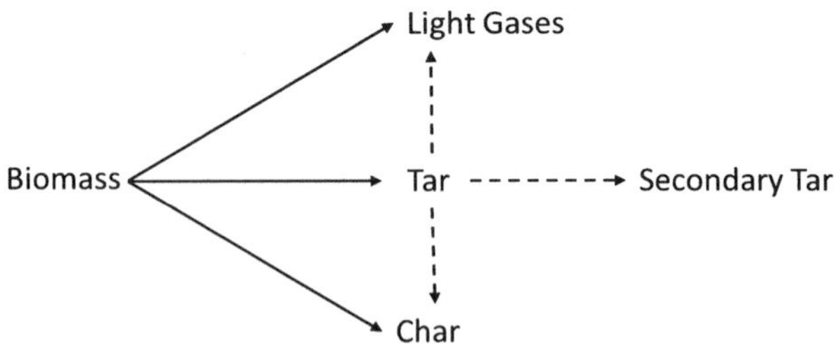

Fig. 15.7 Illustration of the successive and parallel mechanisms of biomass pyrolysis

Gasification

Overview of Gasification

Gasification is a partial combustion process that converts carbonaceous materials like biomass into useful gaseous fuels with a useable heating value or chemical feedstock. Unlike combustion where the product flue gas has no residual heating value, the product flue gas from gasification still carries valuable heating value. Gasification stores part of the energy from the fuel into chemical bonds while combustion releases it. The gasification process adds hydrogen to and strips carbon away from the feedstock to produce gases with a higher hydrogen-to carbon (H/C) ratio, while combustion oxidizes the hydrogen and carbon into water and carbon dioxide, respectively. Gasification takes place in reducing (oxygen-deficient) environments requiring heat; combustion takes place in an oxidizing environment giving off heat. Therefore gasification in its broader sense includes pyrolysis, partial oxidation and hydrogenation.

The nomenclature of gasification is sometimes not clearly defined. Producer gas refers to the low heating value gas mixture of carbon monoxide (CO), hydrogen (H_2), carbon dioxide (CO_2), methane (CH4) and other low molecular weight hydrocarbons and nitrogen (N_2) produced from gasification of carbonaceous feedstocks in air. Historical applications of producer gas have included heat and electricity production, as well as the production of synthetic liquid fuels. Synthesis gas (syngas) refers to a gas mixture of predominantly CO and H_2 produced from gasification of carbonaceous feedstocks in oxygen and steam

Gasification has the advantage of converting solid fuels into gases that are much easier to handle than solids. But the most relevant reason justifying gasification lies in a more chemical/thermodynamic consideration. The oxidation of carbon is illustrative:

$$
\begin{array}{ll}
C + \tfrac{1}{2} O_2 === CO & -111\ \mathrm{Mj/kmol} \\
CO + \tfrac{1}{2} O_2 === CO_2 & -283\ \mathrm{Mj/kmol} \\
C + O_2 === CO_2 & -394\ \mathrm{Mj/mol}
\end{array}
$$

The above three reaction proves that by sacrificing 28 % of the heating value of pure carbon when converting solid carbon to gas, 72 % of the heating value of the carbon can be stored in the gas.

Gasification started as a source for lighting (production of town gas) and heating. However, the water gas process which is an endothermic reaction producing a gas consisting of equal amount of carbon monoxide and hydrogen became significant for the chemical industry as of the 1900s.

$$C + H_2O ===== CO + H_2 \quad + 131\ \mathrm{MJ/kmol}$$

Subsequent conversion of carbon monoxide into hydrogen by the Water Gas Shift Reaction (WGSR) made it possible to convert carbonaceous feedstock to synthesis gas (mixture of hydrogen and carbon monoxide) or pure hydrogen which are necessary chemicals for methanol and ammonia synthesis, respectively.

$$CO + H_2O ============ CO_2 + H_2 \quad \Delta H = -41.1\ kJ/mol$$

Fischer-Tropsch synthesis of hydrocarbons and synthesis of acetic acid anhydride are other application of synthesis gas in the chemical industry.

A typical biomass gasification process may include the following steps:

- Drying.
- Devolatilization or pyrolysis.
- Volatile combustion.
- Char gasification and combustion.

Figures 15.8 and 15.9 illustrate the various steps involved in the gasification process.

The volatile/char combustion steps of the gasification process are exothermic reactions releasing heat. The syngas leaves the gasifier at higher temperatures, thus is sent to the heat recovery boiler to generate steam that is sent to the steam turbine or used as process heat. The waste heat from the gas turbine is also recovered to generate steam for the same purposes (Figs. 15.8 and 15.9).

As already mentioned, the WGSR is a further step for converting the syngas into hydrogen used for ammonia production in the fertilizer industry, petroleum refineries for a variety of operations and recently as fuel for power generation and transportation. The use of gasification for power generation has also increased the use of water gas shift reactors multifold. The earliest recording of the reaction dates back to 1888, and its prominence came with the Haber ammonia synthesis process and development of catalyst by Bosch and Wilde in 1912. The catalyst developed

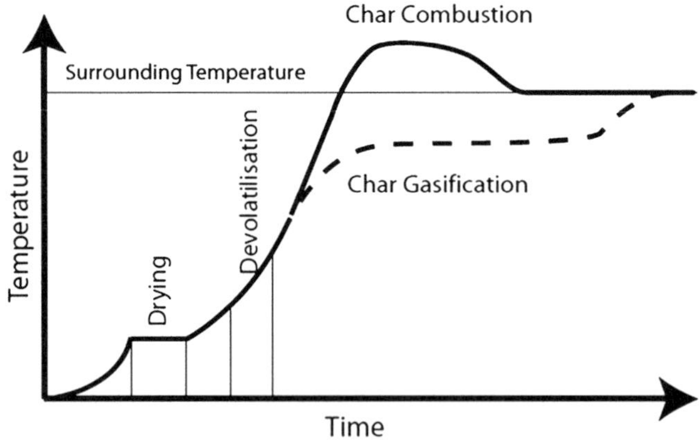

Fig. 15.8 Principle stages of conversion of a small solid fuel particle in a hot surrounding versus particle temperature and time. The extension of drying and devolatilisation are marked with thin lines (Thunman and Bo Leckner 2007)

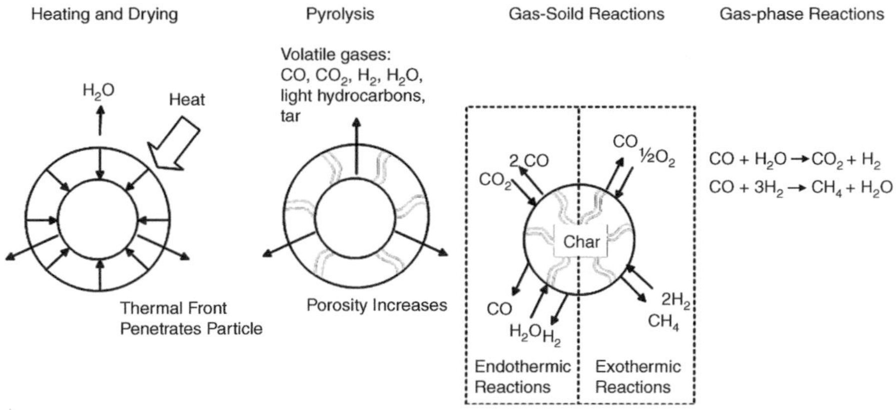

Fig. 15.9 The process of thermal gasification (Brown 2011)

containing iron and chromium was capable of catalyzing the reaction at 400–500 °C and reduced the exit carbon monoxide content to around 2 % Smith R J et al. (2010).

Water gas shift reaction is a moderately exothermic reversible reaction and is expressed by

$$CO + H_2O ============= CO_2 + H_2 \quad \Delta H = -41.1\ \mathrm{kJ/mol}$$

Due to its moderate exothermicity, the WGSR is thermodynamically unfavorable at elevated temperatures. However, in the gas phase the rate of the WGSR is almost negligible until reaching temperature exceeding 800 °C. Conversion at these temperatures is still lower compared to thermodynamic equilibrium expectation; therefore the use of catalysts for these reactions becomes a primary research concern.

The shift reaction will operate with a variety of catalysts between 200 and 500 °C. The effect of pressure on the reaction is minimal since the reaction keeps the number of reactants and products moles equal. However, the equilibrium for H_2 production is favored by high moisture content and low temperature for the exothermic reaction. Normally, excess moisture present in the syngas from slurry-fed gasifiers is sufficient to drive the shift reaction to achieve the required H_2-to-CO ratio. On the other hand, additional steam injection before the shift may be needed for syngas output by dry-fed gasifiers. In any case, the scrubber syngas feed is normally reheated to 0–10 °C above saturation temperature to avoid catalyst damage by condensation of liquid water in the shift reactor. Shifted syngas is cooled in the low temperature gas cooling (LTGC) system by generating low pressure steam, preheating boiler feed water, and heat exchanging against cooling water before going through the acid gas removal system for sulfur removal.

Commercially, the WGSR is a two stage process, operating first under a high temperature regime to maximize the initial rate, followed by cooling, and finally a lower temperature regime to maximize total conversion of carbon monoxide.

The high temperature bed is usually an iron oxide catalyst promoted with chromium oxide. Chromium is added as a structural stabilizer to reduce the sintering commonly observed in high temperature iron oxides. The low temperature bed is usually a copper based catalyst generally promoted by zinc oxide and alumina in 1:1:1 ratio.

Fundamentals of Gasification

A deep understanding of the gasification process as well as how the feedstock and operating parameters influence the performance of the plant is crucial to the design and operation of the plant. A good apprehension of the fundamental reactions taking place during gasification is essential to the planning, design, operation, troubleshooting, and process improvement of the plant. As already mentioned, gasification proceed in the following steps: heating and drying, devolatilization or pyrolysis, homogeneous reactions (including volatile combustion), and heterogeneous reactions (including char combustion and gasification) (Fig. 15.10).

Heating and Drying

The fuel particle is heated as it enters the reactor. Once the particle temperature reaches 100 °C (depending on the reactor pressure, 100 °C at 1 atm), moisture escapes the particle and surface drying takes place. The particle temperature remains there until all the surface moisture have been removed. Following surface moisture evaporation, upon subsequent heating of the particle, inherent moisture evaporates at about 110–120 °C. No chemical reaction occurs during heating and drying. Every kilogram of moisture in the biomass takes away a minimum of 2,260 kJ of extra energy from the gasifier to vaporize water, and that energy is not recoverable. The time required for drying depends on the particle size and the

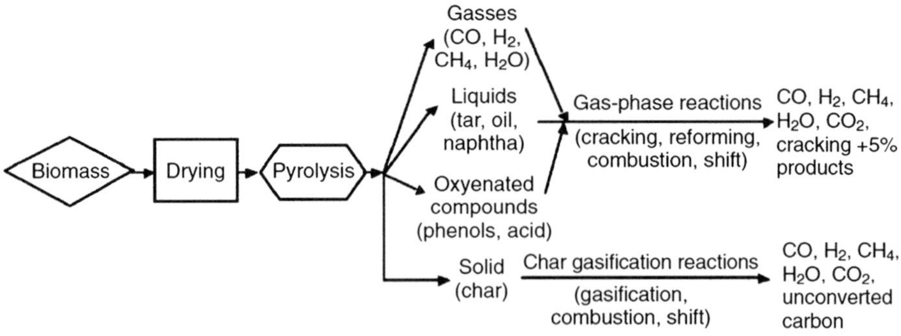

Fig. 15.10 Reaction sequence for gasification of biomass (Higman and Burgt 2008)

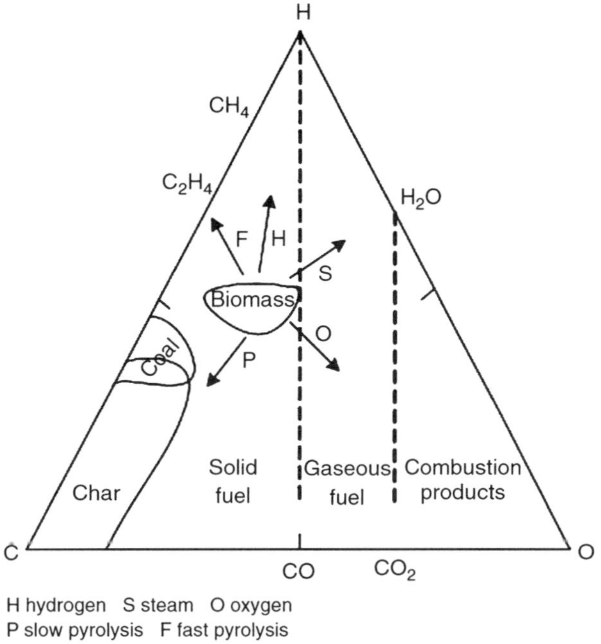

Fig. 15.11 C-H-O diagram of the gasification process (Basu 2010)

ignition temperature of the biomass fuel. Once the drying process is completed, the temperature of the particle increases. At about 200–300 °C, decomposition of biomass begins (Fig. 15.11).

Pyrolysis and Gas-Solid Reactions

Biomass is less resistant to heat since its constituents are polymers of cellulose, hemicelluloses and lignin joined together with relatively weak ether bonds (R-O-R, bond energy ~380–420 kJ/mol). The different components of biomass (cellulose, hemicellulose and lignin) decompose at different temperatures. Hemicellulose is the first component to break down at temperatures between 225 °C and 325 °C. Lignin is the next component to break down but its decomposition is not completed until above 500 °C. Cellulose is the last component to initiate break down (between 300 and 350 °C) but its decomposition completes faster than the other two components (Fig. 15.12). This observation shows that the chemical bonds in cellulose have a consistent structure while lignin exhibits high dissimilarity in its chemical bonds. Weak bonds of lignin (ether bonds) break easily at lower temperatures while strong bonds (aromatic structure) break at higher temperatures. Decomposition of lignin also leaves a high amount of residues behind while cellulose and hemicellulose leave almost no residue. The amount of residues left is important for the design of boilers and gasifiers since it represents the amount of char produced for the next phase of gasification or combustion.

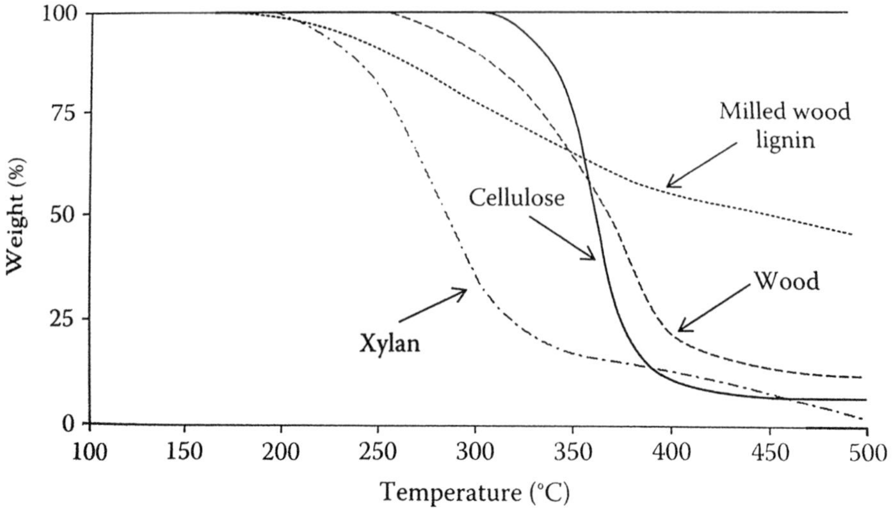

Fig. 15.12 Degradation of biomass components as a function of temperature (Shafizadeh 1982)

The major gases released during pyrolysis are hydrogen, methane, carbon dioxide, carbon monoxide, water vapor, light hydrocarbons and high molecular weight compounds (tar). More than 80 % of biomass conversion takes place during the devolatilization stage. The lower temperature at which these volatiles are released does not favor the cracking of higher molecular weight compounds and somehow justifies why tar is usually observed during biomass gasification. However they don't pose any problem in combustion as there are easily burned in the presence of oxygen.

Devolatilization is as well an endothermic process like drying. The devolatilization temperature dictates the composition of volatiles gases. Dehydration and depolymerization are the two mechanisms by which cellulose is decomposed. Dehydration leads to the formation of anydrocellulose which later decomposes to char, tar, CO, CO_2, and water. Levoglucosan is formed as the main intermediate of depolymerization which later decomposes into char and combustible volatiles. Low temperatures enhance dehydration while depolymerization occurs mostly at higher temperatures. Lignin degradation produces char (50–65 %), tar (10 %), organic acids (formic, acetic, propionic, etc), phenolic compounds (phenol, cresol, guaiacol, etc) and catechols. Devolatilization is a very fast process when compared to the overall gasification. As gasification proceeds, volatile gases formed can either react with the char or react among themselves in the gas phase (Fig. 15.13).

Char–gas reactions convert the solid char to H_2, CO, CO_2, and CH_4. The followings are the solid-gas reactions that occur during gasification:

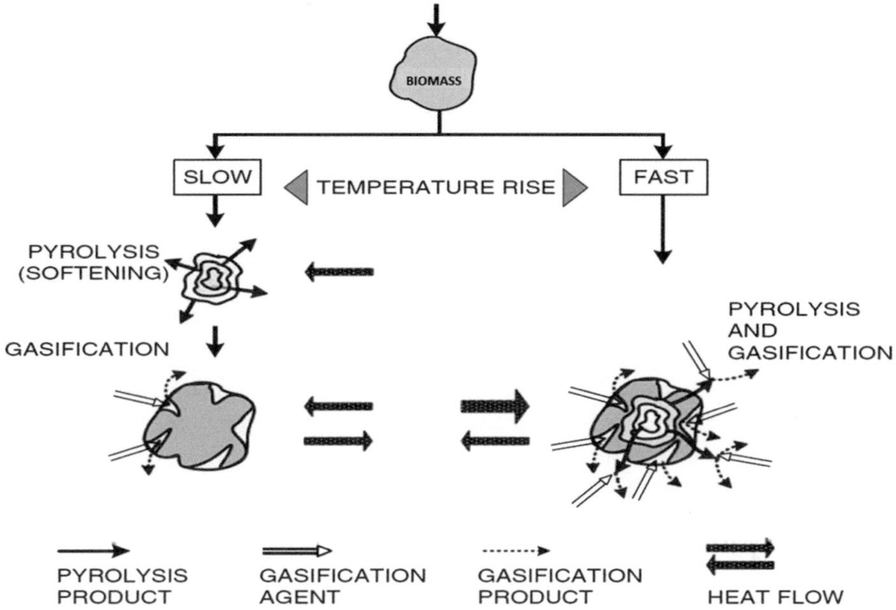

Fig. 15.13 Influence of heating rate on gasification process (Higman and Burgt 2008)

Carbon–oxygen reactions

$$C + \frac{1}{2} O_2 === CO \quad -111\ Mj/kmol$$
$$C + O_2 === CO_2 \quad -394\ Mj/mol$$

Boudouard reaction

$$C + CO_2 === 2CO \quad 172\ Mj/kmol$$

Water gas reaction

$$C + H_2O === H_2 + CO \quad 131\ Mj/kmol$$

Hydrogenation reaction

$$C + 2H_2 === CH_4 \quad -75\ Mj/kmol$$

Reactions of char with oxygen are usually complete. These reactions are so fast that they quickly consume the oxygen, leaving hardly any free oxygen for any other reactions. This highly exothermic reaction is important for supplying the energy to drive the endothermic processes of heating, drying, pyrolysis and char gasification. The hydrogenation reaction also helps to provide some energy for the endothermic reactions, although the relatively low concentration of H_2 in the gasification environment makes it a small contribution compared with the carbon–oxygen reaction.

Except the Boudouard and water-gas reactions, the remaining reactions are exothermic; therefore these two reactions will be given more attention as gasification reactions. The carbon conversion is dictated by the extend of the above reactions. The rate of the water–gas reaction is about 2–5 times faster than that of the Boudouard reaction (Di Blasi 2009). The hydrogeneation reaction is the slowest of the heterogeneous reactions. The relative rates of the various heterogeneous reactions reaction follow the following order:

$$R_{C+O2} >> R_{C+H2O} > R_{C+CO2} >> R_{C+H2}$$

Since the main gasification reactions are endothermic (Boudouard and water-gas) the gas temperature in the reduction section will decrease. Typical temperatures in this section range from 800 to 1,100 °C. If complete gasification takes place, all the carbon is converted to carbon monoxide and CO_2. The remaining solid residue exiting this section consists of ash and unburned carbon.

Several investigations have been performed on the Boudoard and water-gas reactions. Ergun (1956) developed one of the widely used model for the Boudouard reaction, he proposes a two-steps process:

$$\text{Step 1: } C_{fas} + CO_2 ===== C(O) + CO$$
$$\text{Step 2: } C(O) \rightarrow CO + C_{fas}$$

In the first step, CO_2 dissociates at a free carbon active site (C_{fas}), releasing carbon monoxide and forming an oxidized surface complex (C(O)). In the second step, the carbon–oxygen complex produces a molecule of CO and a new free active site. The rate-limiting step is the desorption of the carbon–oxygen surface complex.

The model for the water gas reaction is basically similar:

$$\text{Step 1: } C_{fas} + H_2O ===== C(O) + H_2$$
$$\text{Step 2: } C(O) \rightarrow CO + C_{fas}$$

In this case, the first step is the dissociation of a water molecule at a carbon free active site (C_{fas}), releasing hydrogen and forming an oxidized surface complex (C(O)). In the second step, the carbon–oxygen complex produces a molecule of CO and a new free active site. In some models the rate-limiting step is the desorption of the carbon–oxygen surface complex, as for the Boudouard reaction. Other models include the possibility of hydrogen inhibition by the inclusion of a third step:

$$\text{Step 3a: } C_{fas} + H_2 ===== C(H)_2$$
Or
$$\text{Step 3b: } C_{fas} + ½\, H_2 ===== C(H)$$

whereby some of the sites can become blocked by hydrogen. The presence of hydrogen has a strong inhibiting effect on the char gasification rate in H2O. For example, 30 % hydrogen in the gasification atmosphere can reduce the gasification rate by a factor as high as 15 (Basu 2010). So an effective means of accelerating the water–gas reaction is continuous removal of hydrogen from the reaction site.

Gas-Phase Reactions

Volatile gases released during devolatilization react among themselves in the gas phase. Multiple reactions take place simultaneously during gasification. The following reactions are a simplified set of reactions, limited to carbon, hydrogen and oxygen containing compounds, happening during gasification.

Oxidation reactions

$$H_2 + ½\,O_2 \;=\!=\!=\!=\!=\; H_2O \qquad -241.2\ \text{kJ/mol}$$
$$CH_4 + 2O_2 \;=\!=\!=\!=\!=\; 2H_2O + CO_2 \qquad -241\ \text{kJ/mol}$$

Water-gas shift

$$CO + H_2O \;=\!=\!=\!=\!=\; CO_2 + H_2 \qquad -41.1\ \text{kJ/mol}$$

Steam reforming

$$CH_4 + H_2O \;=\!=\!=\!=\!=\; CO + 3\,H_2 \qquad +206.1\ \text{kJ/mol}$$
$$C_nH_m + n\,H_2O \;=\!=\!=\!=\!=\; nCO + (n + m/2)\,H_2O$$

Dry reforming

$$CH_4 + CO_2 \;=\!=\!=\!=\!=\; 2CO + 2\,H_2 \qquad +247.1\ \text{kJ/mol}$$
$$C_nH_m + CO_2 \;=\!=\!=\!=\!=\; 2nCO + m/2\,H_2$$

Among the gas phase reactions, the two most important ones in terms of determining the final composition of the gas are the water-gas shift and steam reforming reactions. The water-gas shift reaction is slightly exothermic, and its equilibrium yield decreases slowly with temperature. Depending on temperature, it may be driven in direction—that is, products or reactants. The equal amount of moles in the reactants side as well as in the product side makes this reaction insensitive to pressure. Above 1,000 °C the water-gas shift reaction rapidly reaches equilibrium, but at a lower temperature it needs heterogeneous catalysts. This reaction has a higher equilibrium constant at a lower temperature, which implies a higher yield of H_2 at a lower temperature. With increasing temperature, the yield decreases but the reaction rate increases. Optimum yield is obtained at about 225 °C.

Because the reaction rate at such a low temperature is low, catalysts like chromium-promoted iron, copper-zinc, and cobalt-molybdenum are used. At higher temperatures (350–600 °C) Fe-based catalysts may be employed.

The water-gas shift reaction is important in increasing the H_2 content of syngas (important for synthetic fuels production), while the steam reforming reaction can strongly influence the CH_4 content of syngas (important for production of synthetic natural gas). Both of these reactions are exothermic, which means they are thermodynamically favored at low temperatures. However, low temperatures reduce the rates at which these reactions occur. Thus, a better strategy for promoting hydrogen formation is to add steam to the gasifier, while methane formation can be promoted by increasing partial pressures of hydrogen in the gasifier.

Although the ratios of nitrogen and sulfur in biomass are relatively low, compounds containing these elements participate in the solid as well as gas phase reactions. Biomass has a higher content of oxygen, which enhances the direct release of NO and isocyanic acid (HNCO) as part of the devolatilizing gases during pyrolysis. HNCO can then react with steam to give ammonia and carbon dioxide:

$$HNCO + H_2O \rightarrow NH_3 + CO_2$$

Biomass also usually has a higher moisture content. Moisture can react with hydrogen cyanide to give ammonia and carbon monoxide, this reaction is known to be enhanced by the presence of CaO, which is present in most biomass:

$$HCN + H_2O \rightarrow NH_3 + CO \tag{15.5}$$

Some trace elements found in biomass fuels dictate the reaction mechanisms involving thermal decomposition of sulfur. Among these elements, K, Ca, Cl and silicon (Si) have the most significant effect. A better understanding of these elements' inclusion in the plant is necessary to gain a strong insight on how sulfur will behave. It is generally accepted that the stability of organically-bound sulfur in biomass is low, resulting in its decomposition at lower temperatures (673 K) during the devolatilization. Due to their high stabilities, the inorganic sulfates will not be released during the devolatilization stage.

Mineral matters in biomass can significantly influence the reaction rate of biomass gasification. Alkali metals, potassium, and sodium are active catalysts in reactions with oxygen-containing species (CO_2 and H_2O). Dispersed alkali metals in biomass contribute to the high catalytic activity of inorganic materials in biomass. Inorganic matter also affects pyrolysis, giving char of varying morphological characteristics. Potassium and sodium catalyze the polymerization of volatile matter, increasing the char yield; at the same time they produce solid materials that deposit on the char pores, blocking them. During subsequent oxidation of the char, the alkali metal catalyzes this process. Polymerization of volatile matter dominates over the pore-blocking effect. A high pyrolysis temperature may result in thermal annealing or loss of active sites and thereby loss of char reactivity.

Char gasification takes place on the surface of solid char particles, which is generally taken to be the outer surface area. Given that char particles are highly porous, and that the surface areas of the inner walls are far higher than the external surface of the particle, without any physical restriction, reaction gases can potentially enter the pores and react internally at the inner particle surface. The reaction rate expressed with an account of the inner pore surface area is the intrinsic reaction rate whereas the reaction expressed without account of particle surface is usually referred to as the apparent reaction rate. The reacting gas molecules must enter the pore of the char particle for the gasification reaction to take place within on the inner surface. Limited availability of the reacting gas i.e. entire consumption of the gas by reactions happening at the outer surface of the char restricts the gasification reactions to take place only at the outer surface of the particle. Mass transfer limitation can be one of the causes of this restriction. Pore surface reactions require that the gasification agent (CO_2 or H_2O) diffuses through to the inner char surface in order to react with the active sites.

Diffusion is a finite rate mechanism. If the kinetic rate of the reaction is much faster than the diffusion rate of the gasifying agent to the char surface, all of the gas molecules transported are consumed on the external surface of the char, leaving none to enter the pores and react on their surfaces. In this case, overall reaction is controlled by diffusion and this is termed the diffusion- or mass-transfer-controlled regime of reaction. However, if the kinetic rate of reaction is slow compared to the transport rate of the gasifying agent gas molecules, then the gas molecules will diffuse into the pores and react on the char walls. The reaction in this situation is "kinetically controlled." Between the two extremes lie intermediate regimes. The relative rates of chemical reaction and diffusion determine the gas concentration profile in the vicinity of the char particle; how the reaction progresses; and how char size, pore distribution, reaction temperature, char gas relative velocity, and so forth, influence overall char conversion (Fig. 15.14).

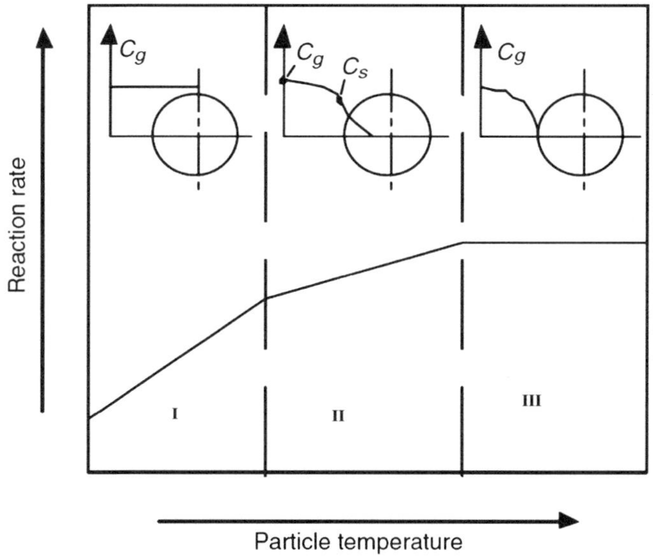

Fig. 15.14 Char—gas reaction regime (Basu 2010)

Gasifiers

Air-Blown Gasifiers

Various combinations of nitrogen, oxygen, air, steam, and carbon dioxide are possible for gasification of biomass; however air offer the simplest and cost effective option. Therefore air-blown biomass gasification systems have been thoroughly investigated. They are the most represented type of gasifiers in commercial-scale application. Air-blown gasification can be carried out in any type of gasifiers including fixed-bed, fluidized-bed both circulating and bubbling and entrained flow reactor. At an equivalence ratio of 0.25 (required for optimized output) the product gas is highly diluted by nitrogen from air, reducing the calorific value of the gas. Air gasification provides a greater volume of syngas per mass of biomass, due to the nitrogen in the air. Consequently, air blown gasifier must be bigger in size compared to an equivalent oxygen blown gasifier. The high volume of nitrogen gas in the syngas is troublesome when it comes to maintaining engine efficiency and avoiding a substantial de-rating of the engines. There is substantial loss of energy in air blown gasifiers as the high volume of nitrogen in the air must be heated to gasification temperature and then cooled down after exiting the gasifier. Figure 15.15 below shows a comparison of the efficiency of air blown and oxygen blown gasifiers. It can be observed that the efficiency of an air blown gasifier operating at 800 °C is equivalent to that of an oxygen blown gasifier operating at 1,400 °C; However, the tarry gas produced at this condition in the air blown gasifier is detrimental. Plasma assisted air gasification has been introduced to take care of the tar; however the inefficiency associated with heating the inert nitrogen remains.

Despite the above mentioned shortcomings of air blown gasification, producer gas from these systems has been successfully used in furnaces, boilers and internal

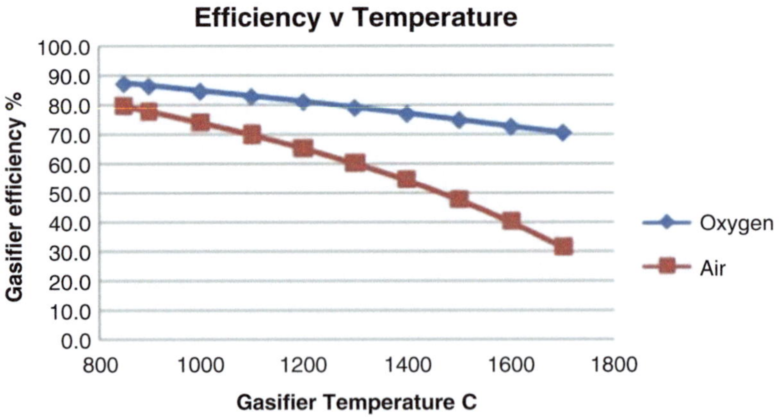

Fig. 15.15 Comparison of the efficiency of air blown and oxygen blown gasifiers

combustion engines. However, for applications involving chemical or fuel synthesis, the nitrogen in syngas becomes more problematic as the equipments must take into account the unreactive nitrogen gas volume.

Combustion

Overview of Biomass Combustion

Combustion is the dominant technology among all the thermal processes used to convert biomass to energy and fuels. It is one of the oldest forms of biomass conversion known by mankind as it is well known that humans have burnt biomass to obtain heat and sometimes light thousands of years ago. Combustion is an exothermic reaction during which a fuel (biomass in this case) reacts with oxygen (from air or supplied) to form carbon dioxide and water vapor. The exothermic nature of the reaction implies that heat is generated during this process. The amount of chemical energy released usually as radiant energy or thermal energy is dictated by the enthalpy of combustion of biomass. Some heat is normally supplied to start up the reaction; once started the reaction continues spontaneously by the heat generated by the reaction itself. A simplified combustion equation for cellulose which is one of the constituents of non-food biomass, can be illustrated as:

$$(C_6H_{10}O_5)_n + 6nO_2 \rightarrow 6nCO_2 + 5nH_2O + \text{Heat}$$

Beside carbon, hydrogen and oxygen, biomass also contains nitrogen and sulfur. These five elements forms the organic constituents of biomass. Determining the percentage of the five elements in biomass is one of the common method for characterizing a solid fuel called elemental or ultimate analysis. Biomass also contains inorganic elements which form ash upon combustion.

Combustion of biomass proceeds in various forms: evaporation combustion, decomposition combustion, surface combustion and smoldering combustion. Evaporation combustion takes place when volatiles contained in the biomass escape the fuel following heating and react with oxygen in the gas phase and burn. Decomposition combustion happens when the products of devolatilization (H_2, CO, CmHn) react with oxygen in the gas phase, form flame and burn. Usually, biomass itself does not directly undergo decomposition combustion but instead the tar derived from biomass during devolatilization. After devolatilization, the solid char containing mostly carbon burns by surface combustion. In this process, oxygen diffuses into the pores created by the escape of volatiles and react at the surface of the char and carbon dioxide (and water vapor) produced diffuses out. When the temperature of combustion is lower than the ignition temperature of the volatile components, smoldering combustion occurs. Decomposition and surface combustion are the main forms of biomass combustion encountered in industrial as well as residential application of biomass combustion.

The main technologies used for combustion of biomass are:

- Fixed bed Combustion.
- Stoker Combustion.
- Suspension Combustion.
- Fluidized Bed Combustion.

Fixed-Bed Combustion: Pile Combustion

Pile combustion is the simplest form of direct firing. The furnace in these systems has a fixed grate inside the combustion chamber. The biomass fuel is fed onto the grate; air passes through the grate from below (under-fire air) and burns the fuel. The bulk of the combustion process takes place on the grate of a pile burner known as the primary combustion chamber. At this stage, combustion is still incomplete; a considerable amount of unburned carbon as well as combustible carbon monoxide is carried over to the secondary combustion chamber, located above the first. Air is further injected (over-fire air) to complete the combustion. The boiler is located above the second combustion chamber in order to absorb the heat generated during combustion. The steam produced in the boiler drives a steam turbine. The steam exiting the turbine is condensed and sent back to the boiler, therefore recycled throughout the system. For Combined Heat and Power systems, the steam taken from the turbine outlet is used for heat recovery prior to its condensation.

The biomass fuel is usually introduced from above the grate. The pile burner has the ability to handle wet and dirty fuels, but the efficiency drops significantly in these cases. One disadvantage of pile burners is that there is no mean of removing the ash unless the furnace is shut down. This makes continuous operation of these systems impossible. Automatic control of pile burners is difficult and their response to energy input is slow. Therefore it becomes difficult to modify the electricity output to match the demand.

Fixed-Bed Combustion: Grate Combustion

Grate combustion is a common technology for burning solid fuels. Biomass combustion systems using grates are still widely used for hot water boilers and production of steam in small scale plants. Despite their less tolerance to fuel quality variations compared to fluidised bed boilers, progress resulting research and development have improve grate combustion technologies such that they are able to compete with modern combustion systems. Burning of wet fuels such as sawdust and bark residues is made possible in the new improved grade firing technologies. Grate combustion systems are simple in construction, therefore adequate and very competitive in small scale settings.

Suspension Combustion

In suspension combustion (also called pulverized combustion or entrained flow combustion) the biomass fuel, ground to small sizes, is blown in the combustion chamber with air using specially designed burners. The biomass powder and the air are mixed by the burner during injection and the mixture burns rapidly into a flame. These systems require very fine particle sizes and low moisture content fuels, usually less than 15 % moisture. The efficiency of suspension fired boiler can reach 80 %, thus they allow for smaller size furnace for a given output. The main drawback of these systems is that they require extensive biomass drying and grinding.

Fluidized Bed Combustion

Fluidized bed combustion is one of the most efficient methods of directly burning biomass. Fluidized bed combustion systems are certainly the most versatile as these systems can easily cope with diverse fuels (type and size) a wide range of moisture content. In these systems, biomass is burned in a self-mixing suspension of gas and solid bed particles (the bed material is usually silica or dolomite). Air enters the combustion chamber from below the bed and fluidizes the mixture of biomass and bed particles. The fluidization velocity dictates the behavior of the system between bubbling and circulating fluidized bed. The bubbling fluidized bed (BFB) systems are offer a simplified construction and lower investment, thus considered a better option for small-scale applications.

Fundamentals of Biomass Combustion

Global Mechanism of Biomass Combustion

Combustion involves simultaneous heat and mass transfer mechanisms, with chemical reaction and fluid flow. This makes it a very complex phenomenon to predict. The basic reaction of biomass complete combustion in air can be described by a single global reaction:

$$\begin{aligned} \mathrm{CuHvOwNxSy} - \mathrm{M} + \left(\mathrm{u} + \frac{\mathrm{v}}{4} - \frac{\mathrm{w}}{2} + \mathrm{y}\right)\ (\mathrm{O_2} + 3.76\ \mathrm{N_2}) \rightarrow \\ \mathrm{uCO_2} + \frac{\mathrm{v}}{2}\mathrm{H_2O} + \left[3.76\left(\mathrm{u} + \frac{\mathrm{v}}{4} - \frac{\mathrm{w}}{2} + \mathrm{y}\right) + \frac{\mathrm{x}}{2}\right]\mathrm{N_2} + \mathrm{y}\ \ \mathrm{SO_2} + \mathrm{Ash} \end{aligned}$$

Where $C_uH_vO_wN_xS_y$-M represents the composition of biomass, based on elemental analysis with inorganic elements of biomass (Na, K, Cl, Si, Ti, Zn, Cu, etc.) all lumped into M and converted to ash after combustion. The above reaction is an ideal state, usually incomplete combustion occurs with release of unburnt pollutants such as CO, C_xH_y, PAH, tar, soot, unburnt carbon, H_2, HCN, NH_3, and N_2O. Ash and contaminants such as ash particles (KCl, NaCl, etc.), HCl, PCDD/F, Cu, Pb, Zn, Cd, etc and other pollutants from complete combustion NO, NO_2, SO_2 can be observed.

Air-Fuel Ratio

The stoichiometric Air to fuel ratio is the ratio of the ratio of the mass flow of air to the mass flow of fuel at stoichiometric combustion.

$$\left(\frac{A}{F}\right)_{stoic} = \frac{4.76*\left(u + \frac{v}{4} - \frac{w}{2} + y\right)}{1\text{ mole of fuel}} * \frac{MW_{air}}{MW_{fuel}}$$

Where MW stands for the molecular weight.

The equivalence ratio is the ratio of the stoichiometric air to fuel ratio to the actual air to fuel ratio.

$$\phi = \frac{\left(\frac{A}{F}\right)_{stoic}}{\left(\frac{A}{F}\right)} = \frac{\left(\frac{F}{A}\right)}{\left(\frac{F}{A}\right)_{stoic}} = \frac{1}{1 + \frac{E\%}{100}}$$

Where E% is the percentage of excess air, $E\% = \frac{(1-\phi)}{\phi} * 100\%$

If $\phi > 1$ we have a fuel rich mixture.

If $\phi < 1$ we have a fuel lean mixture.

References

Abdullah, H., & Wu, H. (2009). Biochar as a fuel: 1. Properties and grindability of biochars produced from the pyrolysis of mallee wood under slow-heating conditions. *Energy & Fuels, 23*(8), 4174–4181.

Acharjee, T. C., Coronella, C. J., & Vasquez, V. R. (2011). Effect of thermal pretreatment on equilibrium moisture content of lignocellulosic biomass. *Bioresource Technology, 102*(7), 4849–4854.

Agar, D., & Wihersaari, M. (2012). Bio-coal, torrefied lignocellulosic resources—Key properties for its use in co-firing with fossil coal—Their status. *Biomass and Bioenergy, 44*, 107–111.

Antal, M. J., Jr., & Varhegyi, G. (1995). Cellulose pyrolysis kinetics: The current state of knowledge. *Industrial & Engineering Chemistry Research, 34*(3), 703–717.

Arias, B., Pevida, C., Fermoso, J., Plaza, M. G., Rubiera, F., & Pis, J. J. (2008). Influence of torrefaction on the grindability and reactivity of woody biomass. *Fuel Processing Technology, 89*(2), 169–175.

Azeez, A. M., Meier, D., & Odermatt, J. (2011). Temperature dependence of fast pyrolysis volatile products from European and African biomasses. *Journal of Analytical and Applied Pyrolysis, 90*(2), 81–92.

Banyasz, J. L., Li, S., Lyons-Hart, J., & Shafer, K. H. (2001). Gas evolution and the mechanism of cellulose pyrolysis. *Fuel, 80*(12), 1757–1763.

Basu, P. (2010a). *Biomass gasification and pyrolysis: Practical design and theory*. Burlington, MA: Elsevier.

Basu, P. (2010b). Chapter 3—Pyrolysis and torrefaction. In P. Basu (Ed.), *Biomass gasification and pyrolysis* (pp. 65–96). Boston: Academic.

Basu, P. (2013). Chapter 4 - Torrefaction. In P. Basu (Ed.), *Biomass gasification, pyrolysis and torrefaction* (2nd ed., pp. 87–145). Boston: Academic.

Bellur, S. R., Coronella, C. J., & Vásquez, V. R. (2009). Analysis of biosolids equilibrium moisture and drying. *Environmental Progress & Sustainable Energy, 28*(2), 291–298.

Bergman, P. C. A., Boersma, A. R., Kiel, J. H. A., Wilberink, R. W. A., Bodenstaff H., & Heere P. G. T. (2003). *Torrefaction for entrained flow gasification of biomass*. Petten, The Netherlands, Energy Research Centre of the Netherlands (ECN) and TU/e, ECN-C-05-067.

Bergman, P. C. A., Boersma, A. R., Zwart R. W. R., & Kiel, J. H. A. (2005). *Torrefaction for biomass co-firing in existing coal-fired power stations "BIOCOAL"*. Petten, The Netherlands, Energy Research Centre of the Netherlands (ECN), ECN-C–05-013.

Bhuiyan, M. T., Hirai, N., & Sobue, N. (2000). Changes of crystallinity in wood cellulose by heat treatment under dried and moist conditions. *Journal of Wood Science, 46*(6), 431–436.

Bhuiyan, M. T., Hirai, N., & Sobue, N. (2001). Effect of intermittent heat treatment on crystallinity in wood cellulose. *Journal of Wood Science, 47*(5), 336–341.

Boerjan, W., Ralph, J., & Baucher, M. (2003). Lignin biosynthesis. *Annual Review of Plant Biology, 54*(1), 519–546.

Bradbury, A. G. W., Sakai, Y., & Shafizadeh, F. (1979). A kinetic model for pyrolysis of cellulose. *Journal of Applied Polymer Science, 23*(11), 3271–3280.

Brebu, M., & Vasile, C. (2010). Thermal degradation of lignin—A review. *Cellullose Chemical Technology, 44*(9), 353–363.

Bridgeman, T. G., Jones, J. M., Shield, I., & Williams, P. T. (2008). Torrefaction of reed canary grass, wheat straw and willow to enhance solid fuel qualities and combustion properties. *Fuel, 87*(6), 844–856.

Bridgeman, T. G., Jones, J. M., Williams, A., & Waldron, D. J. (2010). An investigation of the grindability of two torrefied energy crops. *Fuel, 89*(12), 3911–3918.

Bridgwater, A. V. (2012). Review of fast pyrolysis of biomass and product upgrading. *Biomass and Bioenergy, 38*, 68–94.

Broido, A., Javier-Son, A. C., Ouano, A. C., & Barrall, E. M. (1973). Molecular weight decrease in the early pyrolysis of crystalline and amorphous cellulose. *Journal of Applied Polymer Science, 17*(12), 3627–3635.

Broido, A., & Nelson, M. A. (1975). Char yield on pyrolysis of cellulose. *Combustion and Flame, 24*, 263–268.

Brown, R. C. (2011). *Thermochemical processing of biomass: Conversion into fuels, chemicals and power*. West Sussex, UK: John Wiley & Sons.

Brunow, G., & Lundquist, K. (2011). Functional groups and bonding patterns in lignin (including the lignin-carbohydrate complexes). In C. Heitner, D. Dimmel, & J. Schmidt (Eds.), *Lignin and lignans: Advances in chemistry* (pp. 167–299). Boca Raton, FL: CRC Press.

Chan, W.-C. R., Kelbon, M., & Krieger, B. B. (1985). Modelling and experimental verification of physical and chemical processes during pyrolysis of a large biomass particle. *Fuel, 64*(11), 1505–1513.

Chen, D., Zhou, J., Zhang, Q., Zhu, X., & Lu, Q. (2014). Upgrading of rice husk by torrefaction and its influence on the fuel properties. *BioResources, 9*(4), 5893–5905.

Chen, W.-H., Cheng, W.-Y., Lu, K.-M., & Huang, Y.-P. (2011). An evaluation on improvement of pulverized biomass property for solid fuel through torrefaction. *Applied Energy, 88*(11), 3636–3644.

Chen, W.-H., Hsu, H.-C., Lu, K.-M., Lee, W.-J., & Lin, T.-C. (2011a). Thermal pretreatment of wood (Lauan) block by torrefaction and its influence on the properties of the biomass. *Energy, 36*(5), 3012–3021.

Chen, W.-H., Hsu, H.-C., Lu, K.-M., Lee, W.-J., & Lin, T.-C. (2011b). Thermal pretreatment of wood (Lauan) block by torrefaction and its influence on the properties of the biomass. *Energy, 36*(5), 3012–3021.

Chew, J. J., & Doshi, V. (2011). Recent advances in biomass pretreatment – Torrefaction fundamentals and technology. *Renewable and Sustainable Energy Reviews, 15*(8), 4212–4222.

Collard, F.-X., & Blin, J. (2014). A review on pyrolysis of biomass constituents: Mechanisms and composition of the products obtained from the conversion of cellulose, hemicelluloses and lignin. *Renewable and Sustainable Energy Reviews, 38*, 594–608.

Collard, F.-X., Blin, J., Bensakhria, A., & Valette, J. (2012). Influence of impregnated metal on the pyrolysis conversion of biomass constituents. *Journal of Analytical and Applied Pyrolysis, 95*, 213–226.

Deng, J., Wang, G.-j., Kuang, J.-h., Zhang, Y.-l., & Luo, Y.-H. (2009). Pretreatment of agricultural residues for co-gasification via torrefaction. *Journal of Analytical and Applied Pyrolysis, 86*(2), 331–337.

Di Blasi, C. (1998). Comparison of semi-global mechanisms for primary pyrolysis of lignocellulosic fuels. *Journal of Analytical and Applied Pyrolysis, 47*(1), 43–64.

Di Blasi, C. (2009). Combustion and gasification rates of lignocellulosic chars. *Progress in Energy and Combustion Science, 35*(2), 121–140.

Di Blasi, C., & Russo, G. (1993). Modeling of transport phenomena and kinetics of biomass pyrolysis. In A. V. Bridgwater (Ed.), *Advances in thermochemical biomass conversion* (pp. 906–921). Netherlands: Springer.

Drift, A. V. D., & Boerrigter, H. (2006). *Synthesis gas from biomass for fuels and chemicals*. Pettel, The Netherlands, Energy Research Center of the Netherlands (ECN).

Drift, A. V. D., Boerrigter, H., Coda, B., Cieplik, M. K., & Hemmes, K. (2004). *Entrained flow gasification of biomass: Ash behaviour, feeding issues, and system analyses*. Pettel, The Netherlands, Energy Research Center of the Netherlands (ECN).

Duca, D., Riva, G., Pedretti, E. F., Toscano, G., Chiara, Mengarelli., & Rossini, G. (2014). Solid biofuels production from agricultural residues and processing by-products by means of torrefaction treatment: the case of sunflower chain. *Journal of Agricultural Engineering XLV* (416): 97–102.

Ergun, S. (1956). Kinetics of the reaction of carbon with carbon dioxide. *The Journal of Physical Chemistry, 60*(4), 480–485.

Feldman, D. (1985). Wood—chemistry, ultrastructure, reactions, by D. Fengel and G. Wegener, Walter de Gruyter, Berlin and New York, 1984, 613 pp. Price: 245 DM. *Journal of Polymer Science, Polymer Letters Edition, 23*(11), 601–602.

Felfli, F. F., Luengo, C. A., Suárez, J. A., & Beatón, P. A. (2005). Wood briquette torrefaction. *Energy for Sustainable Development, 9*(3), 19–22.

Ferro, D. T., Vigouroux, V., Grimm, A., & Zanzi, R. (2004). *Torrefaction of agricultural and forest residues*. Guantánamo, Cuba, Cubasolar: Cubasolar.

Garcia-Perez, M., Metcalf, J., U. Washington State, Extension, P. Energy, U. Washington State, et al. (2009). *The formation of polyaromatic hydrocarbons and dioxins during pyrolysis a review of the literature with descriptions of biomass composition, fast pyrolysis technologies and thermochemical reactions*. Retrieved from http://www.pacificbiomass.org/documents/TheFormationOfPolyaromaticHydrocarbonsAndDioxinsDuringPyrolysis.pdf.

Girard, P., & Shah, N. (1989). *Recent developments on torrefied wood – An alternative to charcoal for reducting deforestation*. FAO/CNRE Workshop, Poros, Norway, FAO.

Henrich, E., Dahmen, N., & Dinjus, E. (2009). Cost estimate for biosynfuel production via biosyncrude gasification. *Biofuels, Bioproducts and Biorefining, 3*(1), 28–41.

Henrich, E., Dahmen, N., Raffelt, K., Stahl, R., & Weirich, F.(2007). *The Karlsruhe "BIOLIQ" process for biomass gasification*. 2nd European summer school on renewable motor fuels, Warsaw, Poland.

Henrich, E., & Dinjus, E. (2002). *Tar-free, high pressure synthesis gas from biomass*. Expert meeting on pyrolysis and gasification of biomass. Strasbourg, France.

Henrich, E., & Weirich, F. (2004). Pressurized entrained flow gasifiers for biomass. *Environmental Engineering Science, 21*(1), 53–64.

Henriksson, G., Li, J., Zhang, L., & Lindstrom, M. E. (2010). Lignin utilization. In M. Crocker (Ed.), *Thermochemical conversion of biomass to liquid fuels and chemicals* (pp. 222–262). Cambridge, UK: The Royal Society of Chemistry.

Higman, C., & Burgt, M. V. D. (2008). *Gasification*. New York, USA: Elsevier.

Husain, A., & Kelman, A. (1959). Tissue is disintegrated. In J. G. Horsfall & A. E. Dimond (Eds.), *Plant pathology* (pp. 143–188). New York: Academic Press.

Ibrahim, R. H. H., Darvell, L. I., Jones, J. M., & Williams, A. (2013). Physicochemical characterisation of torrefied biomass. *Journal of Analytical and Applied Pyrolysis, 103*, 21–30.

Jakab, E., Faix, O., Till, F., & Székely, T. (1995). Thermogravimetry/mass spectrometry study of six lignins within the scope of an international round robin test. *Journal of Analytical and Applied Pyrolysis, 35*(2), 167–179.

Jerrold, E. W., & Roger, M. R. (2005). *Chemistry of wood strength* (Handbook of wood chemistry and wood composites). Boca Raton, FL: CRC Press.

Kim, Y.-H., Lee, S.-M., Lee, H.-W., & Lee, J.-W. (2012). Physical and chemical characteristics of products from the torrefaction of yellow poplar (Liriodendron tulipifera). *Bioresource Technology, 116*, 120–125.

Kobayashi, H. (1976). *Devolatilization of pulverized coal at high temperatures*. Cambridge, MA: Dept. of Mechanical Engineering, Massachusetts Institute of Technology.

Koukios, E. (1993). Progress in thermochemical, solid-state refining of biofuels — From research to commercialization. In A. V. Bridgwater (Ed.), *Advances in thermochemical biomass conversion* (pp. 1678–1692). Netherlands: Springer.

Lédé, J., Blanchard, F., & Boutin, O. (2002). Radiant flash pyrolysis of cellulose pellets: products and mechanisms involved in transient and steady state conditions. *Fuel, 81*(10), 1269–1279.

Lee, J., Buchanan, A. C., III, Evans, B., & Kidder, M. (2013). Oxygenation of biochar for enhanced cation exchange capacity. In J. W. Lee (Ed.), *Advanced biofuels and bioproducts* (pp. 35–45). New York: Springer.

Li, H., Qu, Y., & Xu, J. (2015). Microwave-assisted conversion of lignin. In Z. Fang, J. R. L. Smith, & X. Qi (Eds.), *Production of biofuels and chemicals with microwave* (3rd ed., pp. 61–82). Netherlands: Springer.

Liang, B., Lehmann, J., Solomon, D., Kinyangi, J., Grossman, J., O'Neill, B., et al. (2006). Black carbon increases cation exchange capacity in soils. *Soil Science Society of America Journal, 70* (5), 1719–1730.

Lomax, J. A., Commandeur, J. M., Arisz, P. W., & Boon, J. J. (1991). Characterisation of oligomers and sugar ring-cleavage products in the pyrolysate of cellulose. *Journal of Analytical and Applied Pyrolysis, 19*, 65–79.

Lu, J.-J., & Chen, W.-H. (2013). Product yields and characteristics of corncob waste under various torrefaction atmospheres. *Energies, 7*(1), 13–27.

Lu, Q., Yang, X.-C., Dong, C.-q., Zhang, Z.-f., Zhang, X.-M., & Zhu, X.-F. (2011). Influence of pyrolysis temperature and time on the cellulose fast pyrolysis products: Analytical Py-GC/MS study. *Journal of Analytical and Applied Pyrolysis, 92*(2), 430–438.

Lv, G.-J., Wu S.-B., & Lou R. (2010). *Kinetic study for the thermal decomposition of hemicellulose isolated from corn stalk.*

Madorsky, S. L., Hart, V. E., & Straus, S. (1956). Pyrolysis of cellulose in a vacuum. *Journal of Research of the National Bureau of Standards, 56*(6), 343–354.

Mamleev, V., Bourbigot, S., Le Bras, M., & Yvon, J. (2009). The facts and hypotheses relating to the phenomenological model of cellulose pyrolysis: Interdependence of the steps. *Journal of Analytical and Applied Pyrolysis, 84*(1), 1–17.

McGrath, T. E., Chan, W. G., & Hajaligol, M. R. (2003). Low temperature mechanism for the formation of polycyclic aromatic hydrocarbons from the pyrolysis of cellulose. *Journal of Analytical and Applied Pyrolysis, 66*(1–2), 51–70.

Medic, D. (2012). *Investigation of torrefaction process parameters and characterization of torrefied biomass*. Iowa: Agricultural Engineering; Biorenewable Resources and Technology Ames, Iowa State University.

Mei, Y., Liu, R., Yang, Q., Yang, H., Shao, J., Draper, C., et al. (2015). Torrefaction of cedarwood in a pilot scale rotary kiln and the influence of industrial flue gas. *Bioresource Technology, 177*, 355–360.

Mukherjee, A., & Lal, R. (2013). Biochar impacts on soil physical properties and greenhouse gas emissions. *Agronomy, 3*(2), 313–339.

Mullen, C. A., & Boateng, A. A. (2011). Characterization of water insoluble solids isolated from various biomass fast pyrolysis oils. *Journal of Analytical and Applied Pyrolysis, 90*(2), 197–203.

Nachenius, R. W., Ronsse, F., Venderbosch, R. H., & Prins, W. (2013). Biomass pyrolysis. In M. Dmitry Yu (Ed.), *Advances in chemical engineering* (42nd ed., pp. 75–139). New York: Academic Press.

Naik, S. N., Goud, V. V., Rout, P. K., & Dalai, A. K. (2010). Production of first and second generation biofuels: A comprehensive review. *Renewable and Sustainable Energy Reviews, 14*(2), 578–597.

Neves, D., Thunman, H., Matos, A., Tarelho, L., & Gómez-Barea, A. (2011). Characterization and prediction of biomass pyrolysis products. *Progress in Energy and Combustion Science, 37*(5), 611–630.

Nunn, T. R., Howard, J. B., Longwell, J. P., & Peters, W. A. (1985a). Product compositions and kinetics in the rapid pyrolysis of milled wood lignin. *Industrial & Engineering Chemistry Process Design and Development, 24*(3), 844–852.

Nunn, T. R., Howard, J. B., Longwell, J. P., & Peters, W. A. (1985b). Product compositions and kinetics in the rapid pyrolysis of sweet gum hardwood. *Industrial & Engineering Chemistry Process Design and Development, 24*(3), 836–844.

Offrion, V. F. O. (1900). *Improvements in the process of and apparatus for rationally and continuously treating or torrefying coffee*. Great Britain, Offrion, V. F. O.

Ohliger, A., Förster, M., & Kneer, R. (2013). Torrefaction of beechwood: A parametric study including heat of reaction and grindability. *Fuel, 104*, 607–613.

Ouyang, L., Wang, F., Tang, J., Yu, L., & Zhang, R. (2013). Effects of biochar amendment on soil aggregates and hydraulic properties. *Journal of soil Science and Plant Nutrition, 13*(4), 991–1002.

Pastorova, I., Botto, R. E., Arisz, P. W., & Boon, J. J. (1994). Cellulose char structure: A combined analytical Py-GC-MS, FTIR, and NMR study. *Carbohydrate Research, 262*(1), 27–47.

Pettersen, R., Tshabalala, M. A., Han, J. S., Rowell, M. R., & Rowell, J. S. (2005). *Cell wall chemistry* (Handbook of wood chemistry and wood composites). Boca Raton, FL: CRC Press.

Phanphanich, M., & Mani, S. (2011). Impact of torrefaction on the grindability and fuel characteristics of forest biomass. *Bioresource Technology, 102*(2), 1246–1253.

Pimchuai, A., Dutta, A., & Basu, P. (2010). Torrefaction of agriculture residue to enhance combustible properties. *Energy & Fuels, 24*(9), 4638–4645.

Pohlmann, J. G., Osório, E., Vilela, A. C. F., Diez, M. A., & Borrego, A. G. (2014). Integrating physicochemical information to follow the transformations of biomass upon torrefaction and low-temperature carbonization. *Fuel, 131*, 17–27.

Probstein, R. F., & Hicks, R. E. (2006). *Synthetic fuels*. Mineola, NY: NY, Dover Publications Inc.

Raffelt, K., Henrich, E., Koegel, A., Stahl, R., Steinhardt, J., & Weirich, F. (2004). The BTL2 process of biomass utilization entrained-flow gasification of pyrolyzed biomass slurries. *Applied Biochemistry and Biotechnology, 129*(1-3), 153–164.

Ralph, J., Peng, J., Lu, F., Hatfield, R. D., & Helm, R. F. (1999). Are lignins optically active? *Journal of Agricultural and Food Chemistry, 47*(8), 2991–2996.

Reed, T., & Gaur, S. (1997). The high heat of fast pyrolysis for large particles. In A. V. Bridgwater & D. G. B. Boocock (Eds.), *Developments in thermochemical biomass conversion* (pp. 97–103). Netherlands: Springer.

Repellin, V., Govin, A., Rolland, M., & Guyonnet, R. (2010). Energy requirement for fine grinding of torrefied wood. *Biomass and Bioenergy, 34*(7), 923–930.

Rowell, R. M. (2005). *Moisture properties* (Handbook of wood chemistry and wood composites). Boca Raton, FL: CRC Press.

Saleh, S. B., Hansen, B. B., Jensen, P. A., & Dam-Johansen, K. (2013). Efficient fuel pretreatment: Simultaneous torrefaction and grinding of biomass. *Energy & Fuels, 27*(12), 7531–7540.

Samolada, M. C., & Vasalos, I. A. (1991). A kinetic approach to the flash pyrolysis of biomass in a fluidized bed reactor. *Fuel, 70*(7), 883–889.

Satpathy, S. K., Tabil, L. G., Meda, V., Naik, S. N., & Prasad, R. (2014). Torrefaction of wheat and barley straw after microwave heating. *Fuel, 124*, 269–278.

Scheirs, J., Camino, G., & Tumiatti, W. (2001). Overview of water evolution during the thermal degradation of cellulose. *European Polymer Journal, 37*(5), 933–942.

Schobert, H. H. (2013). *Chemistry of fossil fuels and biofuels*. Cambridge, UK: Cambridge University Press.

Shafizadeh, F. (1982). Introduction to pyrolysis of biomass. *Journal of Analytical and Applied Pyrolysis, 3*(4), 283–305.

Shang, L., Ahrenfeldt, J., Holm, J. K., Sanadi, A. R., Barsberg, S., Thomsen, T., et al. (2012). Changes of chemical and mechanical behavior of torrefied wheat straw. *Biomass and Bioenergy, 40*, 63–70.

Silakul, T., & Jindal, V. K. (2002). Equilibrium moisture content isotherms of mungbean. *International Journal of Food Properties, 5*(1), 25–35.

Smith, B. R. J., Loganathan, M., & Shantha, M. S. (2010). A review of the water gas shift reaction kinetics. *International Journal of Chemical Reactor Engineering, 8*(1).

Smith, H. (1977). *The molecular biology of plant cell wall*. Berkeley: University of California Press.

Sohi, S., Gaunt, J., & Atwood, J. (2013). *Biochar in growing media: A sustainability and feasibility assessment*. Edinburgh: UK Biochar Research Center.

Stelte, W. (2014). *Optimization of product specific processing parameters for the production of fuel pellets from torrefied biomass*. Taastrup, Denmark: Center for Biomass and Biorefinery, Danish Technological Institute.

Sule, I. (2012). *Torrefaction behaviour of agricultural biomass*. ON, Canada: School of Engineering Guelph, University of Guelph.

Svoboda, K., Pohořelý, M., Hartman, M., & Martinec, J. (2009). Pretreatment and feeding of biomass for pressurized entrained flow gasification. *Fuel Processing Technology, 90*(5), 629–635.

Tchapda, A. H., & Pisupati, S. V. (2014). A review of thermal co-conversion of coal and biomass/ waste. *Energies, 7*(3), 1098–1148.

Thiel, F. C. (1897). *New or improved roaster or torrefier for coffee and other vegetable substances*. Great Britain, Thiel, F. C.

Thunman, H., & Leckner, B. (2007). *Thermo chemical conversion of biomass and wastes. Nordic graduate school BiofuelGS-2*. Göteborg, Sweden: Chalmers.

Thurner, F., & Mann, U. (1981). Kinetic investigation of wood pyrolysis. *Industrial & Engineering Chemistry Process Design and Development, 20*(3), 482–488.

Tumuluru, J. S., Hess, J. R., Boardman, R. D., Wright, C. T., & Westover, T. L. (2012). Formulation, pretreatment, and densification options to improve biomass specifications for co-firing high percentages with coal. *Industrial Biotechnology, 8*(3), 113–132.

Van de Velden, M., Baeyens, J., Brems, A., Janssens, B., & Dewil, R. (2010). Fundamentals, kinetics and endothermicity of the biomass pyrolysis reaction. *Renewable Energy, 35*(1), 232–242.

van der Stelt, M. J. C., Gerhauser, H., Kiel, J. H. A., & Ptasinski, K. J. (2011). Biomass upgrading by torrefaction for the production of biofuels: A review. *Biomass and Bioenergy, 35*(9), 3748–3762.

Varhegyi, G., Jakab, E., & Antal, M. J. (1994). Is the Broido-Shafizadeh model for cellulose pyrolysis true? *Energy & Fuels, 8*(6), 1345–1352.

Wang, G., Luo, Y., Deng, J., Kuang, J., & Zhang, Y. (2011). Pretreatment of biomass by torrefaction. *Chinese Science Bulletin, 56*(14), 1442–1448.

Wannapeera, J., Fungtammasan, B., & Worasuwannarak, N. (2011). Effects of temperature and holding time during torrefaction on the pyrolysis behaviors of woody biomass. *Journal of Analytical and Applied Pyrolysis, 92*(1), 99–105.

Wardrop, A. B. (1964). The structure and formation of the cell wall in xylem. In M. H. Zimmermann (Ed.), *The formation of wood in forest trees* (pp. 87–134). New York: Academic Press.

Werner, K., Pommer, L., & Broström, M. (2014). Thermal decomposition of hemicelluloses. *Journal of Analytical and Applied Pyrolysis, 110*, 130–137.

Wichman, I., & Melaaen, M. (1993). Modeling the pyrolysis of cellulosic materials. In A. V. Bridgwater (Ed.), *Advances in thermochemical biomass conversion* (pp. 887–905). Netherlands: Springer.

Wooten, J. B., Seeman, J. I., & Hajaligol, M. R. (2003). Observation and characterization of cellulose pyrolysis intermediates by 13C CPMAS NMR. A new mechanistic model. *Energy & Fuels, 18*(1), 1–15.

Xue, G., Kwapinska, M., Kwapinski, W., Czajka, K. M., Kennedy, J., & Leahy, J. J. (2014). Impact of torrefaction on properties of Miscanthus giganteus relevant to gasification. *Fuel, 121*, 189–197.

Yan, W., Acharjee, T. C., Coronella, C. J., & Vásquez, V. R. (2009). Thermal pretreatment of lignocellulosic biomass. *Environmental Progress & Sustainable Energy, 28*(3), 435–440.

Yang, H., Yan, R., Chen, H., Lee, D. H., & Zheng, C. (2007). Characteristics of hemicellulose, cellulose and lignin pyrolysis. *Fuel, 86*(12–13), 1781–1788.

Zaror, C. A., & Pyle, D. L. (1986). Competitive reactions model for the pyrolysis of lignocellulose: A critical study. *Journal of Analytical and Applied Pyrolysis, 10*(1), 1–12.

Zulfiqar, M., Moghtaderi, B., & Wall, T. F. (2006). Flow properties of biomass and coal blends. *Fuel Processing Technology, 87*, 281–288.

Part III
Bioprocess Technology

Chapter 16
Dynamic Enzymatic Kinetic Resolution of NSAIDS

A.H. Kamaruddin, M.H. Uzir, F.N. Gonawan, and S.Y. Lau

Introduction

Non-steroidal anti-inflammatory drugs (NSAIDs) are commonly consumed by people and are often very effective to reduce pain. NSAIDs also possess antipyretic, analgesic and anti-inflammatory effects by reducing fever, alleviating pain and reducing inflammation, as their name implies (Litalien and Jacqz-Aigrain 2001). NSAIDs are chiral drug and each compound consists of an asymmetric carbon center and tends to form two non-superimposable mirror images. These mirror images or optical isomers of a chiral molecule are called (*R*)-enantiomer and (*S*)-enantiomers respectively. In nature, the chiral NSAIDs exists in a racemic mixture where both (*R*)-enantiomer and (*S*)-enantiomers are in equal composition. In racemic mixture, both enantiomers have identical chemical and physical properties. However, the optically pure enantiomer belong to NSAID class, (*S*)-ibuprofen differs significantly in its physical properties from the racemic mixture. The melting point of (*S*)-ibuprofen is 52 °C, which is 25 °C lower than the racemate (melting point = 75 °C) (Perry and Green 1997). It has low solubility in water but highly soluble in organic solvents such as acetone, acetonitrile, isooctane and n-hexane.

Furthermore, they possess different aroma and flavor characteristics and more importantly the compounds possess different toxicity and biological activity (Sheldon 1996; Rouhi 2003). Thus, each of the enantiomers of an NSAID exhibits different pharmacodynamics and pharmacokinetic properties. In particular, each

A.H. Kamaruddin (✉) • M.H. Uzir • F.N. Gonawan
School of Chemical Engineering, Universiti Sains Malaysia,
Engineering Campus, Penang, Malaysia
e-mail: chazlina@usm.my

S.Y. Lau
Department of Chemical Engineering, Curtin University,
Sarawak Campus CDT 250, Miri, Sarawak, Malaysia

P. Ravindra (ed.), *Advances in Bioprocess Technology*,
DOI 10.1007/978-3-319-17915-5_16

enantiomers of the same drug will result in specific health effect when consumed by a patient. For instance, the interaction of beta-blockers with beta-adrenoceptors is highly stereoselective (Mehvar and Brocks 2001). Beta-blocker is an antihypertensive agent for lowering the blood pressure that includes propranolol, atenolol and metoprolol. The stereoselective action of propranolol was exerted by beta blocking activity of (*S*)-propranolol instead of (*R*)-propranolol (Stoschitzky and Lindner 1990). Administration of racemic drugs could lead to undesirable serious side effects. The most tragic case was reported in 1960 due to the use of thalidomide in the treatment of symptom associated with morning sickness among pregnant woman (Ito et al. 2011). The desired anti-nausea effects were associated by (*R*)-isomer, while the (*S*)-form exhibits teratogenic effect and causes birth defects as well as death to babies (Knoche and Blaschke 1994). It is suggested that, the two enantiomers in racemate drugs should be marketed and labelled as two separate drugs with different properties (Ariëns 1991).

Following the thalidomide tragedies, the introduction of new drugs in the market was significantly affected. The pharmaceutical company has realized, that the racemate drugs possess high risk to the consumer. In the early 1990s, the characterization of each enantiomer is required for all new marketed drugs that regulated by the Food and Drugs Administration (FDA) and the European Committee for Proprietary Medicinal Product (Shah and Branch 2003). These new regulations encouraged the application of chiral technology in the development of new drugs. Nowadays, most of all marketed drugs have been sold as a single isomeric form. The increasing awareness in using optically pure enantiomer in the drug formulation has stirred a growing interest for research and development for these purposes (Ward and Ward 2011). Numerous methods have been established by researcher in the production of optically pure drugs. Presently, single enantiomers of drugs can be synthesised through chiral pool technology, separation of racemate via classical resolution, crystallization technique or enzymatic kinetic resolution, biological asymmetric methods and chemical asymmetric techniques (Rouhi 2003; Sheldon 1996). However, these methods need to be improved in order to meet the current needs in reducing production cost and zero environmental implication. Thus, new and advanced approach were introduced for the last 10 years which include; dynamic enzymatic kinetic resolution (DEKR), chromatography separation and selective complex-metal catalysis. Has been foreseen that there will be impressive growth for such technologies in future (Rouhi 2003).

Dynamic Enzymatic Kinetic Resolution (DEKR) of NSAIDs

Enzymatic Kinetic Resolution

The use of enzymes as biocatalysts has become a valuable tool for optically pure NSAIDs synthesis. This method is preferable since it gives more opportunities and benefits in obtaining safer compound and shorter synthesis route (Ansari and Husain 2012). Therefore, the catalyzed-synthesis of optically pure NSAIDs are more preferable through enzymatic reaction compared to that of chemical approach. The chemical synthesis approach has been found to be outdated for a single enantiomer synthesis due to several drawbacks (Contesini et al. 2010); (1) chemical catalyst is not regio-selective; (2) the chemical reaction routes require higher energy and need to be performed in more extreme conditions and (3) chemical synthesis technique requires various chemical solvents. In contrast, enzyme-catalyzed reactions are well-established as a very useful mean to prepare optically active compounds. Some enzymes are very specific to a particular reaction involving only certain substrates. The degree of specificity of an enzyme is depending on its biological role (Ghanem et al. 2010). Additionally, enzymatic process offered more advantages over chemical process where the reactions are often highly enantio- and regio-selective towards the substrate (Ghanem 2007; Pollard and Woodley 2007; Patel 2002). In fact, they do not require high reaction temperature and normally carried out at atmospheric pressure thus, gives less problem with undesired side reactions (Schulze and Wubbolts 1999; Patel 2002).

In general, enzymatic kinetic resolution (EKR) is a type of separation process that depends on enzyme selectivity towards the enantiomers of a racemic substrate. During a resolution process, an enantiomer of a racemate will be left untouched by the enzyme, meanwhile the desired enantiomer is converted into optically pure product (Patel 2002). The resolution is based on the distinct kinetic rate of enzyme-catalyzed reaction towards the racemic substrate. General illustration of the mechanism is shown in Fig. 16.1. The symbols *R* and *S* denotes the (*R*)- and (*S*)-type of substrate, whereas P and Q signify the product and by-product produced during the reaction, respectively. Meanwhile, the symbols k_S and k_R correspond to the rate reaction forming product P and by-product Q, respectively. The enzymatic kinetic resolution takes place when $k_S \neq k_R$ and eventually the reaction will remain at 50 % conversion of the racemic substrate. Ideally, only the selected enantiomer is catalyzed to form a product at high conversion rate, while its mirror image would catalyze at relatively lower rate.

The enzymatic kinetic resolution of the optically pure NSAIDs can be accomplished using several approaches, these include; (1) enzymatic esterification and chemical hydrolysis, (2) chemical esterification and enzymatic hydrolysis and (3) enzymatic esterification and enzymatic hydrolysis. For the first approach, an optically pure (*S*)-substrate ester (desired configuration) is prepared by enzyme-

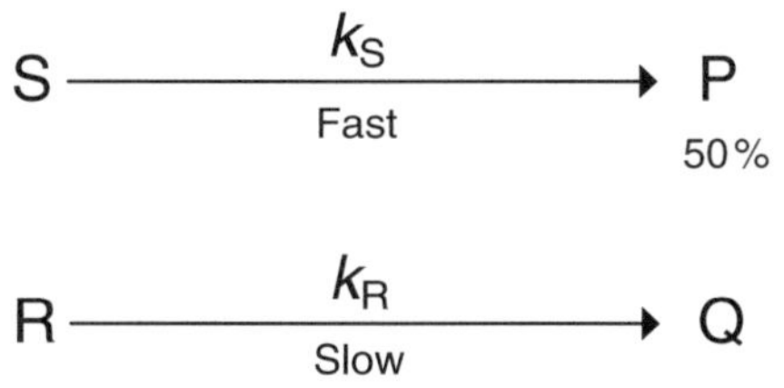

Fig. 16.1 General mechanisms for kinetic resolution

catalyzed esterification of the racemic substrate. The (*S*)-substrate ester is then chemically hydrolysed to the (*S*)-product by acid-catalyzed reaction. However, this approach causes racemization of side reaction by the acid catalyst. The drawback of the first method was overcome by converting the racemic substrate into racemic ester by acid-catalyzed esterification using enzyme as hydrolysing agent to give an optically pure (*S*)-product. In the latter approach, both esterification and hydrolysis are selectively catalyzed by enzyme. However, this process is non-economical and time consuming due to coupled enzymatic reaction.

Lipase (triacylglycerol acyl hydrolases, EC 3.1.1.3) is the most well-known type of enzymes used in the resolution of racemic compound (Patel 2002). It belongs to the hydrolase's family and frequently used in enantioselective hydrolysis, transesterification, esterification and interesterification reactions (Effenberger et al. 1997; Brady et al. 2004; Jin et al. 2003). The active centre of lipases constitutes a distinctive environment which is capable of distinguishing between the enantiomers (Muralidhar et al. 2001). The used of lipase from various species as a catalyst to kinetically resolve racemic compound exhibited high enantiospecificity for NSAIDs. Several esterification and hydrolysis processes involving different origins of lipases as catalyst is summarized in Table 16.1.

Soluble enzymes have been proven to be the effective biocatalysts in the enzymatic kinetic resolution (Mustranta 1992). In spite of the advantages of enzyme, its application in industrial has been limited since the soluble enzymes are relatively unstable, high cost for enzyme extraction from natural resources, and lacking technology in the recovery of active enzyme from the reaction mixture. Thus, the used of soluble lipase in the batch operation is uneconomic in most applications as the activity of the enzymes tend to constantly decrease with reaction times. Numerous efforts were taken in order to overcome the limitation of soluble lipase, which include lipase immobilization by numerous novel methods to hydrophobic or hydrophilic surfaces (Brady and Jordaan 2009; Sheldon 2007; Johnson et al. 2008). The introduction of immobilized lipase in kinetic resolution is encouraging since the bio-catalyst is reusable and thus, will increase the productivity.

Conceptual Strategies of DEKR

Enzymatic kinetic resolution (EKR) of a racemate is a promising methods for the production of single enantiomer of NSAIDs with high optical purity for pharmaceutical industry. However, this approach suffers serious drawback with only 50 % theoretical conversion of the substrate, since the undesired enantiomer remained untouched. The limitation of EKR has been a point of research interest in the past decades in order to improvise the strategies and achieves 100 % conversion of the racemic substrate. Several strategies have been proposed that includes; the dynamic kinetic asymmetric transformation (DYKAT), stereo-inversion, cyclic deracemization, enantio-convergent transformation and dynamic enzymatic kinetic resolution (DEKR) (Steinreiber et al. 2008; Carnell 1999; Wang et al. 2013; Lee 2008; Martín-Matute and Bäckvall 2007). Amongst these methods, DEKR is perhaps the one that has been intensively investigated in recent years.

Table 16.1 Lipase-catalyzed esterification and hydrolysis reaction of NSAIDs substrate (Fazlena 2005)

Drugs	Lipases	Process	Conversion	Optical purity	References
Flurbiprofen	*Candida rugosa*	Esterification	79.0	93.0	Bhandarkar and Neau (2000)
Ibuprofen	*Candida rugosa*	Esterification	42.6	n.d	Ikeda and Kurokawa (2002)
Suprofen	*Candida rugosa*	Esterification	95.0	98.0	Tsai and Huang (1999)
Ibuprofen	*Candida antartica* (type B)	Esterification	n.d.	n.d.	Ducret et al. (1998)
Naproxen	*Lipase MY*	Esterification	43.0	n.d	Chang and Tsai (1997)
Naproxen	*Candida cylindracea*	Hydrolysis	25.0	95.0	Battistel et al. (1991))
Flurbiprofen	*Candida antartica Novozym® 435*	Esterification	5.0	91.0	Morrone et al. (1995)
Flurbiprofen	*Candida antarticaNovozym® 435*	Transesterification	34.9	82.0	Morrone et al. (1995)
Naproxen	*Carica papaya*	Hydrolysis	46.1	>99.0	Chen and Tsai (2005)
Naproxen	*Candida rugosa*	Hydrolysis	38.4	98.0	Gyo Lee et al. (2001)
Ibuprofen	*Candida rugosa*	Hydrolysis	33.0	96.0	Madhav and Ching (2001)
Ibuprofen	*Candida rugosa*	Esterification	>45.0	n.d.	Kin and Lee (1996)
Ketoprofen	*Candida rugosa*	Hydrolysis	22.3	94.0	Liu et al. (2004)
Naproxen trifluoroethylthioester	*Candida rugosa*	Hydrolysis	18.7	75.5	Chang et al. (1998)
Ibuprofen phenyl thioester	*Candida rugosa*	Hydrolysis	37.8	66.6	Chang et al. (1998)
Ketoprofen	*Candida rugosa*	Hydrolysis	>99	>45	Kim et al. (2000)
Ketoprofen	*Candida antarticaNovozym® 435*	Hydrolysis	22.5	63.7	Jin et al. (2003)
Naproxen	*Candida rugosa*	Hydrolysis	45.0	>96.0	Xin et al. (2000)

The DEKR is generally introduced to overcome the disadvantages encountered in EKR. The method of DEKR, can take place by combining the standard EKR and *in-situ* racemization process of the chiral NSAID substrates. In the standard EKR, the hydrolysis process takes place by means of enantioselective biocatalyst to hydrolyze racemic substrate and left with the unreacted enantiomer. Once the EKR is coupled with the *in-situ* racemization, the unreacted enantiomer will be recycled as a racemic mixture. As a result, the desired enantiomer is continuously replenished by the other enantiomer during the course of the resolution process. Hence, it is possible to reach 100 % theoretical conversion of the substrate because both enantiomers are fully utilized. Substrate racemization may be achieved either by chemical or biochemical reaction or even spontaneously. The racemization process condition must satisfy the kinetic resolution process, thus the performance of the enzyme is improved. The compatibility between both the kinetic resolution and racemization processes is the key determining factors for the success of the DEKR. The overall mechanism for DEKR of ibuprofen is illustrated in Fig. 16.2.

In the EKR process, the product is formed at the very beginning of the reaction but decreases when the reaction runs at about 50 % conversion. This behaviour is due to gradual depletion of the desired enantiomer from the racemic substrate. In contrast, the depletion of the desired enantiomer can be overcome with continuous racemization of the unreacted enantiomer throughout the course of the reaction. In other words, in DEKR, the selection of the desired enantiomer is always available for the enzyme because the composition of the racemic substrate remain constant throughout the reaction. Thus, the production of the optically pure product will not be interrupted and 100 % conversion of racemic substrate can be achieved. Figure 16.3 demonstrates that high activity and optically selective lipase enzyme should be used to provide a maximum catalytic capability in order to achieve high rate of (*S*)-NSAID ester hydrolysis. When the (*R*)-NSAID ester concentration builds up in the system, highly efficient base catalyst should be used to provide a rapid racemization of (*R*)- and (*S*)-NSAID esters. The rate of (*S*)-NSAID ester hydrolysis should overtake the rate of racemization, so that an efficient DEKR is achieved.

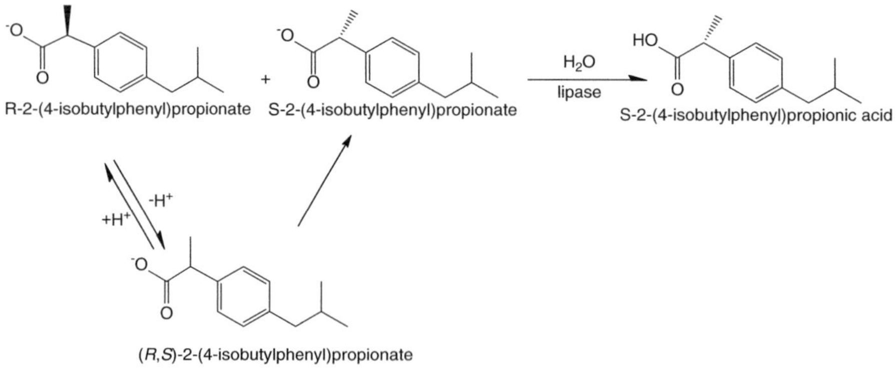

Fig. 16.2 Reaction route for DEKR of racemic ibuprofen ester (Xin et al. 2001)

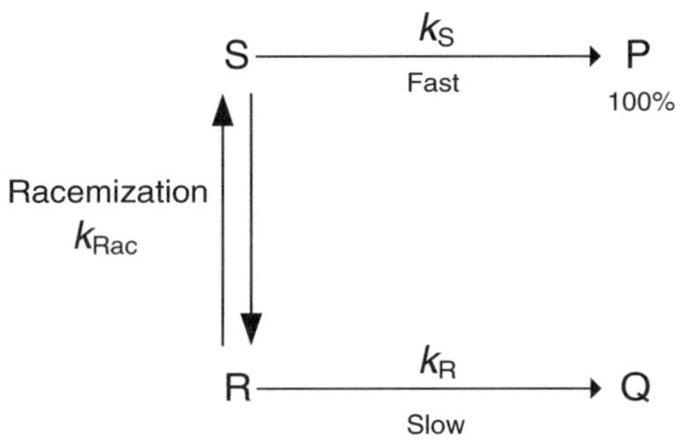

Fig. 16.3 Schematic representation of a DEKR process. (*R*) and (*S*) represent the two NSAID enantiomers; *P* and *Q* represent the two NSAID enantiomers of the product. k_{rac} denotes rate constant for racemization; k_R and k_S denotes rate constant for hydrolysis reaction of (*R*)-isomer and (*S*)-isomer respectively

Base-Catalyzed Racemization of NSAIDs

The racemization of the NSAIDs compound gaining much interest due to it significant contribution for DEKR process. The racemization of NSAIDs, an α-alkyl carboxylic acid can be racemized by base, acid, metal complex, and enzyme catalysis. Base-catalyzed racemization is the most commonly used method for racemization of NSAIDs compound that bears hydrogen at the chiral center or α-carbon with low pKa value. In spite of that, metal complex catalyst has started to be acknowledged as a potential replacement for catalyzing racemization process. However, due to high cost and a concern on the use of toxic substances during the preparation of the metal complex catalysts has deviated some interests into such systems especially for the production of pharmaceutical base products. In fact, bases are the most preferred catalyst in racemization of α-alkyl carboxylic acid as well as NSAID compounds due to their lowest costs, which became an added advantage.

The racemization of NSAID compounds by base-catalyzed reaction can be described by keto-enol tautomer mechanism as illustrated in Fig. 16.4. The reaction starts with the removal of hydrogen ion from the α-carbon by the base catalyst to form an enolate. Enolate is an unstable form of intermediate of the NSAIDs and rapidly transforms to more stable form, thus it present at very low concentration during the inter-conversion process (Yuchun et al. 2000). The enolates is stabilized by either readdition of the hydrogen, adjacent groups or other functionalities. These process occur non-stereospecifically, with simultaneous inversion and retention of configuration, leading to racemization (Wiberg and Wiberg 2001). Racemization depends on several parameters that includes; reaction media, strength of the bases, and type of substituents of the substrate. The removal of the hydrogen from the α-carbon of NSAIDs requires activated hydroxide ion (OH^-) or bases (Gonawan et al. 2013a). Hydroxide-catalyzed racemization of (*R*)-ibuprofen ester was significantly slower in organic or organic-aqueous media. In fact, the racemization was

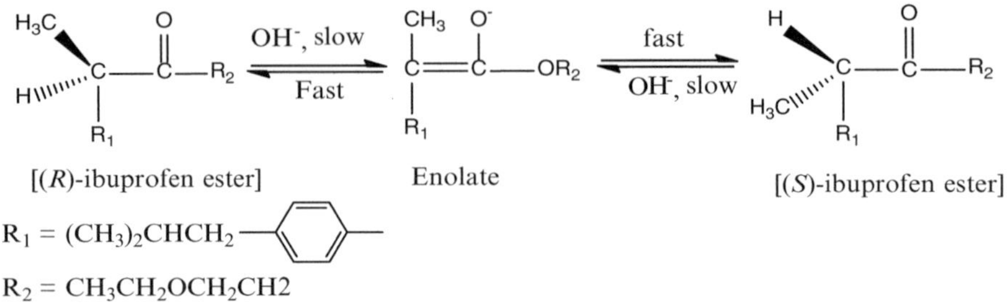

Fig. 16.4 Racemization of ibuprofen enantiomers by base-catalyzed keto-enol tautomerism (Yuchun et al. 2000)

proceeded rapidly in the organic-DMSO media. It is highlighted that the reactivity of the OH^- was hindered by partial positive charge of water and lead to slower racemization. Meanwhile, in the presence of DMSO, the OH^- dissipated better in the reaction medium compared to water due to partial negative charges of DMSO dominated the media. As a result, the OH^- is more reactive in the racemization reaction of the (*R*)-substrate.

In addition, the concentration of the base can strongly influence the racemization process. The high amount of hydroxyl ion (OH^-) could actually facilitate the racemization faster. The racemization of (*R*)-ibuprofen ester increased as the concentration of the base catalyst was increased (Gonawan et al. 2013a). Besides, the action of the nucleophile is significantly affected in the present of cation due to ionic interaction. Hence, cation free media is more preferable for racemization. In this case, anion resin is more suitable to be applied as the base catalyst compared to using base in salt form, such as sodium hydroxide. Base catalyst in salt form consist cation and needs to be ionized in order to execute racemization reaction. The ionization of the base is influenced by their dissociation constant in that particular media. Bases with low dissociation constant will reduce the effectiveness of the catalysis, resulted in slow racemization process. It can be compared to anion resin that readily available in the form of anion, thus, can spontaneously execute racemization reaction.

The rate of racemization can be probed by either looking at the inter-conversion rate of enantiomer or the formation rate of the racemate (Ebbers et al. 1997). The muta-rotation of optically pure enantiomer to a racemate can be described by a reversible reaction. This racemate mixture contains equi-molar proportion of each enantiomer. The rate of reaction for the excess enantiomer can be described by (16.1). The base catalyst is nonselective toward the enantiomers and randomly racemize the enantiomer. The racemization process can only be terminated by the removal of the base catalyst from the reaction media. The process will eventually reach an equilibrium condition where, $k_1 = k_{-1}$. The variation of the enantiomeric excess of the substrate (ee_S) can be determined by solving the differential equation (16.1) with conditions at $[R]_o = [S]_t + [R]_t$ and $[S]_o = [R]_t + [S]_t - [R]_o$, and it is given by (16.2). Where, k_{int} is defined as the inter-conversion constant and the equation follows the initial condition when, at $t=0$; $ee_S = ee_{So}$. The terms ee_S and ee_{So} are given by (16.3) and (16.4).

$$\frac{d[R]}{dt} = k_{-1}[S] - k_1[R] \quad (16.1)$$

$$ee_S = ee_{So}e^{-k_{int}t} \quad (16.2)$$

$$ee_S = \frac{[R] - [S]}{[R]_o + [S]_o} \quad (16.3)$$

$$ee_{So} = \frac{[R]_o - [S]_o}{[R]_o + [S]_o} \quad (16.4)$$

It was found that, the inter-conversion constant is directly proportional to the initial substrate concentration at a constant base concentration. Equation (16.5) describes the racemization process with the influence of base and initial substrate concentrations. The term k_{abs} represents the absolute rate constant of racemization in a specific system that involves operational parameters such as reaction environment and the reactor configuration (Gonawan et al. 2013a).

$$ee_S = ee_{So}exp\left(\frac{-k_{abs}[OH]}{[R]_o + [S]_o}t\right) \quad (16.5)$$

$$k_{int} = \frac{-k_{abs}[OH]}{[R]_o + [S]_o} \quad (16.6)$$

DEKR of NSAIDs

The DEKR can be carried out in two different modes of operations; (1) kinetic resolution and racemization in separate units and (2) one-pot or *in-situ* process. For a separated system, the EKR and racemization are carried out individually in two separate vessels. The resolution of substrate is carried out in a biphasic batch reactor until the desired conversion is achieved. The product is then extracted and the remaining unreacted substrate will undergo racemization in the subsequent vessel. Furthermore, the DEKR can be improved by using immobilized enzyme and fixed-bed of base catalyst in series as illustrated in Fig. 16.5. The immobilized enzyme can be physically separated from the reaction media with the fine-meshed filter. The substrate in the organic phase can be continuously discharged into the racemization reactor with fixed ionic resin as the base catalyst. The substrate is fed into the fixed-bed reactor to undergo racemization and subsequently return back into the

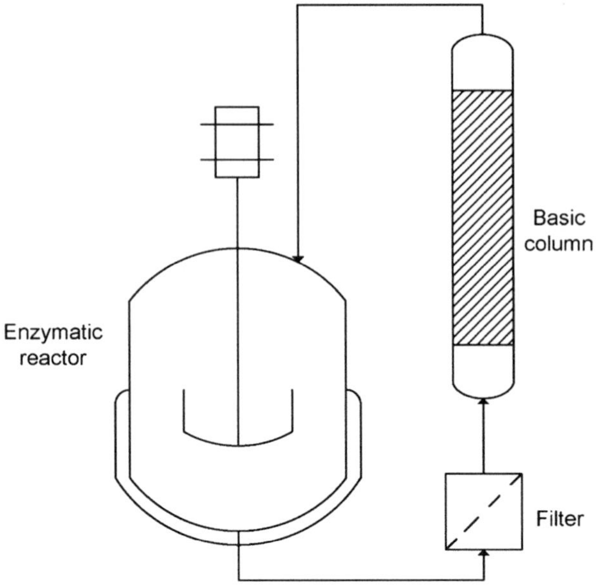

Fig. 16.5 Continuous DEKR system by batch EKR and fixed bed racemization

EKR system. This operation can be conducted in a continuous mode until the desired conversion and optical purity are achieved. In addition, the conversion and the optical purity can be controlled by varying the enzyme and base loading in both reactors until the equilibrium is reached in the DEKR system.

In-situ DEKR system in a single unit is rather complicated since the enzymatic hydrolysis and the base-catalyzed inter-conversion are perform simultaneously. In-depth consideration should be made on the compatibility of the two processes involved that distinctly differs in terms of reaction requirements. Several issues have been highlighted on the early development of *in-situ* DEKR which includes; the sensitivity (activity and stability) of enzyme towards the base catalyst, the possibility of side (racemization of the product) as well as reverse reactions, and complicated procedure for product recovery. Hence, the operation of *in-situ* DEKR in classical batch reactor is unfavourable due to lack of control on the reaction performance. Furthermore, base-catalyzed racemization requires reactive anion to strip the hydrogen at the chiral centre of NSAIDs (ibuprofen, naproxen and etc.) for the formation of enolates. The reactivity of the base (anion) is significantly reduced with the presence of buffering agent during one-pot DEKR. The reactivity of the base is believed to be blocked by the ionic interaction with positively charged buffering agent as well as the partial positive-charged of water.

A novel approach has been suggested in order to overcome the limitation of the conventional batch reactor. An enzyme-mediated at hollow-fiber membrane reactor perhaps the most suitable system to be used for the *in-situ* enzyme reaction. Figure 16.6 illustrates the hollow-fiber membrane fabricated with porous and dense layers are being utilized for enzyme immobilization and selective separation. Enzyme is immobilized on a porous layer of hydrophilic membrane module with

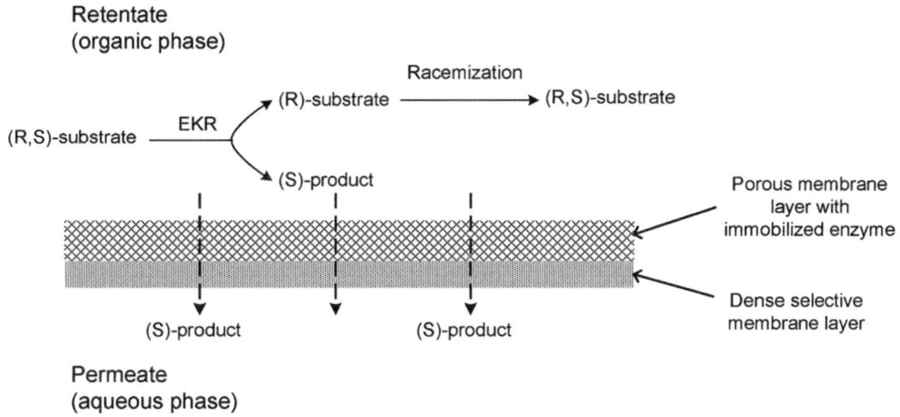

Fig. 16.6 Mechanism of enzyme-mediated membrane reactor for DEKR process

aqueous phase on the permeate side and organic phase on the retentate side. Weak base is used in the organic phase to catalyze the racemization of the excess substrate. The hydrolysis of substrate is carried out in the membrane matrix and the product is separated physically (molecular size, transmembrane pressure) or chemically (hydrophilicity, solubility, concentration gradient) into the lumen side of the membrane. Meanwhile the racemization reaction is taking place at the organic phase in the presence of base catalyst. Immobilized enzyme has been found to be chemically resistant and thermally stable. The addition of co-solvent for base activation is required in order to improve the equilibrium state between both reactions (Fazlena et al. 2006). However, the addition of co-solvent could cause problem to the catalytic activity of the lipase employed. This is due to the fact that some enzyme molecules are unable to exhibit maximum performance in some polar solvents (Fu and Vasudevan 2010). Meanwhile the addition of dimethyl sulfoxide has slightly decreased the activity of lipase during the hydrolysis of ibuprofen ester (Gonawan et al. 2013b). An intense research is in progress to achieve maximum activity of enzyme in the immobilized form as well as in the extreme media environment.

In-Situ *DEKR in Hollow-Fiber Membrane Reactor*

The DEKR process has been proposed to take place in a membrane-mediated reactor (Wang et al. 2004; Lau et al. 2010, 2011). The reactor is divided into two compartments by a semi-permeable membrane. Besides being a selective barrier, the membrane itself can be utilized as a support for the enzyme, in which the enzyme can be immobilized in the porous matrix or on the surface of the membrane. The reaction media in the enzymatic membrane reactor (EMR) can be of aqueous, organic and two-liquid phases (organic-aqueous). The versatility of the two-liquid phases in hollow-fiber membrane reactor for DEKR for the production of (*S*)-ibuprofen acid has been investigated (Sie Yon et al. 2013; Lau et al. 2010). Polyacrylonitrile (PAN) has been used as membrane material due to its superior resistance to hydrolysis and oxidation. The used of membrane in hollow-fiber configuration is an added advantage in particular for its high surface area for maximum enzyme loading which consequently provides higher mass transfer rate. The reactant is continuously fed into the lipase-immobilized membrane matrix where the reaction is taking place and at the same time, the product is discharged. The main advantage in adopting membrane technology in the production of enantiomer is that continuous mass production is guaranteed.

Effect of Substrate Concentration

The concentration of racemic ibuprofen ester used in the EMR system was in the range of 10–200 mM. Figure 16.7 shows a plot that captured the production of (*S*)-ibuprofen acid for the period of 10 h of the DEKR reaction in the EMR at different initial racemic ibuprofen ester concentrations. It can be seen that throughout the reaction time, the (*S*)-ibuprofen acid concentrations for 10, 50 and 80 mM initial substrate increased gradually until it reaches a steady-state at the end of the reaction. However, higher concentrations of initial substrate i.e. 150 and 200 mM gave different trends, where lower production at the beginning of the reaction was observed with a dramatic increase after fourth hour until the reaction completed.

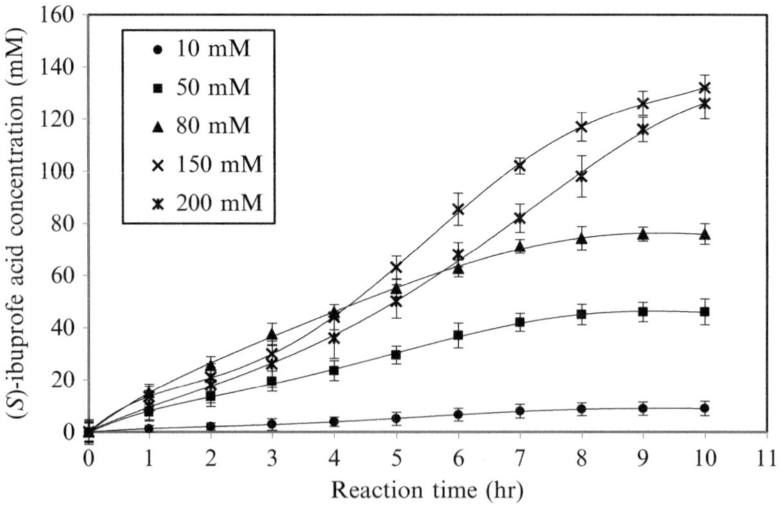

Fig. 16.7 Time-course concentration profiles of the (*S*)-ibuprofen acid at substrate concentrations of 10, 50, 80, 150 and 200 mM at 45 °C and 2 g/L enzyme loading

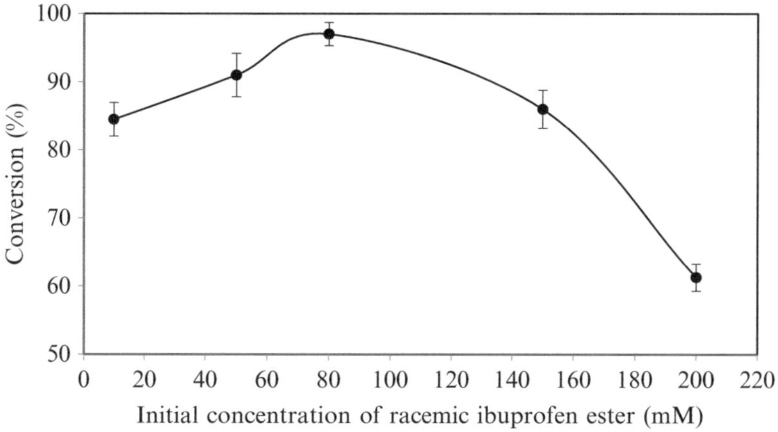

Fig. 16.8 Conversion profile of the racemic ibuprofen ester at the variation of the initial racemic ibuprofen ester concentrations

Figure 16.8 apparently shows the process performance of a DEKR. It can be observed that the percentage conversion gradually increased when the substrate concentration was increased until a level where the conversion rate started to decrease as the substrate concentration continuously increased.

It has found that, 80 mM ester is the most favourable substrate concentration due to the optimum performance of hydrolysis reaction with 97.3 % conversion. However, the conversion dropped to 63.4 % as the substrate concentration was further increased to 200 mM.

In addition, Fig. 16.9 depicts the enantiomeric excess of desired product and unreacted substrate over a variation of conversions for 10, 50, 80, 150 and 200 mM racemic ibuprofen ester. It was observed that the overall product enantiomeric excess (ee_p) was maintained at high values (>90 %) for all of the substrate concentrations. However, the specific value of ee_p for each substrate concentrations was obtained at different conversion values. For instance, 150 mM of substrate could give 95 % of ee_p and remain constant for the conversion above 15 %, but for the lower substrate concentration such as 10 mM, only 90 % ee_p could be attained when the conversion is above 65 %.

On the other hand, the ee_s stayed at low level for all substrate concentrations in the DEKR system. It can be seen that 200 mM substrate give the highest ee_s which is not preferable for a DEKR system. The smaller ee_s values are preferable as it indicates a relatively low amount of unreacted substrate. The result shows that the substrate concentration in the range of 50–80 mM was optimal for the resolution of racemic ibuprofen ester, gives low ee_s and high ee_p were obtained.

The conversion stopped at 65 % is due to the fact that poor resolution was obtained at high substrate concentration i.e. 200 mM. When the reaction started to diminish, there was 36 % of substrate remained unreacted. This condition may be related to the substrate inhibition, which resulted in low efficiency of DEKR

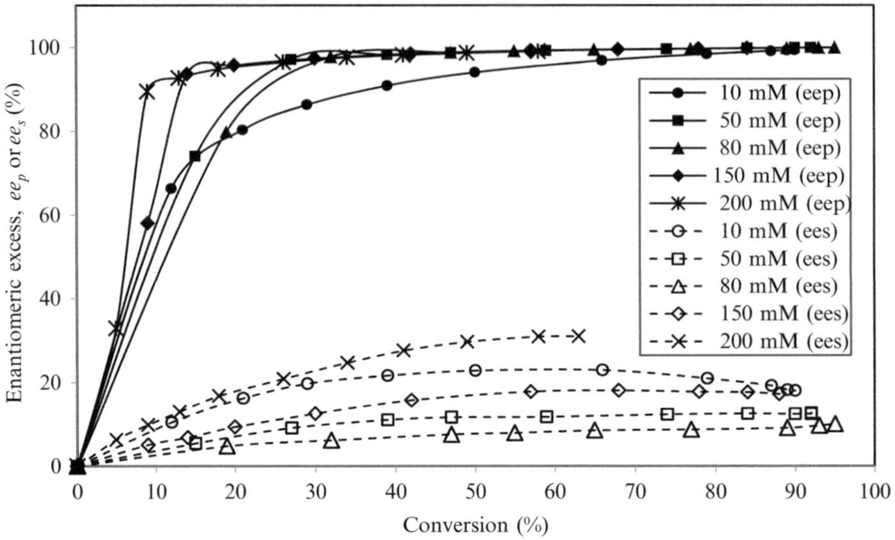

Fig. 16.9 Product and substrate enantiomeric excesses at different of conversions for different substrate concentrations used in the EMR

process when high concentration of substrate was used (Fazlena et al. 2006). Generally, only adequate amount of racemic ibuprofen substrate was able to bind to the active site of the enzyme. When very high concentrations of substrate were used, the active sites of the enzyme became saturated and therefore, unavailable for further binding with other substrate molecules. The substrate molecules are either competing for the active sites on the enzyme surfaces which was blocked by the substrate molecules themselves (competitive inhibition) or binds to the enzyme-substrate complex and therefore, reduces the concentration of enzyme-substrate complex (uncompetitive inhibitor). Both conditions automatically reduced the affinity of the enzyme towards the substrates and thus, lowering the hydrolysis rate of a reaction.

Effect of Base Concentration

The choice of trioctylamine concentration is one of the important factors affecting the *in-situ* racemization and hydrolysis in the lipase mediated membrane reactor. Trioctylamine is a base normally used to catalyze keto-enol tautomerism in the DEKR. In the recent investigation, the effect of trioctylamine concentration on the enzymatic hydrolysis reaction in the EMR was studied at 30, 80 and 100 mM at 45 °C, 0.84 g/m^2 enzyme loading, 50 mM initial concentration, 100 ml/min organic flow rate and 200 ml/min aqueous flow rate. Figure 16.10 demonstrates the effect of trioctylamine concentration on the resolution of the ibuprofen racemate.

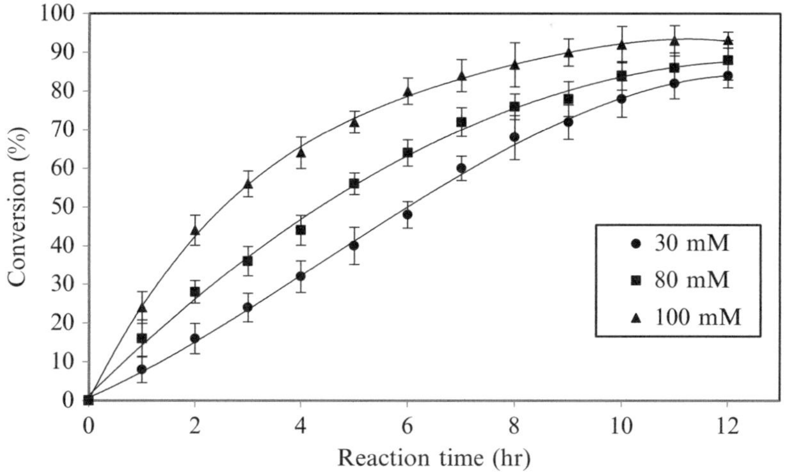

Fig. 16.10 Time-course conversion of racemic ibuprofen ester for three different base concentrations

It was observed that in the first few hours of the reaction, the conversion profiles increased linearly with time and later started to approach a plateau after about 10 h. The organic phase with an added 100 mM trioctylamine gave the highest conversion, where 90 % of the ester was converted to product at the eighth hour of reaction and gave a maximum conversion of 94 % at the tenth hour. The ester with 30 mM trioctylamine showed the lowest conversion where only 72 % ester was converted to product at the end of the reaction. Meanwhile, 80 % conversion was achieved for the substrate with 80 mM trioctylamine.

A time-course variation of substrate enantiomeric excess (ee_s) was depicted in Fig. 16.11. The average ee_s values at 30, 80 and 100 mM concentration of trioctylamine are 38.32 % (average error of ±3.52 %), 20.86 % (average error of ±3.05 %) and 11.18 % (average error of ±3.20 %) respectively. It can be seen that higher base concentrations favored the DEKR as the lower ee_s values were obtained.

The substrate enantiomeric excess was used as an indicator to maintain the success of an asymmetric synthesis (Patel 2006). For an optimal DEKR system, the lower ee_s value is preferred as it implies that the remaining amount of the unreacted enantiomer in the substrate is low and would result in higher optical purity. The slight decrease of the ee_s trends over the reaction time is due to the coupling effect of an *in-situ* racemization and hydrolysis in the DEKR system. The excess unreacted enantiomer i.e. (*R*)-ibuprofen ester underwent keto-enol tautomerism to balance the amount of racemate, meanwhile the reactive enantiomer i.e. (*S*)-ibuprofen ester was hydrolyzed at a higher rate than racemization. This resulted in the decrease of overall substrate and enantiomeric excess in the DEKR system.

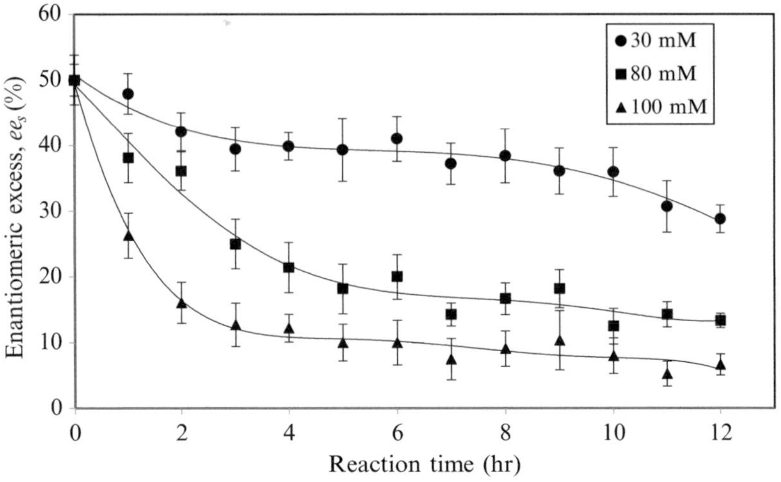

Fig. 16.11 Time-course variations of enantiomeric excess in the dynamic kinetic resolution of racemic acid at 30, 80 and 100 mM concentrations of trioctylamine

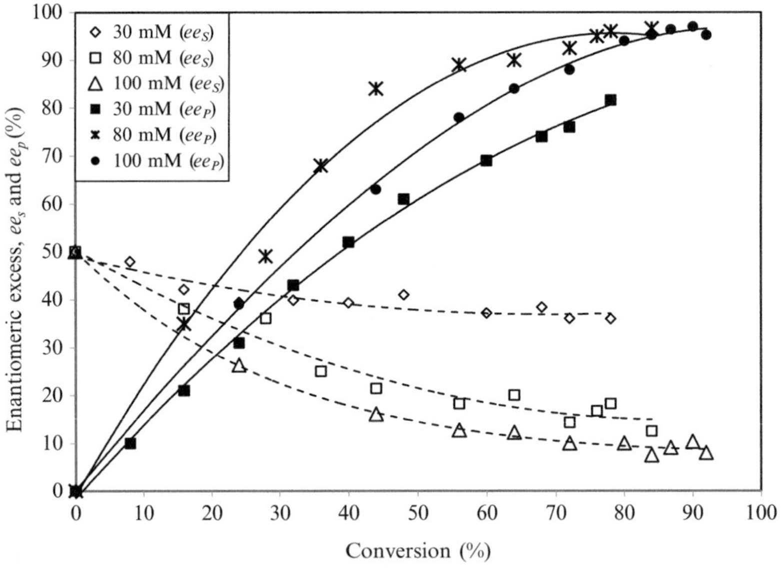

Fig. 16.12 Enantiomeric excess of ibuprofen acid against different conversions at 30, 80 and 100 mM trioctylamine concentrations

In addition, the relationship between the substrate and product enantiomeric excesses against the conversion at three different base concentrations (30, 80 and 100 mM) were further determined and are presented in Fig. 16.12. The plot apparently shows the quality of the DEKR at three different trioctylamine doses.

It was observed that the ee_s profiles reduced when the conversion was increased. In the meantime, the product enantiomeric excess (ee_p) shows the opposite trend, where the ee_s profiles increased proportionally with the conversion. For an ideal DEKR, high conversion is preferable, which as the same time gives high ee_p value with low ee_s value.

The investigation shows that the organic phase at 100 mM base gave the best results where 97.1 % ee_p and 7.3 % ee_s were obtained at 94.5 % conversion. On the other hand, the organic phase with the lowest trioctylamine dose (30 mM), only gave 78.4 % ee_p and 39.3 % ee_s at the maximum conversion of 79.1 %. In the DEKR, trioctylamine has the ability to extract a proton (H^+) from the α-carbon atom of (*R*)-ibuprofen ester (keto-molecule).

As a result, enolate (intermediate) was formed during the reaction and it is highly reactive towards electrophiles (Yuchun et al. 2000). Once the enolate has been attacked by the electrophiles, the hydrogen atom's position of the keto-molecule changes and (*S*)-ibuprofen ester is formed. This keto-enol tautomerism (Fig. 16.4) is directly dependent on the base concentration, where the higher amount of base could enhance the racemization of the unreacted (*R*)-ibuprofen ester.

Effect of Organic Feed Flow Rate

In an EMR, the immobilized lipase on the membrane surface is in direct contact with the organic phase and the flow rate is an important parameter which could affect the efficiency of substrate diffusion in the membrane reactor. Hence, the performance of the EMR was investigated at organic flow rate in the range between 50–200 ml/min at a fixed aqueous flow rate of 200 ml/min. Figure 16.13 shows the overall concentration profiles for all the organic flow rates at a given range.

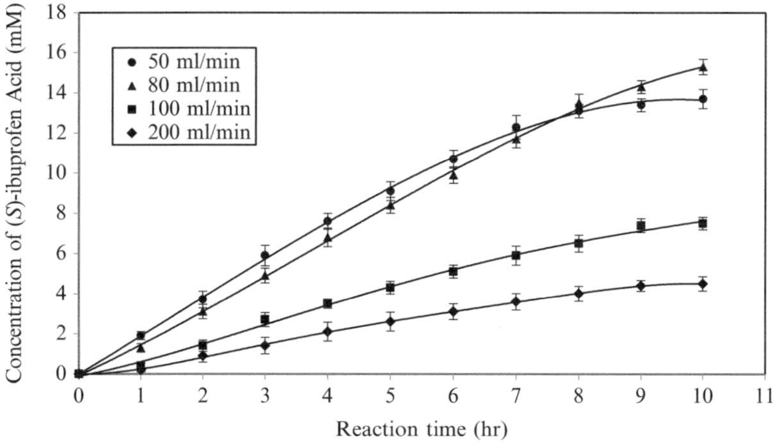

Fig. 16.13 The effect of organic flow rates on (*S*)-ibuprofen acid concentration

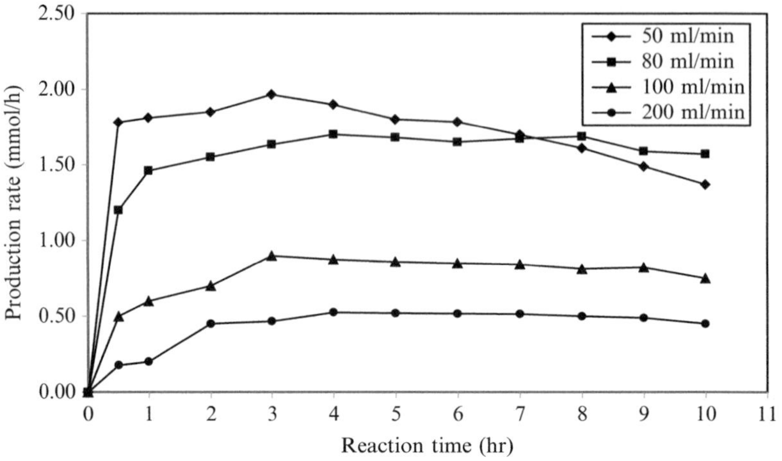

Fig. 16.14 Effect of organic phase flow rates on the (*S*)-acid production rate with lipase immobilized on the porous layer

A steady formation of the (S)-ibuprofen acid with respect to time can be observed with an increasing organic flow rates resulted in lower overall product concentration profiles. For instance, the EMR operated at the organic flow rate of 50 ml/min could give higher product concentration (12.5 mM) as compared to the organic flow rate of 200 ml/min, which only gave approximately 4.5 mM of (*S*)-ibuprofen acid.

The result also shows that although the EMR operated at organic flow rate of 50 ml/min which give higher concentration at the beginning of the reaction, it started to reach steady-state after 7 h of operation with a product concentration of 12.5 mM. On the other hand, the product concentration for EMR operated at organic flow rate of 80 ml/min continued to increase after the eighth hour and surpassed the amount obtained in the former case with 15.6 mM of (*S*)-ibuprofen. Therefore, 80 ml/min of organic flow rate is preferable for a long-run of the DEKR reaction in an EMR. Furthermore, at low organic flow rate will give sufficient retention time for the lipase-catalyzed hydrolysis reaction at the reaction layer (Long et al. 2005).

In addition, the performance of EMR for hydrolysis reaction was also investigated in terms of (*S*)-ibuprofen acid production rate at different organic flow rates. Figure 16.14 shows the effect of organic flow rate towards the reactor performance. It can be observed from the results that lower the organic flow rate, gave a higher (*S*)-ibuprofen rate of production. An average production rate of 1.74 mmol/h was obtained at the organic flow rate of 50 ml/min, but the operation of EMR at organic flow rate of 200 ml/min gave only an average production of 0.43 mmol/h. This is most likely due to the diffusion limitation of the substrate molecules into membrane porous (catalytic) layer and therefore, resulted in a poor resolution at higher organic flow rates.

Although the highest initial production rate (1.89 mmol/h) was observed at the organic flow rate of 50 ml/min, the production rate gradually decreased throughout the operation until it reached 1.37 mmol/h at the tenth hour of process. On the other hand, the organic flow rate of 80 ml/min portrayed a stable production rate as compared to the former case. Therefore, the optimum organic flow rate of 80 ml/min was chosen due the consistent operation, which then provides a relatively high production rate.

Effect of Aqueous Feed Flow Rate

The productivity of DEKR was greatly affected by the volumetric flow rate of the aqueous phase (Mak et al. 2009). Therefore, the effect of aqueous flow rate was investigated with the range of flow rates of 50–350 ml/min. Figure 16.15 shows the product concentration of (*S*)-ibuprofen acid as a function of time. It can be observed from the plot that at higher aqueous phase flow rates, the concentration of optically pure (*S*)-ibuprofen acid is also higher. The desired product concentration up to 13.1 mM was obtained at 350 ml/min, whereas there was only 7.5 mM of (*S*)-ibuprofen produced at 50 ml/min after 10 h of operation. The average error of this experiment was in the range of $\pm 0.27 - 0.35$ mM.

In the EMR, both the (*R*)- and (*S*)-ester are hydrophobic compounds, whereas the product ((*S*)-acid) is hydrophilic and could easily diffuse into the aqueous phase at the lumen side. Basically, the permeability of a membrane not only depends on its pore size, structure and trans-membrane pressure but also its hydrophilicity. As a result, the less reactive (*R*)-ester (non-polar) in the organic phase was retained at the shell side.

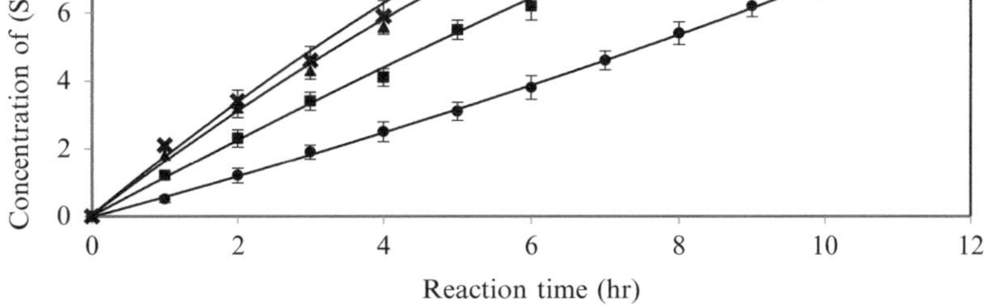

Fig. 16.15 The effect of aqueous flow rates on (*S*)-acid concentration with lipase immobilized on porous layer

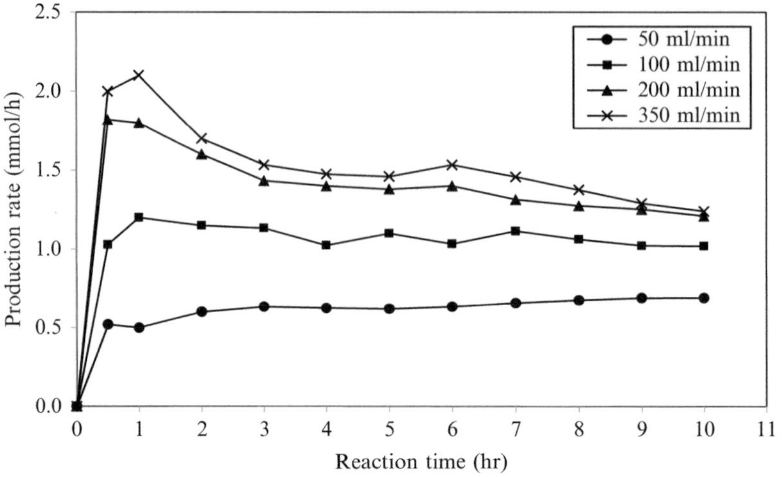

Fig. 16.16 Effect of aqueous phase flow rates on membrane productivity with lipase immobilized on porous layer

The effect of aqueous flow rate on the (*S*)-ibuprofen production rate was also investigated and the results are presented in Fig. 16.16. The results show that the aqueous flow rates of 50 and 100 ml/min gave constant (*S*)-ibuprofen production rates with average values of 0.634 and 1.08 mmol/h respectively. For the increased flow rates i.e. 200 and 350 ml/min, the average production rates increased to 1.40 and 1.51 mmol/h respectively. However, the production rates for both 200 and 350 ml/min showed a slight decrease throughout the process. This phenomenon is probably due to the formation of aqueous turbulence flow near the membrane interface, which might cause conformational change to the enzyme is active site.

In general, the production rates of (*S*)-acid increased with the increase in aqueous flow rates. This is due to the role of hydrophilic membrane in the hydrolysis process. During the reaction, the mass transfer of the accumulated (*S*)-ibuprofen molecules in the aqueous phase would increase proportionally to the convective driving force induced by the trans-membrane pressure as well as nominal flow rate. At higher aqueous flow rate will facilitate the diffusion of polar products from the membrane to the aqueous stream, i.e. the flux of (*S*)-ibuprofen will increase, especially when hydrophilic membrane is employed (Long et al. 2005).

In the recent investigation, the aqueous flow rate of 200 ml/min was selected as the optimum flow rate since it gave more stable production rate if compared to that of 350 ml/min. In fact, there is only a small difference in the average production rate (0.11 mmol/h) even though 75 % of the flow rate was increased. Therefore, pump operated at lower flow could reduce the power consumption in long term operation.

Process Modelling of DEKR in Membrane Reactor

Basically, the membrane module of EMR is immobilized with enzyme on either dense or porous layers of membrane. Lau and co-workers immobilized the lipase on the porous layer of polyacrylonotrile (PAN) hollow-fiber membrane via ultrafiltration. The mechanistic action of the process is divided into two distinctive zones in the immobilized hollow fiber membrane; (1) diffusion layer and (2) a biocatalytic layer. The cross-section of an asymmetric hollow fiber membrane is illustrated in Fig. 16.17 (Lau et al. 2010). In the diffusion layer, the racemization of the unreacted substrate takes place spontaneously by the base-catalyzed reaction. The biocatalytic side on the porous layer represents the boundary between the aqueous phase at the lumen and the organic phase at the shell side of the membrane. The development of the theoretical model is divided into two parts; (1) the overall rate of enzymatic and racemization reactions that occur inside and outside of the membrane matrix support respectively, and (2) diffusion of substrate through the membrane surface. In this particular study, the mass convection at the lumen side as well as the formation of concentration polarization layer were neglected. The mechanistic model was developed based on the combination of kinetic resolution and dynamic racemization for both (*R*)- and (S)-enantiomers. Meanwhile, the diffusion of the substrate was modelled based on convective mass transfer of the substrate in the membrane module. Theoretical analysis was carried out in order to study the performance of the system model.

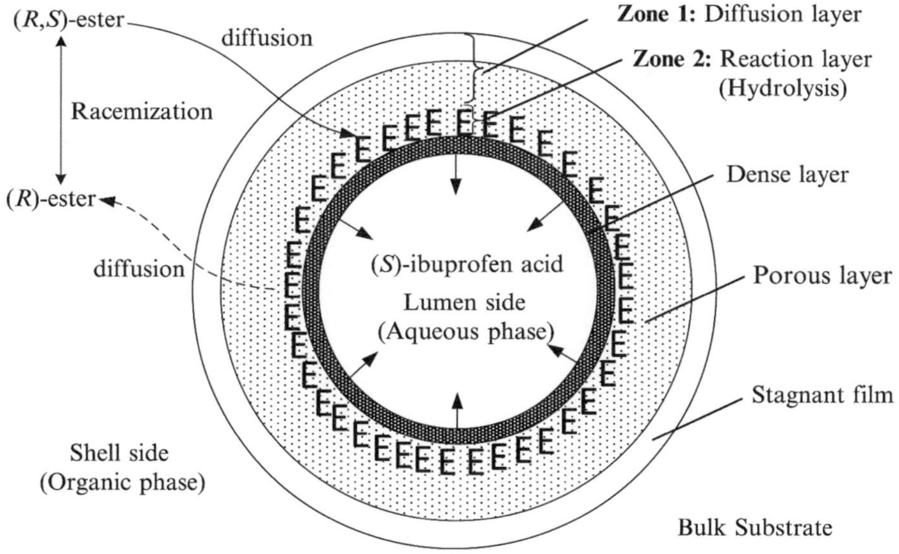

Fig. 16.17 Cross-section diagram of an asymmetric hydrophilic membrane with immobilized lipase on the porous layer (Lau et al. 2010)

Enzymatic Hydrolysis and Racemization

The kinetics of an enzyme-catalyzed hydrolysis of racemic ibuprofen ester in the membrane porous layer was developed base on the steady-state assumption. The substrate ((*S*)-ibuprofen ester) and the by-product (alcohol) were found to be uncompetitive and competitive inhibitors, respectively (Lau 2013; Sie Yon et al. 2013). The reaction mechanism of the enzyme-catalyzed conversion of substrate (*S*) to product (P′) is clearly illustrated in Fig. 16.18. The derivation of the mechanism is based on a rapid equilibrium assumption in which the rates of formation and dissociation of the enzyme-substrate complexes (denoted as ES, ES*, ESS*, EI or ESI) reach the equilibrium state instantly compared to that of the rate formation of the free enzyme (denoted as E) and product.

The reaction rate of the EKR process for both (*R*)- and (*S*)-enantiomers is represented by the following kinetic model (Lau et al. 2010; Lau 2013);

$$\upsilon_A = \frac{dS_A}{dt} = \frac{\upsilon_{Amax}[S_A]}{(K_{mA} + [S_A])\left(1 + \frac{[S_I]}{K_{nI1}} + \frac{[S_B]}{K_{uS1}}\right)} \tag{16.7}$$

$$\upsilon_B = \frac{dS_B}{dt} = \frac{\upsilon_{Bmax}[S_B]}{(K_{mB} + [S_B])\left(1 + \frac{[S_I]}{K_{nI1}} + \frac{[S_A]}{K'_{uS1}}\right)} \tag{16.8}$$

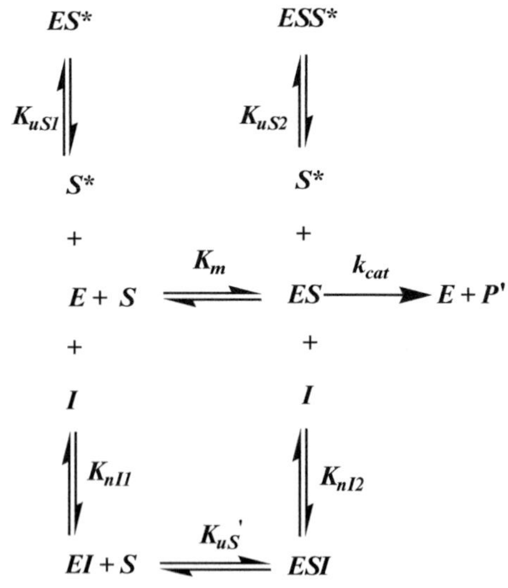

Fig. 16.18 Kinetic mechanism of enzymatic hydrolysis of racemic ibuprofen ester with uncompetitive substrate inhibition and non-competitive product inhibition. S* denoted as the isomer of (S)-substrate; K_m is the Michaelis-Menten constant for substrate ester; k_{cat} represents the enzymatic hydrolysis constant; K_{uS}, K_{uS}' and K_{nI} are the substrate and the by-product inhibition constants, respectively (Lau et al. 2010)

where ν_A, ν_{Amax}, S_A, K_{mA} and K'_{uSI} are defined as the reaction rate, maximum reaction rate, substrate concentration, Michaelis-Menten constant and uncompetitive substrate inhibition constant respectively for (S)-enantiomer; while, ν_B, ν_{Bmax}, S_B, K_{mB} and K_{uSI} are that for (R)-enantiomer respectively. S_I refers to the concentration of non-competitive alcohol inhibitor. The model equations are suitable for the hydrolysis of ibuprofen ester in any media. Some additional assumptions were made for hydrolysis process in enzyme-mediated membrane which include; (1) evenly distributed enzyme molecules in the porous membrane support; (2) evenly distributed concentration of substrate (3) constant enzyme activity throughout the reaction (Lau et al. 2010).

It was observed during the racemization of ibuprofen in the presence of base catalyst that the rate of dynamic equilibrium (V_{eq}) after racemization reaction time (t) by the (R)- and (S)-enantiomers respectively, is proportional to their concentrations. At the steady-state condition, the variation in concentrations of (R)- and (S)-enantiomers are defined as follows (Yuchun et al. 2000);

$$\frac{dS_A}{dt} = \frac{V_{eq}}{2}\left[\frac{S_B \quad S_A}{S_A + S_B}\right] \tag{16.9}$$

$$\frac{dS_B}{dt} = \frac{V_{eq}}{2}\left[\frac{S_A - S_B}{S_A + S_B}\right] \tag{16.10}$$

Catalytic Model of DEKR in Membrane Reactor

A coupled enzymatic hydrolysis and first order racemization model were employed in order to describe the rate for DEKR of racemic ibuprofen ester. The resulted model equations were rendered dimensionless using the constants provided in Table 16.2. Hence, the dimensionless form of the coupling reaction of EKR and dynamic racemization for (R)- and (S)-isomers were given as follows;

$$\upsilon_A = -\frac{dS_A}{d\tau} = \frac{\psi_A S_A}{(\Theta_A + S_A)(1 + \varphi\xi_{IP} + S_B\xi_{IS})} + \frac{V_{eq}}{2}\left[\frac{S_A - S_B}{S_A + S_B}\right] \tag{16.11}$$

$$\upsilon_B = -\frac{dS_B}{d\tau} = \frac{\psi S_B}{(\Theta_B + S_B)(1 + \varphi\xi_{IP} + S_A\xi_{IS})} + \frac{V_{eq}}{2}\left[\frac{S_B - S_A}{S_A + S_B}\right] \tag{16.12}$$

It is assumed that the (S)-isomer is the dominant substrate for a selective lipase-catalyzed hydrolysis reaction as compared to (R)-isomer counterpart (Long et al. 2005). Meanwhile, the inter-conversion of the unreacted substrate ((R)-isomer) to a racemic form dominates at the organic phase of the system. Thus, the V_{eq} and ψ_B were insignificant for (S)- and (R)-isomers respectively. In order to simplify the

Table 16.2 Dimensionless terms defined for reaction rate model (Lau et al. 2010)

Dimensionless term	(*S*)-enantiomer	(*R*)-enantiomer
Enzymatic DEKR constant	$\psi_A = \frac{\Theta t_o \upsilon_{max}}{K_{mA}}$	$\psi_B = \frac{\Theta_B t_o \upsilon_{max}}{K_{mB}}$
Michaelis-Menten constant	$\Theta_A = \frac{K_{mA}}{S_{Ao}}$	$\Theta_A = \frac{K_{mB}}{S_{Bo}}$
Substrate inhibition constant	$\xi_{IS} = \frac{S_{Ao}}{K'_{uS1}}$	$\xi_{IS} = \frac{S_{Bo}}{K_{uS1}}$
By-product inhibition constant	$\xi_{IP} = \frac{S_{by}}{K_{nI1}}$	$\xi_{IP} = \frac{S_{by}}{K_{nI1}}$
Product inhibition fraction	$\varphi = \frac{S_{Io}}{S_{Ao}}$	$\varphi = \frac{S_{Io}}{S_{Bo}}$

mathematical equations, let $V_{eq} = 0$ in (16.11) and $\psi_B = 0$ in (16.12) which results in elimination of the racemization and enzymatic hydrolysis terms, respectively. Therefore, the model equations were then reduced into (16.13) and (16.14).

$$\upsilon_A = -\frac{dS_A}{d\tau} = \frac{\psi_A S_A}{(\Theta_A + S_A)(1 + \varphi\xi_{IP} + S_B\xi_{IS})} \tag{16.13}$$

$$\upsilon_B = -\frac{dS_B}{d\tau} = +\frac{V_{eq}}{2}\left[\frac{S_B - S_A}{S_A + S_B}\right] \tag{16.14}$$

Incorporation of Mass Transfer into the DEKR Model

The radial mass transport through the hollow-fiber membrane is induced by both the pressure difference across the membrane and the convection mass transfer. Several assumptions related to the mass transfer in the system were made for the coupled reaction-diffusion model; (1) the effect of axial flow gradient on the radial diffusion is negligible; (2) Fick's law is applied for the diffusion of substrate; (3) the fluid properties is constant throughout the boundary condition of the membrane; and (4) Eddy diffusion is insignificant (Long et al. 2003; Nagy and Kulcsár 2009; Hu et al. 2008; Lau 2013). The DEKR model derived in the previous section was incorporated together with the diffusion mechanism in order to obtain a complete DEKR model. The generalized mass transfer equation with that of reaction terms for (*S*)-ibuprofen ester in cylindrical coordinate was modelled by integration of differential equation presented by (16.15) (Long et al. 2003):

$$u(r)\frac{dS_i}{dr} + D_{eff}\left[\frac{d^2S_i}{dr^2} + \frac{1}{r}\frac{dS_i}{dr}\right] = \alpha_P \upsilon \tag{16.15}$$

Where, S_i denotes the substrate concentration of racemic ester; α_P is a measure of membrane characteristic (tortuosity, porosity, etc.); D_{eff} is the effective diffusivity of the ibuprofen ester and v is the rate of enzymatic reaction. The model equation comprises the expressions of mass transport for the diffusive-convective fluxes and the rate equation for the enzymatic hydrolysis reaction (Lau 2013). The first term on the left-hand side of (16.15) represents the radial diffusive mass transfer, where $u(r)$ is the radial flow velocity in a bundle of hollow-fiber membrane expressed as;

$$u(r)\frac{F}{2\pi LrN} \tag{16.16}$$

Where, F is the total volumetric flow rate of substrate through the membrane pore channel; L is the effective length of fiber; r is the fiber radius and N is the number of fibers in a membrane bundle.

The term ν on the right hand side of (16.15) represents the EKR and dynamic racemization of the substrate given by (16.13) and (16.14) for (S)- and (R)-isomers respectively. The reaction-diffusion model of DEKR with mass transfer of the substrate on the membrane was developed by substituting the reaction rate (16.13 and 16.14) derived in the previous section into (16.15) which then resulted into (16.17) and (16.18) (Lau et al. 2010);

$$u(r)\frac{dS_A}{dr} + D_{eff}\left[\frac{d^2S_A}{dr^2} + \frac{1}{r}\frac{dS_A}{dr}\right] = \alpha_P\left\{\frac{\upsilon_{max}[S_A]}{(K_{mA} + [S_A])\left(1 + \frac{[S_I]}{K_{nI1}} + \frac{[S_B]}{K_{uS1}}\right)}\right\} \tag{16.17}$$

$$-\left\{u(r)\frac{dS_A}{dr} + D_{eff}\left[\frac{d^2S_A}{dr^2} + \frac{1}{r}\frac{dS_A}{dr}\right]\right\} = \frac{V_{eq}}{2}\left[\frac{S_B - S_A}{S_A + S_B}\right] \tag{16.18}$$

Equation (16.17) represents the concentration distribution of (S)-ibuprofen ester at both shell-side and membrane matrix during the hydrolysis reaction. Meanwhile, (16.18) represents the concentration distribution of (R)-ibuprofen ester during the base-catalyzed reaction. During the DEKR, the concentration of (S)-ibuprofen decreases while the concentration of (R)-ibuprofen ester builds up at the membrane layer. Subsequently, the (R)-ibuprofen ester gradually dissipates into the bulk flow due to convective mass transfer. The presence of base catalyst on the bulk layer promotes the formation of enolates by base-catalyzed keto-enol tautomerism and leads to racemization (Yuchun et al. 2000). Thus, the excess of (R)-ibuprofen ester is continuously converted to racemic form and the resulted (S)-isomer is recycled for EKR process.

Equations (16.17) and (16.18) were simplified, rearranged and transformed into the dimensionless form (Lau et al. 2010; Lau 2013);

$$\frac{d^2S_A}{dR^2} + \frac{1}{R}(1 + B_o)\frac{dS_A}{dR} = \frac{\Phi^2 S_A}{\left(1 + \frac{S_A}{\Theta_A}\right)(1 + \varphi\xi_{IP} + S_B\xi_{IS})} \tag{16.19}$$

$$\frac{d^2S_B}{dR^2} + \frac{1}{R}(1 + B_o)\frac{dS_B}{dR} = \gamma\left[\frac{S_A - S_B}{S_A + S_B}\right] \tag{16.20}$$

Where, B_o is the Bodenstein number which is defined as the ratio of substrate feed flow rate to the effective diffusivity as expressed by (16.21). Φ^2 is the Thiele Modulus for enzymatic hydrolysis of ibuprofen ester and γ represents the racemization characteristics.

$$B_o = \frac{u(r)r}{D_{eff}} = \frac{F}{2\pi LND_{eff}} \tag{16.21}$$

The defined dimensionless groups are given in (16.22)–(16.27).

$$\Phi^2 = \frac{a^2\alpha_P \upsilon_{max}}{D_{eff}K_{mA}} \tag{16.22}$$

$$\varphi = \frac{S_{Io}}{S_{Ao}} \tag{16.23}$$

$$\xi_{IP} = \frac{S_I}{K_{nI1}} \tag{16.24}$$

$$\xi_{IS} = \frac{S_{Bo}}{K_{uS1}} \tag{16.25}$$

$$\gamma = \frac{a^2k_{rac}S_{base}}{S_{To}D_{eff}} \tag{16.26}$$

$$R = \frac{x}{r} \tag{16.27}$$

Where, R is the dimensionless coordinate at radial direction normalized with radius of the membrane, r. The equations are associated with the boundary conditions at the fiber external radius ($x = c$) and internal radius ($x = b$). Thus, the boundary conditions in dimensionless form are given by (Lau 2013; Lau et al. 2010);

$$\frac{dS_A}{dR} = 0; \frac{dS_B}{dR} = 0; \text{ at } R = \frac{b}{r} \tag{16.28}$$

$$\frac{dS_A}{dR} = \frac{B_o}{R}(1 - S_A); \frac{dS_B}{dR} = \frac{B_o}{R}(1 - S_B), \text{ at } R = \frac{c}{r} \tag{16.29}$$

The substrate concentration at the external radius is assumed equal to the substrate concentration at the bulk region and the polarization layer is neglected. The boundary conditions demonstrate the diminishing of substrate concentration gradient at the interface between the porous layer and the shell-side of the membrane module (Lau 2013).

Process Parameters Study of DEKR in the EMR

The performance of enzymatic membrane reactor in operating the DEKR of (*R*,*S*)-ibuprofen ester was studied based on several dimensionless analyses of the model (Lau 2013). In the following section, the DEKR process is hypothetically studied by manipulation of process variables incorporated as dimensionless constants in the developed models. The input parameters are presented in Table 16.3 (Gonawan et al. 2013b; Lau et al. 2010; Lau 2013).

Effect of Michelis-Menten Constant (Θ_A)

In the DEKR process, both substrate isomers are utilized and selectively converted into desired (*S*)-product. Simultaneously, the unreacted (*R*)-substrate is converted to its racemic form and thus, provide continuous supply of (*S*)-substrate until 100 % conversion of the substrate is obtained. The performance of this reaction was evaluated by the mathematical model developed (16.13 and 16.14) in the previous section. The study on the effect of dimensionless Michaelis-Menten constant (Θ_A)

Table 16.3 Input parameters for the simulation of DEKR reaction rate in hollow fiber membrane reactor

Properties	Unit	Present study
Initial racemic substrate concentration, S_{To}	mM	30–60
Effective diffusivity, D_{eff}	$cm^2\ min^{-1}$	$1–10 \times 10^{-4}$
Thiele Modulus, Φ^2	–	0.1–65
Dimensionless racemization constant, γ	–	0.1–2
Substrate inhibition constant for (*R*)-ester, K_{uSI}	$mM\ L^{-1}$	500–1,000
Product inhibition constant for alcohol, K_{nII}	$mM\ L^{-1}$	200–400
Dimensionless Michaelis-Menten constant, Θ_A	–	0.1–5
Maximum reaction rate, V_{max}	$mM\ h^{-1}$	2–7
Dimensionless Substrate inhibition constant, ξ_{IS}	–	0.1–10
Dimensionless by-product inhibition constant, ξ_{IP}	–	0.1–10
Racemization reaction constant, k_{rac}	$mM\ h^{-1}$	1–5
Michaelis-Menten constant for (*S*)-enantiomer, K_{mA}	mM	5–50

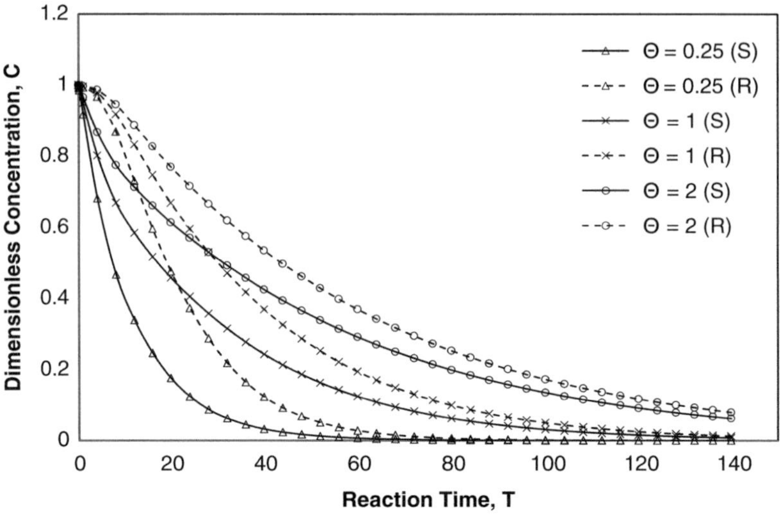

Fig. 16.19 Simulated concentration profiles at different dimensionless Michaelis-Menten constants, Θ_A

in the DEKR process was carried out at Θ_A between 0.25 and 2. Other parameters such as substrate and base concentration were remained constant. The value of Θ_A in (16.13) represents the effect of enzyme-substrate interaction during the DEKR. At lower value of Θ_A, better enzyme-substrate formation is observed and the product is rapidly released from the resulted enzyme-substrate complex.

It was observed in Fig. 16.19 that, both (*R*)- and (*S*)-substrate profiles decrease with the reaction time. This implies that, the racemization of (*R*)-substrate is an effective way to achieve high conversion of the racemic substrate. At lower value of Θ_A, the utilization of both substrates is rapid and thus the DEKR process completes at shorter period of time. This is due to the lower K_m which gives better performance in the hydrolysis of (*S*)-substrate and results in high amount of excess (*R*)-substrate. Subsequently, the unreacted (*R*)-substrate is simultaneously recovered to provide continuous supply of (*S*)-substrate for the hydrolysis reaction. The difference between dimensionless concentration of (*R*)- and (*S*)-substrate is small at high Θ_A values. This implies that, the conversion of the (*R*)-substrate to the (*S*)-substrate is dependent on the concentration of the excess (*R*)-substrate present in the bulk flow. At larger value of Θ_A, only small amount of excess (*R*)-substrate was formed due to slow conversion of (*S*)-substrate to the product. This behaviour also shows that, the decreasing trend of the (*R*)-substrate is controlled by the consumption rate of the (*S*)-substrate. The concentration of the substrates diminished rapidly and reach to a convergent point at specific reaction time, *T*. In the presence of DEKR process, K_m value should be as minimum as possible in order to improve the conversion of the racemic substrate.

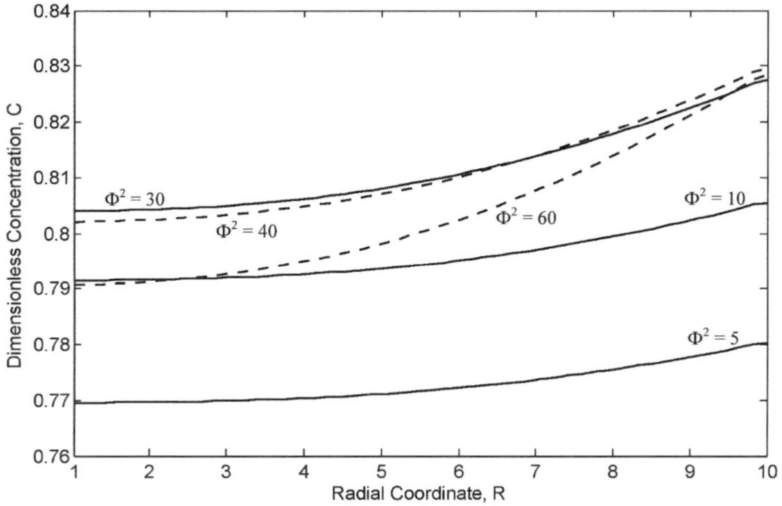

Fig. 16.20 Effect of Φ^2 on the dimensionless concentration profiles at $B_0 = 17.34$; $\gamma = 0.3$

Effect of Thiele Modulus (Φ^2)

The Thiele Modulus constant, Φ^2 represents the maximum reaction rate with the influence of mass transfer characteristic through the membrane. Figure 16.20 demonstrates the predicted concentration profile across the membrane during the DEKR process. The Φ^2 constant was manipulated between Φ^2 range of 5–60, with constant values of $B_o = 17.36$, racemization constant, $\gamma = 0.30$. In general, the concentration profiles in the membrane were highly dependent on the value of Φ^2. The increase of Φ^2 values at 5–30 results in steady increase of the dimensionless concentration profile. However, further increase in Thiele modulus ($\Phi^2 = 40$) shows less significance in increasing reaction. Further increase the Φ^2 to 60 shows low concentration at the module inlet but increase dramatically throughout the membrane module.

According to (16.22), the value of Φ^2 is proportionally correlated to the maximum reaction rate (V_{max}) but inverse proportional to the solute diffusivity (D_{eff}) and Michaelis constant (K_m). The effect of Φ^2 on the concentration profiles is mainly due to the V_{max} and K_m, it is assumed that the porosity (α_P) and the D_{eff} are constant for the same bio-catalytic membrane system. Thus, a larger value of Φ^2 is attributed by broader range of V_{max} and K_m values which indicates a better hydrolysis performance on the catalytic membrane. In the present study, the performance of enzymatic hydrolysis is increased between $5 < \Phi^2 < 30$. As a result, the dimensionless concentration on the radial coordinate of the membrane is increased. Theoretically, the concentration profiles in the membrane should decrease with the increase of enzymatic hydrolysis performance, since the substrate is consumed over time. However, an opposite behavior is observed during the DEKR process. For a rapid hydrolysis reaction should result in more excess (R)-substrate. However, the base-catalyzed racemization takes place in the bulk flow has isomerized the (R)-substrate into equi-molar substrate. At Φ^2 value between 5 and 30, the racemization is a prevailing process in bulk flow. As a result, the concentration profile of the reacting substrate is increased. The DEKR process is controlled by the hydrolysis reaction at $\Phi^2 > 30$, indicated by the lower concentration profiles as the value of Φ^2 is increased. Furthermore, at $\Phi^2 > 30$ the concentration build up at radial coordinate, $R = 10$ was at the maximum level due to reverse transfer of the substrate to the bulk solution. This behavior was attributed to the high mass transfer resistance in the membrane pore channel due to high feed concentration of the (S)-substrate. This study has highlighted the effect of the Φ^2 constant on the DEKR process. In practice, the performance of the hydrolysis reaction can be improved by enhancing the catalytic activity of the immobilized lipase on the porous matrix.

Effect of Racemiztion Constant (γ)

The racemization constant (γ) represents the rate of racemization with the influence of mass transfer parameters through the membrane. In general the rate of racemazation show slight increase as the concentration of the base is increased. This statement is valid with an assumption that the ionization of the base was not hindered by partial positive charge in the bulk flow. Figure 16.21 shows the effect of γ on the substrate concentration profiles within the observed boundary at radial

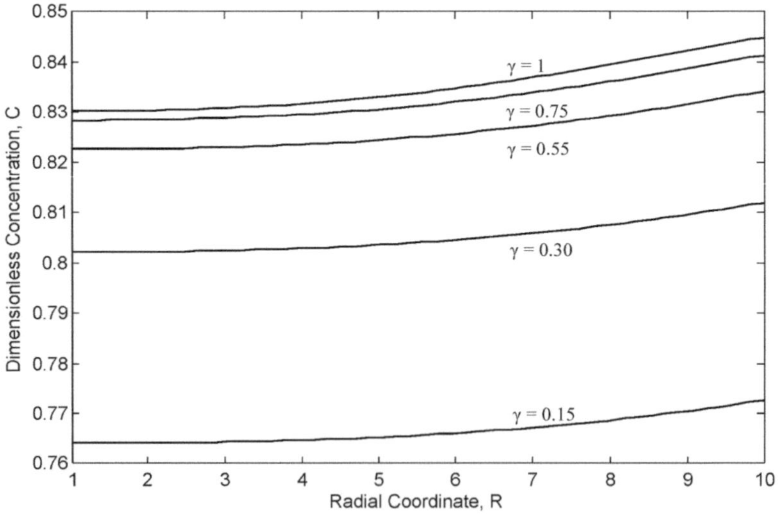

Fig. 16.21 Effect of dimensionless racemization constant, γ on the dimensionless concentration profiles at $B_0 = 17.34$; $\Phi^2 = 1$

direction of the bio-catalytic membrane. The racemization of the unreacted (*R*)-substrate provides continuous supply of (*S*)-substrate for hydrolysis reaction in the porous membrane matrix.

The value of γ is dependent on the inter-conversion rate of the racemization (k_{rac}) and the concentration of the base. In the present study, the value of γ is varied between 0.15 and 1 and the membrane characteristic parameters are assumed as constant values. It is found that, the substrate concentration profiles in the membrane are greatly dependent on the γ value. The concentration of substrate in the membrane increases with the increase of γ values. The increasing trend demonstrates that the racemization of unreacted substrate proceed much faster compared to the hydrolysis reaction. Consequently, gradual increase in (*S*)-substrate concentration profile is observed along the membrane. The increment of the concentration profile is gradually decreases as the value of γ is increased from 0.15 to 1. The racemization reaction will only proceed at the presence of an excess amount of (*R*)-substrate in the media. At larger values of γ ($\gamma > 0.75$), the racemization rate is rapid and results in smaller ee_S values. Consequently, the concentration of (*R*)- and (*S*)-substrates are close to equilibrium (equi-molar of enantiomers) and thus no significant rise in (*S*)-substrate concentration profile was observed when the γ value is increased from 0.75 to 1.

Effect of Bodenstein Number ($\mathbf{B_o}$)

The effect of volumetric flux at the shell side of the membrane on the concentration profile was studied by manipulating the Bodenstein number (B_o). The B_o constant represents feed flow rate at the sell side at constant membrane parameters (α_P and D_{eff}). According to (16.21), B_o is directly proportional to the feed flow rate. The effect of feed flow rate was studied at B_o values between 2 and 12 and the result is depicted in Fig. 16.22. It is clearly shows that, the flow rate at the shell side of the membrane contributes significant effect on the concentration profile in the membrane. The concentration profiles of the substrate increase gradually as the B_o values are increased. The concentration of the substrate in the membrane shows steady profile at $B_o > 8$. This is due to the high diffusion resistance at high substrate concentration in the membrane pore channel. On top of that, external factor such as bulk flow rate and trans-membrane pressure could give significant effect on the diffusive force of the substrate. For instant, higher feed flow rate will results in lower thickness of thin diffusion barrier at the surface of the membrane. Thus, the diffusive force of the substrate through the membrane is improved. As a result, the concentration profile of the substrate in the membrane is increased with increases value of B_o as shows in Fig. 16.22.

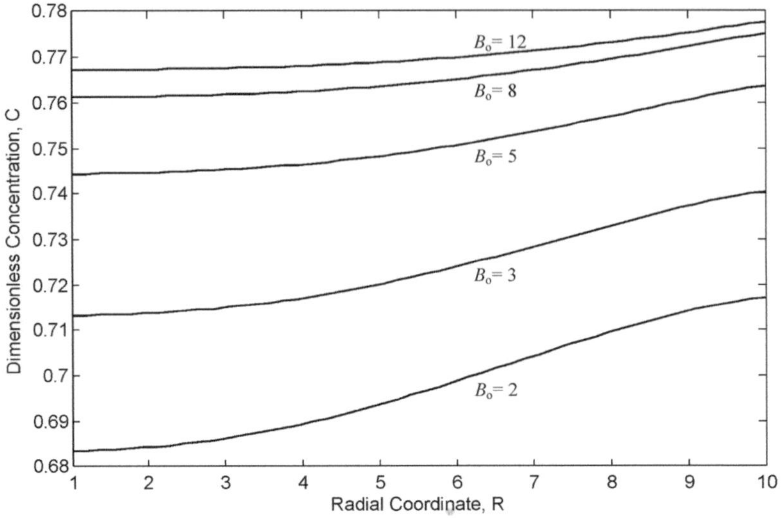

Fig. 16.22 Effect of dimensionless γ on enantiomeric excess at $B_0 = 17.34$; $\Phi^2 = 1$; $\gamma = 0.3$

Pilot Scale Production and Toxicological Analysis

A pilot scale process was developed based on the optimal results obtained from the bench scale process. The plant consists of three sections; (1) chemical esterification of (*R*,*S*)-ibuprofen acid; (2) enzymatic dynamic kinetic resolution of (*R*,*S*)-ibuprofen acid, and (3) crystallization and purification of (*S*)-ibuprofen acid. In this novel process, only five main units are crucially involved. Among the unit operations, the EMR is described as the "heart of the process" where the hydrolysis, by-product recovery as well as product separation carry out simultaneously. A general process flow of each units operation is presented in a block flow diagram (BFD) shown in Fig. 16.23.

The chemical esterification of the racemic ibuprofen acid was performed in batch reflux reactor. The reactant mixture consists of 2-ethoxyethanol, racemic ibuprofen acid, isooctane and P-toluene acid as catalyst. The ibuprofen ester was purified and collected in a holding tank for further resolution process. In membrane reactor, the optimal condition obtained in the bench scale studies were employed. After the resolution process completed, the (S)-ibuprofen acid was collected in the aqueous tank.

For the downstream processing, the product solution was handled carefully in order to preserve the quality of the (*S*)-Ibuprofen, which could degrade at elevated temperature. The product solution was concentrated in a vacuum evaporator below 0.07 bar which is capable to evaporate excess water at 40–42 °C. The crystallization of the concentrated (*S*)-ibuprofen acid was carried out at temperature below 5 °C. In order to promote the nucleation of (*S*)-ibuprofen crystal, the solubility of the (*S*)-ibuprofen was reduced by adding sulfuric acid (H_2SO_4). The crystallization unit was operated at 5 °C for 4 days.

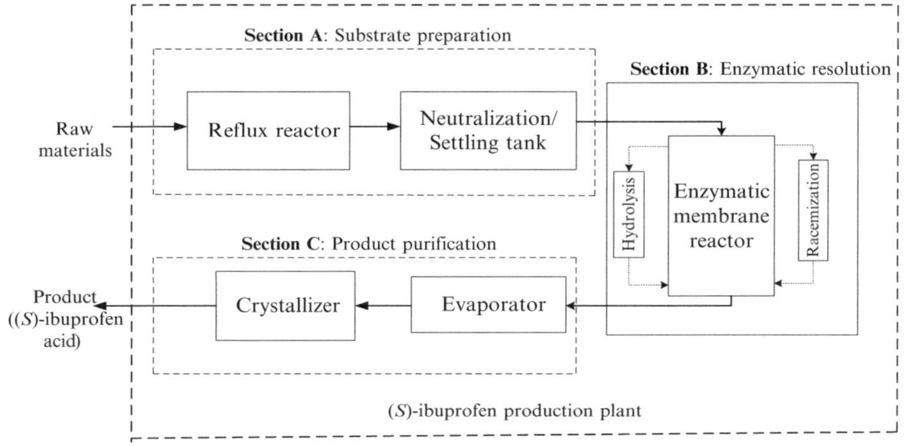

Fig. 16.23 Block flow diagram (BFD) of the (*S*)-ibuprofen production plant (Lau et al. 2011)

Table 16.4 The effect of acid on the recovery of (*S*)-ibuprofen crystal

Batch no.	H_2SO_4 (ml)	(*S*)-ibuprofen acid (g)	Recovery (%)
SIA-04-01	1.0	0.3	15
SIA-04-02	2.0	0.7	35
SIA-04-03	3.0	1.2	60
SIA-04-04	4.0	1.5	75
SIA-04-05	5.0	1.6	80

Investigation on the effect of H_2SO_4 on the recovery of (*S*)-ibuprofen acid from the solution was carried out and the result is tabulated in the Table 16.4. It has been found that, the yield of (*S*)-ibuprofen acid from the solution was higher with the increases amount of H_2SO_4. The addition of 5 ml of H_2SO_4 has provide more acidic environment compared to 1 ml of H_2SO_4, where the solubility of the (*S*)-ibuprofen acid is greatly reduced and large amount of (*S*)-ibuprofen acid crystal was able to form. As a result, 80 % of (*S*)-ibuprofen acid was recovered from a solution with the addition of 5 ml H_2SO_4.

The (*S*)-ibuprofen acid crystal obtained from the Pilot plant was further purified to remove possible residues and any other ionic salt. It is a common practice in pharmaceutical industry to re-crystallize the products in order to obtained highly pure product without the presence of any contaminants that might get carried out from the production line. In addition, re-crystallization is primarily important to achieved desirable crystal structure. This is due to the fact that, the structure or shape of (*S*)-ibuprofen could be changed through crystallization in organic solvents. The (*S*)-ibuprofen obtained was re-crystalized in acetone and acetonitrile. The resulted crystal of (*S*)-ibuprofen acid was analyzed for preliminary toxicological study.

Table 16.5 The cytotoxicity level of ibuprofen acid obtained from MTS Cytotoxicity Assay on cell culture originated from Chinese hamster lung

Samples	Solvent	IC_{50} (mM) (STDEV <5 %)
SIA-00	–	5.4
SIA-03	–	3.0
SIA-04	Acetone	3.5
SIA-05	Acetone	3.4
SIA-06	Acetonitrile	4.4

The preliminary toxicological study of (*S*)-ibuprofen was carried out in order to investigate the presence of toxic effect of (*S*)-ibuprofen that being produced from the (*S*)-ibuprofen pilot plant. In the presence investigation, MTS Cytotoxicity assay of several (*S*)-ibuprofen were tested on V79B Cell originated from Chinese hamster lung. The study has been performed in compliance with the Organization for Economic Co-operation and Development (OECD) Principles of Good Laboratory Practice (GLP). The study is conducted systematically according to the Study Plan and relevant Standard Operating Procedures. The results were tabulated in Table 16.5.

The samples SIA-00 represents standard sample of (*S*)-ibuprofen acid. However, the cytotoxicity effect of (*S*)-ibuprofen acid produced from the Pilot Plant was slightly over than the standard sample (SIA-00). The (S)-ibuprofen of Sample SIA-04 to SIA-06 were prepared with different solvents. It has been found that, the (*S*)-ibuprofen acid purified and crystallized from acetonitrile ($IC_{50} = 4.4$ mM) solution gives lower cytotoxicity effect rather than (*S*)-ibuprofen acid obtained in acetone ($IC_{50} = 3.4$ mM). Meanwhile, the (*S*)-ibuprofen acid that was not purified and directly obtained from the pilot plant results in higher cytotoxicity level with IC_{50} of 3.0 mM. This results has suggested that, the traceable amount of organic solvents and inorganic ionic compounds from the EMR as well as crystallizer units were still present in the sample thus affect the lethal toxicity of the (*S*)-ibuprofen acid produced. Nevertheless, the lethal dosage deviation of the prepared samples from the standard was not significant (IC_{50} 3.0–5.4) with average standard deviation of less than 5 %. In this particular study the samples were recrystallized twice and further crystallization cycle is proposed to increase the purity and quality of the final product. This result implied that, the production of optically pure (*S*)-ibuprofen acid by DEKR process in the EMR is a feasible approach and comparable to existing technologies.

Technical Challenges and Future Prospect

The increasing popularity of optically pure NSAIDs such as (*S*)-ibuprofen, has motivated the interest of developing a high efficiency, productive and environmental-friendly synthesis method. This could be possible by the introduction of the lipase-catalyzed membrane reactor, which applies the DEKR technology. This innovative chiral separation technology becomes more potent and viable owing to its excellent theoretical yield (nearly 100 %) when it is integrated with racemization process. The EMR has been widely used in the field of food processing, fine chemicals synthesis especially in the pharmaceutical area as well as for water treatment and purification (Rios et al. 2004). The main goal of this technology is to guarantee maximum enzyme selectivity and activity of the whole reaction system. As a result, the product quality is preserved and product inhibition is mitigated. Besides, the advantage of this integrated process with single-step reaction and simultaneous separation is the reduction of a step in the downstream process. However, there are still a few challenges that the EMR should be able to tackle. The enzyme activity which decreases with time has to be minimized by reducing the loss of catalyst and the shear stress effect. The membrane fouling and polarisation layer should be avoided in order to increase the EMR performance. Rios and co-workers have listed down a few types of commonly used membrane reactors for different applications (Rios et al. 2004).

To date, the EMR are still employed in bench-scale rather than systems with scaled-up applications. This signified that, the current progress in DEKR has not caught the interest of the producer. In fact, some technical aspects in the application of EMR in DEKR process remain unexplored. For instant, a change in characteristics of the membrane due to immobilization of biocatalyst cannot be rule out. There is no related studies carried out on the permeation flux, separation factor, and rejection of the solute of the bio-catalytic membrane. These parameters are crucial in the evaluation of the membrane performance that primarily work for separation purposes instead of employed as the catalytic layer. Furthermore, the membrane material that was employed for DEKR process was a commercial grade membrane designed to be used in a conventional separation processes. Hence, a new membrane material that is capable to bind the enzyme in a single step process is required. Ideally, the immobilization of the enzyme on the membrane surface is a reversible process, hence the membrane can be reused even after the deactivation of the immobilized enzyme. This approach will significantly improve the cost benefits of the EMR application. Nevertheless, the development of the chiral resolution technologies is still in progress and more ideas have to be explored.

Acknowledgements The authors would like to acknowledge Universiti Sains Malaysia for funding the current study through the Research University Grant Scheme (1001/PJKIMIA/814114). The financial support from the Malaysian Technology Development Corporation (MTDC) and research facilities provided by Universiti Sains Malaysia are also duly acknowledged.

References

Ansari, S. A., & Husain, Q. (2012). Potential applications of enzymes immobilized on/in nano materials: A review. *Biotechnology Advances, 30*(3), 512–523. doi:10.1016/j.biotechadv.2011.09.005.

Ariëns, E. J. (1991). Racemic therapeutics – Ethical and regulatory aspects. *European Journal of Clinical Pharmacology, 41*(2), 89–93. doi:10.1007/bf00265897.

Battistel, E., Bianchi, D., Cesti, P., & Pina, C. (1991). Enzymatic resolution of (S)-(+)-naproxen in a continuous reactor. *Biotechnology and Bioengineering, 38*(6), 659–664. doi:10.1002/bit.260380611.

Bhandarkar, S. V., & Neau, S. H. (2000). Lipase-catalyzed enantioselective esterification of flurbiprofen with n-butanol. *Electronic Journal of Biotechnology, 3*(3), 1–7. doi:10.2225/vol3-issue3-fulltext-3.

Brady, D., & Jordaan, J. (2009). Advances in enzyme immobilisation. *Biotechnology Letters, 31*(11), 1639–1650. doi:10.1007/s10529-009-0076-4.

Brady, D., Steenkamp, L., Skein, E., Chaplin, J. A., & Reddy, S. (2004). Optimisation of the enantioselective biocatalytic hydrolysis of naproxen ethyl ester using ChiroCLEC-CR. *Enzyme and Microbial Technology, 34*(3–4), 283–291. doi:10.1016/j.enzmictec.2003.11.002.

Carnell, A. (1999). Stereoinversions using microbial redox-reactions. In K. Faber (Ed.), *Biotransformations* (Advances in Biochemical Engineering/Biotechnology, Vol. 63, pp. 57–72). Berlin: Springer. 10.1007/3-540-69791-8_3.

Chang, C.-S., & Tsai, S.-W. (1997). A facile enzymatic process for the preparation of (s)-Naproxen ester prodrug in organic solvents. *Enzyme and Microbial Technology, 20*(8), 635–639. doi:10.1016/S0141-0229(96)00222-0.

Chang, C.-S., Tsai, S.-W., & Lin, C.-N. (1998). Enzymatic resolution of (RS)-2-arylpropionic acid thioesters by Candida rugosa lipase-catalyzed thiotransesterification or hydrolysis in organic solvents. *Tetrahedron: Asymmetry, 9*(16), 2799–2807. doi:10.1016/S0957-4166(98)00296-1.

Chen, C.-C., & Tsai, S.-W. (2005). Carica papaya lipase: A novel biocatalyst for the enantioselective hydrolysis of (R, S)-naproxen 2,2,2-trifluoroethyl ester. *Enzyme and Microbial Technology, 36*(1), 127–132. doi:10.1016/j.enzmictec.2004.07.004.

Contesini, F. J., Lopes, D. B., Macedo, G. A., Nascimento, M. G., & Carvalho, P. O. (2010). Aspergillus sp. lipase: Potential biocatalyst for industrial use. *Journal of Molecular Catalysis B: Enzymatic, 67*(3–4), 163–171. doi:10.1016/j.molcatb.2010.07.021.

Ducret, A., Trani, M., & Lortie, R. (1998). Lipase-catalyzed enantioselective esterification of ibuprofen in organic solvents under controlled water activity. *Enzyme and Microbial Technology, 22*(4), 212–216. doi:10.1016/S0141-0229(97)00180-4.

Ebbers, E. J., Ariaans, G. J. A., Houbiers, J. P. M., Bruggink, A., & Zwanenburg, B. (1997). Controlled racemization of optically active organic compounds: Prospects for asymmetric transformation. *Tetrahedron, 53*(28), 9417–9476. doi:10.1016/S0040-4020(97)00324-4.

Effenberger, F., Graef, B. W., & Oßwald, S. (1997). Preparation of (S)-naproxen by enantioselective hydrolysis of racemic naproxen amide with resting cells of Rhodococcus erythropolis MP50 in organic solvents. *Tetrahedron: Asymmetry, 8*(16), 2749–2755. doi:10.1016/S0957-4166(97)00335-2.

Fazlena, H. (2005). *Chemoenzymatic dynamic kinetic resolution of ethoxyethyl ibuprofen ester*. Penang: Universiti Sains malaysia.

Fazlena, H., Kamaruddin, A. H., & Zulkali, M. M. D. (2006). Dynamic kinetic resolution: alternative approach in optimizing S-ibuprofen production. *Bioprocess and Biosystems Engineering, 28*(4), 227–233. doi:10.1007/s00449-005-0024-1.

Fu, B., & Vasudevan, P. T. (2010). Effect of solvent – Co-solvent mixtures on lipase-catalyzed transesterification of canola oil. *Energy & Fuels, 24*(9), 4646–4651. doi:10.1021/ef901176h.

Ghanem, A., Aboul-Enein, M. N., El-Azzouny, A., & El-Behairy, M. F. (2010). Lipase-mediated enantioselective kinetic resolution of racemic acidic drugs in non-standard organic solvents: Direct chiral liquid chromatography monitoring and accurate determination of the enantiomeric excesses. *Journal of Chromatography A, 1217*(7), 1063–1074. doi:10.1016/j.chroma.2009.10.080.

Ghanem, A. (2007). Trends in lipase-catalyzed asymmetric access to enantiomerically pure/enriched compounds. *Tetrahedron, 63*(8), 1721–1754. doi:10.1016/j.tet.2006.09.110.

Gonawan, F. N., Sie Yon, L., Kamaruddin, A. H., & Uzir, M. H. (2013a). Rapid base-catalyzed racemization of (R)-ibuprofen ester in isooctane–dimethyl sulfoxide medium with improved kinetic model. *Industrial & Engineering Chemistry Research, 53*(2), 635–642. doi:10.1021/ie403070u.

Gonawan, F. N., Yon, L. S., Kamaruddin, A. H., & Uzir, M. H. (2013b). Effect of co-solvent addition on the reaction kinetics of the lipase-catalyzed resolution of ibuprofen ester. *Journal of Chemical Technology & Biotechnology, 88*(4), 672–679. doi:10.1002/jctb.3885.

Gyo Lee, E., Soon Won, H., & Hyun Chung, B. (2001). Enantioselective hydrolysis of racemic naproxen methyl ester by two-step acetone-treated Candida rugosa lipase. *Process Biochemistry, 37*(3), 293–298. doi:10.1016/S0032-9592(01)00213-8.

Hu, Y., Wang, Y., Luo, G., & Dai, Y. (2008). Modeling of a biphasic membrane reactor catalyzed by lipase immobilized in a hydrophilic/hydrophobic composite membrane. *Journal of Membrane Science, 308*(1–2), 242–249. doi:10.1016/j.memsci.2007.09.064.

Ikeda, Y., & Kurokawa, Y. (2002). Enantioselective esterification of racemic ibuprofen in isooctane by immobilized lipase on cellulose acetate-titanium iso-propoxide gel fiber. *Journal of Bioscience and Bioengineering, 93*(1), 98–100. doi:10.1016/S1389-1723(02)80062-7.

Ito, T., Ando, H., & Handa, H. (2011). Teratogenic effects of thalidomide: molecular mechanisms. *Cellular and Molecular Life Sciences, 68*(9), 1569–1579. doi:10.1007/s00018-010-0619-9.

Johnson, A., Zawadzka, A., Deobald, L., Crawford, R., & Paszczynski, A. (2008). Novel method for immobilization of enzymes to magnetic nanoparticles. *Journal of Nanoparticle Research, 10*(6), 1009–1025. doi:10.1007/s11051-007-9332-5.

Jin, J. N., Lee, S. H., & Lee, S. B. (2003). Enzymatic production of enantiopure ketoprofen in a solvent-free two-phase system. *Journal of Molecular Catalysis B: Enzymatic, 26*(3–6), 209–216. doi:10.1016/j.molcatb.2003.06.004.

Kin, M. G., & Lee, S. B. (1996). Enzymatic resolution of racemic ibuprofen by lipase-catalyzed esterification reaction: Effects of water content and solid supports. *Journal of Fermentation and Bioengineering, 81*(3), 269–271. doi:10.1016/0922-338X(96)82221-5.

Kim, M. G., Lee, E. G., & Chung, B. H. (2000). Improved enantioselectivity of Candida rugosa lipase towards ketoprofen ethyl ester by a simple two-step treatment. *Process Biochemistry, 35*(9), 977–982. doi:10.1016/S0032-9592(00)00129-1.

Knoche, B., & Blaschke, G. (1994). Investigations on the in vitro racemization of thalidomide by high-performance liquid chromatography. *Journal of Chromatography A, 666*(1–2), 235–240. doi:10.1016/0021-9673(94)80385-4.

Long, W. S., Bhatia, S., & Kamaruddin, A. (2003). Modeling and simulation of enzymatic membrane reactor for kinetic resolution of ibuprofen ester. *Journal of Membrane Science, 219*(1–2), 69–88. doi:10.1016/S0376-7388(03)00181-9.

Long, W. S., Kamaruddin, A., & Bhatia, S. (2005). Chiral resolution of racemic ibuprofen ester in an enzymatic membrane reactor. *Journal of Membrane Science, 247*(1–2), 185–200. doi:10.1016/j.memsci.2004.09.019.

Lau, S. Y., Gonawan, F. N., Bhatia, S., Kamaruddin, A. H., & Uzir, M. H. (2011). Conceptual design and simulation of a plant for the production of high purity (S)-ibuprofen acid using innovative enzymatic membrane technology. *Chemical Engineering Journal, 166*(2), 726–737. doi:10.1016/j.cej.2010.11.072.

Lau, S. Y. (2013). *Dynamic kinetic resolution of (R, S)-ibuprofen ester in an enzymatic hollow fiber membrane reactor system: Process design, modeling and simulation studies*. Penang: Universiti Sains Malaysia.

Lau, S. Y., Uzir, M. H., Kamaruddin, A. H., & Bhatia, S. (2010). Lipase-catalyzed dynamic kinetic resolution of racemic ibuprofen ester via hollow fiber membrane reactor: Modeling and simulation. *Journal of Membrane Science, 357*(1–2), 109–121. doi:10.1016/j.memsci.2010.04.008.

Lee, E. (2008). Epoxide hydrolase-mediated enantioconvergent bioconversions to prepare chiral epoxides and alcohols. *Biotechnology Letters, 30*(9), 1509–1514. doi:10.1007/s10529-008-9727-0.

Liu, Y.-Y., Xu, J.-H., Wu, H.-Y., & Shen, D. (2004). Integration of purification with immobilization of Candida rugosa lipase for kinetic resolution of racemic ketoprofen. *Journal of Biotechnology, 110*(2), 209–217. doi:10.1016/j.jbiotec.2004.02.008.

Litalien, C., & Jacqz-Aigrain, E. (2001). Risks and benefits of nonsteroidal anti-inflammatory drugs in children: A comparison with paracetamol. *Paediatric Drugs, 3*(11), 817–858. doi:10.2165/00128072-200103110-00004.

Madhav, M. V., & Ching, C. B. (2001). Study on the enzymatic hydrolysis of racemic methyl ibuprofen ester. *Journal of Chemical Technology & Biotechnology, 76*(9), 941–948. doi:10.1002/jctb.466.

Martín-Matute, B., & Bäckvall, J.-E. (2007). Dynamic kinetic resolution catalyzed by enzymes and metals. *Current Opinion in Chemical Biology, 11*(2), 226–232. doi:10.1016/j.cbpa.2007.01.724.

Mak, X. Y., Laurino, P., & Seeberger, P. H. (2009). Asymmetric reactions in continuous flow. *Beilstein Journal of Organic Chemistry, 5*, 19. doi:10.3762/bjoc.5.19.

Mehvar, R., & Brocks, D. R. (2001). Stereospecific pharmacokinetics and pharmacodynamics of beta-adrenergic blockers in humans. *Journal of Pharmacy and Pharmaceutical Sciences, 4*(2), 185–200.

Morrone, R., Nicolosi, G., Patti, A., & Piattelli, M. (1995). Resolution of racemic flurbiprofen by lipase-mediated esterification in organic solvent. *Tetrahedron: Asymmetry, 6*(7), 1773–1778. doi:10.1016/0957-4166(95)00223-C.

Muralidhar, R. V., Marchant, R., & Nigam, P. (2001). Lipases in racemic resolutions. *Journal of Chemical Technology and Biotechnology, 76*(1), 3–8. 10.1002/1097-4660(200101)76:1<3::AID-JCTB336>3.0.CO;2-8.

Mustranta, A. (1992). Use of lipases in the resolution of racemic ibuprofen. *Applied Microbiology and Biotechnology, 38*(1), 61–66. doi:10.1007/bf00169420.

Nagy, E., & Kulcsár, E. (2009). Mass transport through biocatalytic membrane reactor. *Desalination, 245*(1–3), 422–436. doi:10.1016/j.desal.2009.02.005.

Patel, R. N. (2002). Microbial/enzymatic synthesis of chiral intermediates for pharmaceuticals. *Enzyme and Microbial Technology, 31*(6), 804–826. doi:10.1016/S0141-0229(02)00186-2.

Patel, R. N. (Ed.). (2006). *Biocatalysis in the pharmaceutical and biotechnology industries. A biocatalysis for synthesis for chiral pharmaceutical intermediates* (1st ed.). New York, NY: CRC Press.

Perry, R. H., & Green, D. W. (1997). *Perry's chemical engineers' handbook. Heat and mass transfer* (7th ed.). New York, NY: Mc-Graw Hill.

Pollard, D. J., & Woodley, J. M. (2007). Biocatalysis for pharmaceutical intermediates: The future is now. *Trends in Biotechnology, 25*(2), 66–73. doi:10.1016/j.tibtech.2006.12.005.

Rios, G. M., Belleville, M. P., Paolucci, D., & Sanchez, J. (2004). Progress in enzymatic membrane reactors – A review. *Journal of Membrane Science, 242*(1–2), 189–196. doi:10.1016/j.memsci.2003.06.004.

Rouhi, A. M. (2003). Chiral business. *Chemical and Engineering News Archive, 81*(18), 45–61. doi:10.1021/cen-v081n018.p045.

Schulze, B., & Wubbolts, M. G. (1999). Biocatalysis for industrial production of fine chemicals. *Current Opinion in Biotechnology, 10*(6), 609–615. doi:10.1016/S0958-1669(99)00042-7.

Shah, R. R., & Branch, S. K. (2003). Regulatory Requirements for the Development of Chirally Active Drugs. In M. Eichelbaum, B. Testa, & A. Somogyi (Eds.), *Stereochemical aspects of drug action and disposition. Handbook of experimental pharmacology* (Vol. 153, pp. 379–399). Berlin: Springer. doi:10.1007/978-3-642-55842-9_16.

Sheldon, R. A. (1996). Chirotechnology: Designing economic chiral syntheses. *Journal of Chemical Technology & Biotechnology, 67*(1), 1–14. doi:10.1002/(sici)1097-4660(199609)67:1<1::aid-jctb531>3.0.co;2-l.

Sheldon, R. A. (2007). Enzyme immobilization: The quest for optimum performance. *Advanced Synthesis & Catalysis, 349*(8-9), 1289–1307. doi:10.1002/adsc.200700082.

Sie Yon, L., Gonawan, F. N., Kamaruddin, A. H., & Uzir, M. H. (2013). Enzymatic deracemization of (R, S)-ibuprofen ester via lipase-catalyzed membrane reactor. *Industrial & Engineering Chemistry Research, 52*(27), 9441–9453. doi:10.1021/ie400795j.

Steinreiber, J., Faber, K., & Griengl, H. (2008). De-racemization of enantiomers versus de-epimerization of diastereomers—Classification of dynamic kinetic asymmetric transformations (DYKAT). *Chemistry – A European Journal, 14*(27), 8060–8072. doi:10.1002/chem.200701643.

Stoschitzky, K., & Lindner, W. (1990). Lindner W (1990) [Specific and nonspecific effects of beta receptor blockers: stereoselectively different properties exemplified by (R)- and (S)-propranolol]. *Wiener Medizinische Wochenschrift, 140*(6-7), 156–162.

Tsai, S. W., & Huang, C. M. (1999). Enantioselective synthesis of (S)-suprofen ester prodrugs by lipase in cyclohexane. *Enzyme and Microbial Technology, 25*(8–9), 682–688. doi:10.1016/S0141-0229(99)00108-8.

Wang, L. W., Cheng, Y. C., & Tsai, S. W. (2004). Process modeling of the lipase-catalyzed dynamic kinetic resolution of (R, S)-suprofen 2,2,2-trifluoroethyl thioester in a hollow-fiber membrane. *Bioprocess and Biosystems Engineering, 27*(1), 39–49. doi:10.1007/s00449-004-0379-8.

Wang, S.-Z., Wu, J.-P., Xu, G., & Yang, L.-R. (2013). Chemo-enzymatic asymmetric synthesis of S-citalopram by lipase-catalyzed cyclic resolution and stereoinversion of quaternary stereogenic center. *Bioprocess and Biosystems Engineering, 36*(8), 1031–1037. doi:10.1007/s00449-012-0855-5.

Ward, T. J., & Ward, K. D. (2011). Chiral separations: A review of current topics and trends. *Analytical Chemistry, 84*(2), 626–635. doi:10.1021/ac202892w.

Wiberg, E., & Wiberg, N. (Eds.). (2001). *Inorganic chemistry. Stereochemistry of molecules* (35th ed.). Academic Press: Florida.

Xin, J.-Y., Li, S.-B., Chen, X.-H., Wang, L.-L., & Xu, Y. (2000). Improvement of the enantioselectivity of lipase-catalyzed naproxen ester hydrolysis in organic solvent. *Enzyme and Microbial Technology, 26*(2–4), 137–141. doi:10.1016/S0141-0229(99)00147-7.

Xin, J.-Y., Li S-b, S., Xu, Y., & Chui, C.-g. (2001). Dynamic enzymatic resolution of Naproxen methyl ester in a membrane bioreactor. *Journal of Chemical Technology & Biotechnology, 76*(6), 579–585. doi:10.1002/jctb.413.

Yuchun, X., Huizhou, L., & Jiayong, C. (2000). Kinetics of base catalyzed racemization of ibuprofen enantiomers. *International Journal of Pharmaceutics, 196*(1), 21–26. doi:10.1016/S0378-5173(99)00438-X.

Chapter 17
Catgut Waste Utilization for Protease Production Using *Bacillus subtilis*

Jerusha Kezia Clement, Raja Mohan, and Jegannathan Kenthorai Raman

Introduction

The field of industrial enzymes is now experiencing major R&D initiative, resulting both in the development of a number of new products and in improvement in the process and performance of several existing products. Currently, new and emerging applications are driving demand and the industry is responding with a continuous stream of innovative products. Proteases occupy a pivotal position with respect to their applications in both physiological and commercial fields. Today proteases account for approximately 60 % of the total enzyme sales in various industrial market sectors in India and 30 % worldwide. World enzyme demand will rise 6.8 % per year to $8.0 billion in 2015 (World Enzymes 2011).

Enzymes are obtained from plants, animals or microbes. The inability of the plant and animal proteases to meet current world demands has led to an increased interest in microbial proteases. Microbial proteases account for approximately 40 % of the total worldwide enzyme sales (Pandey et al. 2006). The choice of a particular source depends on the efficiency, degree of control, steady production, composition of enzyme and degree of purity of the preparation. The isolation of enzymes from plant cells or animal sources involves a high cost factor and low yield as compared to that from microbial sources.

Several microbial strains including fungi (*Aspergillus flavus*, *Aspergillus melleu*, *Aspergillus niger*, *Chrysosporium keratinophilum*, *Fusarium graminarum*, *Penicillium griseofulvin*, *Scedosporium apiosermum*) and bacteria (*Bacillus licheniformis*, *Bacillus firmus*, *Bacillus alcalophilus*, *Bacillus amyloliquefaciens*, *Bacillus*

J.K. Clement • R. Mohan
School of Biotechnology and Health sciences, Karunya University, Karunya University, Coimbatore 641114, India

J.K. Raman (✉)
BPE, EPFL, GC A3 444, Station 18, EPFL, Lausanne 1015, Switzerland
e-mail: jeganbio@gmail.com

P. Ravindra (ed.), *Advances in Bioprocess Technology*,
DOI 10.1007/978-3-319-17915-5_17

proteolyticus, *Bacillus subtilis*, *Bacillus thuringiensis*) (Ellaiah et al. 2001) are reported to produce proteases. Of these, protease produced by *Bacillus* sp. is most widely used and studied (Priest 1977).

The use of cheap sources of carbon and nitrogen like wheat bran, rice bran, marine products waste, gelatin, soybean, ossein waste, agro based products are important as they can significantly reduce the cost of valuable products like enzymes and proteins (Anupama and Ravindra 2000; Naidu and Devi 2005; Wen-Teish et al. 2003; El-Safey and Abdul-Raouf 2004; Shaheen and Shah 2008; Jegannathan and Viruthagiri 2009; Boominadhan and Rajakumar 2009). At present, the overall cost of enzyme production is very high (due to high cost of substrates and mediums used) and therefore, development of novel processes to increase the yield of proteases with respect to their industrial requirements coupled with lowering down the production cost is highly appreciable from the commercial point of view (Mukherjee et al. 2008).

Sutures can be divided into two types: Absorbable surgical suture and non absorbable surgical suture. Absorbable sutures are made of materials which are broken down in tissue after a given period of time. They are used therefore in many of the internal tissues of the body and no need for the patient to have the sutures removed. Non-absorbable must be manually removed and are used in the external skin. Catgut is the natural absorbable surgical suture made from collagen that is derived from the submucosal layer of small intestine of sheep or the serosal layer of small intestine of cattle (Miller et al. 1964). Collagen is a group of naturally occurring proteins. In nature, it is found exclusively in animals, especially in the flesh and connective tissues of mammals. It is the main component of connective tissue, and is the most abundant protein in mammals, making up about 25–35 % of the whole-body protein content (Neuberger 1955).

Various products and effluents have been used as a nutrient source for the production of protease enzyme. Widely used substrates are gelatin (El-Safey and Abdul-Raouf 2004), soybean (Shaheen and Shah 2008), agro based products like soy cake, groundnut cake, coconut cake, wheat bran, brewery spent grain (Boominadhan and Rajakumar 2009), glucose (Ul Qadar et al. 2009), starch (Almeida do Nascimento and Martins 2004), lactose (Abdelnasser et al. 2009), soybean meal (Joo et al. 2002), rice bran (Naidu and Devi 2005), dextrose (Das and Prasad 2010), green gram husk (Prakasham et al. 2006), yeast extract, glucose and asparagines (Asokan and Jayanthi 2010) and ossein waste (Jegannathan and Viruthagiri 2009).

Catguts that do not meet the physical properties like consistent diameter, sufficient initial strength and firm needle (Kristensen 2005), unused catguts in hospitals, clinics and pharmacies that gets expired and catgut strings which are interrupted by fat sequences are treated as biomedical waste and are disposed by incineration, steam sterilization or chemical decomposition. Effluent obtained from catgut manufacturing industry is recycled to produce manure and fat (Verheijen et al. 1996). Thus far, there is no research done on the catgut waste. Therefore the aim of this study is to utilize catgut waste rich in carbon and nitrogen sources as a novel substrate for protease production and optimization of the process conditions.

Materials and Methods

Materials

Chemicals ($MgSO_4.7H_2O$, KH_2PO_4, $FeSO_4{\cdot}7H_2O$, casein, phosphate buffer, trichloro acetic acid, sodium carbonate, glucose, BSA solution, Anthrone, sulphuric acid, Na_2CO_3, NaOH, NaK Tartrate, $CuSO_4$, Folin–Ciocalteau's reagent, tyrosine) used for the production and quantification of protease enzyme were purchased from Himedia Laboratries Pvt. Ltd., Mumbai, India. *Bacillus subtilis* B 456 was obtained from ARS, New York. Expired catgut used as catgut waste was obtained from various hospitals and pharmacies.

Methods

The catgut was soaked in water overnight and dried using a hot air oven for an hour to make the catgut brittle. It was then ground in a grinder to form fine powder. Uniform size of powder was obtained by sieving. For the organism to use catgut waste as a substrate, it is necessary to analyze the catgut waste for the occurrence of carbohydrate and protein which can be used as a total nutrient.

Carbohydrate Estimation

Various concentrations of glucose were prepared to a final volume of 1 ml with water. Test sample was prepared by digesting 100 mg of catgut waste in 1 ml of 10 mM phosphate buffer. 4 ml of Anthrone solution (200 mg anthrone in 100 ml of ice cold 95 % H_2SO_4) was added in all test tubes. The tubes were heated for 8 min in a boiling water bath and rapidly cooled. The green to dark green colour was read at 630 nm (Hedge and Hofreiter 1962).

Protein Estimation

Various concentrations of BSA solution were prepared to a final volume of 1 ml with water. Test sample was prepared by digesting 100 mg of catgut waste in 1 ml of 10 mM phosphate buffer. 5 ml of copper solution was added, mixed and allowed to stand for 10 min. Later 0.5 ml of Folin–Ciocalteau's reagent was added and again incubated at room temperature for 30 min in dark. The absorbance was then measured using a spectrophotometer at 660 nm (Lowry et al. 1951).

Inoculum Preparation

Nutrient broth was prepared and was sterilized in an autoclave at 121 °C for 15 min. The sterilized broth was cooled and the lyophilized form of *Bacillus subtilis* was inoculated. It was then incubated at 30 °C for 24 h at 200 rpm in incubator shaker.

Production of Protease

The production mixture was prepared using catgut waste powder, 0.5 % $MgSO_4{\cdot}7H_2O$, 0.5 % KH_2PO_4 and 0.01 % $FeSO_4{\cdot}7H_2O$. The production mixture was autoclaved at 121 °C for 15 min and then cooled to room temperature. Inoculum was added and the production medium was incubated in an incubator shaker. Process parameters were maintained appropriately to the optimization design. Medium was then centrifuged at 10,000 rpm at 4 °C for 15 min and the clear supernatant crude enzyme was analyzed for its protease activity using enzyme assay.

Test for Protease

The protease enzyme is said to liberate tyrosine when it digests casein. Standardization of tyrosine method is used to determine the unknown concentration of tyrosine liberated from protease. Various concentrations of tyrosine were prepared and 0.2 ml of it was taken in 15 ml centrifuge tubes. Add 2 ml of Lowry's reagent to each vial and incubate at room temperature for 10 min. Later 0.2 ml of Folin–Coalteau's reagent was added and again incubated at room temperature for 30 min in dark. The absorbance was then measured using a spectrophotometer at 660 nm (Lowry et al. 1951). 0.5 ml of the enzyme solution was added to 2.5 ml of 0.65 % casein solution and in 50 mM phosphate buffer. The mixture was incubated for 30 min at 37 °C and the reaction was stopped with equal volume of 110 mM trichloro acetic acid (TCA) solution. After 10 min, the mixture was centrifuged for 10 min at 5,000 rpm. To 0.5 ml of the filtrate obtained, 5 ml of 0.5 M sodium carbonate and 1 ml of threefold diluted Folin's reagent was added, and stored at 30 °C for 30 min. Later the colour formed was measured at 660 nm using spectrophotometer (Folin and Ciocalteau 1927; Anson 1938).

Medium Optimization

Medium optimization is necessary to determine the set of optimal conditions to give high productivity of the product. Various process parameters influencing protease production during submerged fermentation were optimized. Plackett Burman

Table 17.1 Experimental variables of Plackett-Burman design

Order	Variables	High (+)	Low (−)
A	Substrate concentration (w/v%)	1	5
B	Inoculum level (v/v%)	2	6
C	Inoculum age (h)	6	30
D	pH	5	9
E	Temperature (°C)	30	50
F	Duration (h)	24	120
G	Mixing intensity (rpm)	100	180

design was used to screen the most important variables in the medium and Response Surface design was used to determine optimum level of each key independent variable (Stanbury et al. 1999).

Plackett Burman design: The first step of optimization is to identify which ingredients of the medium have a significant effect on protease production. Plackett Burman design does not consider the interaction effects between variables but to screen the important variable affecting the enzyme production (Plackett and Burman 1946). This technique allows the evaluation of X − 1 variables where X is the number of experiments to be carried out. X must be a multiple of 4. Any factor not assigned to a variable can be designated as dummy variable which is helpful in estimating the variance (experimental error) of an effect. The variables to be optimized at their levels are given in Table 17.1. Each variable is represented at two levels, a high level denoted by "+1" and a low level denoted as "−1". In the variables (A–G), E and G are designated as dummy variables. The high level of each variable is far enough from the low level so that a significant effect, if exists, is likely to be detected.

Response surface design: The next stage of optimization is to determine the optimum level of each key independent variable which has been identified by the Plackett Burman design in the form of a polynomial model. Using Response Surface Methodology (RSM) the relationship among the variables, i.e. substrate concentration, inoculum level, duration were expressed mathematically (Box and Wilson 1951). Design Expert Version 8.0.4 is the software used for the design of experiments, regression analysis and graphical analysis of data obtained.

The most popular response surface method (RSM) design is the central composite design (CCD). A CCD has three groups of design points viz. two-level factorial or fractional factorial design points, axial points (sometimes called "star" points) and center points. CCD's are designed to estimate the coefficients of a quadratic model. All point descriptions will be in terms of coded values of the factors. Each factor in the CCD was studied at five different levels (−1.68, −1, 0, +1, +1.68). A central composite factorial design with three factors leading to a total of 20 sets per experiment was formulated to optimize the Substrate concentration, Inoculum level, Duration. The minimum and maximum ranges of variables were investigated and the full experimental plan with respect to their values in actual and coded form is listed in Table 17.2.

Table 17.2 Ranges of variables in RSM

	Experimental factor		
Coded unit	A: Substrate concentration (w/v%)	B: Inoculum level (v/v%)	C: Duration (h)
−1.68	1	2	24
−1	2	3	48
0	3	4	72
1	4	5	96
1.68	5	6	120

Classical method: Classical method of medium optimization is performed by changing one independent variable while fixing all the others at a certain level. Classical method done after RSM was used to obtain confirmatory results of optimization.

Effect of substrate concentration: Catgut waste was taken in different concentrations ranging from 1 to 5 % (w/v). Other parameters were maintained to be constant with respect to the results of RSM.

Effect of inoculum level: Amount of microorganism used has a large impact on the protease production. Level of inoculum is an important biological factor which determines the biomass production in fermentation. Inoculum level was used at various levels ranging from 2 to 6 % (v/v).

Effect of incubation period: To study the effect of incubation time on protease production, the submerged fermentation was carried out for different duration (24, 48, 72, 96 and 120 h) and the enzyme produced was assayed.

Results and Discussion

Analysis of Catgut Waste

The analysis result shows the catgut waste is rich in carbhohydrate (6.1 %) and protein (23.9 %) source that could influence the production of protease. The industry spends a lot of capital to treat catgut waste. Thus, utilizing this waste as a rich nutrient content saves the treatment cost and also provides additional income to the industry.

Protease Assay

To determine the protease produced by *Bacillus subtilis* using catgut waste as its substrate, the absorbance of various protease samples after enzyme assay was read using spectrophotometer at 660 nm. With the help of the standard curve, the concentration of tyrosine corresponding to the absorbance of the protease sample can be determined. One unit (U) of protease is the amount of enzyme that releases

1 μmol tyrosine/min under the reaction conditions. Hence the protease produced in terms of Unit can be calculated using the following formula:

$$\text{U/ml enzyme} = \frac{\text{tyrosine equivalents }(\mu\text{g})\text{ released} \times \text{total volume (ml) of assay}}{\text{time of assay (min)} \times \text{volume of enzyme (ml)} \times \text{volume (ml) determined}}$$

Plackett Burman Design

Plackett Burman methodology was performed for seven variables. The results were calculated manually to determine the most important variables. The advantage of using Plackett Burman plans lies in the clean and easy selection of variables in determining the important variables. Among the seven factors (Table 17.3), A is the medium constituent, B and C are inoculum constituents, D and F are process parameters and E and G are dummy variables.

The analysis of data was done and the F test values for the variables were, 484.18, 1.24, 0.06, 0.02, 0.69, 39.7 and 1.31. The factors A, B and F show large effects which are significant, whereas C and D show a very low effect which are not significant. Factors E and G were not actually optimized since they were assumed to be dummy variables and hence they are not considered to be significant though they have large effects. Therefore, factors A, B and F i.e. substrate concentration, Inoculum level and duration have been identified as the most important factors.

Response Surface Design

The protease activity of three most important variables obtained from Plackett Burman design was run for 20 runs in a single block (Table 17.4). The 13th run is found to have the greatest enzyme activity. Hence the corresponding optimum condition of the process to be maintained are substrate concentration—4 %, inoculum level—5 % and 96 h of incubation period. The model opted is found to be

Table 17.3 Enzyme yield results of Plackett Burman design

	Variables							
Trials	A	B	C	D	E	F	G	Enzyme yield (U/ml)
1	+	+	+	–	+	–	–	40
2	–	+	+	+	–	+	–	18.1
3	–	–	+	+	+	–	+	13.6
4	+	–	–	+	+	+	–	53.6
5	–	+	–	–	+	+	+	18
6	+	–	+	–	–	+	+	56.6
7	+	+	–	+	–	–	+	42.8
8	–	–	–	–	–	–	–	12.5

Table 17.4 Enzyme yield results of Response Surface Design

Run order	Blocks	A	B	C	Actual enzyme yield (U/ml)	Predicted enzyme yield (U/ml)	Residual value
1	1	0	0	−1.68	94.88	95.02	−0.14
2	1	−1.68	0	0	98.56	98.91	−0.35
3	1	0	−1.68	0	86.88	90.56	−3.68
4	1	0	0	0	89.76	90.24	−0.48
5	1	−1	1	−1	94.56	94.46	0.1
6	1	0	1.68	0	94.88	98.34	−3.46
7	1	−1	1	1	89.12	89.32	−0.2
8	1	0	0	0	90.08	90.24	−0.16
9	1	−1	−1	1	96	95.71	0.29
10	1	0	0	0	90.4	90.24	0.16
11	1	0	0	1.68	96.48	96.84	−0.36
12	1	1	−1	−1	99.84	96.94	2.9
13	1	1	1	1	111.36	111.25	0.11
14	1	0	0	0	90.56	90.24	0.32
15	1	1.68	0	0	111.2	114.67	−3.47
16	1	1	−1	1	102.88	100.28	2.6
17	1	0	0	0	91.04	90.24	0.8
18	1	−1	−1	−1	102.72	100.13	2.59
19	1	1	1	−1	111.04	108.63	2.41
20	1	0	0	0	90.24	90.24	0.0024

Table 17.5 Degree of freedom for evaluation

Terms	Degree of freedom
Model	9
Residuals	10
Lack of fit	5
Pure error	5
Corrected total	19

Response Surface Quadratic Model. Degree of freedom (df) for the model (Table 17.5) is the number of model terms, including the intercept, minus one. A recommendation is a minimum of 3 lack of fit df and 4 df for pure error. This ensures a valid lack of fit test. Fewer df will lead to a test that may not detect lack of fit. The Model F-value of 17.67 implies the model is significant. There is only a 0.01 % chance that a "Model F-Value" this large could occur due to noise. Value of P less than 0.0500 indicate model terms are significant. In this case A, AB, A^2, C^2 are significant model terms (Table 17.6). Values greater than 0.1 indicate the model terms are not significant. If there are many insignificant model terms (not counting those required to support hierarchy), model reduction may improve your model. The "Lack of Fit F-value" of 68.79 implies the Lack of Fit is significant. There is only a 0.01 % chance that a "Lack of Fit F-value" this large could occur due to noise.

Table 17.6 ANOVA for RSM

Source	Sum of squares	Degree of freedom	Mean square	F value	P value
Model	1,060.51	9	117.83	17.67	<0.0001
A	299.72	1	299.72	44.94	<0.0001
B	23.97	1	23.97	3.59	0.0872
C	2.73	1	2.73	0.41	0.5365
AB	150.68	1	150.68	22.59	0.0008
AC	30.11	1	30.11	4.51	0.0596
BC	0.26	1	0.26	0.039	0.8477
A^2	493.49	1	493.49	73.99	<0.0001
B^2	11.73	1	11.73	1.76	0.2144
C^2	97.35	1	97.35	14.6	0.0034
Residual	66.7	10	6.67		
Lack of fit	65.74	5	13.15	68.79	0.0001
Pure error	0.96	5	0.19		
Corrected total	1,127.21	19			

Table 17.7 Standard deviation and correlation coefficients

Terms	Values
Std. dev.	2.58
Mean	96.62
C.v%	2.67
Press	502.14
R-squared	0.94
Adjusted r-squared	0.89
Predicted r-squared	0.55
Adequate precision	13.88

R-squared is the correlation coefficient for the model. For the data, 94.08 % of the variation in activity is explained by the model. Since, $R^2 > 90$ % (goodness of fit), the system is fit. Adjusted R-squared represents the amount of variation that can be explained by the model. Adequate precision measures the signal to noise ratio. A ratio greater than 4 is desirable and ratio of 13.878 indicates an adequate signal (Table 17.7). This model can be used to navigate the design space.

Final Equation in Terms of Coded Factors:

The application of RSM yielded the following regression equation which is an empirical relationship between the activity and variables.

Activity = 90.24 + 4.68 (Substrate concentration) + 1.32 (Inoculum level) − 0.45 (Duration) + 4.34 (Substrate concentration) (Inoculum level) + 1.94 (Substrate concentration) (Duration) − 0.18 (Inoculum level) (Duration) + 5.85 (Substrate concentration) (Substrate concentration) + 0.90 (Inoculum level) (Inoculum level) + 2.60 (Duration) (Duration)

The unusual observations were investigated to determine if the data were recorded correctly and whether or not the data collection process was affected by any other factors. Few data have a standardized residual greater or lesser than 2. Therefore, the response surface model does sufficiently account for the observed level of the response variable. Response surface plots as a function of two factors at a time, maintaining all other factors at a fixed level (zero, for instance), are more helpful in understanding the effect of three factors. The contour plot shows how a response variable relates to two factors based on a model equation. The surface plot which is a three dimensional wireframe graph represents the functional relationship between the response and the experimental factors (Fig. 17.1). The response surface helps to visualize how the response reacts to the changes in the experimental factors. Response Surface methodology not only yielded the optimum values of each factor but also the interaction that occurs between each factor. Few residual values in the Response Surface Methodology seem to be large that also causes variation in the maximum production of protease. This may be due to the experimentation error. Thus, a multifactorial statistical approach that considers the interaction of independent variables provides a basis for the model to search for the nonlinear nature of the response in a short-term experiment.

Medium Optimization

The traditional one-factor-at-a-time technique used for optimizing a variable system is not only time consuming but also often easily confuses the alternative effect between the components and also it requires a larger number of experiments to determine the optimum levels. The drawbacks were eliminated by optimizing all the effecting parameters collectively by response surface methodology (RSM) which includes factorial design and regression analysis. Thus this method was used to confirm the obtained results.

Effect of substrate concentration: Substrate concentration in the production media is an important parameter for microbial growth. Low substrate concentration may not be sufficient for microbial growth whereas, higher substrate concentration may inhibit microbial growth thereby decreasing the production of the product. In this study, protease was influenced by the concentration of the substrate. The maximum production was observed at a substrate concentration of 4 w/v% for 5 v/v% inoculum concentration, 18 h inoculum age, pH 7, 140 rpm, 40 °C and 96 h incubation period (Fig. 17.2).

Effect of inoculum level: The experimental result (Fig. 17.3) shows that there is a significant increase in protease production with an increase in inoculum level up to an optimum level and then the yield is reduced. The decrease seen in the protease production with large amount of inoculum level could be due to the shortage of nutrients available for the faster growth of the culture (Hesseltine et al. 1976). Hence an optimum level should be maintained to obtain maximum yield. Here, the

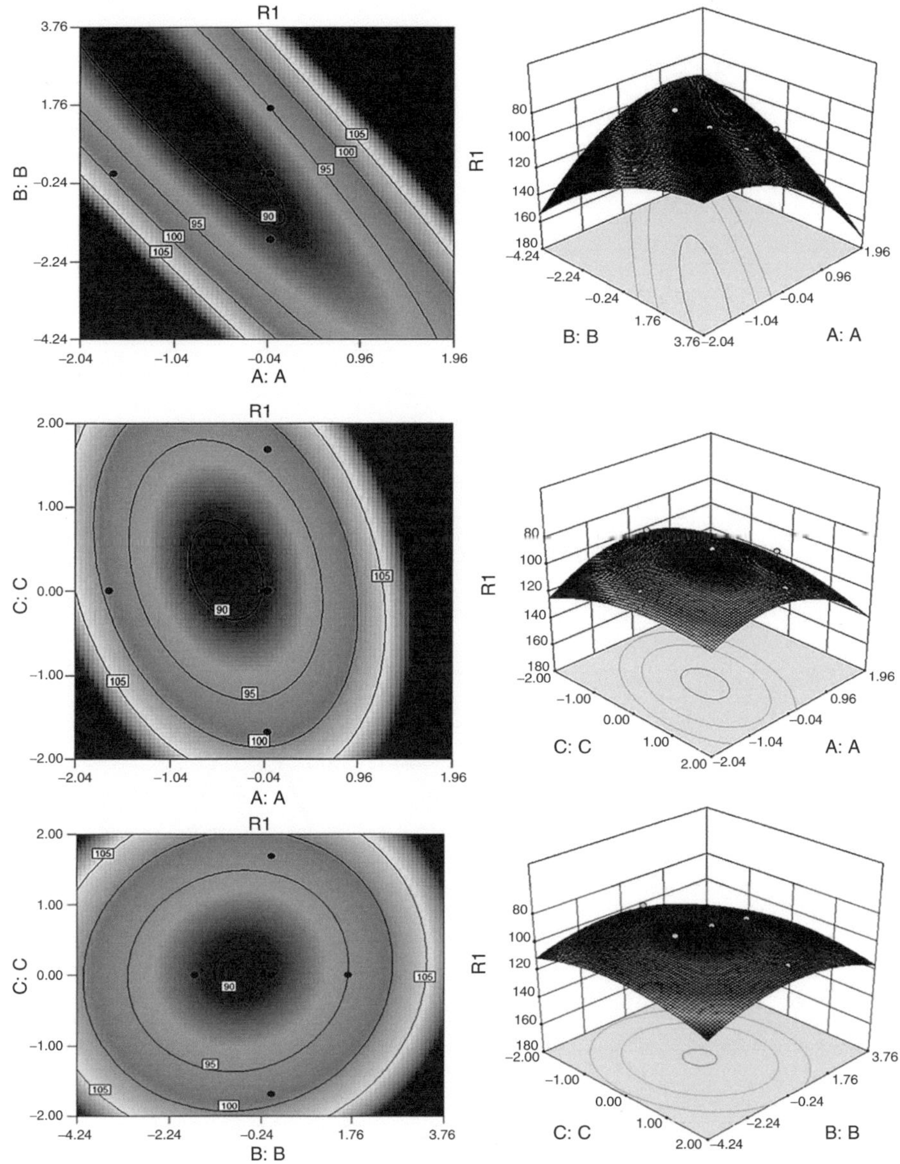

Fig. 17.1 Contour plot and Surface plot showing the interaction between A, B and C

maximum production was obtained at 5 v/v% for 4 w/v% substrate concentration, 18 h inoculum age, pH 7, 140 rpm, 40 °C and 96 h incubation period.

Effect of incubation period: Enzyme production from microorganism is directly correlated to the time period of incubation (Smitt et al. 1996). In the present study, we observed that 96 h of incubation as the optimum growth duration for the

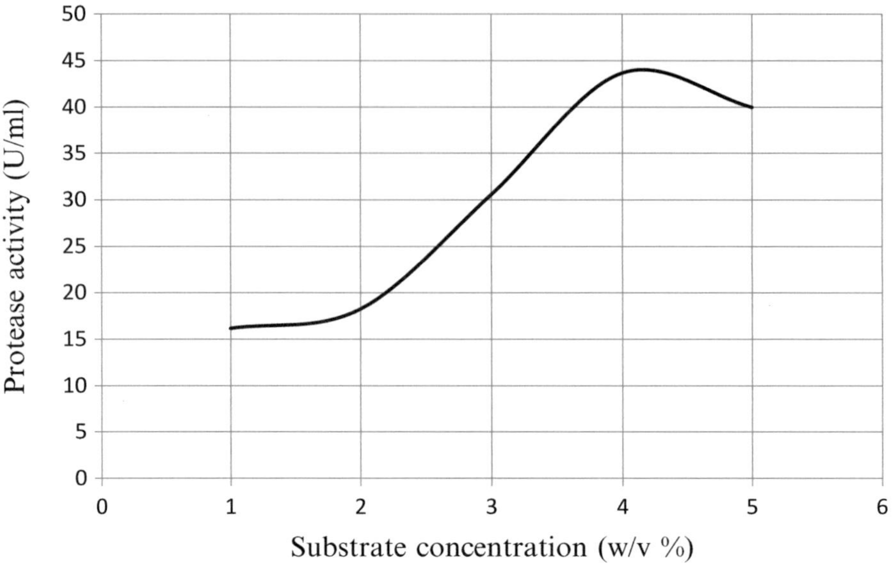

Fig. 17.2 Effect of substrate concentration on protease production

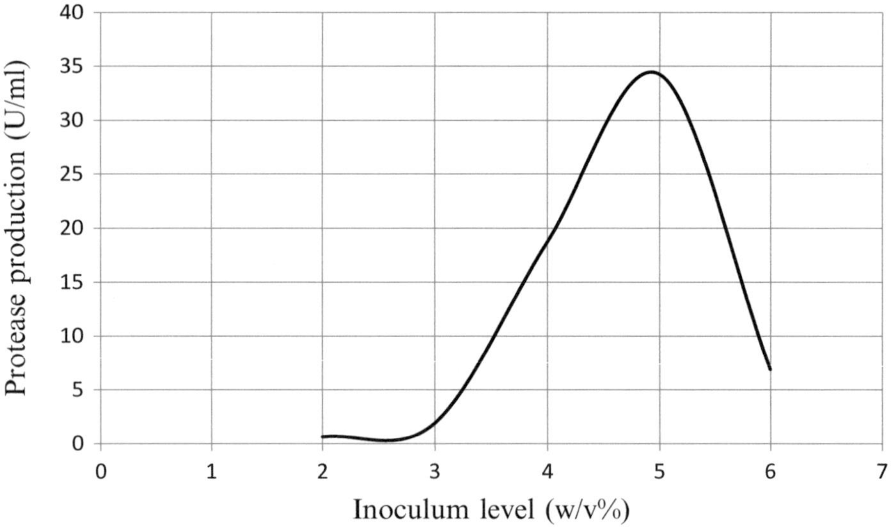

Fig. 17.3 Effect of inoculum level on protease production

presently reported *Bacillus* strain. And longer duration neither supported enzyme production nor maintained the enzyme produced. The protease produced in the sample is found to increase and then decrease as the duration of incubation increases. The steady increase of protease production is due to the utilization of the substrate to produce enzyme. Whereas the decrease in the protease production is due to the reason that when there is a limitation is the substrate, the microbes start utilizing the already produced enzyme in the vessel because of its proteinaceous nature. Thus it leads to the depletion in the produced enzyme. Hence the produced enzyme should be removed for the reactor when the production of maximum protease is reached. Thus optimization of parameters is an important aspect to be considered in the development of fermentation technology.

Conclusion

The effect of catgut waste as an alternative substrate on bacterial protease production under submerged production was studied. Catgut waste rich in protein could be a low cost substrate for protease production. The process optimization studies showed that a maximum amount of protease 111.36 U/ml could be produced at substrate concentration 4 % (w/v), inoculum level 5 % (v/v) and 96 h of incubation period. Thus production of protease from catgut waste could be an economical and eco-friendly approach to handle this waste.

References

Abdelnasser, S., Ibrahim, S., & Al-Salamah, A. L. (2009). Optimization of media and cultivation for alkaline protease production by alkaliphilic Bacillus haludurans. *Research Journal of Microbiology, 4*, 251–259.

Almeida do Nascimento, W. C., & Martins, M. L. L. (2004). Production and properties of an extracellular protease from thermophilic Bacillus sp. *Brazilian Journal of Microbiology, 35*, 1–2.

Anson, M. L. (1938). The estimation of pepsin, trypsin, papain, and cathepsin with hemoglobin. *Journal of General Physiology, 22*, 79–89.

Anupama, P., & Ravindra, P. (2000). Value-added food: Single cell protein. *Biotechnology Advances, 18*, 459–479.

Asokan, S., & Jayanthi, C. (2010). Alkaline protease production by Bacillus licheniformis and Bacillus coagulans. *Journal of Cell and Tissue Research, 10*, 2119–2123.

Boominadhan, U., & Rajakumar, R. (2009). Optimization of protease enzyme production using Bacillus sp. isolated from different wastes. *Botany Research International, 2*, 83–87.

Box, G. E. P., & Wilson, K. B. (1951). On the experimental attainment of optimum conditions. *Journal of the Royal Statistical Society: Series B, 13*, 1–45.

Das, G., & Prasad, M. P. (2010). Isolation, purification & mass production of protease enzyme from *Bacillus subtilis*. *International Research Journal of Microbiology, 1*, 26–31.

Ellaiah, P., Srinivasulu, B., & Adinarayana, K. (2001). A review on microbial alkaline proteases. *Journal of Scientific and Industrial Research, 61*, 690–704.

El-Safey, E. M., & Abdul-Raouf, U. M. (2004) *Production, purification and characterization of protease enzyme from Bacillus subtilis*. International conferences for development and the environment in the Arab World Assiut University, Egypt, March 23–25

Folin, O., & Ciocalteau, V. (1927). On tyrosine and tryptophan determinations in proteins. *Journal of Biological Chemistry, 73*, 627–650.

Hedge, J. E., & Hofreiter, B. T. (1962). Carbohydrate chemistry. In R. L. Whistler & J. N. Be Miller (Eds.), *Methods in carbohydrate chemistry*. New York, NY: Academic.

Hesseltine, C. W., Smith, M., & Wang, H. L. (1976). Product of fungal spore as inocula for oriental fermented food. *Developments in Industrial Microbiology, 17*, 101–115.

Jegannathan, K. R., & Viruthagiri, T. (2009). Ossein waste – A potential raw material for protease production. *International Journal of Chemical Reactor Engineering, 7*, Article 14.

Joo, H. S., Kumar, G. C., Park, G. C., Kim, K.T., Paik, S.R., & Chang, C.S. (2002). Optimization of the production of an extracellular alkaline protease from Bacillus horikoshii. *Process Biochemistry, 3*, 155–159.

Kristensen, G. H. (2005). In: European Pharmacopoeia 5.0, pp 873.

Lowry, O. H., Rosebrough, N. J., Farr, A. L., et al. (1951). Protein measurement with the folin phenol reagent. *Journal of Biological Chemistry, 193*, 265–275.

Miller, J. M., Zoll, D. R., & Brown, E. O. (1964). Clinical observations on use of an extruded collagen suture. *Archives of Surgery, 88*, 167–174.

Mukherjee, A. K., Adhikari, H., & Sudhir, K. R. (2008). Production of alkaline protease by a thermophilic Bacillus subtilis under solid-state fermentation (SSF) condition using Imperata cylindrica grass and potato peel as lowcost medium: Characterization and application of enzyme in detergent formulation. *Biochemical Engineering Journal, 39*, 353–361.

Naidu, K. S. B., & Devi, L. K. (2005). Optimization of thermostable alkaline protease production from species of Bacillus using rice bran. *African Journal of Biotechnology, 4*, 724–726.

Neuberger, A. (1955). Fibrous proteins and their biological significance. *Symposia of the Society for Experimental Biology, 9*, 72–84.

Pandey, A., Webb, C., Soccol, C. R., & Larroche, C. (2006). *Enzyme technology* (p. 320). New York, NY: Springer.

Plackett, R. L., & Burman, J. P. (1946). The design of optimal multifactorial experiments. *Biometrika, 33*, 305–325.

Prakasham, R. S., Subba Rao, C., & Sarma, P. N. (2006). Green gram husk – An inexpensive substrate for alkaline protease production by Bacillus sp. in solid-state fermentation. *Bioresource Technology, 97*, 1449–1454.

Priest, F. G. (1977). Extracellular enzyme synthesis in the genus Bacillus. *Microbiology and Molecular Biology Reviews, 41*, 711–753.

Shaheen, M., & Shah, A. (2008). Influence of culture conditions on production and activity of protease from *Bacillus subtilis*. *Pakistan Journal of Botany, 40*, 2161–2169.

Smitt, J. P., Rinzema, J., Tramper, H., Van, M., & Knol, W. (1996). Solid state fermentation of wheat bran by *Trichoderma reesei* QMQ 414. *Applied Microbiology and Biotechnology, 46*, 489–496.

Stanbury, P. F., Whitaker, A., & Hall, S. J. (1999). *Principles of fermentation technology* (2nd ed., pp. 110–114). Oxford: Butterworth-Heinemann Publisher.

Ul Qadar, S. A., Shireen, E., Iqbal, S., & Anwar, A. (2009). Optimization of protease production form newly isolated strain of Bacillus sp. PCSIR EA-3. *Indian Journal of Biotechnology, 8*, 286–290.

Verheijen, L. A. H. M., Wiersema, D., & Hulshoff Pol, L. W. (1996). *Management of waste from animal product processing*. Roma: Food and Agriculture Organization of United Nations.

Wen-Teish, C., Chin-Shuh, C., & San-Lang, W. (2003). An antifungal chitinase produced by *Bacillus cereus* with shrimp and crab shell powder as a carbon source. *Current Microbiology, 47*, 102–108.

World Enzymes. (2011). *Industry study with forecasts for 2015 & 2020* (p. 44). New York, NY: World Enzymes.

Chapter 18
Membrane Processes for Microalgae in Carbonation and Wastewater Treatment

Rosalam Sarbatly, Emma Suali, Farhana Abd Lahin, and Chel-Ken Chiam

Introduction

Membrane processes have gained an increase attention in many biological, chemical and physical processes because the membrane processes have fulfilled the requirements of process intensification (Drioli and Paul 2007). The process intensification is defined as a strategy to develop novel apparatuses and techniques that, compared to the conventional ones, leads to a substantially shrinking equipment size, reducing energy consumption, boosting plant efficiency, or minimising waste production, which eventually resulting in smaller, cleaner, more energy efficient and higher productive technologies (Stankiewicz and Moulijn 2002; Drioli and Paul 2007). Microalgae on the other hand are photosynthetic microorganism which requires CO_2, light and nutrients such as nitrogen and phosphorus to grow. This makes microalgae suitable for bioenergy production and wastewater treatment. As the world moves toward sustainable and renewable energy resource, microalgae are seen as one of the renewable energy resources that have high potential for biomass production.

Microalgae which is known to have 40 times higher photosynthesis compared to terrestrial plant (Suali et al. 2012) also has a high potential as a biological approach for CO_2 mitigation based on microalgae CO_2 consumption data that were reported by many (Yoo et al. 2010; Suali and Sarbatly 2012). The crucial step to achieve both biomass production and CO_2 mitigation by microalgae is the success of CO_2 transporting process into the bioreactors.

Being a photosynthetic microorganism, microalgae have very basic necessities including CO_2 supply, light and nutrient such as nitrogen and phosphorus. Microalgae are usually seen to thrive in waters with abundant fertilizers, a phenomenon which commonly known as eutrophication. In the 1940s, the potential of

R. Sarbatly (✉) • E. Suali • F.A. Lahin • C.-K. Chiam
Faculty of Engineering, Universiti Malaysia Sabah, Jalan UMS,
88400 Kota Kinabalu, Sabah, Malaysia
e-mail: rslam@ums.edu.my

P. Ravindra (ed.), *Advances in Bioprocess Technology*,
DOI 10.1007/978-3-319-17915-5_18

using microalgae as a medium for nutrient removal in wastewaters has been proposed (Caldwell 1946) and its potential has shown good progress ever since (Oswald and Gotaas 1957; Green et al. 1996). Microalgae has shown to significantly reduce biological oxygen demand (BOD) and chemical oxygen demand (COD) as well as nutrients especially nitrogen and phosphorus.

The utilization of microalgae has been researched to treat several types of wastewater sources including municipal (Kim et al. 2012; Singh and Thomas 2012; Wang et al. 2009; Hammouda et al. 1995; Cho et al. 2013), agro industrial (Gonzalez et al. 1997; Gantar et al. 1991) and industrial (Yun et al. 1997; Hongyang et al. 2011). Microalgae have also been researched in its sorption ability for heavy metal containing wastewater (Chinnasamy et al. 2010; Cai et al. 2013). This has opened up the opportunity to utilize the available unlimited resources of wastewater in order to cultivate microalgae for the dual purpose of removing nutrients and producing biofuels (Kim et al. 2010).

Microalgae based wastewater treatment has several benefits compared to chemical and biological treatment. The microalgae wastewater treatment does not require vigorous aeration that usually required in biological treatment. The biomass produces in microalgae treatment can be further utilized in other industries such as biofuel production and agricultural as feed, compared to the sludge produced in biological treatment that requires further treatment. As for chemical treatment, the additional chemical cost can be avoided. Furthermore, the microalgae treatment can serve a dual purpose of wastewater treatment and CO_2 sequestration at the same time. The nitrogen removed during the treatment process is also able to be recovered in the form of biomass and can be further reused. However, carbonation and nutrient up-take are crucial in microalgae cultivation. The carbonation and nutrient up-take by the microalgae are further discussed in the next sections. The functions of integrated membrane processes with the microalgae are demonstrated by experimental works. Section "Carbonation of Microalgae" discussed the experimental work on microalgae carbonation using membrane and Section "Microalgae Separation Process" discussed the experimental work on wastewater treatment in membrane integrated microalgae cultivation process.

Membrane

A membrane is a thin and permeable partition which separates two different phases. Synthetic membranes can be categorised into hydrophilic and hydrophobic membranes. The hydrophilic membranes are essential for water permeation through the pores or matrixes of the membranes. The hydrophilic membranes are made from polyethersulfone, cellulose acetate, polyacrylonitrile, polyvinylchloride, etc. The application of vapour permeation through the membrane pores or matrixes requires hydrophobic membranes. The examples of polymers which are used to manufacture the hydrophobic membranes are polyethylene, polypropylene, polyvinylidene fluoride and polytetrafluoroethylene. The application of membrane for the purpose of

Table 18.1 Driving forces distinguish the membrane processes

Driving force	Membrane process
Direct pressure gradient	Microfiltration, ultrafiltration, nanofiltration, reverse osmosis
Electrical gradient	Electrodialysis, membrane electrophoresis, membrane electrolysis
Activity gradient	Pervaporation, vapour permeation, gas separation
Thermal gradient	Membrane distillation
Salinity gradient	Pressure retarded osmosis, forward osmosis, osmotic distillation
Thermal and salinity gradients	Osmotic membrane distillation

carbonation and wastewater treatment is relatively new in the industry. Thus, there are still lots of issues that need to be considered before the implementation of membrane processes for wastewater treatment, biomass production and CO_2 mitigation by microalgae, especially the industrial level.

Membrane Processes

Membrane processes are developed in which the driving forces distinguish the processes. Table 18.1 shows an overview of the membrane processes and their respective driving forces. The applications of membrane processes in various industries include concentrating fruit juices, whey protein, ginseng extract, etc.; desalination; removals of CO_2, O_2 and NH_3 from water; removals of alcohols from fermented products; energy production; and wastewater treatments.

Carbonation of Microalgae

Carbonation of microalgae using membrane is one of the crucial steps in using microalgae for bioenergy production and for CO_2 utilization. The others step is known as deoxygenation. Usually, deoxygenation must be conducted using hydrophobic hollow fibre membrane to prevent the risk of clogging by the accumulated microalgae biomass in the membrane module. Carbonation is the process of dispersing or dissolving CO_2 gas into the microalgal media. Carbonation, on the other hand can be conducted by almost all types of membrane as long as the pressure for gas inlet is higher compared to the liquid inlet. The main process of carbonation comprises of three stages: (1) the transfer of gases from the gas phase liquid to the gas-membrane interface; (2) transfer of the gases through the membrane to the liquid-membrane interface; and (3) transfer of the gases into the bulk liquid phase. The reverse of this process is known as degassing, which applied for

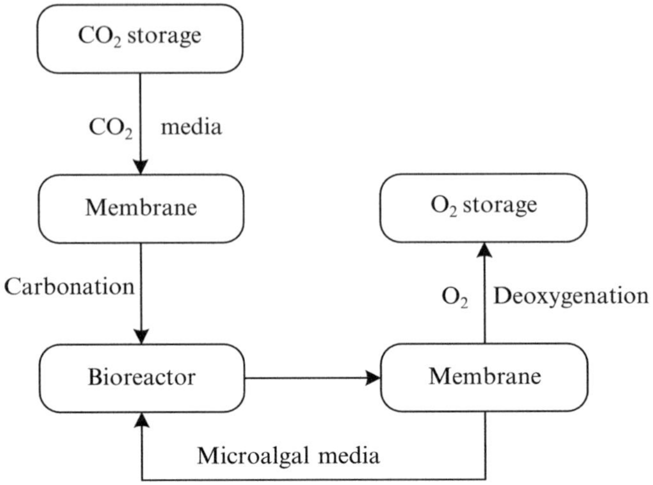

Fig. 18.1 Process diagram of carbonation and degassing in membrane-bioreactor

the removal of O_2 from the bioreactor. Figure 18.1 shows the process diagram of carbonation and degassing of microalgae using membranes. The present work mainly focused only on the carbonation process.

Microalgae Separation Process

The application of membrane in the separation process of microalgae is strictly physical, and the performance only depends on the membrane pore size and its operating parameters (Anselme et al. 1993). Therefore, the need for prior chemical addition is omitted (Wicaksana et al. 2012) and the resultant permeates can be maintained below the specific limit without any further treatment.

In microalgae biomass separation from its medium specifically, the most superior membrane filter has yet to be determined, due to the complex nature of microalgae itself. The performance of membrane filtration varies depending on various factors including hydrodynamic condition, concentration and the properties of culture such as shape, age and debris of algae. Some researcher has mentioned that the selectivity is better with small pore size, as the retention of cells is higher (Rossignol et al. 1999) and pore clogging is minimal (Rossi et al. 2004). The preferred membrane filters are however narrowed between two pore sizes, microfiltration (MF) and ultrafiltration (UF) (Rossi et al. 2004; Castaing et al. 2010; Rossignol et al. 1999; Zhang et al. 2010).

The most significant problem regarding membrane separation in microalgae application is the decline of permeates flux due to fouling. Fouling is mainly due to the formation of the microalgae cake formation, adsorption of organic matter, concentration polarization and eventually pore clogging (Zhang et al. 2010; Chen et al. 1997). Membrane fouling will directly affect the productivity in separation the microalgae cells and is further complicated by the complex character of microalgae itself. The fouling is caused by the combination of the size, morphology characteristic and especially the extracellular organic matter (EOM) attached to the cells (Babel and Takizawa 2010) that will significantly affect the specific cake resistance developed in the membrane filter (Babel et al. 2002). This could lead to permeability decline and higher energy demands for maintaining a constant permeate flux (Drews et al. 2006).

Membrane fouling can be limited by adjustment of the operating parameters of the membrane, among which is the utilization of cross-flow membrane filtration rather than frontal filtration (Chow et al., 1997) and working with the optimum operating parameters such as cross flow velocity and transmembrane pressure (TMP) (Song 1998). Selection of a lower pressure operation and lower cross flow velocity is more suitable for a long term operation of membrane as it presents a gentle condition for the fragile species (Frappart et al. 2011). Other research has reported that the lower fouling rate was achieved at a higher cross flow velocity (Wicaksana et al. 2012) and the flux was also observed to increase with increasing cross flow velocity by Choi and colleagues (Choi et al. 2005). This phenomenon is due to increase of shear velocity which is caused by cross flow velocity elevation. This makes it more difficult for the microalgae cell to be deposited on the membrane, thus inducing flux (Zhang et al. 2010; Ahmad et al. 2012). In the other hand, high cross flow velocity may impose high shear stress to the microalgae cell causing it to break and possibly releasing back the nutrients into the medium (Himberg et al. 1989; Ladner et al. 2010). This condition is also disadvantages to the process where the biomass productivity is required such as microalgae cultivation for biofuel production.

Experimental Study on Carbonation of Microalgae Using Membrane

Introduction

Experimental works were conducted to measure the efficiency of membrane application for carbonation of microalgae cultivation in a bioreactor. In this study, an isolate microalga from Borneo known as *Chlorella* sp. was selected as a test subject for carbonation of microalgal media. The strain was isolated from Tun Fuad Stephens Lake that located at 6°N and 116°E by the research team of the Borneo Marine Research Institutes, Universiti Malaysia Sabah. The investigation on this microalga will give benefit to the local community of Borneo as it was never mentioned in any publication.

The concentration of CO_2 that was applied in this work is limited to a CO_2 concentration within the range of 0.5 kg/m^3 and 3.5 kg/m^3. This range was selected based on the capability of microalgae to grow in water with dissolved CO_2 (DCO_2) less than 3.5 kg/m^3 (3,500 ppm). This CO_2 concentration is about 10 times higher compared to atmospheric CO_2, which is usually less than 350 ppm. The microalgae concentration is fixed at 0.0401 kg/m^3 and cultivated in photobioreactor (PBR). In this study, the term of PBR is used to define a system that consists of mini bioreactors which was made up of tubular acrylic pipes. The bioreactor unit is equipped with white cool fluorescent lamps as a light source in the photosynthesis, stirrer and aeration system. The microalgae were cultivated in the bioreactor. The other unit that made up the PBR system includes gas exchange unit and a CO_2 supply unit. Cultivation of microalgae was conducted phototrophically in Jaworski medium (JM). The experimental work was conducted at controlled temperature at 27 °C under two fluorescent white cool lamps with irradiance at 296 μE/s^1/m^2. The cultivation was conducted under the 12 h light, 12 h dark cycle.

Experimental Procedures and Data Analysis

The DCO_2 was measured using CO_2 metre CGP-31. The CO_2 metre does not measure the total dissolved inorganic carbon (DIC) but total DCO_2. Thus, to measure carbonate and carbonic acid, the pH solution was taken simultaneously. The pH is a measure of the total ratio of hydrogen ion (H^+) to hydroxyl ions (OH^-) which effective way to measure the carbonic acid and carbonate in the media (Chen and Durbin 1994).

The carbonation efficiency was analysed by comparing the outlet and inlet of CO_2 concentration. The CO_2 device measures CO_2 in mg/l and %. The value was converted to SI unit (kg/m^3) during the data analysis. To determine the dissolved CO_2 without interference from atmospheric recarbonation, each tubular bioreactor fitted with rubber stopper drilled with a 20 mm-diameter hole at the top of the reactor.

The carbonation was started from CO_2 supply unit and forwarded to gas exchange unit before entering the bioreactor unit. Carbonation of microalgae was conducted using direct and indirect membrane-based bubbling technique. The direct supply of CO_2 into bioreactor unit was used to compare the effectiveness of membrane for carbonation.

Experimental Setup

The setup that involved in the experimental work is shown in Fig. 18.2. The membrane photobioreactor (PBR) setup that was developed for carbonation consists of three main units. This includes CO_2 supply unit, gas exchange unit and bioreactor unit.

The membrane that was used for the carbonation is a GFD hydrophobic UFS220 and the characteristic of this membrane is shown in Table 18.2. The purpose for using GFD membrane is it was made from polyethylene hollow fibre.

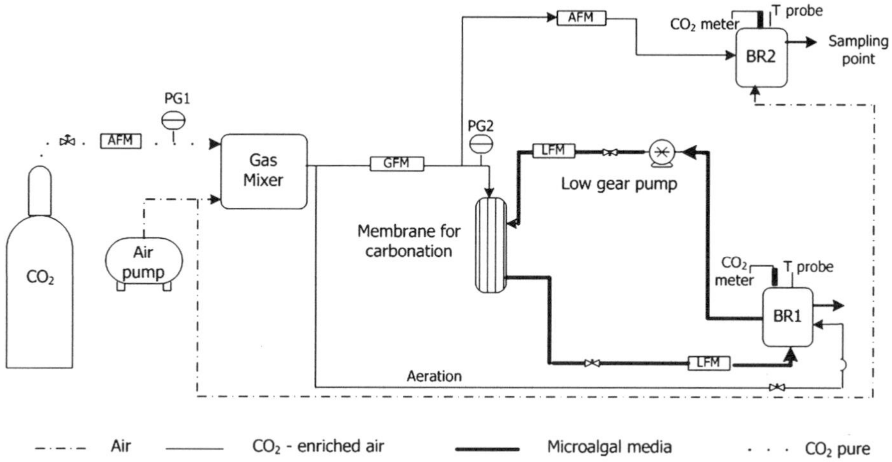

Fig. 18.2 Membrane bioreactor setup for the carbonation process (adopted from Suali, 2014)

Table 18.2 Characteristics of membrane for carbonation

Features	Membrane for carbonation
Number of fibres	2,400
Fibre pore size (μm)	0.01–0.1
Length (mm)	495
Total surface area (m^2)	0.80
Shell internal diameter (mm)	60
Capacity (m^3/s^1) × 10^{-5}	2.7
Membrane thickness (mm)	0.65
Shell side volume (m^3) × 10^{-3}	0.43
Volume occupies by hollow fibre (m^3) × 10^{-3}	2.5

Source: GDP (2013)

Polyethylene is known with an excellent chemical resistance characteristic, which left undamaged even when backwashed using chemicals making this membrane physically suitable for carbonation. In addition, polyethylene is hydrophobic that prevent fouling of microalgae cell in the module. The membrane module was also designed with an inlet for media and an inlet for gas. The outlet gas can also be controlled and closed when necessary, according to experimental purpose. This makes the selected membrane suitable for the carbonation in microalgae media, beside the lower cost compared to other available commercial membrane in the market.

Results and Discussion

The carbonation efficiency of the technique to aid carbonation was evaluated by comparing the amount of CO_2 at the membrane inlet and membrane outlet. The amount of CO_2 at the membrane outlet is the maximum amount of CO_2 that can absorb in the microalgae media in the bioreactor. At a normal equilibrium state, the absorption of gas to liquid is according to Henry's law. However, it is time consuming to reach the equilibrium state. With the use of a membrane, the absorption of CO_2 into the liquid can be achieved in lesser time compared to without membrane. Table 18.3 shows the carbonation efficiency of both direct and indirect membrane-based bubbling technique.

From a carbonation perspective, the indirect membrane-based bubbling only had a slight advantage compared to the direct membrane-based bubbling technique. However, both direct and indirect membrane-based bubbling achieved more than 80 % carbonation compared to only 29 % when applying a direct bubbling technique without membrane (data not shown).

The result also showed that the carbonation can be increased by approximately 4-fold when carbonised with the use of a membrane. The feasibility of carbonation using inoculated microalgae in bioreactors should not be an issue because the carbonation efficiency was not less than that of direct membrane-based bubbling technique. The key to carbonation or CO_2 absorption into the media includes contact area between CO_2 and the liquid phase within the membrane. Thus, inoculated microalgae cells in the bioreactor receive equivalent or higher CO_2 concentrations than direct carbonation microalgae.

Table 18.3 Carbonation efficiency based on direct and indirect membrane-based bubbling

	Eff. of carbonation[a]			pH		Acc. microalgae[b]	
Inlet CO_2 $(kg/m^3) \times 10^{-3}$	Dir. (%)	Ind. (%)	RSD (%)	Dir.	Ind.	Dir. $(kg/m^3) \times 10^{-3}$	RSD (%)
0.49	76	76	3	5.18	5.18	0.056	14
0.67	82	84	2	5.06	5.06	0.056	14
0.95	61	63	3	5.04	5.03	0.052	13
1.25	56	57	1	4.99	4.98	0.048	12
1.58	53	54	1	4.93	4.93	0.040	10
1.87	49	50	3	4.91	4.90	0.020	5
2.55	43	47	8	4.85	4.83	0.020	5
2.8	42	43	3	4.83	4.83	0.012	3
3.11	40	40	1	4.82	4.82	0.008	2

Acc. accumulated, *Diss.* dissolved, *Eff.* efficiency, *Dir.* direct membrane-based bubbling technique, *Ind.* indirect membrane-based bubbling technique
RSD relative standard deviation (to note the similarity and variation of direct membrane-based bubbling value compared to indirect membrane-based bubbling technique)
RSD%: Average difference in percentage of both techniques
[a]Carbonation efficiency based on value of CO_2 dissolved in media compared to inlet value
[b]Acuumulate Microalgae based on value of CO_2 dissolved in media compared to inlet value

Under equilibrium conditions, CO_2 solubility in water according to Henry law is nearly 84 % of the total supply ($DCO_2 = 0.8317 \times CO_2$ inlet) (Davis and Masten 2004). This estimate was the maximum value of gas soluble into the media. However, the equilibrium state of CO_2 solubility in liquid media under natural conditions takes longer to achieve when compared to carbonation using membranes, which reduces the potential of microalgae to capture and use CO_2 for photosynthesis.

During the carbonation, some CO_2 accumulates inside the membrane because of its larger space compared to the space of the liquid phase. Thus, the CO_2 was accumulated within the space instead of direct dissolving into liquid phase. Thus, for (see Table 18.3) future work, the gas space in membrane must be smaller compared to liquid space. The used of superphobic membrane resulted in less accumulated microalgae. The experimental results on the application of membrane as aids to the carbonation processes are proven effective. The cultivated microalgae from the carbonation study were used for further experimental work in wastewater treatment. The same microalgae culture system was modified to suit the experimental work on a wastewater treatment as discussed in Sect. 4.

Experimental Study on Separation of Microalgae Using Membrane

Introduction

The cultivated microalgae for carbonation purpose were reused for the purpose of wastewater treatment. In this work, the membrane was used in the separation process between microalgae biomass and the effluent. Due to the micron size of the microalgae cell, the separation of the resulting biomass during treatment is challenging, however very important. The wastewater treatment process may not be successful or even may lead to future problem such as future eutrophication if the microalgae biomass is not removed from the effluent before discharged into the environment. Studies have shown that with the growth of microalgae, the COD reading will increase (Cho et al. 2013), in fact, every gram of microalgae will contribute to 1.25 g of COD (Craggs et al. 1997; Ozkan 2012). This has led the main focus of microalgae wastewater treatment to the method of separation between the treated water and grown biomass. It is essential that the system design is accountable for a feasible and effective treatment especially to separate the microalgae biomass.

The current techniques for microalgae biomass separation available include floatation, sedimentation, flocculation, filtration and centrifugation. Even though these techniques are established in separation processes and has been used for centuries, there could be drawbacks in terms of microalgae separations especially for wastewater treatment (Lahin et al., 2013). The applied technique should be able to completely separate the biomass without any further contamination in the treated water such as added chemicals that lead to another treatment, which usually happens in the flocculation processes. The technique should also be able to serve its purpose by its own without the need of combination with other processes such as flocculation-sedimentation and flocculation-filtration.

The selected process should be economical and efficient, suitable for and industrial scale operation. The cost of microalgae separation or harvesting either in wastewater treatment or biofuel production can sometimes be the main contributor in the total operational cost. For instance, Grima et al. (2003) reported that algae harvest and dewatering contributed up to 30 % of total costs of the whole system in a microalgae producing plant. In a recent report on the techno-economic analysis of autotrophic microalgae for fuel production, it was estimated that biomass harvesting costs will be 21 % of the total capital cost of an open pond system (Davis et al. 2011).

Experimental Procedures and Data Analysis

A laboratory scale microalgae membrane bioreactor was constructed to study the efficiency of membrane separation process as well as in treatment of nitrogen rich synthetic wastewater. A synthetic ammonia rich wastewater which was prepared daily was used as wastewater for the experiment. Schematic of the microalgae membrane bioreactor is as shown in Fig. 18.3.

The microalgae membrane bioreactor treatment system consisted of one photobioreactor and three additional tanks, influent tank where the untreated

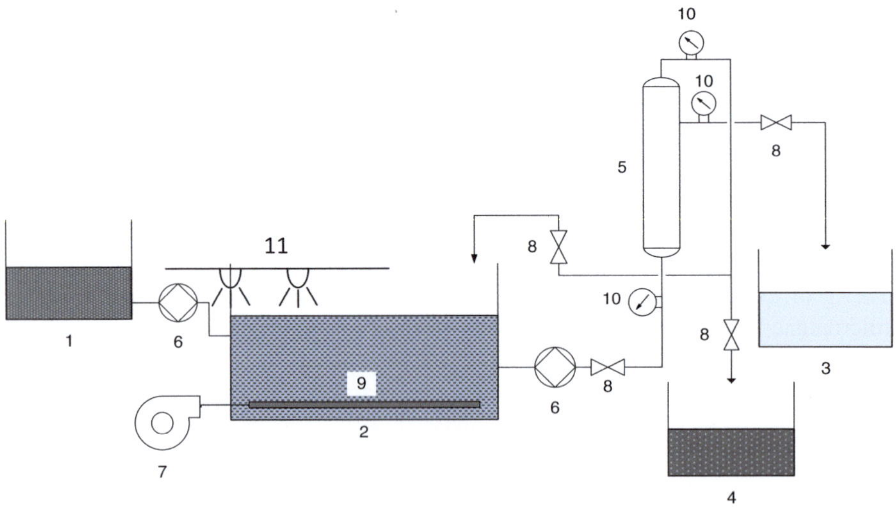

Fig. 18.3 Schematic of microalgae membrane bioreactor treatment (1) untreated synthetic wastewater tank, (2) microalgae photobioreactor, (3) treated water tank, (4) excess algae holding tank, (5) ultrafiltration membrane, (6) pump, (7) air pump, (8) valve, (9) air diffuser, (10) pressure gauge, (11) florescent lamps

Table 18.4 Operating parameter of microalgae membrane bioreactor

Operating parameter	Specifications
Membrane filter unit	Commercial ultrafiltration membrane Molecular weight cut-off: 50 KDa
Ambient temperature	20–24 °C
Synthetic wastewater temperature	22–23 °C
CO_2 supply flow rate	15 L/min
Synthetic wastewater pH	7–7.5
Light supply	Intensity: 4,000 ± 20 Lux Light/dark cycle: 12/12 h
Photobioreactor working volume	70 L
Membrane cleaning	Forward flushing: 30 s Backwashing: 4 min TMP: 0.2 bar

wastewater was stored, the excess algae holding tank and the treated water tank. Two water pumps were used to feed wastewater into the photobioreactor and to flow the mixture of treated wastewater and algae through the membrane. UF membrane with 50 kDa pore size was connected downstream to the photobioreactor for biomass separation and concentration. The clean water and concentrated biomass are flowed into the treated water tank and algae holding tank respectively. Transparent glass (prospect) was used as the photobioreactor material so that light supplies are even throughout the whole reactor.

The wastewater sample prepared was initially stored in the effluent tank. Upon the beginning of the system operation, 30 L of the wastewater was pumped into the photobioreactor where 20 L of microalgae has been cultured beforehand and was prepared in the specified concentration OD680 reading at approximately 0.150. The ratio of microalgae and synthetic wastewater volume was 2:3. The microalgae treatment was operated in two different short retention time (RT) inside the photobioreactor. During this operation, CO_2 and light were supplied to the microalgae by aeration and artificial illumination respectively. The treatment system was conducted in a semi-batch operation where each cycle; only 2/3 of the photobioreactor volume is filtered to maintain the microalgae to synthetic wastewater volume ratio. Membrane cleaning by forward flushing and backwashing was conducted after the membrane filtration operation. The operating parameters for the microalgae membrane bioreactor are summarized in Table 18.4.

Results and Discussions

The performance of the membrane was evaluated based on its capability in removing BOD and COD as well as its ability to retain microalgae biomass which was assessed by the turbidity reading of the membrane permeate. Turbidity was recorded to be below 5 Fau and the BOD and COD were 70 % removed from the wastewater. The result in Table 18.5 shows that the membrane unit was able to retain microalgae biomass that will otherwise contribute to COD and BOD contamination in the permeate water.

The membrane bioreactor was able to remove the ammonia content in the wastewater, where more than 80 % of NH_3 and more than 20 % of PO_4^{3-} was removed. The highest removed nutrient was achieved in operation with low concentration wastewater sample in 2 days RT where 81.9 % of NH_3 and 25.5 % of PO_4^{3-} was removed from the wastewater (refer Table 18.6). Others researcher has also reported that microalgae nutrient removals are preferable in a lower initial concentration (Gantar et al. 1991; Li et al. 2013). The nutrient uptake is also influenced by the initial concentration of microalgae where a higher microalgae concentration will lead to a higher nutrient uptake.

The nutrient uptake by microalgae is a very complex process as it can depend on many parameters that may include nutrient concentration, light intensity, the nitrogen/phosphorus ratio, light/dark cycle (Aslan and Kapdan 2006; Kim et al. 2012; Singh and Thomas 2012; Wang et al. 2009; Hammouda et al. 1995; Gonzalez et al. 1997; Gantar et al. 1991; Yun et al. 1997; Hongyang et al. 2011; Kim et al. 2010). The N:P ratio in the wastewater is an equally important parameter that will affect the efficiency of microalgae wastewater treatment. This has been shown through the lower PO_4^{3-} removal compared to the NH_3 removal. This implies that in the nature of the synthetic wastewater used, the limiting nutrient is nitrogen. The N:P ratio of used wastewater feed is 5:1 which suggested by to be nitrogen limited based on the optimal N: P ratio for microalgae growth graph. This result was also obtained by Gonzalez et al. (1997) and Yun et al. (1997) where phosphorus removal was low even with the exhaustion of nitrogen content where the microalgae of *Chlorella vulgaris* were used. Both of the research were done using medium containing low N:P ratio. From other literature, Ruiz-Martinez et al. (2012), Hultberg et al. (2013) and Li et al. (2013) that used medium with very high N:P ratio, the phosphorus removal was reported to be higher than nitrogen removal.

Table 18.5 Ultrafiltration membrane performance in removing turbidity, COD and BOD

		Final concentration		Percentage removal	
Average value	Initial concentration	3 days RT	2 days RT	3 days RT	2 days RT
Turbidity (Fau)	50 ± 5	<5	<4	83.1	91.1
COD (mg/L)	60 ± 10	11	7	21.7	25.5
BOD (mg/L)	30 ± 5	87.8	92.6	87.8	92.6

Table 18.6 Average nitrogen and phosphorus removal

		Final concentration		Percentage removal	
Average value	Initial concentration	3 days RT	2 days RT	3 days RT	2 days RT
NH_3 (mg/L)	50 ± 10	11 ± 10	9 ± 10	80.5	81.9
PO_4^{3-} (mg/L)	10 ± 5	7.9 ± 5	7.5 ± 5	21.7	25.5

RT retention time

Conclusion

The result of the experimental work on the carbonation of microalgae using membrane is considered effective because of the desired carbonation was achieved in a shorter period compared to the carbonation without the application of membrane. The carbonation of microalgae without a membrane will be depended solely on the equilibrium state which depends on pressure and temperature. The carbonation efficiency is within the ranges from 40 to 83 %. When carbonised at a high CO_2 inlet concentration, the low carbonation efficiency (less than 60 %) indicates that the high CO_2 inlet concentration at static medium volume is not suitable for the purpose of carbonation. A high CO_2 concentration promotes instability of bubbling, thus resulting in high CO_2 escape into the headspace of the bioreactor.

The study on the separation process of biomass and wastewater effluent on the other hand has shown that wastewater treatment using microalgae with the aids of membrane is able to achieve desired nutrient removal efficiency with less required time compared to the biological treatment. Even though microalgae membrane bioreactor system does not allow recovery of ammonia and urea, this can be compensated by the various possible applications of the resulting biomass from the treatment system. It can be established that the microalgae membrane bioreactor system that incorporate UF membrane as the separation unit of microalgae biomass is highly efficient in producing a good quality effluent water, by removing the NH_3, PO_4^{3-}, turbidity, BOD and COD as well as retain the microalgae biomass. However, the operating parameters should be further studied to avoid any unnecessary extra cost for membrane maintenance and damages

Acknowledgments We would like to thank Universiti Malaysia Sabah, Faculty of Engineering and, Minerals and Materials Research Unit for the research facilities. We would like to acknowledge Borneo Marine Research Institute for their provision and supervision on the microalgae related issue.

References

Ahmad, A. L., Mat Yasin, N. H., Derek, C. J. C., & Lim, J. K. (2012). Cross flow microfiltration of microalgae biomass for biofuel production. *Desalination, 302*, 65–70.

Anselme, C., Mandra, V., Baudin, I., & Mallevialle, J. (1993). Optimum use of membrane processes in drinking water treatment. *Water Supply, 12*, 1–2.

Aslan, S., & Kapdan, I. K. (2006). Batch kinetics of nitrogen and phosphorus removal from synthetic wastewater by algae. *Ecological Engineering, 28*, 64–70.

Babel, S., Takizawa, S., & Ozaki, H. (2002). Factors affecting seasonal variation of membrane filtration resistance caused by *Chlorella* algae. *Water Research, 36*, 1193–1202.

Babel, S., & Takizawa, S. (2010). Microfiltration membrane fouling and cake behaviour during algal filtration. *Desalination, 261*, 46–51.

Cai, T., Park, S. Y., & Li, Y. (2013). Nutrient recovery from wastewater streams by microalgae: Status and prospects. *Renewable and Sustainable Energy Reviews, 19*, 360–369.

Caldwell, D. H. (1946). Sewage oxidation ponds-performance, operation and design. *Sewage Works Journal, 18*, 433–458.

Castaing, J. B., Massé, A., Pontie, M., Sechet, V., Haure, J., & Jaouen, P. (2010). Investigating submerged ultrafiltration (UF) and microfiltration (MF) membranes for seawater pre-treatment dedicated to total removal of undesirable micro-algae. *Desalination, 253*, 71–77.

Chen, C. Y., & Durbin, E. G. (1994). Effects of pH on the growth and carbon uptake of marine phytoplankton. *Marine Ecology Progress Series, 109*(83–94), 84–93.

Chen, V., Fane, A. G., Madaeni, S., & Wenten, I. G. (1997). Particle deposition during membrane filtration of colloids: Transition between concentration polarization and cake formation. *Journal of Membrane Science, 125*, 109–122.

Chinnasamy, S., Bhatnagar, A., Hunt, R. W., & Das, K. C. (2010). Microalgae cultivation in a wastewater dominated by carpet mill effluents for biofuel applications. *Bioresource Technology, 101*, 3097–3105.

Cho, S., Lee, N., Park, S., Yu, J., Luong, T. T., Oh, Y. K., & Lee, T. (2013). Microalgae cultivation for bioenergy production using wastewaters from a municipal WWTP as nutritional sources. *Bioresource Technology, 131*, 515–520.

Choi, H., Zhang, K., Doinysiou, D. D., Oether, D. B., & Sorial, G. A. (2005). Influence of cross-flow velocity on membrane performance during filtration of biological suspension. *Journal of Membrane Science, 248*, 189–199.

Chow, C. W. K., Panglisch, S., House, J., et al. (1997). A study of membrane filtration for the removal of cyanobacterial cells. *Journal of Water SRT—Aqua, 46*(6), 324–334.

Craggs, R. J., McAuley, P. J., & Smith, J. V. (1997). Wastewater nutrient removal by marine microalgae grown on a corrugated raceway. *Water Resources, 31*(7), 1701–1707.

Davis, M. L., & Masten, S. J. (2004). *Principles of Environmental Engineering and Science* (p. 704). New York: McGraw-Hill Higher Education.

Davis, R., Aden, A., & Pienkos, P. (2011). Techno-economic analysis of autotrophic microalgae for fuel production. *Applied Energy, 88*, 3524–3531.

Drews, A., Lee, C. H., & Kraume, M. (2006). Membrane fouling—A review on the role of EPS. *Desalination, 200*, 186–188.

Drioli, E., & Paul, D. R. (2007). Preface: Advanced membrane technology III—Membrane engineering for process intensification conference. *Industrial and Engineering Chemistry Research, 46*, 2235.

Frappart, M., Masse, A., Jaffrin, M. Y., Pruvost, J., & Jaouen, P. (2011). Influence of hydrodynamics in tangential and dynamic ultrafiltration systems for microalgae separation. *Desalination, 265*, 279–283.

Gantar, M., Obreht, Z., & Dalmacija, B. (1991). Nutrient removal and algal succession during the growth of *spirulina plantensis* and *scenedesmus quadricauda* on swine wastewater. *Bioresource Technology, 36*, 167–171.

GDP. (2013). Data sheet report. Retrieved from http://gdpfilter.co.id/images/Data%20Sheet%20&%20Drawing%20UF/S-220%20Data%20Sheet.pdf

Gonzalez, L. E., Canizares, R. O., & Baena, S. (1997). Efficiency of ammonia and phosphorus removal from a Colombian agro industrial wastewater by the microalgae *Chlorella vulgraris* and *Scnedesmus dimorphus*. *Bioresource Technology, 60*, 259–265.

Green, F. B., Bernstone, L. S., Lundquist, T. J., & Oswald, W. J. (1996). Advanced integrated wastewater pond systems for nitrogen removal. *Water Science and Technology, 33*, 207–217.

Grima, E. M., Belarbi, E., Fernandez, F. A., Medina, A. R., & Chisti, Y. (2003). Recovery of microalgal biomass and metabolites: Process options and economics. *Biotechnology Advances, 20*, 491–515.

Hammouda, O., Gaber, A., & Abdel-Raouf, N. (1995). Microalgae and wastewater treatment. *Ecotoxicology and Environmental Safety, 32*, 205–210.

Himberg, K., Keijola, A. M., Hiisvirta, L., Pyysalo, H., & Sivonen, K. (1989). The effect of water-treatment processes on the removal of hepatotoxins from Microcystis and Oscillatoria cyanobacteria: A laboratory study. *Water Research, 23*, 979–984.

Hongyang, S., Yalei, Z., Chunmin, Z., Xuefei, Z., & Jinpeng, L. (2011). Cultivation of *Chlorella pyrenoidosa* in soybean processing wastewater. *Bioresource Technology, 102*, 9884–9890.

Hultberg, M., Carlsson, A. S., & Gustafsson, S. (2013). Treatment of drainage solution from hydroponic greenhouse production with microalgae. *Bioresource Technology, 136*, 401–406.

Kim, J., Lingaraju, B. P., Rheaume, R., Lee, J. Y., & Siddiqui, K. F. (2010). Removal of ammonia from wastewater effluent by *Chlorella vulgaris*. *Tsinghua Science and Technology, 15*(4), 391–396.

Kim, S., Jung, C. W., & Lee, B. (2012). Treatment characteristic of persuant algae-bacteria ratio on actual municipal wastewater condition. Proceedings of the 2012 joint conference: Korea Society of Water and Wastewater (KSWW), Korean Society on Water Quality (KSWQ), May 21–22.

Ladner, D. A., Vardona, D. R., & Clark, M. M. (2010). Effects of shear on microfiltration and ultrafiltration fouling by marine bloom-forming algae. *Journal of Membrane Science, 356*, 33–43.

Lahin, F. A., Sarbatly, R., & Suali, E. (2013). Membrane bioreactor for wastewater treatment using microalgae. Fourth International Graduate Conference on Engineering, Science and Humanities (IGCESH) 16–17 April 2013, Univertisi Teknologi Malaysia, Johor Bahru, Johor, Malaysia.

Li, S., Luo, S., & Guo, R. (2013). Efficiency of CO_2 fixation by microalgae in a closed raceway pond. *Bioresource Technology, 136*, 267–272.

Oswald, W. J., & Gotaas, H. B. (1957). Photosynthesis in sewage treatment. *Transactions of the American Society of Civil Engineers, 122*, 73–105.

Ozkan, A. (2012). Development of a novel algae biofilm photobioreactor for biofuel production. A dissertation, The University of Texas at Austin.

Rossi, N., Jaouen, P., Legentilhomme, P., & Petit, I. (2004). Harvesting of cynobacterium *Arthrospira plantesis* using organic filtration membranes. *Food and Bioproducts Processing, 82*(C3), 244–250.

Rossignol, N., Vandanjon, L., Jaouen, P., & Quéméneur, F. (1999). Membrane technology for the continuous separation microalgae/culture medium: compared performances of cross-flow microfiltration and ultrafiltration. *Aquacultural Engineering, 20*, 191–208.

Ruiz-Martinez, A., Garcia, N. M., Romero, I., Seco, A., & Ferrer, J. (2012). Microalgae cultivation in wastewater: Nutrient removal from anaerobic membrane bioreactor effluent. *Bioresource Technology, 126*, 247–253.

Singh, G., & Thomas, P. B. (2012). Nutrient removal from membrane bioreactor permeate using microalgae and in a microalgae membrane photoreactor. *Bioresource Technology, 117*, 80–85.

Song, L. (1998). Flux decline in crossflow microfiltration and ultrafiltration: Mechanisms and modelling of membrane fouling. *Journal of Membrane Science, 139*, 183–200.

Stankiewicz, A., & Moulijn, J. A. (2002). Process intensification. *Industrial and Engineering Chemistry Research, 41*, 1920–1924.

Suali, E. (2014). Carbon dioxide utilisation by integrated microalgae cultivation process in membrane photobioreactor. PhD thesis, Universiti Malaysia Sabah.

Suali, E., & Sarbatly, R. (2012). Microalgae conversion to biofuel. *Renewable and Sustainable Energy Reviews, 16*(6), 4316–4342.

Suali, E., Sarbatly, R., & Shaleh, S. R. M. (2012). Characterisation of local *Chlorella sp.* toward biofuel production. International conference on applied energy, 5–8 July, Suzhou, China.

Wang, B., Li, Y., Wu, N., & Lan, C. Q. (2009). CO_2 bio-mitigation using microalgae. *Bioresource Technology, 102*, 35–42.

Wicaksana, F., Anthony, G. F., Pharima, P., & Robert, F. (2012). Microfiltration of algae (*Chlorella sorokiniana*): Critical flux, fouling and transmission. *Journal of Membrane Science, 387–388*, 83–92.

Yoo, C., Jun, S., Lee, J., Ahn, C., & Oh, H. M. (2010). Selection of microalgae for lipid production under high levels carbon dioxide. *Bioresource Technology, 101*, 71–74.

Yun, Y. S., Lee, S. B., Park, J. M., Lee, C. I., & Yang, J. W. (1997). Carbon dioxide fixation by algal cultivation using wastewater nutrients. *Journal of Chemical Technology and Biotechnology, 69*, 451–455.

Zhang, X., Qiang, H., Sommerfeld, M., Puruhito, E., & Chen, Y. (2010). Harvesting algal biomass for biofuels using ultrafiltration membranes. *Bioresource Technology, 101*, 5297–5304.

Chapter 19
A Systems View of Lignocellulose Hydrolysis

Deepti Tanjore and Tom L. Richard

Introduction

For 475 million years terrestrial plants have waged an evolutionary battle with microbes, insects, and most recently large mammals to develop tissues that can protect and retain their hard earned energy and nutrients in the face of incessant external attack. This evolutionary pressure has resulted in a complex plant cell wall structure with inherent resistance to degradation by mechanical force (physics), water (chemistry) and enzymes (biology). The most recent combatant in this battle is *Homo sapiens*, which has tackled this problem of recalcitrance using a variety of practical and technical means, applying knowledge from many fields of science and engineering. Yet the deconstruction of plant cell wall polymers into water soluble molecules, a process termed lignocellulose hydrolysis, remains one of the fundamental challenges for agriculture (livestock feed), forestry (pulp and paper) and the energy industry (biofuels).

Lignocellulosic biomass is the term used to collectively refer to the stem, stalks, leaves, and other non-fruiting parts of a plant. On a dry matter basis, most of this biomass is composed of long-chain polysaccharides, cellulose and hemicellulose, which are assembled from various sugar molecules and provide structure and function in plant cell walls. These polysaccharides are protected by a more complicated polymer, lignin. Over the past century researchers have developed numerous strategies to extract sugars for fermentation or chemical conversion to fuel precursors, fuels, chemicals, and several other bio-based products. Much of this

D. Tanjore
Advanced Biofuels (and Bioproducts) Demonstration Unit (AB-PDU), Lawrence Berkeley National Laboratory, Berkeley, CA 94608, USA

T.L. Richard (✉)
Biological Engineering, Penn State University, University Park, State College, PA 16801, USA
e-mail: trichard@psu.edu

P. Ravindra (ed.), *Advances in Bioprocess Technology*,
DOI 10.1007/978-3-319-17915-5_19

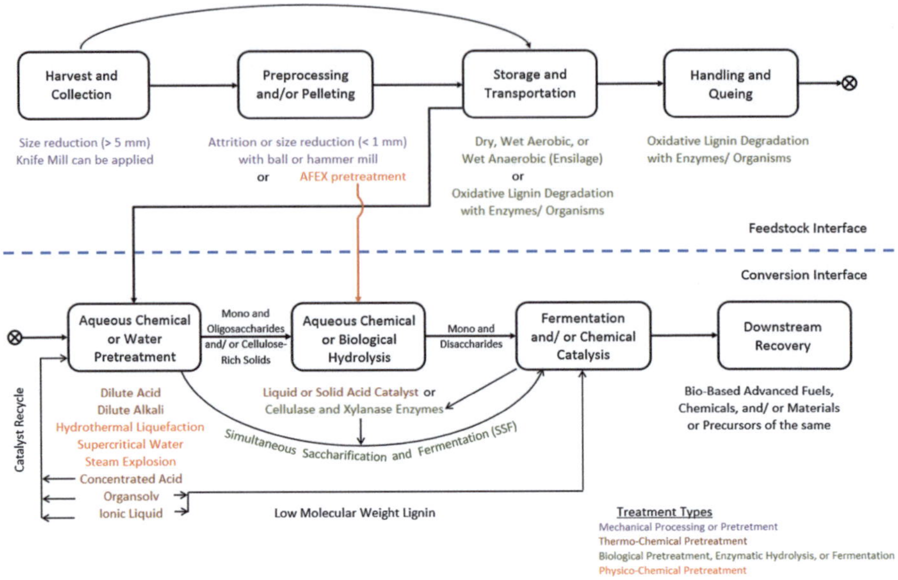

Fig. 19.1 Summary figure (Vidal et al. 2011)

research has focused on process chains that are designed to recover sugars using two hydrolysis stages: pretreatment and saccharification. Both of these stages will be covered in this chapter. In the last decade, recognition of the high costs associated with sugar recovery has accelerated research in this area, and novel process chains are being designed to integrate two or more unit operations into a single process, or perform multi-stage hydrolysis in a single reactor in an attempt to cut the costs. The individual unit processes involved in hydrolysis can be broadly categorized as (1) mechanical, (2) biological, or (3) chemical. In this chapter, we will focus on several of these processes with a review of historical technology development, sugar release kinetics, and reactor design and scale-up (see Fig. 19.1). We will further compare a few standard lignocellulose hydrolysis strategies to explore a very popular question: which treatment strategy is the best?

Mechanical Treatment (Size Reduction)

Mechanical treatment is the minimalist's approach to break down biomass. Often, mechanical pretreatment is confused with size reduction. Reducing biomass to sizes less than 5 mm is commonly employed to improve handling of the low-bulk density biomass. Smaller biomass particles, with higher surface area, improve the mass transfer of catalysts and water during subsequent unit operations. However, such size reducing mechanical treatments are not normally categorized as pretreatment because particle size reduction on its own, without additional chemical or biological processing, rarely has much impact on downstream sugar hydrolysis and thereby product yields. However, some intensive size reduction processes can result in extremely small biomass particles and/or physiological changes in the plant cell wall that do increase sugar release during downstream saccharification process; such technologies can be considered mechanical pretreatments. During intensive mechanical pretreatments, the heat generated from friction between the biomass and size reduction equipment may also contribute towards the pretreatment effect.

Knife Mills Minimize Structural Changes

A knife mill has been very commonly employed for studies examining chemical pretreatments. Cutting with a sharp knife has little impact on the overall structure of the cell wall and the minimal heat generated during the process, so this type of size reduction does not cause a pretreatment effect and is not expected to interfere when evaluating the impacts of chemical pretreatment processes. Several makes of knife mills are available for bench and pilot research, including the Thomas Wiley mills (Thomas Scientific, Swedesboro, NJ), and are accompanied with various screens of less than 2 mm. The National Renewable Energy Laboratory (NREL) released a series of Laboratory Analytical Protocols (LAPs) that establish methods for biomass characterization and conversion, including particle size reduction to improve subsample homogeneity. The protocol on performing compositional analysis of various lignocellulosic biomass types (NREL/TP-510-42618) recommends that biomass be knife-milled and sieved to obtain biomass particles between 0.177 and 0.841 mm (80 and 20 standard mesh sizes, respectively) prior to performing gravimetric and chromatographic analytical tests (Hames et al. 2008; Sluiter et al. 2008). Apart from preparing biomass for characterization or other research studies there is no particular advantage in employing knife mills, especially because they do not scale very well to commercial facilities that may need to process several hundred tons of biomass per day. If a mechanical pretreatment is to be included in a process chain, it is preferable to choose one that has a substantial impact at the physiological level and may reduce the required severity of downstream treatments and/or improve sugar yields.

Ball and Hammer Milling Reduce Cellulose Crystallinity

A typical ball mill pounds steel balls against biomass and each other in a gyrating cylindrical container (Barakat et al. 2014). The impact and friction between the balls not only reduces the size of biomass but also substantially reduces its crystallinity. Crushing cell walls and reducing particle size and crystallinity can lead to improved sugar yields during saccharification (Lin et al. 2010) and thus clearly counts as mechanical pretreatment. Hammer mills also breakdown biomass with shear and friction. Several hammers rotate on a central axis to pound biomass against the mill's rough cylindrical surface until the particles are small enough to pass through the screen. Barakat et al. (2014) hammer milled miscanthus to particle sizes less than 5 mm. These biomass particles were further ball milled for 24 h and sieved to obtain four ranges of particle sizes: (1) 250–355, (2) 150–250, (3) 63–150, and (4) <63 μm. Each sieve size of biomass was separately enzymatically hydrolyzed with the same treatment conditions. For the three particle size ranges larger than 63 μm the sugar yields were lower than 15 % of theoretical, while for the smallest particle size range, <63 μm, there was a significant increase in sugar yield of up to 30 % of theoretical. Similar results have been observed for several other feedstocks as well. Particle sizes must typically be less than 100 μm to significantly increase the sugar yield from mechanically pretreated biomass. However, such mechanical pretreatments can be energy intensive, as below 1 mm energy consumption increases exponentially with decreasing particle size. Mani et al. (2004) and Blanch et al. (2011) investigated the variation of this increase in energy consumption with biomass type. Among six combinations of biomass types and milling options that ground biomass to less and 1 mm particle size, switchgrass and wheat straw consumed the highest and lowest amounts of energy, respectively (Blanch et al. 2011; Mani et al. 2004). To achieve the small particle sizes required to significantly increase hydrolysis, the overall energy consumed by conventional size reducing equipment can be higher than the energy contained in the biomass (Barakat et al., 2013). To avoid these high energy costs, mechanical treatments are often combined with chemical or biological treatments.

Extrusion Integrated with Chemical Pretreatment

Combining mechanical pretreatments with a chemical or a thermal treatment process can not only reduce energy requirements, but also increase yield. Yoo et al. (2011) performed extrusion of soybean hulls at temperatures up to 80 °C and optimized the process with varying moisture contents and extrusion times. The extrusion process hydrolyzed cellulosic sugars up to 95 % of theoretical yield, higher than corresponding alkali (93 % of theoretical) and acid treatments (70 % of theoretical) without extrusion (Yoo et al. 2011). However, the authors noted that this process was not as effective on corn stover, where the extrusion process led to a sugar yield of only 60 % of theoretical. While biomass types with low recalcitrance can benefit from integrated processes at lower temperatures to reduce generation of inhibitors, the more recalcitrant feedstocks such as corn stover and switchgrass may still require stronger, more invasive pretreatments to break down the lignin structure of the plant cell wall to achieve high sugar yields through hydrolysis.

Biological Pretreatment and Enzymatic Saccharification (Bio-catalysis)

The deconstruction of lignin structures in the cell wall using microbes and/or enzymes as catalysts is usually referred to as biological pretreatment, and is classified with other pretreatment processes in the first stage of hydrolysis. The use of cellulase enzymes to convert cellulose into sugar monomers and oligomers is referred to as enzymatic saccharification and is classified in the second stage of hydrolysis. While it is convenient to separate these biological processes conceptually, it is important to realize that many of the relevant microbes hydrolyze both lignin and cellulase simultaneously to obtain carbon and energy from the biomass. Effective biological pretreatment may require multiple enzymes and chemical mediators to address physical as well as biochemical barriers to hydrolysis; mixtures of enzymes can work synergistically to expand small pores, open up the cell wall matrix and increase access (Jeremic et al. 2014). Some of the first studies on the impact of fungal enzymes on wood decay and cellulose hydrolysis were published in the early 1900s (Duggar and Davis 1914; Schmitz and Zeller 1919; Spaulding 1906). While bacteria such as Actinomycetes were found to be effective on grasses, fungi have been the most popular sources of commercial plant cell wall degrading enzymes, producing a variety of lignin degrading as well as cellulose and hemicellulose hydrolyzing enzymes (Saritha et al. 2012).

Fungal Enzymes Are Capable of Lignin Removal But at a Slow Rate

While several fungi are capable of producing lignin degrading enzymes, the most popular species for application in biofuels are the white-rot fungi. The white rot fungi are capable of selectively metabolizing hemicellulose and some low molecular weight lignin and leave cellulose relatively unaffected. Amongst the white rot species, *Phanerochaete chrysosporium* has been the most well studied fungus for the production of lignin degrading enzymes (Orth et al. 1993; Tien and Kirk 1983). These aerobic fungi can grown on biomass using the solid-state fermentation technologies familiar in the mushroom industry, or in simple bench-scale laboratory systems as illustrated in Fig. 19.2. Lignin degradation with enzymes is chiefly an oxidative process involving the production of manganese peroxidases (MnP), lignin peroxidases (LiP), and laccases. Typically, the activity of peroxidases is dependent on several isoenzymes produced by white rot fungi that remain attached to the fungal cells when oxidizing lignin to monolignols (Higuchi 1985), so that cell-free enzyme preparations may not be effective. Moreover, even with the presence of living fungi, the kinetics of lignin degradation were observed to be slow, with only 10–15 % of theoretical delignification occurring after 7–14 days

Fig. 19.2 Biological pretreatment of corn stover (particle size >1 mm) with white rot fungi

(Tanjore et al. 2009). The relatively slow kinetics of lignin degradation with fungal enzymes may not be conducive to commercial biorefinery operations that process over 1,000 tons per day, and where pretreatment is expected to be complete within a couple of hours.

Microbial pretreatment of biomass has the potential to reduce the severity required for downstream thermal and chemical pretreatment processes, but has not been shown to be effective on its own (Keller et al. 2003). Gui et al. (2013) observed that pretreatment with *P. chrysosporium* followed by 2.5 % (v/v) sulfuric acid on Chinese liquorice increased the sugar yields by 20 % (w/w), especially when treated at lower temperatures (Gui et al. 2013). Lignin is an extremely complicated phenylpropanoid compound with more than 11 types of bonds among several types of molecules. Compared to cellulose that has only a single type of bond (the β-1,4 glycosidic bond connecting glucan molecules) and a simple 180° rotation that can be decoupled with one or two enzymes, biological cleaving of lignin using the standard enzymatic paradigm of specific molecule recognition might require a dozen or more enzymes, all of which would be different to produce and isolate. However, several lines of evidence suggest that lignin degradation occurs by chemical intermediaries and reaction mechanisms that are much less specific than typical enzymatic biocatalysis. A better understanding of these mechanisms may inspire future biomimetic pretreatment strategies that are both rapid and effective.

Cellulolytic Enzymes

In the early 1900s, some studies identified and grouped cellulases under the broad class of carbohydrases, along with maltases, pectinases, lactases, etc., all of which are enzymes that disintegrate plant material including the fruit (Jacobson and Holmes 1914; Schmitz 1920). In 1920s, cellulases were observed in shipworms and molds, mushrooms, and other fungi that could digest saw dust and filter paper (Boynton and Miller 1927; Diehl 1949; Euler 1912; Miyaji 1930; Wolfgang and Hans 1937). Reese et al. (1957) correlated the drop in degree of polymerization (DP) of cellulose with change in intrinsic viscosities when enzymatic hydrolysis was conducted with enzymes from *Trichoderma viride* QM 6a. They noted that three main parameters influenced enzymatic hydrolysis (1) Particle size (40 mesh Solka Floc powdered cellulose was ball milled to various sizes), (2) hydratability (or wet-ability, which is generally lower in crystalline materials), and (3) pore size (for improved mass transfer of the bio-catalyst) (Reese et al. 1957). Almin et al. (1967) also used viscometric measurements to evaluate activity of cellulase from *Penicillium chrysogenum notatum* on carboxy-methyl cellulose or CMC. Several more activity assays followed, with those of Wood and Bhat (1988) and Ghose (1987) being the most popularly applied (Béguin 1983; Ghose 1987; Wood and Bhat 1988).

Meanwhile, several researchers reported that cellulase was not a single enzyme but a mixture of component enzymes, and that each component enzyme had unique activity on cellulosic materials (Li et al. 1965; Selby and Maitland 1967; Wood 1968). Reese et al. (1950) identified two distinct enzymes from *Aspergillus fumigatus* that were effective on crystalline and amorphous cellulose respectively (Reese et al. 1950). They postulated that one enzyme was only able to hydrolyze the β-1,4 bonds in amorphous cellulose and thus could not hydrolyze crystalline cellulose without the presence of the other enzyme. Li et al. (1965) reported two enzymes from *Trichoderma viride*, β-1,4-D-glucanglucohydrolase and β-1,4-D-glucan 4-glucano hydrolase, along with β-1,4 glucosidase, and termed them as exo and endo-glucanase. This study was followed by investigations by several other researchers who postulated cellulase mechanisms and attempted to increase enzyme production (Berghem and Pettersson 1973; Berghem et al. 1976; Hurst et al. 1978; Sternberg 1976; Wood 1968). From this body of research we now know that endoglucanase, exoglucanase, and β-glucosidase work in a serial fashion to break down long chains of tightly packed glucan molecules. The two enzymes originally described by Reese et al. (1950) were an endoglucanase, which randomly breaks down long chains of crystalline cellulose to soluble oligomers, and an exoglucanase that converts the oligomers to cellobiose, a dimer of two glucose units. This dimeric sugar is then broken down to monomeric sugar by β-glucosidase (Beckham et al. 2010).

The late 1970s saw the expansion of biofuel cellulase applications from model cellulose compounds to more complex forms of lignocellulosic biomass (Brooks et al. 1979; Perez et al. 2013). Very early on, Selby and Maitland observed that alkali pretreatment increased the porosity of cellulose and thereby its enzymatic digestibility (Selby and Maitland 1967). Researchers later observed that a chemical pretreatment was required for effective use of cellulases for lignocellulosic biomass hydrolysis (Brooks et al. 1979; Himmel et al. 2007; Zhang and Lynd 2004). Cellulases are now almost exclusively used for the second-stage of hydrolysis process in commercial scale lignocellulosic biorefineries (Energy.Gov 2014). Cellulases are currently being produced at an industrial scale by Dupont Industrial Biosciences (formerly Genencor), Novozymes Inc., and several other companies. In the last few years there have been considerable advances in our understanding of the parameters that influence cellulase action, including lignin deposition, xylo-oligomer concentration, and cellulose crystallinity. Each chemical pretreatment strategy affects the biomass in different ways, with both favorable and unfavorable impacts on enzymatic hydrolysis. These interactions of chemical pretreatment with enzymatic hydrolysis will be discussed in the following section.

Aqueous Phase Thermochemical Treatment (or Chemical Catalysis)

Thermochemical treatment is the most well studied type of biomass hydrolysis, dating back to earlier investigations conducted in the 1930s and 1940s. In this chapter we focus only on aqueous phase thermochemical treatments and do not consider dry phase treatments, such as pyrolysis and torrefaction. Aqueous treatments can utilize either homogeneous catalysts (acids, alkalis, ionic liquids) or heterogeneous catalysts (solid catalysts or gases, in a different phase than the water). In some cases, just water is used as the catalyst due to its enhanced solvent properties at high pressures and supercritical conditions. The oldest documented studies of homogenous catalysis of biomass employed sulfuric acid and calcium chloride (Kressman 1922; Simonsen 1898). Early alkali treatments were designed to unlock the lignin-hemicellulose matrix (the function designated "pretreatment" today) to access cellulose for the production of cellulose esters, while acid treatments were focused on recovering cellulosic sugars (today called "saccharification") for ethanol production (Kressman 1922; Simonsen 1898).

Homogenous Catalysts

Acid treatments Can Be Used for Both Pretreatment and Saccharification

Dilute Acids Are Effective at High Temperature and Pressure

One of the first publications on dilute acid treatment of lignocellulosic biomass described the Scholler process, which treated sawdust chips with 0.5 wt% H_2SO_4 at

190–193 °C and a pressure of 165–180 psi for 45 min (Faith 1945). In contrast to several previous attempts, in the Scholler process repeated hydrolysate removal reduced the degradation of sugars normally observed with such intensive pretreatment (Katzen and Othmer 1942). The Scholler process was implemented in a commercial plant at Tornesch, Germany in 1931 that used large reactors (13.5 × 2.5 m) to treat 24 batches between shutdowns for cleaning. These shutdowns were used to remove lignin plugs that often formed in the ceramic-lined reactors, replace broken bricks in the reactor, and test for leaks. The sugar solutions from the reactors were cooled in heat exchangers and neutralized before being fed to a fermenter. In addition to the lignin plugs, resins accumulated in the heat exchangers and were another source of interruption that required physical cleaning. Even with high retention times and availability of physical labor for cleaning, sugar yields were observed to be only 49 % of theoretical, much lower than needed to be commercially viable for the German economy at the time. However, by 1941 this technology was used to establish at least 20 additional plants in Germany as a hedge against wartime disruptions of oil imports.

About this time, the Madison wood sugar process was established in the United States of America, removing hydrolysate by continuously flowing dilute acid through a packed bed of wood chips at temperatures ranging 150–180 °C. Harris and Beglinger (1945 and 1946) found that an extended heating period of 2.8 h with immediate removal of sugars improved sugar yields to 54 % of theoretical, and was thus an improvement on the Scholler process (Harris and Beglinger 1946; Harris et al. 1945). The Madison process was used to design and build a 4–6 million gallon per year ethanol plant in Springfield, Oregon (Bioenergy in Oregon 2010). Both the Scholler and Madison processes were based on a single-step hydrolysis process. Since these processes used repeated flushing with dilute acid to limit the formation of furfuraldehyde, treatment was less than optimal for total sugar hydrolysis. For more complete conversion of biomass with high sugar and low furfural yields, two stage hydrolysis was developed to separate hemicellulose hydrolysis from hydrolysis of recovered cellulose. Dunning and Lathrop (1945) developed a low temperature (98 °C), long retention time (50–185 min) pretreatment process with more concentrated sulfuric acid (4.9–9.8 %) that primarily released hemicellulosic sugars, which were then separated from biomass (Dunning and Lathrop 1945). The remaining biomass was re-treated at much higher temperature (120–130 °C) for 7–10 min to hydrolyze cellulose to glucose. This two-stage process led to 90 % of theoretical sugar yields. Since these findings, biomass hydrolysis has typically been implemented in two-stages, with a "pretreatment" stage designed to break down the lignin-hemicellulose matrix and a "saccharification" stage to hydrolyze the glucan in the residual solids (Tanjore et al. 2011).

Even though the studies conducted in the 1940s laid a strong foundation for acid pretreatment technologies of the day, much of the research was empirical and did not greatly increase scientific understanding of the fundamentals of sugar release. In the past couple of decades several more mechanistic acid pretreatment studies have established the pathways and kinetics of both hemicellulose and cellulose hydrolysis in the presence of lignin. Lee et al. (1999) noted that hemicellulose can convert to xylose and then to degradation products, such as furfuraldehyde and acetic acid, when biomass is treated at temperature higher than 180 °C (see 19.1) (Lee et al. 1999). Jacobsen and Wyman (2000) established that when biomass is treated with dilute acid at temperatures lower than 180 °C, hydrolysis occurs at two separate rates due to slow and fast-hydrolyzing hemicellulose fractions, and this asynchrony mitigates xylose release due to low conversion or high dehydration rates (see 19.2) (Jacobsen and Wyman 2000). Jacobsen and Wyman (2000) further introduced xylo-oligomers to the kinetic analysis; hemicellulose first converts to soluble xylo-oligomers before converting to monomeric xylose (see 19.3).

$$\text{Hemicellulose} \xrightarrow{k_1} \text{Xylose} \xrightarrow{k_2} \text{Degradation Products} = \text{Furfuraldehyde} \quad (19.1)$$

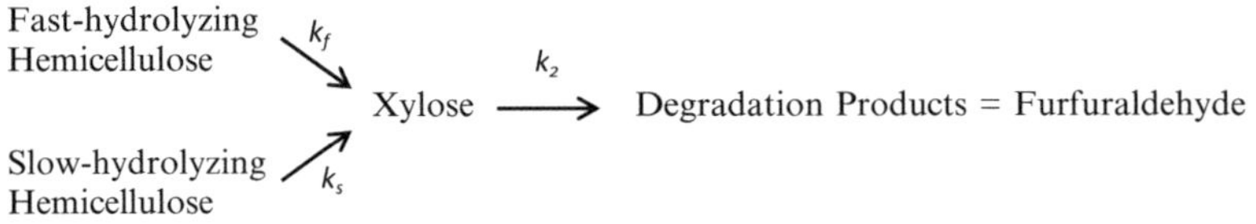

(19.2)

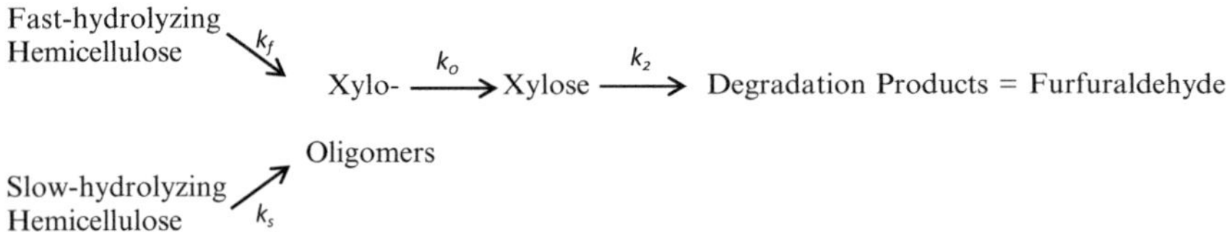

(19.3)

Pretreated biomass can be further hydrolyzed with either chemical or biological catalysts. If saccharification is accomplished by an additional acid hydrolysis, especially at higher severity, it can generate a number of decomposition products along with glucose. Several of these decomposition products have been shown to inhibit subsequent microbial fermentations. HydroxyMethylFurfural (HMF), an aldehyde of glucose, is formed when glucose in high severity treatments is dehydrated by one water molecule. HMF is a relatively unstable compared to its acid, levulinic acid, especially in the presence of water. Levulinic acid is formed when HMF rehydrates with two water molecules, and is co-produced along with acetic acid. High severity acid treatments can also lead to direct dehydration of glucose to levoglucosan, a sugar anhydride (see 19.4–19.7). Since the late 1970s, enzymatic hydrolysis of pretreated biomass with cellulases has proven to be an effective alternative to strong acids, minimizing production of inhibitors during the saccharification stage. High sugar yields (>95 % of theoretical) can be achieved with enzymatic treatment at 50 °C for 4–5 days of retention time (Brooks et al. 1979; Perez et al. 2013).

Glucan hydrolysis to Glucose:

$$C_6H_{12}O_5(162\,g) + H_2O\,(18\,g) \longrightarrow C_6H_{12}O_6(180\,g) \quad (19.4)$$

Glucose dehydration to HydroxyMethylFurfural (HMF):

$$C_6H_{12}O_6(180\,g) \longrightarrow C_6H_6O_3(126\,g) + 3H_2O\,(54\,g) \quad (19.5)$$

HMF rehydration to Levulinic and Acetic acids:

$$C_6H_6O_3(126\,g) + 2H_2O\,(36\,g) \longrightarrow C_5H_8O_3(116\,g) + CH_2O_2(46\,g) \quad (19.6)$$

Glucose dehydration to Levoglucosan:

$$C_6H_{12}O_6(180\,g) \longrightarrow C_6H_{10}O_5(162\,g) + H_2O(18\,g) \quad (19.7)$$

Both the hydrolysis of lignocellulose and the formation of inhibitors are a function of pretreatment severity, also called the Combined Severity (CS) factor, which is dependent of acid concentration, reaction temperature, and retention time. This CS factor can be calculated using an equation proposed by Chum et al. (1990), which is in-turn based on a P-factor proposed by Overend and Chronet (1987). Equation (19.8) indicates these relationships:

$$\text{Combined Severity (CS)} = \log R_0 - \text{pH}, \text{ where } R_0 = t^{\exp\left[\left(T_R - T_H\right)/14.75\right]} \quad (19.8)$$

where, pH = pH of the final slurry, t = reaction time, T_R = reaction temperature, T_H = reference temperature (100 °C)

Increasing pretreatment severity increases the susceptibility of acid pretreated biomass to cellulase treatments and leads to high, nearly theoretical glucose yields. For corn stover biomass, Lloyd and Wyman (2005) observed that increasing the combined severity (log CS) of acid pretreatment from 0.5 to 2.2 substantially increased glucose release after enzymatic saccharification, from 32 to 57 % (mass glucan/mass untreated biomass). On a theoretical basis, the overall sugar yield at log CS 2.2 was 82 % of theoretical, while the maximum overall sugar yield was observed at log CS 1.4 and was 92 % of theoretical. The reason a lower overall sugar yield was observed at a severity of 2.2 was due to a drop of xylose yield from 32 % (at log CS = 1.4) to 19 % of theoretical with increasing severity. Higher pretreatment severity led to the dehydration of xylose to furfuraldehyde, formic acid, and other degradation products leading to a reduction of yield, and this trend continued up to a the highest severity tested in this study (log CS = 2.5). While optimal treatment combinations for other feedstocks can be developed by similar empirical studies of sugar yields and kinetics, incomplete understanding of the fundamental physical and chemical interactions between hemicellulose, cellulose, and lignin during acid pretreatment and enzymatic saccharification continues to limit our ability to innovate and achieve a rational design.

As one step toward a more mechanistic understanding, Donohoe et al. (2008) observed that the lignin structure is modified during acid pretreatment to create detached lignin droplets instead of a continuous lignin matrix within the secondary cell wall. These droplets coalesce and migrate to the surface of the cell wall for re-deposition (Donohoe et al. 2008). Selig et al. (2007) showed that the re-deposited lignin droplets inhibited cellulase enzymes. In controlled experiments pretreating filter paper combined with lignified corn stem rind, sugar yields dropped 5–20 % when lignin droplets were re-deposited on the filter paper cellulose. An additional solvent extraction, with dioxane, was able to remove the lignin droplets that caused the enzymatic inhibition (Selig et al. 2007). Sannigrahi et al. (2010) determined that even cellulose forms degradation products with lignin after high severity acid treatments. These products were detected in the Klason lignin measurements, and were thereby named pseudo-lignin (Sannigrahi et al. 2011). Kumar et al. (2013) have shown that high severity pretreatment of Avicel microcrystalline cellulose, even in the absence of any lignin, can lead to the production of pseudo-lignin that also inhibits cellulase enzymes.

Several research teams have designed lab-scale acid pretreatment reactors in order to test hypotheses, improve process designs and generate pretreated feedstock for downstream processing. Such systems often rely on steam injection to obtain the required fast heating rates, which can be challenging to introduce in a controlled fashion at volumes as small as 10 ml or less. To mimic the very rapid heating rates of steam injection at a much smaller scale, researchers have used Hastelloy tube reactors with sealed ends to maintain the pressure in the reactors (Lee et al. 1999). Tubes of 1.3 cm (½ in.) diameter, when heated in preheated fluidized sand baths, were able to deliver a heating rate similar to that of steam injection. Until recently these were the smallest reactors used in the research community, but a high throughput system with 96 Hastelloy wells is now available for screening purposes (Studer et al. 2010). Several bench scale pretreatment reactors (10 ml to 10 L) were developed either from independent designs or by customizing off-the shelf reactors, such as Dionex accelerated solvent extraction and Parr reactors, that are built to withstand high temperatures and pressures (Narani et al. 2014). Reactors at the 100–1,000 L scale have been developed to process kilograms to hundreds of kilograms of biomass in both batch and continuous phases (AdvanceBioSystems 2012). Although building and operating high-pressure vessels can be expensive, the volume of catalyst required in the aqueous phase of dilute acid pretreatments is relatively low, 0.5 % (w/w) sulfuric acid. Apart from economic benefits, such low concentrations of acids minimize the need for neutralizing alkali and can also have environmental benefits (Shill et al. 2011). Concentrated acids can be employed at lower temperatures and pressures, but these larger quantities of acids have to be recovered and recycled to obtain environmental and economic benefits (Shill et al. 2011).

Concentrated Acids Can Be Effective at Low Pressures But Need to Be Recycled

Concentrated acids have been used for both laboratory-scale research and industrial-scale production (Miller and Hester 2007). Saeman et al. (1945) reported one of the first facilities to commercialize 72 % (w/w) sulfuric acid hydrolysis of wood and cellulose. They demonstrated that more than 70 % of theoretical sugar yields could be obtained from concentrated acid hydrolysis of wood, but only after a primary hydrolysis, or pretreatment, of the wood with a dilute acid (<1 % sulfuric acid) (Saeman et al. 1945). In later studies, concentrated acid was used primarily for pretreatment followed by saccharification with dilute acid. Biomass was first subjected to pretreatment with 72 % (w/w) sulfuric acid at 30 °C for 3 h. This was followed by thermochemical saccharification with 4 % (w/w) sulfuric acid in an autoclave at 121 °C for 1 h. This two-stage hydrolysis process is similar to the Klason lignin test, the gravimetric method applied in determining lignin concentration in biomass (Ritter et al. 1932; Sluiter et al. 2008). The fermentability of the sugar streams obtained from this process was observed to be up to 88.2 % of theoretical, much higher than the 60–70 % of theoretical reported for the Scholler process, possibly due to a lower chance of inhibitor formation at this lower temperature (Bergius 1937). This concentrated acid technology, or a version of it, is being evaluated for commercialization (Bakker et al. 2004; Lane 2012). Virdia, previously HCl Technologies and now acquired by Stora Enso, is currently looking to build a $60 million biochemical plan in Lafourche Parish, LA and extend this enzyme-free technology for commercial scale production (Young 2014).

A major advantage of employing concentrated acids for pretreatment is that high-pressure vessels can be avoided. Lower temperature reactors may not even need to be fabricated from an expensive corrosion-resistant metal; an acid resistant plastic, such as polyethylene, is resistant to 98 % (w/w) sulfuric acid up to 60 °C (Garverick 1994). While these advantages provide an economic benefit upfront in capital expenses, without highly efficient recovery and recycling of the mineral acids this technology can quickly become operationally expensive (Tuck et al. 2012). All biomass feedstocks are highly hygroscopic, and hemicellulose is the most hygroscopic amongst the three major polymer constituents (Acharjee et al. 2011). While most of the hemicellulose is removed during concentrated acid pretreatment, cellulose remaining in the residual solids undergoes a substantial change in crystallinity, making it more hygroscopic. As a result, it is quite difficult to recover the concentrated acid, and separating the solubilized pentoses from the acid can add several downstream unit operations. Although H_2SO_4 is much cheaper than HCl, the low volatility of HCl makes it more economical for recovery through distillation and recycle after condensation (Brown and Brown 2013). While several concentrated acids are being tested for application at low temperatures, other catalysts, such as dilute alkalis that are less hazardous, can also be effective at low temperatures and concentrations.

Dilute Alkalis Can Be Applied at Lower Temperatures But Require Longer Reaction Times

Alkali pretreatment has a long history as a chemical catalyst for lignocellulose hydrolysis. Dreyfus (1936) was issued a U.S. Patent for the invention of alkali pretreatment of lignocellulosic biomass to produce cellulose esters. Dreyfus treated biomass with 0.5–1 % (w/w) NaOH at 120–130 °C for up to 1 h. Similar to acid treatments, the technology has not undergone major changes, although our understanding of the changes in the chemical composition of biomass has improved substantially since then. Taherzadeh and Karimi (2008), in their review of the NaOH treatments indicate that this base can cause (1) disruption of lignin structure, (2) lignin removal, (3) reduced degree of polymerization (DP) of cellulose, (4) cellulose swelling and increased internal surface for improved mass transfer of catalyst during saccharification, (5) decreased cellulose crystallinity, and (6) no production of fermentation-inhibiting aldehydes (Taherzadeh and Karimi 2008). While dilute acid treatments can also disrupt lignin structure and reduce DP of cellulose, they do not have these other effects on biomass.

In addition to NaOH, several other catalysts including the less expensive lime ($Ca(OH)_2$) have been tested for pretreatment of lignocellulose (Chang et al. 1997; Kaar and Holtzapple 2000; Kim and Holtzapple 2006). Lime is more efficient in the presence of an oxidizing agent such as hydrogen peroxide or oxygen itself (Ahmadi et al. 2013). Lime costs less than NaOH and KOH and is easily recoverable by reaction with CO_2. Lime has also been effective in removing acetyl groups from hemicellulose, thereby reducing steric hindrance of enzymes during saccharification (Mosier et al. 2005b). Acetic acid itself is an inhibitor during fermentation, so removal of this and other carboxylic acids is an additional benefit of alkali pretreatment (Li et al. 2013b). Terrabon Inc. demonstrated lime treatment with intermittent pumping of oxygen at 55 °C, enzymatically hydrolyzed the pretreated biomass, and successfully fermented the sugars to produce carboxylic acids. However, the reaction time for this pretreatment was long, about 24 h. Li et al. (2013b) performed alkaline hydrogen peroxide (AHP) pretreatments at various solids loading and found that the yield of sugars dropped substantially with dry matter loadings beyond 30 % (w/w) biomass in total slurry. They found that catalyzing the treatment further with Copper bipyridine at a 2.0 mM concentration improved sugar yields by about 10 % to a total of 80 % of theoretical. The reaction time for this study was also 24 h. In general, acceptable alkali pretreatments can span hours to days, a reaction time that is not favorable for a commercial scale facility.

Ammonium hydroxide treatments have typically shown to effectively treat lignocellulosic biomass at shorter reaction times, between 5 min and 1 h (Hennessey et al. 2009). DuPont demonstrated this technology at their DDCE Biorefinery at Vonore, Tennessee and is further scaling it to their first commercial lignocellulosic ethanol plant in Nevada, Iowa. In this system ammonia is introduced along with steam to simulate NH_4OH pretreatment at 121 °C. The ammonia is captured and recycled to improve economics of the pretreatment process. The ammonia that is absorbed by the biomass can be repurposed as a nutrient, $(NH_4)_2SO_4$, and used for fermentation by neutralizing the biomass with H_2SO_4. Although such process intensification steps improve the economics of NH_4OH pretreatment, high-pressure vessels will be required to accommodate high vapor pressure from NH_4OH compared to other alkalis at 121 °C. After pretreatment, moisture must be removed or lignin may not be separated, potentially inhibiting downstream enzymatic and microbial treatments. Recently researchers have investigated bioconversion of lignin removed after alkali treatments, looking toward advanced fuels as a potential product (Linger et al. 2014). Lignin conversion has long been a topic of great interest but little success in the field of biomass conversion. Several chemical studies have been performed to recover lignin from biomass in a manner that can support downstream conversions to useful products (Adler 1977; Björkman 1956; Holladay et al. 2007). However, only a few pretreatments, such as those involving organic solvents, have been able to release lignin in a "useful" form.

Organosolv Treatments Produce Lignin That Can Potentially Be Converted to Useful Products

Organosolv pretreatment was invented by Theoder Kleinert as an improved method to dissolve lignin and cellulose with an organic solvent in the aqueous phase (Kleinert 1971). The organic solvent, often a lower aliphatic alcohol such as methanol or ethanol, was applied at a concentrations ranging from 20 to 75 % (v/v in water) and reaction temperatures around 180 °C (Kleinert 1971). Several other organic solvents, including dioxane, had previously been tested at room temperatures by researchers such as Björkman and Adler in the late 1950s (Adler 1977; Björkman 1956, 1957; Pepper et al. 1959). In the earliest reported attempt, wood suspended in toulene was finely milled in a ball mill and lignin was extracted with dioxane in water (9:1) (Björkman 1956). Such lignin, referred to as "Björkman lignin" or "milled wood lignin" contains very little carbohydrates, making it an attractive pretreatment and separation process to obtain both sugars and lignin for downstream conversion (Adler 1977).

New variations of organic solvent strategies continue to be developed. Cateto et al. (2011) showed that pretreatment of switchgrass with an ethanol-water mixture (75:25 v/v) along with sulfuric acid (0.9 % w/w final slurry) could not only substantially remove lignin but also render the cellulose-rich solids highly susceptible to enzymatic hydrolysis (Cateto et al. 2011). As with dilute acids, pretreatment with organic solvents reduces the degree of polymerization of glucan in the pretreated biomass. But organic solvents have additional effects that dilute acid pretreatments do not, including reducing the crystallinity of the recovered glucan to accelerate saccharification. Whereas dilute acid pretreatment must be followed by 4–5 days of enzymatic hydrolysis to achieve 95 % of theoretical glucan conversion, organosolv researchers have saccharified 70 % of theoretical glucan within just 8 h (Hallac and Ragauskas 2011), and achieved glucose and xylose yields greater than

90 and 50 % of theoretical (respectively) after 24 h of enzymatic hydrolysis (Cateto et al. 2011). Depending on the fermentation strategy (simultaneous with saccharification or sequential) and kinetics of downstream conversion, faster hydrolysis could reduce the reactor size required during enzymatic treatment and considerably decrease both capital and operational costs. More efficient recycling strategies for the organic solvent may further decrease the costs of process operations (Kautto et al. 2014). However, even with rapid and efficient processing, the investment costs associated with high-pressure corrosion resistant reactors are substantial. A combination of high yields of commodity chemicals and fuels from the sugars, along with value-added coproducts from the lignin, may be necessary for organosolv pretreatment to be economically competitive. Under the auspices of an EU public-private research consortium CIVM and ICN have been investigating such options for carboxylic acids (citric and acetic) (Snelders et al. 2014) and ethanol (Wildschut et al. 2013) respectively, and are making progress toward cost-effectively obtaining chemically convertable lignin and rapidly hydrolysable polysaccharides (Biocore 2013).

Water Alone Has Catalytic Capabilities at High Temperatures and Pressures

Water has autocatalytic properties at high temperatures and pressures, and this phenomena has been exploited through three distinct pretreatment approaches: (1) both temperatures and pressures above the supercritical point, which for water is 374 °C and 22.1 MPa (Peterson et al. 2008); (2) temperatures in the range of 180–220 °C with pressures sufficient to maintain hot water in the liquid state (Mosier et al. 2005a); and (3) pressures ranging up to 40 MPa, with temperatures often in the upper subcritical range (Elliott et al. 2013). All three of these approaches can be classified as hydrothermal processing (Peterson et al. 2008), although in common terminology the first is called supercritical fluid extraction, the second is called liquid hot water pretreatment, and the third approach is called hydrothermal liquefaction. While the first and second approaches are typically followed by saccharification to produce sugar hydrolysates, hydrothermal liquefaction produces a complex "biocrude" suitable for insertion in conventional petroleum refinery processes. Hydrothermal liquefaction is currently under active development by the Pacific Northwest National Laboratory, where high liquefaction rates have been observed at 350 °C and 20 MPa with the resulting biocrude upgraded by catalytic hydrotreating to produce liquid hydrocarbon fuels (Elliott et al. 2013).

In supercritical aqueous pretreatment water can not only dissolve biomass constituents but also separate them into sugar- and lignin-rich phases during cooling and depressurization. Antal et al. (1987) identified hemolysis of biomass in supercritical water as an alternative to high temperature pyrolysis treatments in that both can be non-specific and lead to several intermediates. Renmatix is commercializing supercritical water pretreatment, with Kilambi et al.'s patent application (2013) documenting over 85 % xylan release into the aqueous phase and more than 90 % of theoretical glucans and lignin in the solid phase. The solids recovered from this pretreatment were subjected to dilute acid (0.2 % w/w H_2SO_4) saccharification for only 3 s, after which more than 90 % of theoretical glucose was realized (Kilambi and Kadam 2013). The glucose and xylose streams can be combined for downstream conversion of mixed sugars, while the lignin remains available for other value-added processing.

Liquid hot water pretreatment can also be performed at temperatures and pressures similar to that of dilute acid pretreatment. At temperatures between 190 and 230 °C plant cell wall acetate is converted to acetic acid (contributing to the dilute acid effect), 35–60 % of lignin is dissolved, and significant amounts of xylan are released into the liquid phase (Mosier et al. 2005a; Qing and Wyman 2011; Qing et al. 2010; Yang and Wyman 2004). Most of the xylan is still in the form of xylo-oligomers, which need to be hydrolyzed further with xylanases to ferment them to fuels. Similarly, xylan remaining in the solid fraction also requires xylanases to release the pentoses into the liquid phase. The aforementioned production of acetic acid lowers the pH of the hydrolysates after hot water pretreatment to 4.0 and 4.5. However, this pH does not mandate an expensive corrosion-resistant alloy, but still requires a high-pressure vessel (Saha et al. 2013). Mosier et al. (2005b) optimized hot water pretreatment of corn stover in a flow-through reactor at 190 °C for 15 min, with subsequent enzymatic saccharification achieving 90 and 80 % of theoretical conversion of glucose and xylose respectively (Mosier et al. 2005a).

Ionic Liquids Can Be Applied for Hydrolysis at Atmospheric Pressures

Dilute acid, dilute alkali, organosolv and hydrothermal pretreatments all require high-pressure vessels rated to pressures from 0.103 MPa ($Ca(OH)_2$ or NaOH treatments at 121 °C) up to 0.90–1.24 MPa (H_2SO_4 or water pretreatments at 180–193 °C) (McNeill 1998). The cost of fabricating and operating pressure vessels is substantial, increasing exponentially with increasing temperature and linearly with increasing pressure of steam (Chen et al. 2014). At these temperatures and pressures, continuous pretreatment with high solids slurries has also been found to be operationally difficult, with possible downtime due to repairs (Chen et al. 2014). To address these material flow challenges as well as reactor capital costs, several groups have been investigating the use of room temperature ionic liquids for biomass pretreatment. Ionic liquids (IL), are liquid salts at room temperatures that can be applied on lignocellulosic biomass at atmospheric pressures when heated up to 160 °C to produce rapidly hydrolysable polysaccharides and "useful" lignin (Moens and Khan 2002; Rogers and Seddon 2002; Swatloski et al. 2002).

Li et al. (2010) have shown that 80 % of glucan in Emim-Acetate pretreated solids can be enzymatically hydrolyzed within 8 h of treatment. This is primarily due to significant reductions in the crystallinity of cellulose. During a 3 h pretreatment at 160 °C the crystallinity index dropped from 26.2 in untreated biomass to 2.6 in IL pretreated biomass, whereas for acid treated biomass (1.2 %

w/w sulfuric acid for 1 h) the crystallinity index increased to 39.1 (Kumar et al. 2009). IL pretreatment also led to the release of 69.2 % of the biomass lignin into the liquid phase (Li et al. 2010). Li et al. (2013a, b) was the first IL pretreatment scale-up study, and was conducted at 6 L. They established that the process is scalable to a bench scale and defined the amount of washing required to avoid inhibition of commercial enzymes by the residual IL in pretreated solids (Li et al. 2013a).

Sun et al. (2013) developed an enzyme-free acidolysis process to not only avoid washing but also provide the opportunity to perform both pretreatment and saccharification in one-pot. In this process, IL was used for pretreatment and dilute HCl for saccharification, with over 83 and 52 % of theoretical glucose and xylose respectively released into the liquid phase (Sun et al. 2013). Another one-pot approach was investigated by developing an IL resistant cellulase cocktail, JTherm (Park et al. 2012). Shi et al. (2013) performed a one-pot study with the JTherm enzymes after diluting IL to 20 % (w/w slurry). They achieved 80 % theoretical glucose yields that were recovered using a boronic acid based system. They were also able to recover both the dissolved lignin in IL and the IL itself by tangential flow filtration and sequential filtration followed by vacuum evaporation, respectively (Shi et al. 2013).

Ionic liquids are expensive and very high recovery and recycling efficiencies will be required for this pretreatment to be commercially viable. Shi et al. (2013) have shown that IL used for treatment of lignocellulosic biomass can be recovered at the rate of 90.8 % of theoretical. Shill et al. (2011) also observed a similar recovery yield but only after diluting the IL in water (Shill et al. 2011). In general, recovery and recycling of homogenous pretreatment catalysts are challenging, especially due to the hygroscopic nature of biomass. Easier separation of catalyst from biomass can often be achieved when the catalyst is heterogenous.

Heterogenous Catalysts May Lead to Mass Transfer Issues But Can Be Recycled

Biomass and crystalline cellulose are insoluble substrates that are difficult to dissolve in water. When suspended uniformly in an aqueous phase, homogeneous catalysts in the same phase are capable of penetrating into these insoluble substrates. In the case of heterogeneous catalysts, especially solid catalysts, surface area contact between the solid biomass substrate and solid catalyst is usually not sufficient for effective pretreatment of biomass. Also, with precipitated or residual biomass solids in the system after catalysis, recycling and regenerating the solid catalysts becomes a challenge, nullifying the economic benefit of applying the catalyst. While several researchers and companies have investigated ways to enhance the efficiency and performance of solid heterogeneous catalysts for cellulose hydrolysis after pretreatment of biomass (Huber et al. 2006; Stöcker 2008), difficult challenges remain. Although the enzymes used for hydrolysis of cellulose can also considered solid heterogeneous catalysts (Wald et al. 1984), here and elsewhere they are considered a separate category. An additional group of heterogeneous catalysts act on solid biomass from a gas phase, where mass transfer and recovery challenges are often easier to address.

Gaseous Catalysts Provide the Option of Performing Physico-chemical Treatments

Several gaseous catalysts including sulfur dioxide, carbon dioxide, ozone, etc. have been evaluated for pretreatment of lignocellulosic biomass (Neely 1984; Shi et al. 2011a; Zheng et al. 1998). While several of these catalysts can chemically break down the lignin-hemicellulose matrix, most were only successful when the catalysis was combined with a physical explosion-like process, i.e. rapid release of pressure of the reactor and releasing the biomass to atmospheric conditions within a few seconds. Such rapid release leads to cell wall expansion and makes cellulose available for further hydrolysis.

Steam Explosion

Steam Explosion was developed as the Masonite Process and as early as 1930 (Boehm 1930). The original steam guns, much like those used today, were used to pressurize wood chips to up to 1,000 lb of steam pressure. When the pressure was released the fibers separated and were then used for paper production. One important advantage of applying steam explosion is the ability to produce pretreated biomass at moisture contents as low as approximately 50 % (w/w wet basis). Such solids lead to higher concentrated sugar solutions, up to 100 g/l, which is preferable for many downstream conversions and recovery of fuel. However, such high solids treatments with only water in the gas phase typically lead to low yields. Wu et al. (2013) observed only a 73.6 % (theoretical) conversion of cellulose after treating corn stover at 30 % (w/w) solid loading and enzymatically hydrolyzing the pretreated solids. Including sulfur dioxide during steam explosion had substantially improved the overall yield and kinetics of cellulose conversion during subsequent cellulose hydrolysis (Kang et al. 2013). Kang et al. observed more than 80 % (theoretical) cellulose conversion within 6 h of enzymatic hydrolysis after treating loblolly pine at 215 °C for 5–10 min with sulfur dioxide concentrations of 3–5 % (g/g biomass).

Recent studies have shown that such rapid release combined with acidic catalysis causes a morphological change at the cellular level (Ciesielski et al. 2014; Wang et al. 2014). Slow release of pressure when pretreating corn stover at 2 % (w/w dry biomass) sulfuric acid loading at 160 °C for 5 min in a zipper clave reactor led to statistically negligible reduction in cell wall thickness compared to untreated material. However, the same treatment in a batch steam gun reactor led to halving cell wall thickness and considerable morphological changes as detected by confocal scanning laser microscopy. This effect was further pronounced when the biomass was treated using a horizontal screw reactor, which is a continuous reactor that can also be configured to release biomass pressure rapidly (Ciesielski et al. 2014). While steam explosion has been proven to scale very well, there is still a chance of diluting the sugar solutions that will be produced. To further avoid such dilution, there are advantages to using a gaseous catalyst other than steam that can enhance the hydrolysis process, although such catalysts will need to be recovered and recycled.

Ammonia Fiber EXpansion (AFEX)

Initiated as Ammonia Freeze Explosion in 1982, AFEX was initially applied to treat biomass with ammonia at high pressures of 1.7 MPa for 30 min but at low temperature of 27 °C (Reilly 1987). This process evolved to applying ammonia on moist biomass at 90 °C at 2.1 MPa for 1 h. The ammonia released from this reactor was further recycled. However, a significant amount of ammonia was absorbed onto the biomass. This was converted to ammonium sulfate, a nutrient in fermentation, by neutralizing the alkaline moist biomass with sulfuric acid (Sendich et al. 2008). AFEX treatment does not cause the production of common pretreatment inhibitors such as furfural and HMF. However, minimal removal of lignin due to the “dry to dry” treatment has been suggested to have inhibited enzyme adsorption during hydrolysis (Shi et al. 2011b). Recent efforts in AFEX have used biomass moisture to release lignin from the reactor in an aqueous stream separate from the polysaccharide-rich solid phase. A pilot-scale AFEX system is shown in Fig. 19.3.

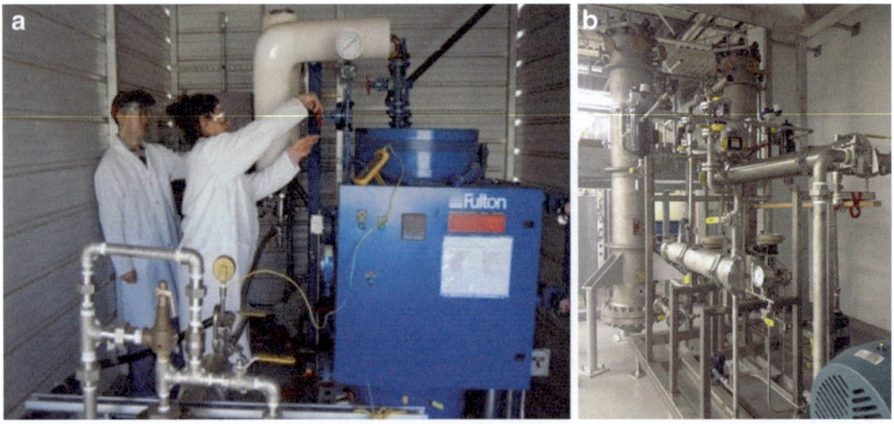

Fig. 19.3 (**a**) Steam Gun at UC-Riverside and (**b**) AFEX Pretreatment Reactor at MBI (*Courtesy*: http://www.cert.ucr.edu/Cellulosic%20Biomass.pdf and http://www.mbi.org/technologies-3/afex/)

AFEX has been promoted as being particularly suitable for decentralized biomass pretreatment process. Small depots that would use the AFEX process to pretreat and pelletize biomass have been proposed, marketing products as both a value-added and highly digestible animal feed and as a biofuels feedstock. AFEX pellets have been shown to perform as well as loose pretreated biomass (Bals et al. 2014). Decentralization and densification promise to reduce transportation related issues that can inflate the supply chain costs of biorefineries, thereby posing a threat to commercialization (Carolan et al. 2007).

Radiant Heat Transfer May Reduce Energy Requirements for Thermal Pretreatment

As is by now apparent, many of the most effective pretreatment processes are operated at elevated temperatures. Heating biomass to those temperatures, especially in an aqueous media, requires a major investment of energy and comes at a considerable operational cost. Many of these processes have been developed at laboratory scale where conductive thermal transfer through vessel walls is sufficient, but this heat transfer mechanism can be both inefficient and uneven at larger scales. Instead, pilot- and commercial-scale systems more commonly use convection, especially of high heat capacity fluids like water, compressed steam, or ammonia. These fluids can be heated externally and then injected to provide more rapid and uniform heat transfer than is possible with conduction. Biomass can also be heated by electromagnetic radiation, and this third mechanism may offer advantages for certain pretreatment processes. While most biomass feedstocks are opaque to visible light and other mid-range wavelengths, these materials are much more translucent to wavelengths in the microwave range (~10^{-2} m), so for some reactor materials and configurations microwaves may penetrate sufficiently to offer practical potential.

Over the last few decades there has been a revolution in thermal transfer technology in the food industry, with the microwave oven now ubiquitous in both commercial and personal kitchens. Thus it is not surprising that several research groups have applied this alternative energy transfer strategy to lignocellulosic biomass (Azuma et al. 1984; Hu and Wen 2008; Intanakul et al. 2003; Ooshima et al. 1984; Zhu et al. 2005). While most of these studies have been exploratory investigations without extensive optimization for specific feedstocks or saccharification systems and with relatively low sugar yields, there have been several indications that these technologies can provide equal or improved hydrolysis to conventional thermal heating methods. Importantly, these studies have demonstrated the potential to reduce the energy input required to achieve a particular degree of pretreatment severity. As lignocellulosic biomass pretreatment technologies are commercialized and operational costs become the dominant cost factor for biofuels, it will become increasingly important to invest in efficient means of thermal energy transfer.

Identifying the Best Treatment Strategy

Several comparative studies have been performed to identify the most appropriate deconstruction strategy for a given feedstock or to establish a novel technology (Li et al. 2010; Sathitsuksanoh et al. 2012; Zhu et al. 2009). The Biomass Refining Consortium for Applied Fundamentals and Innovation (CAFI) is one such group of leading researchers that studied the effects of several pretreatment technologies on enzymatic hydrolysis of corn stover (Mosier et al. 2005b; Wyman et al. 2005) and later poplar. Perhaps the most important conclusion from the comparative research thus far is that no one particular pretreatment is the best for all feedstocks. This is due to the diverse compositions and biochemical makeup of the feedstocks, benefits and limitations of the pretreatment technologies, and complex interactions with downstream saccharification and/or conversion technologies (see Fig. 19.4).

Table 19.1 is a summary of the advantages and disadvantages that can be realized from adopting some prominent pretreatment technologies for the hydrolysis of lignocellulosic biomass. In attempting to identify which pretreatment is the best for any specific situation, it is first essential to recognize that any given pretreatment technology may not be applicable for all situations. The restrictions associated with a particular situation will mandate the course of selecting a deconstruction process. A series of topics are provided below that need to be addressed prior to choosing a

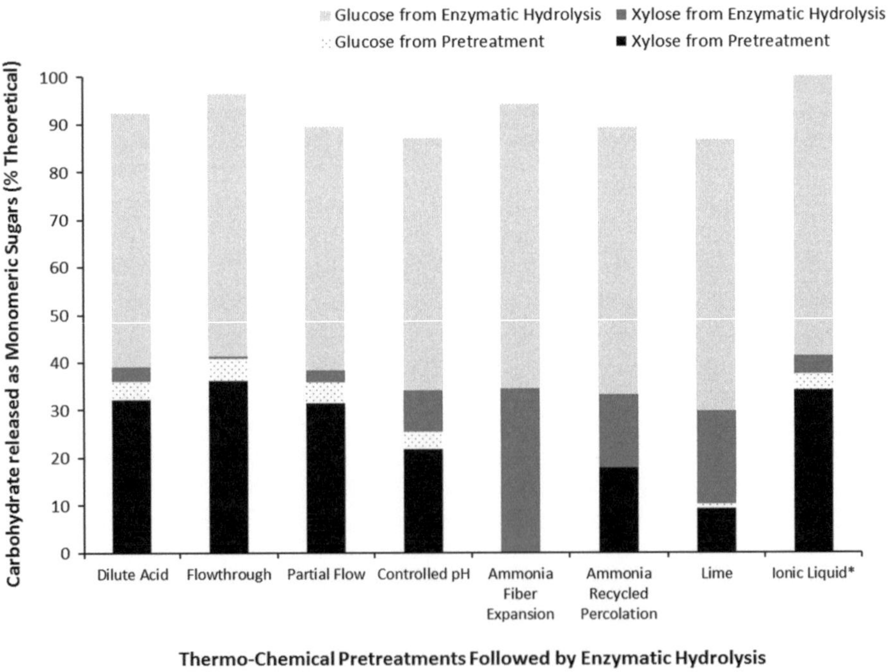

Fig. 19.4 Impact of various thermo-chemical pretreatments on sugar yields after enzymatic hydrolysis (Elander et al. 2009). Ionic liquid* pretreatment data obtained from elsewhere (Li et al. 2011)

Table 19.1 Comparison between some prominent pretreatment

Pretreatment	Advantage	Disadvantage
Dilute acid	Shorter reaction times Scaled to commercial production Xylan conversion to xylose	Inhibitor production Slower enzymatic hydrolysis Corrosion-resistant, high-pressure vessel required
Steam explosion	Shorter reaction times Scaled to commercial production No hazardous catalyst required	Xylo-oligomer conversion Slower enzymatic hydrolysis Corrosion-resistant, high-pressure vessel required
Dilute alkali	Lignin removal Scaled to commercial production Lower pressure treatment	Longer reaction time Slower enzymatic hydrolysis Pressure vessel required
Supercritical water	Shorter reaction times Scaled to pilot-level production Prospects of "Useful" lignin	High-pressure vessel required Inhibitors may require separation
Hydrothermal liquefaction	Shorter reaction times Scaled to pilot-level production Petro-refinery feedstock	High-pressure vessel required High-pressure feed required Limited scale up data
AFEX	Shorter reaction times Scaled to pilot-level production De-centralization prospects	Concerns of ammonia safety Slower enzymatic hydrolysis High-pressure vessel required
Organosolv	Shorter reaction times Faster enzymatic hydrolysis Prospects of "Useful" lignin	Flammability issues Cost of organic solvent mandates recovery Corrosion resistant high-pressure vessel required
Ionic liquids	Faster enzymatic hydrolysis Prospects of "Useful" lignin Lower enzyme loading required	Longer reaction time Very little scale-up data Cost of ionic liquids mandates recovery

pretreatment technology for a specific lignocellulosic biorefinery. While these topics can be addressed in any order, careful selection of a technology may require that this set of topics be addressed several times, iteratively, with frequent detailed interaction with a technology developer or supplier.

1. Identifying the feedstock and local resource constraints for the biorefinery.
   Geographical location will largely determine available feedstocks as well as other input opportunities and constraints. These are probably the most important aspects in choosing a deconstruction process. The following aspects are inherently dependent on these parameters.
   (a) Feedstock Type(s)—All pretreatment technologies are not applicable for all feedstock types. Understanding the compositions of the available feedstocks and their interaction with several pretreatment technologies is imperative in downsizing the various options to a few prospects.
   (b) Availability of Water—The availability of water is the most likely resource constraint for a lignocellulosic bioconversion process. If this limitation is severe and water cannot be recycled from downstream processing, it can

even reduce the list of possibilities to one pretreatment option, AFEX treatment, a dry process.

(c) Other utility and infrastructure connections—While lignocellulosic biorefineries can be net power generators, different pretreatment approaches have different energy requirements for natural gas and electricity. Pretreatment will also influence biofuel yield and the volume and characteristics of coproducts and residues that will need to be transported and marketed or disposed.

2. Assessing downstream compatibility.

   (a) Compatibility with saccharification strategy (Enzymes or Enzyme-Free)—If enzyme treatment is the choice of second-stage hydrolysis, it will be necessary to choose a pretreatment that produces biomass with a low crystallinity and degree of polymerization. If simultaneous saccharification and fermentation is planned, are reaction kinetics and temperature requirements compatible?
   (b) Robustness of Fermentation Organism/Chemical Catalyst—If biomass hydrolysates are to be converted by a fermentation organism or a chemical catalyst that can be poisoned by presence of lignin, it will be essential to delignify the biomass during deconstruction. Alkali pretreatments may be more advantageous than acid pretreatments in such cases.
   (c) Lignin-based Co-Products—If one of the downstream processes is designed to convert low molecular weight lignin into value-added products, it will be necessary to ensure that the deconstruction process is compatible with this objective.

3. Choosing between operational cost and capital investment.

   (a) Choice of reactor metal—If it is not economically or technically feasible to fabricate a reactor out of expensive metallurgy such as Hastelloy or Incoloy, acid pretreatment or ionic liquid acidolysis are not likely to be viable options. In such cases it may be preferable to choose more expensive catalysts such as ionic liquids and enzymes that can be applied at atmospheric pressures and low temperatures and lead to higher OPEX but lower CAPEX.
   (b) Catalyst requirements, recovery, and lifetime. Both biocatalysts (such as enzymes) and chemical catalysts have a limited lifetime before they degrade or are poisoned. These can represent a large fraction of operational costs, and so must be well understood.

4. Availability of Scale-Up Data.

   To make an informed decision for a commercial plant, a deconstruction strategy must absolutely have been tested at mini-pilot, pilot, and demonstration scales. Materials handling before and after pretreatment, mass transfer of catalyst and water during the treatment, choice of impellers, reactor designs, solid–liquid separation, etc. are all issues that arise at different levels when the technology is tested at the various scales. The mass balance closure for scale-up studies should also be high (between 90 and 110 %) to be able to establish the commercial viability of a process (see Fig. 19.5).

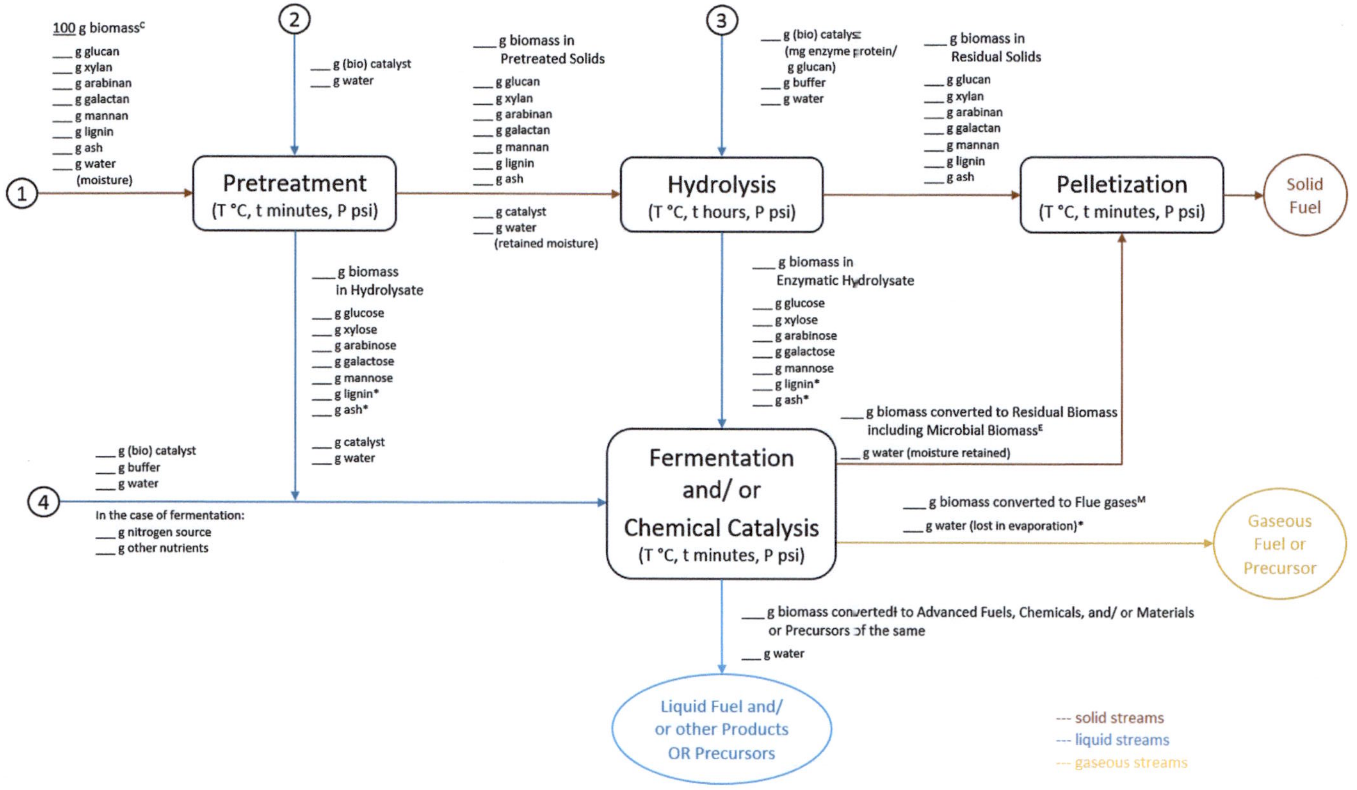

Fig. 19.5 An example mass balance flow sheet with four input streams and three product streams from biomass (Zhang et al. 2009). (*C*) Only stream 1 or the biomass stream carries carbon as an input; *Typically calculated after subtracting from measured values assuming 100 % recovery; (*E*) May require elemental analysis to primarily complete carbon balance; (*M*) May require mass spectrometry to identify the flue gases

5. Techno-Economic Analysis, which should incorporate:

   (a) Mass balance data and energy requirements from scale-up studies.
   (b) Possibility of catalyst recycle or re-use in another unit operation.
   (c) Energy requirements for solid–liquid handling, separation, mixing, etc.
   (d) Co-product value and residue disposal costs at scale.

References

Acharjee, T. C., Coronella, C. J., & Vasquez, V. R. (2011). Effect of thermal pretreatment on equilibrium moisture content of lignocellulosic biomass. *Bioresource Technology, 102*(7), 4849–4854.

Adler, E. (1977). Lignin chemistry—Past, present and future. *Wood Science and Technology, 11* (3), 169–218.

AdvanceBioSystems. (2012). AdvaceBio supplied SuPR2G reactor to Brazilian laboratory. In AdvanceBioSystems (Ed.), *Biomass magazine* (Vol. 2014). University Park, PA: BBI International.

Ahmadi, F., Rad, A. R., Holtzapple, M. T., & Zamiri, M. J. (2013). Short-term oxidative lime pretreatment of palm pruning waste for use as animal feedstuff. *Journal of the Science of Food and Agriculture, 93*(8), 2061–2070.

Almin, K. E., Eriksson, K.-E., & Jansson, C. (1967). Enzymic degradation of polymers: II. Viscometric determination of cellulase activity in absolute terms. *Biochimica et Biophysica Acta (BBA) Enzymology, 139*(2), 248–253.

Antal, M. J., Brittain, A., DeAlmeida, C., Ramayya, S., & Roy, J. C. (1987). *Heterolysis and homolysis in supercritical water. Supercritical fluids: Chemical and engineering principles and applications* (pp. 77–86). Washington, DC: American Chemical Society.

Azuma, J., Tanaka, F., & Koshijima, T. (1984). Enhancement of enzymatic susceptibility of lignocellulosic wastes by microwave irradiation. *Journal of Fermentation Technology, 62*(4), 377–384.

Bakker, R., Gosselink, R., Maas, R., De Vrije, T., De Jong, E., Van Groenestijn, J., et al. (2004). Biofuel production from acid-impregnated willow and switchgrass. *Proceedings 2nd World Conference on Biomass for Energy, Industry and Climate Protection*. 10 May 2008, Rome, Italy.

Bals, B. D., Gunawan, C., Moore, J., Teymouri, F., & Dale, B. E. (2014). Enzymatic hydrolysis of pelletized AFEX™-treated corn stover at high solid loadings. *Biotechnology and Bioengineering, 111*(2), 264–271.

Barakat, A., de Vries, H., & Rouau, X. (2013). Dry fractionation process as an important step in current and future lignocellulosic biorefineries: A review. *Bioresource Technology, 134*, 362–373.

Barakat, A., Chuetor, S., Monlau, F., Solhy, A., & Rouau, X. (2014). Eco-friendly dry chemo-mechanical pretreatments of lignocellulosic biomass: Impact on energy and yield of the enzymatic hydrolysis. *Applied Energy, 113*, 97–105.

Beckham, G. T., Matthews, J. F., Bomble, Y. J., Bu, L., Adney, W. S., Himmel, M. E., et al. (2010). Identification of amino acids responsible for processivity in a family 1 carbohydrate-binding module from a fungal cellulase. *The Journal of Physical Chemistry B, 114*(3), 1447–1453.

Béguin, P. (1983). Detection of cellulase activity in polyacrylamide gels using Congo red-stained agar replicas. *Analytical Biochemistry, 131*(2), 333–336.

Berghem, L. E., & Pettersson, L. G. (1973). The mechanism of enzymatic cellulose degradation. *European Journal of Biochemistry, 37*(1), 21–30.

Berghem, L. E. R., Pettersson, L. G., & Axiö-Fredriksson, U.-B. (1976). The mechanism of enzymatic cellulose degradation. *European Journal of Biochemistry, 61*(2), 621–630.
Bergius, F. (1937). Conversion of wood to carbohydrates. *Industrial and Engineering Chemistry, 29*(3), 247–253.
Biocore. (Ed). (2013). *Newsletter Issue 3*. In: http://www.biocore-europe.org/file/Newsletter%20BIOCORE%20Issue%203.pdf. Paris, France.
Bioenergy in Oregon. (2010). *Report to the oregon legislature, August 2010: Biofuels impact study* (Vol. 2014). Salem, OR: Bioenergy in Oregon. Oregon.gov.
Björkman, A. (1956). *Svensk Papperstidn, 59*, 477.
Björkman, A. (1957). Svensk Papperstidn, 59, 477, 1956; A. Bjorkman and B. Person. *Svensk Papperstidn, 60*(158), 285.
Boehm, R. M. (1930). The masonite process 1. *Industrial and Engineering Chemistry, 22*(5), 493–497.
Boynton, L. C., & Miller, R. C. (1927). The occurrence of a cellulase in the ship-worm. *Journal of Biological Chemistry, 75*(2), 613–618.
Brooks, R. E., Su, T. M., Brennan, M. J. J., Frick, J., & Lynch, M. (1979). *Bioconversion of plant biomass to ethanol*. Final report, 15 December 1976–31 December 1978. COO-4147-7 United States10.2172/5141284Thu Aug 22 07:33:20 EDT 2013NTIS, PC A03/MF A01.ERA-05-034658; EDB-80-105542English.
Brown, R. C., & Brown, T. R. (2013). *Biorenewable resources: engineering new products from agriculture*. New York, NY: Wiley.
Carolan, J. E., Joshi, S. V., & Dale, B. E. (2007). Technical and financial feasibility analysis of distributed bioprocessing using regional biomass pre-processing centers. *Journal of Agricultural and Food Industrial Organization, 5*(2).
Cateto, C., Hu, G., & Ragauskas, A. (2011). Enzymatic hydrolysis of organosolv Kanlow switchgrass and its impact on cellulose crystallinity and degree of polymerization. *Energy & Environmental Science, 4*(4), 1516–1521.
Chang, V. S., Burr, B., & Holtzapple, M. T. (1997). Lime pretreatment of switchgrass. *Applied Biochemistry and Biotechnology, 63–65*, 3–19.
Chen, X., Shekiro, J., Pschorn, T., Sabourin, M., Tao, L., Elander, R., et al. (2014). A highly efficient dilute alkali deacetylation and mechanical (disc) refining process for the conversion of renewable biomass to lower cost sugars. *Biotechnology for Biofuels, 7*(1), 98.
Chum, H., Johnson, D., Black, S., & Overend, R. (1990). Pretreatment-catalyst effects and the combined severity parameter. *Applied Biochemistry and Biotechnology, 24–25*(1), 1–14.
Ciesielski, P. N., Wang, W., Chen, X., Vinzant, T. B., Tucker, M. P., Decker, S. R., et al. (2014). Effect of mechanical disruption on the effectiveness of three reactors used for dilute acid pretreatment of corn stover. Part 2: Morphological and structural substrate analysis. *Biotechnology for Biofuels, 7*(1), 1–12.
Diehl, W. W. (1949). Concerning the identity of Iterson's cellulolytic mycogone. *Mycologia, 277–279*.
Donohoe, B. S., Decker, S. R., Tucker, M. P., Himmel, M. E., & Vinzant, T. B. (2008). Visualizing lignin coalescence and migration through maize cell walls following thermochemical pretreatment. *Biotechnology and Bioengineering, 101*(5), 913–925.
Dreyfus, H. (1936). Lignocellulosic material. Patent US2047314 A.
Dunning, J., & Lathrop, E. C. (1945). Saccharification of agricultural residues. *Industrial and Engineering Chemistry, 37*(1), 24–29.
Duggar, B. M., & Davis, A. R. (1914). Enzyme action in Fucus vesiculosus L. *Annals of the Missouri Botanical Garden, 1*(4), 419–426.
Elander, R., Dale, B., Holtzapple, M., Ladisch, M., Lee, Y. Y., Mitchinson, C., et al. (2009). Summary of findings from the biomass refining consortium for applied fundamentals and innovation (CAFI): Corn stover pretreatment. *Cellulose, 16*(4), 649–659.

Elliott, D. C., Hart, T. R., Schmidt, A. J., Neuenschwander, G. G., Rotness, L. J., Olarte, M. V., et al. (2013). Process development for hydrothermal liquefaction of algae feedstocks in a continuous-flow reactor. *Algal Research, 2*(4), 445–454.

Energy.Gov. (2014). *Project LIBERTY biorefinery starts cellulosic ethanol production.* E.E.a.R. Energy. Retrieved from http://energy.gov/articles/project-liberty-biorefinery-starts-cellulosic-ethanol-production.

Euler, H. (1912). Zur kenntnis der cellulase. *Angewandte Chemie, 25*(6), 250–251.

Faith, W. L. (1945). Development of the scholler process in the United States. *Industrial and Engineering Chemistry, 37*(1), 9–11.

Garverick, L. (1994). *Corrosion in the petrochemical industry*. Geauga, OH: ASM International.

Ghose, T. K. (1987). Measurement of cellulase activities. *Pure and Applied Chemistry, 59*(2), 257–268.

Gui, X., Wang, G., Hu, M., & Yan, Y. (2013). Combined fungal and mild acid pretreatment of glycyrrhiza uralensis residue for enhancing enzymatic hydrolysis and oil production. *BioResources, 8*(4), 5485–5499.

Hallac, B. B., & Ragauskas, A. J. (2011). Analyzing cellulose degree of polymerization and its relevancy to cellulosic ethanol. *Biofuels, Bioproducts and Biorefining, 5*(2), 215–225.

Hames, B., Ruiz, R., Scarlata, C., Sluiter, A., Sluiter, J., & Templeton, D. (2008). Preparation of samples for compositional analysis. *Laboratory Analytical Procedure, 1617.*

Harris, E. E., & Beglinger, E. (1946). Madison wood sugar process. *Industrial and Engineering Chemistry, 38*(9), 890–895.

Harris, E. E., Beglinger, E., Hajny, G. J., & Sherrard, E. C. (1945). Hydrolysis of wood – Treatment with sulfuric acid in a stationary digester. *Industrial and Engineering Chemistry, 37*(1), 12–23.

Hennessey, S. M., Friend, J., Elander, R. T., & Tucker, M. P. (2009). Biomass pretreatment. Google Patents.

Higuchi, T. (1985). Biosynthesis of lignin. Biosynthesis and biodegradation of wood components. pp. 141–160.

Himmel, M. E., Ding, S.-Y., Johnson, D. K., Adney, W. S., Nimlos, M. R., Brady, J. W., et al. (2007). Biomass recalcitrance: engineering plants and enzymes for biofuels production. *Science, 315*(5813), 804–807.

Holladay, J., Bozell, J., White, J., & Johnson, D. (2007). *Top value-added chemicals from biomass.*

Hu, Z., & Wen, Z. (2008). Enhancing enzymatic digestibility of switchgrass by microwave-assisted alkali pretreatment. *Biochemical Engineering Journal, 38*(3), 369–378.

Huber, G. W., Iborra, S., & Corma, A. (2006). Synthesis of transportation fuels from biomass: chemistry, catalysts, and engineering. *Chemical Reviews, 106*(9), 4044–4098.

Hurst, P. L., Sullivan, P. A., & Shepherd, M. G. (1978). Substrate specificity and mode of action of a cellulase from Aspergillus niger. *Biochemistry Journal, 169*, 389–395.

Intanakul, P., Krairiksh, M., & Kitchaiya, P. (2003). Enhancement of enzymatic hydrolysis of lignocellulosic wastes by microwave pretreatment under atmospheric pressure. *Journal of Wood Chemistry and Technology, 23*(2), 217–225.

Jacobsen, S., & Wyman, C. (2000). Cellulose and hemicellulose hydrolysis models for application to current and novel pretreatment processes. In M. Finkelstein & B. Davison (Eds.), *Twenty-first symposium on biotechnology for fuels and chemicals* (pp. 81–96). New York, NY: Humana.

Jacobson, C. A., & Holmes, A. (1914). Enzymes present in Alfalfa. Alfalfa investigation. V. *Journal of the American Chemical Society, 36*(10), 2170–2182.

Jeremic, D., Goacher, R. E., Yan, R., Karunakaran, C., & Master, E. R. (2014). Direct and up-close views of plant cell walls show a leading role for lignin-modifying enzymes on ensuing xylanases. *Biotechnology for Biofuels, 7*(1), 496.

Kaar, W. E., & Holtzapple, M. T. (2000). Using lime pretreatment to facilitate the enzymic hydrolysis of corn stover. *Biomass and Bioenergy, 18*(3), 189–199.

Kang, Y., Bansal, P., Realff, M. J., & Bommarius, A. S. (2013). SO2-catalyzed steam explosion: The effects of different severity on digestibility, accessibility, and crystallinity of lignocellulosic biomass. *Biotechnology Progress, 29*(4), 909–916.

Katzen, R., & Othmer, D. F. (1942). Wood hydrolysis. A continuous process. *Industrial and Engineering Chemistry, 34*(3), 314–322.

Kautto, J., Realff, M. J., Ragauskas, A. J., & Kässi, T. (2014). Economic analysis of an organosolv process for bioethanol production. *BioResources, 9*(4), 6041–6072.

Keller, F. A., Hamilton, J. E., & Nguyen, Q. A. (2003). Microbial pretreatment of biomass. *Applied Biochemistry and Biotechnology, 105*(1-3), 27–41.

Kilambi, S., & Kadam, K. L. (2013). Production of fermentable sugars and lignin from biomass using supercritical fluids. Google Patents

Kim, S., & Holtzapple, M. T. (2006). Effect of structural features on enzyme digestibility of corn stover. *Bioresource Technology, 97*(4), 583–591.

Kleinert, T.N. (1971). Organosolv pulping and recovery process. Google Patents.

Kressman, F. W. (1922). *The manufacture of ethyl alcohol from wood waste*. Washington, DC: US Department of Agriculture.

Kumar, R., Mago, G., Balan, V., & Wyman, C. E. (2009). Physical and chemical characterizations of corn stover and poplar solids resulting from leading pretreatment technologies. *Bioresource Technology, 100*(17), 3948–3962.

Kumar, R., Hu, F., Sannigrahi, P., Jung, S., Ragauskas, A. J., & Wyman, C. E. (2013). Carbohydrate derived-pseudo-lignin can retard cellulose biological conversion. *Biotechnology and Bioengineering, 110*(3), 737–753.

Lane, J. (2012). Virdia ups the ante in the Race for the new sugars. *BiofuelsDigest.com*, 2014. California.

Lee, Y., Iyer, P., & Torget, R. W. (1999). Dilute-acid hydrolysis of lignocellulosic biomass. In G. T. Tsao, A. P. Brainard, H. R. Bungay, N. J. Cao, P. Cen, Z. Chen, et al. (Eds.), *Recent progress in bioconversion of lignocellulosics* (pp. 93–115). New York, NY: Springer.

Li, Z., Chen, C. H., Liu, T., Mathrubootham, V., Hegg, E. L., & Hodge, D. B. (2013a). Catalysis with CuII (bpy) improves alkaline hydrogen peroxide pretreatment. *Biotechnology and Bioengineering, 110*(4), 1078–1086.

Li, C., Tanjore, D., He, W., Wong, J., Gardner, J. L., Sale, K. L., et al. (2013). Scale-up and evaluation of high solid ionic liquid pretreatment and enzymatic hydrolysis of switchgrass. *Biotechnology for Biofuels, 6*(154), 1–13.

Li, C., Cheng, G., Balan, V., Kent, M. S., Ong, M., Chundawat, S. P., et al. (2011). Influence of physico-chemical changes on enzymatic digestibility of ionic liquid and AFEX pretreated corn stover. *Bioresource Technology, 102*(13), 6928–6936.

Li, L. H., Flora, R. M., & King, K. W. (1965). Individual roles of cellulase components derived from Trichoderma viride. *Archives of Biochemistry and Biophysics, 111*(2), 439–447.

Li, C., Knierim, B., Manisseri, C., Arora, R., Scheller, H. V., Auer, M., et al. (2010). Comparison of dilute acid and ionic liquid pretreatment of switchgrass: Biomass recalcitrance, delignification and enzymatic saccharification. *Bioresource Technology, 101*(13), 4900–4906.

Li, C., Tanjore, D., He, W., Wong, J., Gardner, J. L., Sale, K. L., et al. (2013b). Scale-up and evaluation of high solid ionic liquid pretreatment and enzymatic hydrolysis of switchgrass. *Biotechnology for Biofuels, 6*(154), 1–13.

Lin, Z., Huang, H., Zhang, H., Zhang, L., Yan, L., & Chen, J. (2010). Ball milling pretreatment of corn stover for enhancing the efficiency of enzymatic hydrolysis. *Applied Biochemistry and Biotechnology, 162*(7), 1872–1880.

Linger, J. G., Vardon, D. R., Guarnieri, M. T., Karp, E. M., Hunsinger, G. B., Frandern, M. A., et al. (2014). Lignin valorization through integrated biological funneling and chemical catalysis. *Proceedings of the National Academy of Sciences, 111*(33).

Lloyd, T. A., & Wyman, C. E. (2005). Combined sugar yields for dilute sulfuric acid pretreatment of corn stover followed by enzymatic hydrolysis of the remaining solids. *Bioresource Technology, 96*(18), 1967–1977.

Mani, S., Tabil, L. G., & Sokhansanj, S. (2004). Grinding performance and physical properties of wheat and barley straws, corn stover and switchgrass. *Biomass and Bioenergy, 27*(4), 339–352.
McNeill, B. (1998). *Saturated steam tables*. S. Tables S (Ed.). Tempe, AZ.
Miller, S., & Hester, R. (2007). Concentrated acid conversion of pine softwood to sugars, part I: Use of a twin-screw reactor for hydrolysis pretreatment. *Chemical Engineering Communications, 194*(1), 85–102.
Miyaji, K. (1930). Note on the presence of Monilia in mother culture of soy sauce. *Bulletin of the College of Agriculture Research Japan, 10*, 5.
Moens, L., & Khan, N. (2002). Application of room-temperature ionic liquids to the chemical processing of biomass-derived feedstocks. In R. D. Rogers, K. R. Seddon, & S. Volkov (Eds.), *Green industrial applications of ionic liquids* (pp. 157–171). New York, NY: Springer.
Mosier, N., Hendrickson, R., Ho, N., Sedlak, M., & Ladisch, M. R. (2005a). Optimization of pH controlled liquid hot water pretreatment of corn stover. *Bioresource Technology, 96*(18), 1986–1993.
Mosier, N., Wyman, C., Dale, B., Elander, R., Lee, Y., Holtzapple, M., et al. (2005b). Features of promising technologies for pretreatment of lignocellulosic biomass. *Bioresource Technology, 96*(6), 673–686.
Narani, A., Tanjore, D., Gardner, J., Li, C., Rasson, J., & Perry, P. (2014). *Design and operation of a "Tea-Cup" reactor for scale-up of aqueous phase heterogenous catalytic reactions*. Atlanta, GA: Annual Meeting of American Institute of Chemical Engineers.
Neely, W. C. (1984). Factors affecting the pretreatment of biomass with gaseous ozone. *Biotechnology and Bioengineering, 26*(1), 59–65.
Ooshima, H., Aso, K., Harano, Y., & Yamamoto, T. (1984). Microwave treatment of cellulosic materials for their enzymatic hydrolysis. *Biotechnology Letters, 6*(5), 289–294.
Orth, A. B., Royse, D., & Tien, M. (1993). Ubiquity of lignin-degrading peroxidases among various wood-degrading fungi. *Applied and Environmental Microbiology, 59*(12), 4017–4023.
Overend, R. P., & Chronet, E. (1987). Fractionation of lignocellulosics by steam-aqueous pretreatments. *Philosophical Transactions of Royal Society of London, A321*, 523–536.
Park, J. I., Steen, E. J., Burd, H., Evans, S. S., Redding-Johnson, A. M., Batth, T., et al. (2012). A thermophilic ionic liquid-tolerant cellulase cocktail for the production of cellulosic biofuels. *PLoS ONE, 7*(5), e37010.
Pepper, J., Baylis, P., & Adler, E. (1959). The isolation and properties of lignins obtained by the acidolysis of spruce and aspen woods in dioxane-water medium. *Canadian Journal of Chemistry, 37*(8), 1241–1248.
Perez, J., Wilke, C.R., & Blanch, H.W. (2013). Economics of sugar production with Trichoderma reesei Rutgers C-30. *Second Chemical Congress of the North American Continent*, Las Vegas, NV, 25–27 August, 1980.
Peterson, A. A., Vogel, F., Lachance, R. P., Fröling, M., Antal, M. J., Jr., & Tester, J. W. (2008). Thermochemical biofuel production in hydrothermal media: A review of sub-and supercritical water technologies. *Energy & Environmental Science, 1*(1), 32–65.
Qing, Q., & Wyman, C. E. (2011). Supplementation with xylanase and b-xylosidase to reduce xylo-oligomer and xylan inhibition of enzymatic hydrolysis of cellulose and pretreated corn stover. *Biotechnology for Biofuels, 4*(1), 18.
Qing, Q., Yang, B., & Wyman, C. E. (2010). Xylooligomers are strong inhibitors of cellulose hydrolysis by enzymes. *Bioresource Technology, 101*(24), 9624–9630.
Reese, E. T., Segal, L., & Tripp, V. W. (1957). The effect of cellulase on the degree of polymerization of cellulose and hydrocellulose. *Textile Research Journal, 27*(8), 626–632.

Reese, E. T., Siu, R. G. H., & Levinson, H. S. (1950). The biological degration of the soluble cellulose derivatives and its relationship to the mechanism of cellulose hydrolysis. *Journal of Bacteriology, 59*(4), 485–497.

Reilly, P. J. (1987). *Proceedings of the seventeenth annual biochemical engineering symposium*. Ames, IA: Department of Chemical Engineering, Iowa State University.

Ritter, G. J., Seborg, R. M., & Mitchell, R. L. (1932). Factors affecting quantitative determination of lignin by 72 per cent sulfuric acid method. *Industrial and Engineering Chemistry Analytical Edition, 4*(2), 202–204.

Rogers, R. D., & Seddon, K. R. (2002). *Ionic liquids (industrial applications for green chemistry). A. C. S. symposium series*. Washington, DC: American Chemical Society.

Saeman, J. F., Bubl, J. L., & Harris, E. E. (1945). Quantitative saccharification of wood and cellulose. *Industrial and Engineering Chemistry Analytical Edition, 17*(1), 35–37.

Saha, B. C., Yoshida, T., Cotta, M. A., & Sonomoto, K. (2013). Hydrothermal pretreatment and enzymatic saccharification of corn stover for efficient ethanol production. *Industrial Crops and Products, 44*, 367–372.

Sannigrahi, P., Kim, D. H., Jung, S., & Ragauskas, A. (2011). Pseudo-lignin and pretreatment chemistry. *Energy & Environmental Science, 4*(4), 1306–1310.

Saritha, M., Arora, A., & Lata. (2012). Biological pretreatment of lignocellulosic substrates for enhanced delignification and enzymatic digestibility. *Indian Journal of Microbiology, 52*(2), 122–130.

Sathitsuksanoh, N., Zhu, Z., & Zhang, Y.-H. P. (2012). Cellulose solvent-based pretreatment for corn stover and avicel: Concentrated phosphoric acid versus ionic liquid [BMIM] Cl. *Cellulose, 19*(4), 1161–1172.

Schmitz, H. (1920). Enzyme action in Echinodontium tinctorium Ellis and Everhart. *The Journal of General Physiology, 2*(6), 613.

Selby, K., & Maitland, C. (1967). The cellulase of Trichoderma viride. *Biochemistry Journal, 104*, 716–724.

Selig, M. J., Viamajala, S., Decker, S. R., Tucker, M. P., Himmel, M. E., & Vinzant, T. B. (2007). Deposition of lignin droplets produced during dilute acid pretreatment of maize stems retards enzymatic hydrolysis of cellulose. *Biotechnology Progress, 23*(6), 1333–1339.

Sendich, E., Laser, M., Kim, S., Alizadeh, H., Laureano-Perez, L., Dale, B., et al. (2008). Recent process improvements for the ammonia fiber expansion (AFEX) process and resulting reductions in minimum ethanol selling price. *Bioresource Technology, 99*(17), 8429–8435.

Shi, J., Ebrik, M. A., Yang, B., Garlock, R. J., Balan, V., Dale, B. E., et al. (2011a). Application of cellulase and hemicellulase to pure xylan, pure cellulose, and switchgrass solids from leading pretreatments. *Bioresource Technology, 102*(24), 11080–11088.

Shi, J., Ebrik, M. A., & Wyman, C. E. (2011b). Sugar yields from dilute sulfuric acid and sulfur dioxide pretreatments and subsequent enzymatic hydrolysis of switchgrass. *Bioresource Technology, 102*(19), 8930–8938.

Shi, J., Gladden, J. M., Sathitsuksanoh, N., Kambam, P., Sandoval, L., Mitra, D., et al. (2013). One-pot ionic liquid pretreatment and saccharification of switchgrass. *Green Chemistry, 15*(9), 2579–2589.

Shill, K., Padmanabhan, S., Xin, Q., Prausnitz, J. M., Clark, D. S., & Blanch, H. W. (2011). Ionic liquid pretreatment of cellulosic biomass: enzymatic hydrolysis and ionic liquid recycle. *Biotechnology and Bioengineering, 108*(3), 511–520.

Simonsen, E. (1898). Spiritus aus cellulose und holz. *Angewandte Chemie, 11*(10), 219–228.

Sluiter, A., Hames, B., Ruiz, R., Scarlata, C., Sluiter, J., Templeton, D., et al. (2008). *Determination of structural carbohydrates and lignin in biomass. Laboratory analytical procedure*. Golden, CO: National Renewable Energy Laboratory.

Snelders, J., Dornez, E., Benjelloun-Mlayah, B., Huijgen, W. J. J., de Wild, P. J., Gosselink, R. J. A., et al. (2014). Biorefining of wheat straw using an acetic and formic acid based organosolv fractionation process. *Bioresource Technology, 156*, 275–282.

Sternberg, D. (1976). A method for increasing cellulase production by Trichoderma viride. *Biotechnology and Bioengineering, 18*(12), 1751–1760.
Stöcker, M. (2008). Biofuels and biomass-to-liquid fuels in the biorefinery: Catalytic conversion of lignocellulosic biomass using porous materials. *Angewandte Chemie International Edition, 47*(48), 9200–9211.
Studer, M. H., DeMartini, J. D., Brethauer, S., McKenzie, H. L., & Wyman, C. E. (2010). Engineering of a high-throughput screening system to identify cellulosic biomass, pretreatments, and enzyme formulations that enhance sugar release. *Biotechnology and Bioengineering, 105*(2), 231–238.
Sun, N., Liu, H., Sathitsuksanoh, N., Stavila, V., Sawant, M., Bonito, A., et al. (2013). Production and extraction of sugars from switchgrass hydrolyzed in ionic liquids. *Biotechnology for Biofuels, 6*(1), 1–15.
Swatloski, R. P., Spear, S. K., Holbrey, J. D., & Rogers, R. D. (2002). Dissolution of cellose with ionic liquids. *Journal of the American Chemical Society, 124*(18), 4974–4975.
Taherzadeh, M. J., & Karimi, K. (2008). Pretreatment of lignocellulosic wastes to improve ethanol and biogas production: A review. *International Journal of Molecular Sciences, 9*(9), 1621–1651.
Tanjore, D., Marshall, M. N., Richard, T. L. (2009). *Biological pretreatments of corn stover with filamentous fungi*. The 31st symposium on biotechnology for fuels and chemicals.
Tanjore, D., Shi, J., & Wyman, C. (2011). Dilute acid and hydrothermal pretreatment of cellulosic biomass. In B. A. Simmons (Ed.), *Chemical and biochemical catalysis for next generation biofuels* (pp. 64–88). London: Royal Society of Chemistry.
Tuck, C. O., Pérez, E., Horváth, I. T., Sheldon, R. A., & Poliakoff, M. (2012). Valorization of biomass: Deriving more value from waste. *Science, 337*(6095), 695–699.
Vidal, B. C., Jr., Dien, B. S., Ting, K., & Singh, V. (2011). Influence of feedstock particle size on lignocellulose conversion—A review. *Applied Biochemistry and Biotechnology, 164*(8), 1405–1421.
Wald, S., Wilke, C. R., & Blanch, H. W. (1984). Kinetics of the enzymatic hydrolysis of cellulose. *Biotechnology and Bioengineering, 26*(3), 221–230.
Wang, W., Chen, X., Donohoe, B. S., Ciesielski, P. N., Katahira, R., Kuhn, E. M., et al. (2014). Effect of mechanical disruption on the effectiveness of three reactors used for dilute acid pretreatment of corn stover. Part 1: Chemical and physical substrate analysis. *Biotechnology for Biofuels, 7*(1), 1–13.
Wildschut, J., Smit, A. T., Reith, J. H., & Huijgen, W. J. J. (2013). Ethanol-based organosolv fractionation of wheat straw for the production of lignin and enzymatically digestible cellulose. *Bioresource Technology, 135*, 58–66.
Wolfgang, G., & Hans, R. (1937). Process of preparing an enzyme from mold. Google Patents.
Wood, T. (1968). Cellulolytic enzyme system of Trichoderma koningii. Separation of components attacking native cotton. *Biochemistry Journal, 109*, 217–227.
Wood, T. M., & Bhat, K. M. (1988). Methods for measuring cellulase activities. In A. Willis & S. T. K. Wood (Eds.), *Methods in enzymology* (Vol. 160, pp. 87–112). Waltham, MA: Academic.
Wu, D., Li, Q., Wang, D., & Dong, Y. (2013). Enzymatic hydrolysis and succinic acid fermentation from steam-exploded corn stalk at high solid concentration by recombinant Escherichia coli. *Applied Biochemistry and Biotechnology, 170*(8), 1942–1949.
Wyman, C. E., Dale, B. E., Elander, R. T., Holtzapple, M., Ladisch, M. R., & Lee, Y. (2005). Coordinated development of leading biomass pretreatment technologies. *Bioresource Technology, 96*(18), 1959–1966.
Yang, B., & Wyman, C. E. (2004). Effect of xylan and lignin removal by batch and flowthrough pretreatment on the enzymatic digestibility of corn stover cellulose. *Biotechnology and Bioengineering, 86*(1), 88–98.
Young, R. D. (2014). Virdia to build $60 million biochemical plant in Lafourche Parish. *NOLA.com*, vol 2014. Louisiana.

Yoo, J., Alavi, S., Vadlani, P., & Amanor-Boadu, V. (2011). Thermo-mechanical extrusion pretreatment for conversion of soybean hulls to fermentable sugars. *Bioresource Technology, 102*(16), 7583–7590.

Zhang, Y. H. P., & Lynd, L. R. (2004). Toward an aggregated understanding of enzymatic hydrolysis of cellulose: Noncomplexed cellulase systems. *Biotechnology and Bioengineering, 88*(7), 797–824.

Zheng, Y., Lin, H. M., & Tsao, G. T. (1998). Pretreatment for cellulose hydrolysis by carbon dioxide explosion. *Biotechnology Progress, 14*(6), 890–896.

Zhu, Z., Sathitsuksanoh, N., Vinzant, T., Schell, D. J., McMillan, J. D., & Zhang, Y. H. P. (2009). Comparative study of corn stover pretreated by dilute acid and cellulose solvent-based lignocellulose fractionation: Enzymatic hydrolysis, supramolecular structure, and substrate accessibility. *Biotechnology and Bioengineering, 103*(4), 715–724.

Zhu, S., Wu, Y., Yu, Z., Liao, J., & Zhang, Y. (2005). Pretreatment by microwave/alkali of rice straw and its enzymic hydrolysis. *Process Biochemistry, 40*(9), 3082–3086.

Part IV
Food Biotechnology

Chapter 20
Innovations in Alcoholic Beverage Production

Julie Kellershohn and Inge Russell

Introduction

Beverage alcohol production is a very old industry, and it is likely the original biotechnology industry. Like all industries it faces the challenges of staying relevant to the current consumer market and of producing products that are of high quality and are environmentally friendly. It is a very large and varied industry with the production of so many different types of beverages ranging from small traditional local products produced in small villages in Africa to modern global mega-breweries and distilleries, able to churn out vast quantities of products standardized to specific quality parameters. This chapter highlights some of the recent developments in the alcoholic beverage industry, especially regarding innovations in the microbial strains used for fermentation and the changes in consumer perceptions of genetically modified (GM) organisms. The chapter also offers a look at changes in the current alcohol market environment in terms of the rise of craft brewing and craft distilling.

The Dramatic Rise in Craft Brewing and Craft Distilling

Beverage alcohol production is a very old and traditional industry that in recent years has seen some exciting new activity. The dramatic rise in craft brewing and craft distilling in North America is now being seen in other countries. Where for over the past 20 years the long-term traditional large brewers and distillers have

J. Kellershohn
Red Dragon Research, 22 Fairfield Road, Toronto, ON, Canada, M4P 1T1

I. Russell (✉)
Russell & Associates, 76 Knights Bridge Road, London, ON, Canada, N6K 3R4
e-mail: Inge.Russell@gmail.com

P. Ravindra (ed.), *Advances in Bioprocess Technology*,
DOI 10.1007/978-3-319-17915-5_20

Table 20.1 Best global selling beers by volume in 2013 (Bloomberg News 2014)

1. Snow
2. Tsingtao
3. Bud light
4. Budweiser
5. Skol
6. Yanjing
7. Heineken
8. Harbin
9. Brahma
10. Coors light

dominated, the landscape is now changing. Where formerly, there appeared to be little need or desire for innovation other than in terms of new marketing campaigns, today, there is a new level of accelerated growth in new markets. The top ten beers in terms of volume sold in 2013 are shown in Table 20.1.

Snow beer is a lager beer, which since 2008 has been the largest selling beer globally in terms of volume and yet it is a beer that outside of China most consumers are not even aware that it exists. The second best-selling beer by volume is Tsingtao beer, also produced in China, with limited brand recognition outside of China.

Globally there is an emergence of the craft beer (and distilled spirits) market, surpassing industry expert expectations and demonstrating a dramatic shift in consumer buying habits. With over 16.1 million barrels shipped in the US in 2013, by volume, craft beer now outsells the once very popular Budweiser beer (Mickle 2014).

Who Is the Target Market for This Boom in Craft Beers?

In 2013, the overall beer market was estimated to be worth $100 billion USD in sales, of which $14.3 billion USD was the craft beer market. Mintel (2014) is now forecasting that the growth of craft beer could surpass $36 billion USD by 2019. Legislation on drinking and driving is also a factor contributing to the growing consumer focus on quality and uniqueness of product over quantity.

Craft beer consumers stretch across all demographics, but the largest category in the US is the millennial craft beer consumer (age 25–34), currently comprising 43 % of craft beer sales. The most popular craft beers are the ‘seasonal beers’—especially the beers produced in the fall of the year with unique fall-associated flavours such as pumpkin beers, and stronger winter seasonal products with higher alcohol levels such as stout beers made with heavily caramelized malts. Spring and summer offerings are usually lighter beers with fruity and citrusy flavours. It is in these seasonal offering that there is a huge surge in product innovation, as a plethora of ingredients are used in brewing beers with non-traditional flavours, with no limit on the ingenuity of ingredients used.

Following closely on the surge in craft brewing, there has been a significant and unexpected swell in interest in the area of craft distilling. In 2003, Bill Owens founded The American Distilling Institute (ADI), an organization of small-batch, independently-owned distillers in the United States. In less than 10 years, it grew from a few niche members to an organization with meetings of national scope that attract audiences of over 1,000 participants. By 2014, the number of craft distilleries stood at over 1,000 in the USA and that number continues to rise and innovation flourishes in these companies (Lyons 2014). Other regions of the globe, especially Europe, are also seeing a similar rise in craft distilling operations.

What makes these companies unique is their ability to manufacture small batches allowing them to experiment with a wide range of raw materials. From locally grown and organic raw materials used to develop unique marketing positions, to niche distinctive ingredients that can only be sourced with great difficulty and offer new flavour and taste opportunities—there is a wide field that has opened for innovative new products and new marketing campaigns (Driver 2014; Gordon 2014) as well as novel packaging presentation opportunities (Mitchell 2014).

Beverage Alcohol Production and the Perceptions on Using a Genetically Modified (GM) Yeast

Over 25 years ago, when the methods to modify industrial yeast using the newly developed molecular technologies of gene splicing were first introduced, industry experts forecasted that soon everyone in the beverage alcohol industry would be using yeasts that had been modified in this way. The designer strains would be able to expand what substrates the yeast could use, fermentation times would be shorter and flavour profiles would be easily adjusted by tweaking the biochemical pathways. In 1994, the Brewing Research Foundation in the United Kingdom produced a 'light' beer using a genetically modified organism (GMO) as the 'test case' for the industry (Baxter 1975). The glucoamylase gene was introduced into a lager brewing strain. This allowed the genetically modified (GM) yeast to ferment some of the residual starch that is normally left in the beer when a non-GM yeast is used and it also lets the brewer avoid the use of added commercial enzymes to remove some of this starch. The beer produced with the GM yeast received government clearance at that time as a safe product. However, 20 years later, to the best of our knowledge, there are no large commercial breweries using GM yeast in the industry to produce mainstream beers. The anti-GMO sentiments in terms of a brewing yeast strain for beer production are as prevalent now, and indeed even stronger, than they were over 25 years ago, especially in Europe. The same consumer concerns apply to the yeast strains employed by the industry for spirit production even though there would be no live yeast present in the spirit after distillation.

Although there is great reluctance from a public relations point of view to employ a GM yeast for brewing or distilling, this does not mean that tremendous leaps in our understanding of how yeast ferment their various substrates has not been made. Thanks to the new technologies in use in the medical arena today, with yeast as the darling organism of choice for medical research, our knowledge of GM yeast and potential uses and applications continues to expand. It is easy to manipulate as it is a eukaryotic organism that bears a closer relation to humans than the other key organism for research, the bacterium *E. coli,* which is a prokaryotic organism. Yeast not only provides beverages that are a pleasure to drink but yeast is an ideal model system to study both human and animal disease.

The yeast genome of a haploid laboratory strain was first sequenced in 1996 (Goffeau et al. 1996). This accomplishment was as significant as the first landing on the moon in terms of how it has advanced our scientific knowledge. Today, over 30 industrial yeast strains (ranging from laboratory, to fuel alcohol, to beverage and food yeasts) have been sequenced as the technology to do this keeps improving and the cost of sequencing continues to become more economical (Stewart et al. 2013; Mattanovich et al. 2014). The knowledge obtained by researchers using the molecular techniques such as genome sequencing has allowed us to better understand the industrial yeast genome and to make adjustments to it that do not require the use of GM organisms (Borneman et al. 2011). The developments in brewing and distilling yeasts over the past 25 years, and examples of yeast that have been produced in this way, are reviewed in detail by Stewart et al. (2013).

Improvements in Wine Yeast Technology

Consumers are much more likely to accept GM organisms being used for their food or beverage products if they can see a direct link to an improvement in human health (or an avoidance of danger). To obtain a balanced wine, bacterial malolactic fermentations are used in wine production to de-acidify high acid grape must. However in this process there is the possibility of producing undesirable biogenic amines in the wine. A specialized yeast, designated ML01, was genetically engineered and received approval from both the United States FDA and Health Canada for use in wine production. This yeast was developed to avoid the problems that can arise when using bacteria for the malolactic fermentation process, which includes the production of undesirable biogenic amines and precursors to carcinogens produced by the bacteria (Husnik et al. 2006, 2007). Canada and the United States allowed the first release of this GM yeast to the commercial market in 2005. The wines in these two countries do not have to be labelled as using a GM yeast since the yeast is classified as a 'processing aid' by the local government regulators—however due to the many countries that have banned the use of GM organisms for food and beverage production and the current consumer anti-GMO sentiments—these yeast, even in North America, are still being used on a very limited scale.

Another danger in the production of some types of beverage ethanol is the presence of ethyl carbamate in the final product. Ethyl carbamate is a carcinogen, which is formed from a reaction between ethanol with urea. By using recombinant

DNA technology, wine yeasts (an example is strain ECM001) have been created that can reduce the formation of ethyl carbamate in the final beverage. Yeast ECM001 has an extra copy of a *Saccharomyces* yeast gene inserted that reduces the formation of ethyl carbamate by degrading the urea precursor. Since this yeast does not contain DNA from any other organism, it has been approved for commercial use both in the United States and Canada as a GRAS organism. By using strain ECM001, it was shown possible to reduce the level of ethyl carbamate by 90 % in chardonnay wine (Coulon et al. 2006; Dahabieh et al. 2009).

A recent publication by researchers at UC Davis (Jarosz et al. 2014) sheds additional light on a key wine fermentation problem and how it can potentially be avoided. They report on a biochemical communication system that crosses from bacteria to yeast, which makes use of prions. This system is responsible for a chronic winemaking problem—the 'stuck fermentation'. They found that the glucose repression circuit could be interrupted when bacteria jump-start the replication of the prions in the membranes of yeast cells. This interference of the prions causes the yeast to process carbon sources other than glucose. As such, the yeast becomes less effective in glucose metabolism resulting in a fermentation that slows down and eventually stops (i.e. a stuck fermentation).

Will Consumer Attitudes to GM Use for Alcoholic Beverage Production Change?

The reader is directed to a recently published review and meta-analysis by Frewer et al. (2013) comparing attitudes to GM use in foods and beverages in different global regions and to a review by De Steur et al. (2014) on the market potential of GM food with health benefits. The Hartman group (2014), a market research firm, published a report called 'GMO perceptions, knowledge and labelling' which concluded that 40 % of consumers now state that they are avoiding or reducing GMOs in their diet, a marked increase from the 29 % of consumers that described avoiding or reducing GMOs in their diet in 2010. It is very important to note that consumer aversion to GM organisms does not mean they understand exactly what a GM product entails. However, a stated key wish of consumers is a desire for labelling where GM organisms have been used in a food or beverage and they voice concern over a lack of transparency. It appears that consumer engagement in the industry conversation about the role of GM will continue to grow.

The Synthetic Yeast Project

Leaping ahead in new technologies and how to improve industrial strains is the 'synthetic yeast project'. Craig Venter's team in 2010 created the first synthetic bacterium (Gibson et al. 2010). Since yeast is a eukaryotic organism with

16 chromosomes, it is much more difficult to construct a synthetic yeast, however, in 2014 researchers successfully constructed an entire synthetic yeast chromosome (Annaluru et al. 2014). Researchers focused on chromosome 3 and using the blueprint that was available for this chromosome, edited it and joined together small snippets of man-made DNA. Once the chromosome was constructed, it was inserted into a *Saccharomyces cerevisiae* yeast strain and it was shown to be functional. This was the first time this was accomplished in a eukaryotic cell and the next goal of the project is now to synthesize all 16 yeast chromosomes.

An entire new world of yeast biotechnology and innovative opportunities is opening with these types of techniques (Church et al. 2014). How the public will perceive an entirely new synthesized organism is something that will unfold in the coming years, but there is little doubt that the coupling of yeast biology and synthetic biology has exciting potential to advance the application of yeast in all industries (Siddiqui et al. 2012).

Natural Diversity of Yeast

Due to the consumer perception issues with GM organisms there is much work at this time on exploiting natural yeast diversity (Steensels et al. 2014) and on examining yeasts for industrial processes that have not been traditionally used for alcoholic beverage production (called "non-conventional" yeast). One such yeast that is attracting much research attention is *Dekkera bruxellensis* (old name was Brettanomyes), a distant relative of *Saccharomyces cerevisiae,* and considered to be an important flavour contributor in Belgium lambic and gueuze beers, as well as in red wines. It is often found in high ethanol habitats and thus has interesting possibilities for industrial beverage alcohol production (Schifferdecker et al. 2014).

Only a small number of the yeast strains in nature have been classified to date and there is a huge unexplored pool of organisms that may prove to be valuable for fermentation without the use of any type of molecular genetic manipulation. However these new molecular techniques are useful in terms of giving us valuable insights into the origin of the key industrial strains currently in use (Gibson and Liti 2014). Dashko et al. (2014) recently published a paper that examines how, when and why yeast evolved alcoholic fermentation. Yeast are very unique in that they can convert sugars to ethanol under both aerobic and anaerobic conditions. Even with all of that we now know about yeast and its metabolic pathways, the gene expression regulatory networks are still not totally understood.

Ale and lager brewing strains both belong to the genus *Saccharomyces*, but are also very different species with different origins. Ale yeast (*Saccharomyces cerevisiae*) can ferment at a relatively high temperature, whereas lager brewing yeast prefer much cooler temperatures. The old name for lager brewing yeast was *Saccharomyces carlsbergensis* but today this yeast is called *Saccharomyces pastorianus*. Lager brewing originated in Bavaria in the 1400s and today lager beers are the most popular beers in the world. From where this cold-loving lager beer yeast originated has been a mystery and an area of debate for researchers for many years but recently using the new molecular techniques, some answers appear to be on the horizon. *Saccharomyces pastorianus* is a hybrid yeast with part of its genome coming from *Saccharomyces cerevisiae* but the original source of the non-ale subgenome (*Saccharomyces eubayanus*), which gives lager yeast its cryotolerance, has been a source of debate. The most recent speculation is that it is likely a species native to the Tibetan plateau and that it was transferred between Asia and Europe through the Silk Road over 2,000 years ago (Bing et al. 2014).

Global Beverage Alcohol Production

The literature in the fermentation area is vast but one important change is there are now many English publications on the production of beverages from countries such as China, India, Africa, and Japan. This surge in the English literature describes in detail the production of unique alcoholic beverages. Examples include advances in the production of innovative strains for sake production (Soejima et al. 2012); the examination of ancient sake production techniques in Japan (Teramoto et al. 2000); the rise in publications on studies of the indigenous flora in mud pit fermentations from China (Luo et al. 2014; Huang et al. 2014); and alcoholic products produced in Africa using traditional local ways (Kubo 2014 and Kubo et al. 2014). Many of these processes are all now being documented in detail in English in publications such as the 'Journal of the Institute of Brewing' published by the Institute of Brewing and Distilling. At one time this information was not easily found in the English literature and there is much to learn from these traditional processes. It is an opportunity for innovation in terms of regional production processes and the use of local raw materials.

The Environment and New Process Techniques

Life cycle assessment is defined as the "compilation and evaluation of the inputs, outputs and potential environmental impacts of a product system throughout its life cycle" (ISO 14040 2006). Life cycle assessment in the alcoholic beverage industry has been conducted in depth for a number of products such as beer, wine and whisky. A life cycle analysis of beer versus wine is presented by Mattila et al. (2012). The paper emphasizes the many complexities in decision making in this area. A very thorough life cycle assessment of whisky production in Scotland was undertaken by The Scotch Whisky Research Institute (SWRI) (2014), in collaboration with The Scotch Whisky Association (SWA). Their assessment covers all the processes within the Scotch Whisky life cycle. A wide ranging look at an overall life cycle assessment of the beverage industry in Finland, including cider and beer, was prepared in 2013 (Karjalainen 2013). There are many more of these assessments that have been conducted over the previous years and not surprisingly the raw materials to produce the beverage and the packaging materials tend to have the highest carbon footprint. Most manufacturers are striving to be more environmentally friendly, decreasing material costs and in doing so also reducing monetary costs.

Leaner manufacturing (also called brewing intensification) is about process optimization and increased efficiencies have been one of the main priorities for both large and small brewing and distilling companies and wineries. The key areas that have been addressed are increased rates of fermentation and yeast selection (Walker et al. 2012), decreased maturation times for the product (Nedovic et al. 2014; Genisheva et al. 2014), enhanced quality and stability of the final product (Bamforth 2015), and reduced capital expenditure (Stewart 2010).

The implementation of high gravity brewing has been one of the great successes in terms of process optimization (Gibson 2011; Stewart 2014). High gravity brewing was originally developed to reduce capital expenditure as it involves producing a wort at a higher than normal concentration and then diluting with water later in the process. This allows a brewhouse to increase production without having to physically expand the plant. This was the primary goal when high gravity brewing was first implemented, but soon other benefits were apparent such as a reduction in water, energy, labour, cleaning and effluent costs. In many cases improvements in consumer panels in terms of a more positive flavour profile of the product produced by high gravity brewing versus standard gravity brewing were also seen.

Process Improvement Using Immobilized Yeast Technology

Immobilized yeast technology is an area that has great potential but has not yet overcome the barriers to entry for an industrial process for mainstream products (Nedovic et al. 2014). Continuous fermentation using immobilized cell technology is an ambitious brewing fermentation process that has been explored for many years as an alternative to batch processes for beverage alcohol production (Kourkoutas et al. 2004). In brewing, the goal is to make a finished beer in 2 days rather than waiting the 5–7 days for the primary fermentation and then 5–21 days for beer maturation. Pilot plants have been constructed to perform long-term trials, but to date the only large scale production use of immobilized cell technology for beer is for the production of low alcohol beers and for maturation and rapid aging of young beer (Mensour et al. 1997). In theory the high volumetric productivity should lead to significant savings in capital, time and operational costs. However, finding a suitable and affordable carrier and addressing the issues of yeast viability, consistent flavour matches for the product and some consumer negative concerns regarding a change from traditional technology, have all hampered the implementation of this technology. Options continue to be investigated on how to overcome these barriers to success (Pires et al. 2014a, b).

The Future

The biggest opportunity for improving the fermentation process currently lies in the use of GM organisms for the fermentation process since this is the heart of beverage alcohol production. However, since so much of an alcohol beverage brand is associated with consumer perceptions, the use of this technology is not likely to be embraced until there is a massive change in consumer perception and widespread consumer acceptance. All of the ground breaking work being carried out using yeast as a model eukaryote for medical research will continue to enrich our knowledge of the processes that take place within the yeast cell and this valuable new knowledge will lead to innovations in other areas of fermentation biochemistry. The vast area of unexplored yeast on our planet also offers a huge resource for exploration that is not yet being fully exploited.

References

Annaluru, N., Muller, H., Mitchell, L., et al. (2014). Total synthesis of a functional designer eukaryotic chromosome. *Science, 344*(6179), 55–58.

Bamforth, C. (2015). *Brewing materials and processes: A practical approach to beer excellence*. UK: Elsevier.

Baxter, E. D. (1975). The application of genetics to brewing. *Ferment, 8*, 307–311.

Bing, J., Han, P. J., Liu, W. Q., Wang, Q. M., & Bai, F. Y. (2014). Evidence for a Far East Asian origin of lager beer yeast. *Current Biology, 24*, 380–381.

Bloomberg News. (2014). http://www.bloomberg.com/news/2014-06-13/the-top-ten-beers-in-the-world-aren-t-what-you-think-.html.

Borneman, A. R., Desany, B. A., Riches, D., Affourtit, J. P., Forgan, A. H., Pretorius, I. S., Egholm, M., & Chambers, P. J. (2011). Whole-genome comparison reveals novel genetic elements that characterize the genome of industrial strains of *Saccharomyces cerevisiae*. *PLoS Genetics, 7*, e1001287.

Church, G. M., Elowitz, M. B., Smolke, C. D., Voigt, C. A., & Weiss, R. (2014). Realizing the potential of synthetic biology. *Nature Reviews Molecular Cell Biology, 15*, 289–294.

Coulon, J., Husnik, J. I., Inglis, D. L., van der Merwe, G. K., Lonvaud, A., Erasmus, D. J., & van Vuuren, H. J. J. (2006). Metabolic engineering of *Saccharomyces cerevisiae* to minimize the production of ethyl carbamate in wine. *American Journal of Enology and Viticulture, 57*, 113–124.

Dahabieh, M. S., Husnik, J. I., & van Vuuren, H. J. H. (2009). Functional expression of the DUR3 gene in a wine yeast strain to minimize ethyl carbamate in Chardonnay wine. *American Journal of Enology and Viticulture, 60*, 537–541.

Dashko, S., Zhou, N., Compagno, C., & Piškur, J. (2014). Why, when, and how did yeast evolve alcoholic fermentation? *FEMS Yeast Research, 14*, 826–832.

De Steur, H., Blancquaert, D., Lambert, W., Van Der Straeten, D., & Gellynck, X. (2014). Conceptual framework for ex-ante evaluation at the micro/macro level of GM crops with health benefits. *Trends Food Science Technology, 39*, 116–134.

Driver, J. (2014). Marketing Scotch Whisky in the 21st century and previously. In I. Russell & G. Stewart (Eds.), *Whisky technology production and marketing* (2nd ed., pp. 343–358). UK: Elsevier. Chapter 19.

Frewer, L. J., van der lans, I. A., & Fischer, A. R. H. (2013). Public perceptions of agri-food applications of genetic modification – A systematic review and meta-analysis. *Trends Food Science Technology, 30*, 142–152.

Genisheva, Z., Teixeira, J. A., & Oliveira, J. M. (2014). Immobilized cell systems for batch and continuous winemaking. *Trends Food Science Technology, 40*, 33–47.

Gibson, B. R. (2011). 125th anniversary review: Improvement of higher gravity brewery fermentation via wort enrichment and supplementation. *Journal of the Institute of Brewing, 117*, 268–284.

Gibson, D. G., Glass, J. I., Lartigue, C., et al. (2010). Creation of a bacterial cell controlled by a chemically synthesized genome. *Science, 329*(5987), 52–56.

Gibson, B., & Liti, G. (2014). *Saccharomyces pastorianus*: Genomic insights inspiring innovation for industry. *Yeast*. doi:10.1002/yea.3033.

Goffeau, A., Barrell, B. G., Bussey, H., et al. (1996). Life with 6000 genes. *Science, 274*, 546–567.

Gordon, G. E. (2014). Marketing Scotch Whisky. In I. Russell & G. Stewart (Eds.), *Whisky technology production and marketing* (2nd ed., pp. 359–398). UK: Elsevier. Chapter 20.

Huang, Y., Wu, Q., & Xu, Y. (2014). Isolation and identification of a black *Aspergillus* strain and the effect of its novel protease on the aroma of Moutai-flavoured liquor. *Journal of the Institute of Brewing, 120*, 268–276.

Husnik, J. I., Delaquis, P. J., Cliff, M. A., & van Vuuren, H. J. J. (2007). Functional analyses of the malolactic wine yeast ML01. *American Journal of Enology and Viticulture, 58*, 42–52.

Husnik, J. I., Volschenk, H., Bauer, J., Colavizza, D., Luo, Z., & van Vuuren, H. J. J. (2006). Metabolic engineering of malolactic wine yeast. *Metabolic Engineering, 8*, 315–323.

ISO 14040. (2006). Environmental management – Life cycle assessment. *Principles and Framework*. http://www.iso.org.

Jarosz, D. F., Brown, J. C. S., Walker, G. A., Datta, M. S., Ung, W. L., Lancaster, A. K., Rotem, A., Chang, A., Newby, G. A., Weitz, D. A., Bisson, L. F., & Lindquist, S. (2014). Cross-kingdom chemical communication drives a heritable, mutually beneficial prion-based transformation of metabolism. *Cell, 158*, 1083–1093.

Karjalainen, P. (2013). The carbon footprint of the Finnish beverage industry for years 2000–2012 as calculated with CCaLC. Master's Thesis, University of Helsinki.

Kourkoutas, Y., Bekatorou, A., Banat, I. M., Marchant, R., & Koutinas, A. A. (2004). Immobilization technologies and support materials suitable in alcohol beverages production: A review. *Food Microbiology, 21*, 377–397.

Kubo, R. (2014). Production of indigenous alcoholic beverages in a rural village of Tanzania. *Journal of the Institute of Brewing, 120*, 142–148.

Kubo, R., Funakawa, S., Araki, S., & Kitabatake, N. (2014). Production of indigenous alcoholic beverages in a rural village of Cameroon. *Journal of the Institute of Brewing, 120*, 133–141.

Luo, Q., Liu, C., Wu, Z., Wang, H., Li, W., Zhang, K., Huang, D., Zhang, J., & Zhang, W. (2014). Monitoring of the prokaryotic diversity in pit mud from a Luzhou-flavour liquor distillery and evaluation of two predominant archaea using qPCR assays. *Journal of the Institute of Brewing, 120*, 253–261.

Lyons, T. P. (2014). North American whiskies: A story of evolution, experience and an ongoing entrepreneurial spirit. In I. Russell & G. Stewart (Eds.), *Whisky technology production and marketing* (2nd ed., pp. 39–48). UK: Elsevier. Chapter 5.

Mattanovich, D., Sauer, M., & Gasser, B. (2014). Yeast biotechnology: Teaching the old dog new tricks. *Microbial Cell Factories, 13*(1), 34. doi:10.1186/1475-2859-13-34.

Mattila, T., Leskinen, P., Soimakallio, S., & Sironen, S. (2012). Uncertainty in environmentally conscious decision making: Beer or wine? *International Journal of Life Cycle Assessment, 17*, 696–705.

Mensour, N. A., Margaritis, A., Briens, C. L., Pilkington, H., & Russell, I. (1997). New developments in the brewing industry using immobilised yeast cell bioreactor systems. *Journal Institute of Brewing, 103*, 363–370.

Mickle, T. (2014). Bud crowded out by craft beer craze. *Wall Street Journal (Business Section), 23*, 2014.

Mintel. (2014). Craft Beer – US – June 2014 http://www.mintel.com/.

Mitchell, M. (2014). Whisky global packaging developments. In I. Russell & G. Stewart (Eds.), *Whisky technology production and marketing* (2nd ed., pp. 315–434). UK: Elsevier. Chapter 18.

Nedovic, V., Gibson, B., Mantzouridou, T. F., Bugarski, B., Djordjevi, V., & Kalu, A. (2014). Aroma formation by immobilized yeast cells in fermentation processes. *Yeast*. doi:10.1002/yea.3042.

Pires, E. J., Teixeira, J. A., Brányik, T., Côrte-Real, M., Brandão, T., & Vicente, A. A. (2014a). High gravity primary continuous beer fermentation using flocculent yeast biomass. *Journal of the Institute of Brewing, 120*, 486–494.

Pires, E. J., Teixeira, J. A., Brányik, T., & Vicente, A. A. (2014b). Carrier-free, continuous primary beer fermentation. *Journal of the Institute of Brewing, 120*, 500–506.

Schifferdecker, A. J., Dashko, S., Ishchuk, O. P., & Piškur, J. (2014). The wine and beer yeast *Dekkera bruxellensis*. *Yeast, 31*, 323–332.

Scotch Whisky Association. (2014). Scotch Whisky Life Cycle Assessment. http://www.scotch-whisky.org.uk/media/12908/lifecycleassessment.pdf.

Siddiqui, M. S., Thodey, K., Trenchard, I., & Smolke, C. D. (2012). Advancing secondary metabolite biosynthesis in yeast with synthetic biology tools. *FEMS Yeast Research, 12*, 144–170.

Soejima, H., Tsuge, K., Yoshimura, T., Sawada, K., & Kitagaki, H. (2012). Breeding of a high tyrosol-producing sake yeast by isolation of an ethanol-resistant mutant from a trp3 mutant. *Journal of the Institute of Brewing, 118*, 264–268.

Steensels, J., Snoek, T., Meersman, E., Nicolino, M. P., Voordeckers, K., & Verstrepen, K. J. (2014). Improving industrial yeast strains: Exploiting natural and artificial diversity. *FEMS Microbiology Review, 38*, 947–995.

Stewart, G. G. (2010). High gravity brewing and distilling – Past experiences and future prospects. *Journal of the American Society of Brewing Chemists, 68*, 1–9.

Stewart, G. G. (2014). *Brewing intensification*. Saint Paul, MN: American Society of Brewing Chemists.

Stewart, G. G., Hill, A. E., & And Russell, I. (2013). 125th anniversary review: Developments in brewing and distilling yeast strains. *Journal of the Institute of Brewing, 119*, 202–220.

Teramoto, Y., Hano, T., & Ueda, S. (2000). Production and characteristics of an ancient form of sake made with Shitogi. *Journal of the Institute of Brewing, 106*, 95–100.

The Hartman Group. (2014). GMO perceptions knowledge and labeling: A consumers perspective. http://www.hartman-group.com/publications/reports/special-report-gmo-perceptions-knowledge-and-labeling/.

Walker, G., Brosnan, J., Bringhurst, T., & Jack, F. (2012). Selecting new distilling yeast for improved fermentation and for sustainability. In G. Walker, I. Goodall, D. Murray, & R. Fotheringham (Eds.), *Distilled spirits: Science and sustainability* (pp. 1–11). Nottingham: Nottingham University Press.

Chapter 21
Starter Culture Technology: Fermented Foods

D. Prema Latha, S.M. Reddy, K.S. Youn, and Pogaku Ravindra

Introduction

Since the ancient time's fermentation technology have been in common practice of man for the production of different products such as curd, wine, alcohol, idly, dosa etc. In the initial stages these food products were produced without the knowledge of scientific principles involved in the process of product formation. It was due to spontaneous microorganisms present in the environment or container. Traditional fermentations were captured to store the food products (Holzapfel 2002). This concept was originated after later half of nineteenth century when the fermented products developed after pasteurization (Klaenhammer and Fitzgerald 1994). The role of microorganisms in food fermentations were realized (Caplice and Fitzerald 1999; Baldino et al. 2003) with passing of time the clear picture of the fermentation and role played by starter cultures or microorganisms.

In the past fermentation process was utilized to enhance the quality of product (Holzapfel 1997, 2002). Final Product was the result of interaction of flora of the raw material and the environmental conditions prevailing such as temperature, oxygen, water activity, pH etc. the microorganisms beside selecting necessary enzymes they also produce beneficial substances which are needed for production of fermented products (Kalenhammer and Fitzgerald 1994) such as lactic acid, alcohol, organic acid etc. (Hansen 2002). These products act as preservatives, enhance flavor and increase shelf life of product (Nout 1991). Preservation of food is man's struggle for survival since prehistoric times was to ensure required food for all the time. Microbial fermentation not only preserves the food but also contributes the taste, aroma, flavor, stability etc. (Hammes 1990).

D.P. Latha • S.M. Reddy • K.S. Youn • P. Ravindra (✉)
Faculty of Engineering, Chemical Engineering Program, University Malaysia Sabah, Sabah, Malaysia
e-mail: dr_ravindra@hotmail.com

P. Ravindra (ed.), *Advances in Bioprocess Technology*,
DOI 10.1007/978-3-319-17915-5_21

Traditional fermentations were spontaneous, uncontrollable and unpredictable used to produce low yields and variable product quality. Spontaneous fermentation provide the road map to produce food products using a small sample from the previous fermented products by adding (inoculum) to fresh media/milk to initiate the fermentation process. Subsequently it was experienced that some of the fermented foods if added to the raw material the food product quality was improved which lead to the development of back slopping (Ali and Mustafa 2009; Smid et al. 2014). This back slopping improved safety and quality and shelf life of the product.

Concept of Starter Culture

These developments in traditional fermentations opened the doors to the development of Starter Cultures. Starter Culture concept developed to have a better control over the fermentation process to produce quality products. First Starter Culture was introduced in 1893 in Denmark when Emil Christian Hansen working with Carlsberg Brewery isolated the first pure starter culture of yeast. He identified and isolated the yeast responsible for different poor and best quality of Beer (Holzapfel 1997).

"Starter Culture can be defined as the stain of microorganism selected with stable features to produce desirable characters of food under controlled conditions (Wakil et al. 2014; Hati et al. 2013; Holzapfel 2002; Leroy and Devuyst 2004; Rattanachaikunsopon and Phumkhachoran 2010)" starter cultures may be isolated from nature/sources tested for desirable character. They were selected and grown in fermenter. Maximum starter cultures used today are developed from lactic acid bacteria.

Microorganisms in Starter Culture

Based on the microorganism employed, starter cultures are

1. Bacteria
2. Yeasts
3. Fungi

Bacteria

Different bacterial members are employed as starters in different food products.

LACTOBACILLUS: It is a large group containing gram positive rod shaped microaerophilic catalase negative bacteria. First pure culture of lactobacillus was

L. lactis by Joseph Lister from sour milk (Axeleson et al. 1998) There are 2 types of lactobacillus species homo fermenters and hetero fermenters (Howaz 2014). Homofermenters produces max lactic acid from glucose with trace amounts of acetic acid and carbon dioxide optimum growth temperature 37 °C Ex. L. bulgaricus L. helviticus L. acidophilus L. thermophilus L. plantarum. Lactobacilli are used as starter in the production of cheese and yogurt. Heterofermenters produce Volatile substances alcohol and carbondioxide along with lactic acid (Widyastuti et al. 2014). Ex. L. fermentum L. brevis.

They are important as starters in food industry especially in dairy industry because of the following features (1) Production of lactose from glucose (2) Production of gas (3) Resistance to heat even at high temperatures such as production of swiss cheese.

L. ACIDOPHILUS: It is a non motile gram positive rod, homo fermenter. Optimum growth temperature 35–38, pH 5.5–6. Based on their wide range of pH and temperature they are used in many food productions to obtain taste, flavor, texture etc. It is used in the production of acidophilus milk. The microorganism secretes lactic acid and bacteriocins hence used as a biopreservative in food preparations.

L. HELVITICUS: Homofermenter optimum growth temperature 42–43 °C even it can grow at 45 °C. It is used in the preparation of Mozzarella and Emmental cheese and yogurt.

L. LACTIS is a gram positive, non motile, non spore former, catalase negative, mesophilic and micro aerophilic bacterium. This organism is involved in natural souring of milk even in cultured butter milk.

LEUCONOSTOC MESENTEROIDES: It is a gram positive rod/cocci based on available nutrients, heterofermenter Facultative anaerobe catalase negative. This organism can tolerate higher concentrations of sugar and salt more than 50 %.

PROPIONIBACTERIUM: They are gram positive rods/cocci, non motile, non spore formers, mesophilic and catalase negative. It is used as a starter in swiss cheese i.e. P. freudenreichii subsp. freudenreichii and P. freudenreichii subsp. shermanii for the production of eyes and flavor.

STREPTOCOCCI: It is spherical, ovoid, gram positive, non motile, non spore former, facultative anaerobe grows at 37 °C. The cultures used in foods include Streptococcus lactis and Streptococcus cremoris. These are used in the preparation of cheese and cultured butter milk.

STREPTOCOCCUS THERMOPHILUS: They are spherical in shape occur in pairs or in chains homofermenter facultative anaerobe. Optimum growth temperature is 45 °C or more. It is used in the preparation of yogurt.

STAPHYLOCOCCI: It is spherical, ovoid, gram positive bacteria.

Yeast

It is a unicellular eukaryotic organism. It grows in rich carbon sources such as glucose, fructose, sucrose etc. It is obligate aerobe grows in neutral or slightly acidic environment. Microbe gets energy using carbon sources releasing carbon dioxide and alcohol. This feature made this organism as a valuable tool in food industry especially in the preparation of beverages and dairy products. Since eighteenth century man knew the production of alcohol using yeast. There are many products produced using yeast such as Beer, alcohol, wine, bread, (Lin-Xu et al. 2014) kefir etc. Yeast Strain selection depends on the type of product and raw material used. A good yeast starter should have high tolerance to, alcohol, sugar, rapid carbon dioxide production and flocculation.

Bakers' yeast is manufactured using specially selected starters of Saccharomyces cerevisiae with the features of stability, viability in dried stored conditions, rapid carbon dioxide production for bread dough leavening. S. pastorianus is used in the production of Lager beer. Debaryomyces hansenii is used in cheese production.

Molds

They are multicellular filamentous fungi. The body of the fungi consists of intertwined filaments known as hyphae, all hyphae together known as mycelia.

Aspergillus, Penicillium, Rhizopus, yeast (saccharomyces kluveromyces and zygosaccharomyces sps) are the important fungal starters used in food industry. Two species of penicillium are used in the manufacture of cheese P. camemberti a white mold grows on the surface of Camembert and Brie cheese. And blue mold P. roquefortii grows in the interior of cheese-Roquefort, Gorgonzola, Stilton. Penicillum rocqueforti produces blue vein in cheese such as Roquefort. Penicillium camemberti, a white mold used in the preparation of Brie and Camembert cheese. P. roqueforti and P. camemberti both molds change the texture, aroma, taste of cheeses by secreting lipases and proteases. Methyl ketones and free fatty acids impart flavor and aroma to the cheese. Aspergillus sojae is the starter culture used in traditional soy sauce, soy pastes and rice wine preparations. Rhizopus is the starter culture for the Tempeh preparation

Starter Culture

Classification of Starter Cultures

Starter cultures can be classified differently based on number, nature and temperature of the organism.

Based on Number

1. Single strain starter culture.
2. Multi strain starter culture.
3. Mixed strain starter culture.

SINGLE STRAIN STARTER CULTURE: This type of starter culture consists of only one species/strain. Pure starter culture may have one or more strains of same species. Butter milk is produced using starter Streptococcus lactis subsp. cremoris or S. lactis subsp. lactis where as Bulgarian buttermilk is prepared with the starters of L. delbrueckii subsp. bulgaricus. Starter culture type, selection depends on the raw material used and the product to be produced. Yeast Saccharomyces cerevisae employed in the bread preparation is meant for the production of carbon dioxide or leavening where as Saccharomyces cerevisae used in the production of alcohol is meant for production of alcohol.

MULTI(PLE) STRAIN STARTERS CULTURE: These cultures contains three or more single strains of defined mixture i.e. these starters are prepared using few strains of different species or different strains of same species. The combination of cultures may be such as Lactococcus lactis ssp. cremoris and/or lactis with Leuconostoc cremoris and L. lactis.

MIXED STRAIN STARTER CULTURE: This starter cultures contain many unknown strains. Ratio of organisms also varies with product to product. Swiss cheese is produced with mixed cultures of Lactobacillus bulgaricus, Streptococcus thermophilus and Propionibacterium shermanii. Lactobacillus bulgaricus and Streptococcus are involved in the formation of cheese and P. shermani imparts flavor and eyes during cheese ripening.

Kefir a dairy product contains the microbial population of Lactic acid bacteria and Yeast like Lactobacillus bulgaricus, L. helviticus, L. acidophilus, L. kefiranofaciens and Lactococcus lactis sbsp. cremoris and Yeast such as Kluyveromyces marxinus, Kluyveromyces lactis var. lactis, Debaromyces hansenii e dekkera anomala, Saccharomyces cerevisae, Torula spora delbrueckii, Debaromyces accidentalis etc. (Leite et al. 2013).

Based on Temperature

Starters are of two types based on growth temperature

1. MESOPHILIC STARTER CULTURE: optimum growth temperature is 20–30 °C Ex: Genera of Lactococcus, Leuconostoc, Streptococci. These starters are useful in milk fermentation either individually or in combination. Mesophilic starters such as Lactococcus lactis is used for cheese production may be following
   (a) O-type: They are homofermenters produces lactic acid generally used organism is Lactococcus lactis subsp. lactic and Lactococcus lactis subsp. cremoris.

(b) D-type: Contain Streptococcus lactis subsp. lactis var diacetyl lactis along with O-type bacteria. D-type organism produces diacetyl-a characteristic flavor compound of butter milk. It also produces carbon dioxide which imparts the blend of delicate flavor.
(c) L-type: this type of starter culture contains Lenconostoc mesenteroides along with O-type of bacteria and produces diacetyl acetic acid, acetaldehyde etc. and less corbon dioxide than D-type.
(d) LD-type: These starter cultures produce fine blend of delicate flavor and aroma. Organisms includes Streptococcus lactis subsp. lactis var. diacetyl lactis along with Leuconostoc mesenteroids

    Mesophilic starters are used in the production of different cheeses such as Brie Camembert, cheddar, Gouda etc.

2. THERMOPHILIC STARTER CULTURE: Optimum growth temperature is 40–45 °C. They are microaerophilic in nature. Thermophilic cultures include species of Streptococcus and Lactobacillus. Streptococcus thermophilus and Lactobacillus bulgaricus is used in the production of Yogurt (Oskar adolfsson et al. 2014; Widyastuti et al. 2014). The organisms will be maintained in 1:1 ratio. The starter culture lactobacillus bulgaricus develops the flavor component of yogurt acetaldehyde and aroma and Streptococci produces mainly acid.

Thermophilic bacterial cultures and products

Microorganisms	Product
Streptococcus thermophilus Lactobacillus delbruekii subsp. bulgaricus ± Lactobacillus helveticus	Yogurt
Streptococcus thermophilus ± Lactobacillus delbruekii subsp. bulgaricus or Lactobacillus helveticus	Mozzarella cheese Pizza cheese Provolone
Streptococcus thermophilus	Emmental comte Gruyere Jarsuserg
± Lactobacillus delbruekii subsp. lactis Lactobacillus helveticus Streptococcus thermophilus	Hartkase Ber-alpkase Pecorino Romano Gorgonzola
Lactobacillus delbruekii subsp. lactis Lactobacillus helveticus	Bleude Breese Bleude grese Fourme d'ambent

Characteristic Features of Good Starter Culture

1. Only one microorganism is present hence product quality will be uniform.
2. Specially designed strains can be used to develop desired flavor, texture etc.
3. Increased volume of starter culture can decrease fermentation time.
4. Quality and stability of fermentation can be maintained.

5. Contamination can be discouraged by adjusting the parameters such as temperature, oxygen, pH, substrate composition etc.
6. Maximum substrate can be used by the organism hence cost of product can be decreased.

Commercial Starter from Labs

There are some advantages of commercial cultures.

1. Cultures will have only desired microorganisms.
2. If it is a mixed culture then the ratio will be defined.
3. The organisms will have desired features such as resistance to phages, heat resistance etc.
4. The organism can be obtained in desired form if log phase is required they provide the culture i.e. late in lag phase.
5. Temperature and time is adopted by the organism.

Properties of different starter culture

Culture	Properties
Diary culture	Ease to manufacture Stability and consistency Rate of Lactic acid production Lag phase times Phage resistance desirable flavor and texture production Preservation of tolerance produce
Meat cultures	Lack of flavours Fast acidification Desired flavor Antimicrobial activity
Beer culture	Rapid fermentation Desired flavor Production Tolerance and stability of preservation Flocculation Lack of off-flavours Growth at wide temperature range
Wine culture	Osmotolerant Ethane tolerant Flocculation, sedimentation Growth at low temperature Produce consistent flavor Malolactic fermentation
Bread cultures	Freeze tolerant Produce desired flavor

Growth Media for Starters

The principle behind the starter culture propagation is to obtain actively growing cells to form the product. It requires different nutrients and controlling the parameters such as pH, temperature, aeration etc. Traditional growth media for the lactic starters are milk and whey. Commonly required nutrients for starter culture production in the following table (White Head et al. 1976).

Ingredients of the bulk starters of lactic acid bacteria

Carbon source	Nitrogen source	Vitamins and minerals	Phage inhibitory agents	Antoxidants	Neutralizers
Lactose, maltose, sucrose, glucose	Milk protein, whey protein, casein hydrolysates	Yeast extract, corn steap liquor	Prosphates, citrates	Ascorbic acid, ferrous sulphate	Carbonates, phosphates, hydroxides and oxides

For a starter culture, media carbon, nitrogen sources, vitamins, minerals, antoxidants, phage inhibitory substances and neutralizers, was required (White Head et al. 1976).

Methods of Propagation

Starter culture can be, isolated from source/nature or obtained commercially. Some of the dairy plants maintain their own cultures. Some of the plants use new cultures obtained from commercial culture labs which are in frozen condition, concentrated or lyophilized.

Obtaining commercial cultures reduces the need for maintenance and sub culturing. Hence, contamination can be eliminated. This reduces the preparation of bulk starter inoculums which requires the transfer of media larger to larger volumes. This can save the time also. Based on the propagation, the starter cultures can be

1. **Commercial cultures**: original cultures obtained directly from labs.
2. **Mother cultures:** Culture prepared using the master cultures.
3. **Intermediate cultures:** In large industries before the propagation of bulk starters, intermediate cultures will be propagated by increasing the volume of media.
4. **Bulk starters:** the starter used for the production of desired product.

Starter culture production: It consists of following steps: (1) Procuring the organism from Raw material/sources. Environment or from commercial sources. (2) Inoculation material Handling. (3) Sterilization of media. (4) Cooling the medium till it reaches required temperature. (5) Inoculation. (6) Incubation/

production of starter culture. (7) Cooling the finished product. (8) Concentration. (9) Freezing. (10) Drying. (11) Packing and storing (Hoier et al. 2010).

Lactose is the major source of carbon, maltose, sucrose or glucose can be added in less concentration to stimulate the growth (Sandine 1996). Nitrogen can be obtained from yeast extract even it supplies vitamins and minerals also. Nitrogen sources can be milk proteins, whey proteins, casein hydrolysates etc. phosphates has a dual role they act as acid neutralizers and phage inhibitors. Phage inhibitors prevent excessive acidity. Neutralizers can be in the form of ammonium or potassium hydroxide.

Starter Culture for Yogurt

Required features of starter cultures for yogurt are: presence of desired organism, ability to produce acid, aroma and flavor Phage resistance, rapid multiplication under desired cultural conditions of the product.

Starter culture for yogurt generally Lactobacillus delbrueckii ssp. bulgaricus and Streptococcus thermophilus are developed in a glass bottle. Yogurt is a coagulated milk product obtained in frozen or freeze dried form and cultured on suitable medium and the cultured medium is known as mother culture (Isleten and Yucer 2006; Madhu et al. 2013).

100 ml of fresh milk which is free of antibiotics or reconstituted milk is taken in a sterile glass bottle up to 3/4th and heated to 90 °C for 1 h which, sterilizes the media and removes dissolved oxygen and decomposes proteins. Add starter when temperature is 20–30 °C if the starter is mesophile or 40–45 °C if the starter is thermophile and incubate at 21–43 °C.

Taking into consideration of amount of inoculum, the incubation temperature and duration of incubation varies with the quantity of inoculums. Generally inoculum ranges 2–3 %. Based on the amount of inoculum used, the incubation time is between 3 and 20 h. During this period organism multiplies and produces lactic acid and different aromatic compounds such as acetic acid, diacetyl, aldehydes, ketones, alcohols, fatty acids, carbon dioxides etc. depending on the organism employed.

Symbiosis of L. bulgaricus and S. thermophilus is important for the formation of yogurt. Acetaldehyde is the major flavoring compound in yogurt but this will be produced when acidification reaches approximately pH 5.0. The maximum at pH 4.2 and stabilizes at 4.0. Generally the ratio of above two organisms will be in 1:1 to 2:1 for yogurt formation but when the incubatuion temperature is 40 °C the ratio reaches to 4:1. At 45 °C the ratio will be 1:2, at 43 °C it is 1:1. Hence incubation temperature is critical and generally kept at 43 °C. Size of inoculum is 2.5–3.0 % and incubation period is 3–4 h. At the end of incubation period it is cooled to 10–12 °C resulting in the concentrate. This is freezed and dried as per requirements and packed to store (Hoier et al. 2010) (Fig. 21.1).

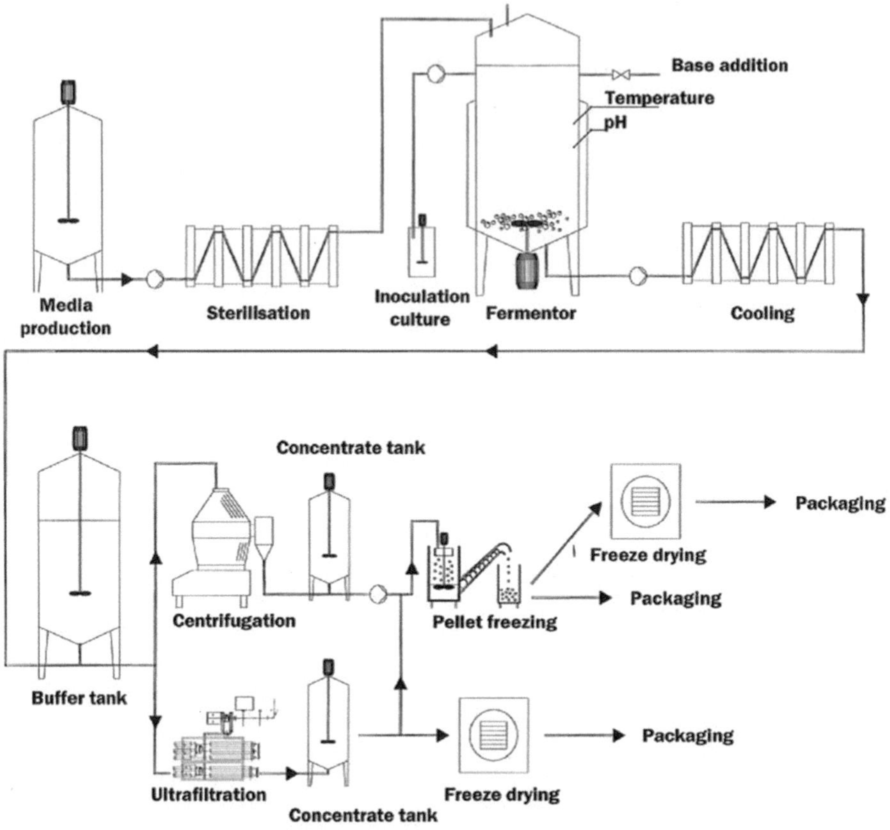

Fig. 21.1 Process flow of a typical starter culture production

Starter Culture for Bread

It consists of preparation of stock culture followed by Mother culture and several intermediate cultures finally seed culture. Generally media contains a carbon source in the form of cane or beet molasses/mineral salts mash having molasses, nitrogen in the form of ammonium salts, urea, malt sprouts etc. inorganic salts and accessory growth substances, vitamin precursors or vitamins. Media is sterilized and Inoculated with starters. Required pH 4.3–4.5 and temperature 30 °C. Medium is aerated rapidly and molasses is added gradually to maintain sugar level 0.5–1.5 %. After 4–5 budding cycles yeast is centrifuged in cream form and passed through filter press to remove excess liquid and made into cakes of yeast with different sizes after addition of vegetable oil.

Starter Cultures for Mold

Mold cultures can be grown on agar slants using the media such as malt extract agar or they can be grown in different ways using, flasks with liquid or solid media such as agar, shallow trays as surface medium, loose media moistened with nutrients such as wheat brawn, Moistened bread, submerged fermentation aerated.

Spores of Penicillium roquefortii is used for blue cheese- Roquefort stilton, Gorgonzola They will be grown on sterile moistened acidified bread after growth, bread and culture is dried and powdered and packed in cans. Moistened sterile crackers also used for the spore production of Pencillium camamberti. Starter cultures used in submerged fermentation generally in the form of pellets or mycelial mats.

Preservation of Starter Cultures

Starter cultures will be stored to make them available for bulk starter production. If starter culture fails the preserved cultures can be used for direct to vat inoculation (DVI). Frequent sub culturing may develop mutants then the ratio of starter cultures may be changed. The starter cultures can be stored in liquid form which is not in use now. It may also be preserved as dried starters in the form of freeze dried concentrates. These freeze dried starters are used in powder form. They can be preserved as frozen starters can be stored at temp −20, −40, −80, −196 °C in liquid nitrogen. Freeze dried starter may be super concentrated to get deep frozen and used directly for the production of product (Hui H.Y.).

Bacterial cultures are stored at room temperature from months to years by preserving, on agar slants incorporated with 1 % sodium chloride.

Or sealing tube cultures with paraffin oil, in the form of dry spore stocks using sterile soil, by lyophilization (freeze drying) and freezing in liquid nitrogen −196 °C. Viability of the starters depends on the type of medium, cryoprotective agents, freezing conditions and method used for concentration.

Fermentation Products

Different starters are involved in formation of different fermented products.

Based on raw material fermented products can be classified as

1. Beverage based fermented products.
2. Cereal based fermented products.
3. Dairy products.
4. Fruits, vegetables and Fish.
5. Legumes.
6. Meats (Campbell Platt).

Bioyogurt

It is produced by using Bifidobacterium along with Streptococcus thermophilus and Lactobacillus bulgaricus. Improvements in the field of molecular biology and bioinformatics revolutionized production of yogurt which stimulates the growth of beneficial microorganisms in the intestine and promotes digestion.

Soy Sauce

It is a dark brown color liquid prepared using soy beans, wheat, salt and microorganisms Mold yeasts and bacteria Aspergillus oryzae, Aspergillus soyae in Koji fermentation and Yeasts and Lactic acid bacteria in Moroni fermentation (Holzapfel et al. 1998). PH of the sauce is 4.6–4.8. The manufacture of soy sauce consists of

Preparation of raw material: Roasting of wheat is followed by coarse grinding. Roasting imparts color, flavor and eliminates undesirable organisms. Soy beans are soaked in water and cooked at high temperature and pressure.

Koji preparation: Coarse wheat flour and cooked soy are used for koji preparation. Now starter is prepared from mold Aspergillus oryzae or A. soyae. Spores are inoculated into mixture. This mixture is spread thinly at 30 °C for 5 days. Koji starter is used to inoculate wheat and soy mixed in 1:1 ratio. This mixture is kept in vats and aerated. Temperature and humidity is maintained. Koji is obtained after 2–3 days as greenish yellow material due to the presence of Aspergillus spores.

Brine fermentation: Deep fermentation tanks are used for koji fermentation. Koji is mixed with equal volume of salt solution 20–23 %. This mixture is called as Moroni. This is fermented for 6–8 months and mixed to create aerobic conditions and to remove unwanted organisms. After formation of Moroni liquid part is separated, filtered (Rowan et al. 1998; Blandini et al. 2003). During fermentation process different starters are grown in three phases. During first phase pediococcus halophilus grows and produce lactic acid resulting a decrease of pH. In second phase Saccharomyces rouxii develops and producing alcohol. In last phase Torulopsis yeasts develop and secretes phenolic compounds essential for flavor of koichuki shoyu flavor, fermentation conditions will select the organisms consistent flavor is possible using pure cultures.

Refining: Fermented Moroni is pressed to release soy sauce. The raw sauce is pasteurized at 70–80 °C to inactivate the enzymes pasteurized and bottled Soy sauce production has been improving since the first production (Blandino et al. 2003).

Yogurt

It is a coagulated milk product, prepared from special starters Lactobacillus delbrueckii ssp. bulgaricus and Streptococcus thermophilus (Isleten and Yucer 2006; Madhu et al. 2013) streptococci grows faster than the rod and leads to the production of acid and the rod adds flavor and aroma. The association of two organisms produces lactic acid maximum and L. bulgaricus produces more acetaldehyde (flavoring component of yogurt) in association with streptococci. Lactic acid produced by them coagulates the milk and confers characters such as aroma, acidity and consistency. Back-slopping was the method used earlier which may not produce consistent quality, a desired feature for industry. The best choice is the utilization of defined cultures.

Tempeh

It is the Indonesian food prepared from fermented soybeans using species of Rhizopus. Traditional tempeh preparation varies little.

Seed coat is removed from soaked and boiled soybeans followed by draining water and cooled. This will be inoculated with species of Rhizopus i.e., R. oryzae. The beans will be packed in small parcels and incubated approximately at 25 °C temperature for 40 h. Finally a white cake will be formed which contains beans bound by mold mycelia. These white cakes will be deep fried for 3–4 min or boiled for 10 min.

Bread

Since centuries human beings has the knowledge of bread preparation. Major ingredients used in bread preparation are Wheat flour, water, salt, microorganism.

Technology brought many changes in traditional bread preparation i.e. Sugar, milk, enzymes (Bacterial and fungal), flavors are added which help in the conversion of starch into simple sugar. Bread preparation includes following steps (1) Preparation of dough (2) Proving: allows the microorganism to grow and leads to the rising of dough. (3) Baking: conversion of dough to bread using high temperature. (4) Cooling: decreasing the temperature to ambient. (5) Slicing and packing.

Gari

It is a food product of West Africa prepared using cassava tubers. The preparation of gari consists of following steps-Cassava tubers will be peeled then grated. The material/mash obtained after peeling will be fermented for 18–48 h in cloth bags, dewatered, sieved and heated. Addition of palm oil is option while heating (Kuye and Sanni 1999).

Idli

It is prepared from separately soaked rice and black gram. After grinding and mixing at ratio of 2:1 by adding little salt and allowed to ferment overnight at room temperature. After that batter is transferred to idli pans and cooked for 5–8 min (Nagaraju and Manohar). The fermentation process will be completed with the lactic acid bacteria such as L. fermentii, L. lactis, L. deleberuekii, Streptococcus faecalis, Leuconastoc misenteroides and pediococcus cerevisiae. It is identified that Leuconastoc misenteroides and streptococcus faecalis are essential for leavening the idli batter (Ramakrishnan 1993; Blandino et al. 2003). Yeasts geotrycum candidum, trychosporon pullulans and torulopsis species were identified in the product (Shrott 1998) it was assessed that fermentation increases all required amino acids and decreases anti nutrients (Blandino et al. 2003).

Dosa

It is prepared from a mixture of ground rice and black gram and prepared as paste and fermented for 8–20 h. During fermentation microbiological, physical, biochemical changes will occur. Nutritive properties of substance will increase due to the activity of different microbes. These fermented substances are spread on a hot plate which is smeared with oil (Blandino et al. 2003; Shrott 1998).

Cheese

Cheese is a proteinaceous material with 30 % fat (Traditional) prepared from milk. This might have originated thousands of years ago. More than 2,000 varieties of cheese are available now. Lactic acid bacteria are used in cheese production starter culture with Resistance to high salt and acidic/alkaline pH is must and desirable as it can use maximum substrate while decreasing the pH by acid production. Ability to compete with natural microflora, dominance over spoilage microorganisms, ability of revitalization and surviving after storage or other desirable characters of cheese fermenting starters.

Lactic acid bacteria used in Cheese are of two types (1) Starter lactic acid bacteria (2) Non starter lactic acid bacteria Starter lactic acid bacteria are involved in acid production during fermentation and contribute to ripening process again starter lactic acid bacteria can be divided as primary starters and secondary starters. Primary starters are involved in acid development and secondary starters provide defined functions for example Propionibacterium shermanii ssp. freudenreichii is involved in gas production. In swiss cheese (journal of dairy science vol 83 (4); 609–619) NSLAB are unique to specific to cheese varieties (Beresford et al. 2001).

Cheese fermentation involves standardization of milk, coagulation by different microbial rennets

Lactobacillus as starter cultures: Lactic acid production is necessary for cheese manufacture. In olden days lactic acid was produced by naturally occurring bacteria. But nowadays starter cultures are used. If cheese prepared at less than 40 °C Lactococcus lactis is used. If cheese is prepared at greater temperatures, streptococcus thermophilus, L. bulgaricus and L. helveticus is used.

Coagulation: Earlier for the process of coagulation rennet enzyme was used obtained from stomach of calves, and young goats Extracts contain pepsin along with rennin (Pallavi et al. 2012). Rennin is responsible for coagulum formation. Pepsin hydrolyses the coagulum and the result is low yield of cheese and bitter taste. Due to high cost of rennet microbial rennet's are used.

Shrinkage of the curd, inoculation of the starter culture of mesophilic lactic acid bacteria Lactococcus lactis or thermophilic lactics (L. bulgaricus, L. helevetics or Streptococcus thermophilus).

Salting of the curd: Salt is added to develop taste and to contribute moisture. It limits the growth of proteolytic bacteria. Curd is pressed into shape before maturation.

Ripening and salting are different stages in cheese product. Enzymes of milk and rennet and those from starter organisms are responsible for the development of special character such as Blue veined cheese—Roqueforti, gorgonzola, Stilton, while swiss cheese from propionibacterium spp., micrococci and Brevibacterium linens confers special flavor to Limburger cheese.

Applications of Starter Culture

1. Starter cultures will preserve the food material. During multiplication they secretes different health promoting substances. Streptococcus thermophles, Lactobacillus bulgaricus regular starters of Yoghurt secretes Folic acid (Wouters et al. 2002). Bifidobacterium spp. secretes Folic acid and biotin (Strozzi and Mogna 2008; Pompei et al. 2007) L. reuteri and other lactobacilli secretes Vitamin B12 (Taranto et al. 2003) Propionibacterium shermani also secretes Vintamin B12 (Burgess et al. 2004) Lactococcus, Lactobacillus, Enterococcus, Leuconostoc and streptococcus secretes Vitamin K (O'Connor et al. 2001; Cooke et al. 2006) Bifidobacterium lactis secretes vitamin B in fermented milk (Beitane and Ciprovica 2012) Enterococcus spp. secretes vitamin B-complex during Tempeh production (Keuth and Bisping 1993).

Different acids and vitamins secreted by starters hence nutritive values of fermented product is superior to the raw foods.

Starters secrets different substances which inhibits the growth of poisonous/unwanted organisms.

1. Lactic acid bacteria has important role in enhancing the properties of fermented foods especially nutritional, technological, organoleptic values in the preparation of fermented foods and beverages (Shah and Prajapati 2013; Patel et al. 2013; Capozzi et al. 2012).

2. The sensory qualities of cheese was improved by lactobacillus lactis subsps. Cremoris (Gued feldt et al. 2001; Gonalez 2010).
3. Dairy starters produce lactic acid from lactose. This lactic acid acts as a preservative and develops desirable texture and characteristic body of fermented milk products (Subriati Hati 2013; Patel et al. 2013).
4. Therapeutic and dietetic products such as Bioghurt, actimel, yakelt are cultured dairy products produced (Paramzit 2011) using starter cultures.
5. Advances in genomics and proteomics of microorganisms providing better understanding and helps in the strain improvement with desired features and also to eliminate undesired features (Del-Cour et al. 1999; Law 2001).
6. Cheese can create required environment for the probiotic organisms hence useful for the people having acidity problem due to the growth of lactobacillus (Karimi et al. 2012).
7. Yogurt is source of nutrients such as proteins, calcium, phosphorous, vitamins B1, B2, B12, folate, Mg, Zn (Baresford et al. 2001; Ahmet et al. 2014).
8. Yogurt can control the growth of unwanted organisms and can cure the diseases like constipation, diarrhea and dysentery (Stuti Agarwal and Renu Prasad 2013).
9. Gene transfer techniques can be used to develop starters to produce high quality and low fat cheese in the place of 30 % fat of traditional cheese causing heart diseases obesity etc. These modified organisms can create opportunities to reduce disease (James 2012).
10. Starters can be used to preserve the product by fermentation process increasing the shelf life and safety of the product. By the secretion of preservatives such as bacteriocins. Starter cultures increases the viscosity of the product. Starter cultures are used for the formation of eyes and colour in cheese.
11. Different starters are involved in the production of vitamins such as folic acid in yogurt (Wouters et al. 2002), bioactive peptides, amino acids and biothickeners (Hati et al. 2013).
12. Folic acid and thiamin and vitamin B3 using starters B. bifidum, B. breve, B. longum longum, B. longum infantis, B. adolescentis (Deguchi et al. 1985).
13. Vitamin B6 and vitamin B12 is produced using the starters B. bifidum, B. breve, B. longum longum, B. infantis (Marks et al. 2010).

List of flavors and products from different microorganisms

Microorganism	Compound	Name of the product
Streptococcus thermophilus, Lactobacillus bulgaricus	Acetaldehyde, Diacetyl, Acetoin, Acetone	Yogurt
Mucor maihi		Cheese
Propionibacterium shermanii	Propionic acid	Swiss cheese
Streptococcus lactis Streptococcus cremoris	Lactic acid, Diacetyl and Acetaldehyde	Buttermilk
Lactobacillus bulgaricus	Acetaldehyde	Buttermilk
Saccharomyces cerevisae	Ethanol	Bread
Mixed cultures of Lb. brevis, Leuconostoc mesenteroids, Lb. plantaram	Acetate, Fatty acids	Sauer kraut
Aspergillus oryzae Lb. S. rouxii	Alkyl phenols and Pyrazines	Soy sauce (Japan)

Table 21.1 Starter cultures for fermented foods and beverages (Jytte josephsen and Lene jesperson 2006)

Fermented foods	Starter cultures
Hard cheeses without eyes Cheeses with small eyes Swiss and Italian type cheeses Butter and buttermilk Yoghurt Fermented, probiotic milk Kefir	Lactococcus lactis ssp. Lactis, L. lactis ssp. Cremoris Lactococcus lactis ssp. Lactis, Lactococcus lactis ssp. Lactis var. diacetylactis, L. lactis ssp. Cremoris, Leuconostoc mesenteroides ssp. Cremoris Lactobacillus delbrueckii ssp. Lactis, L. helveticus, L. casei, L. delbrueckii ssp. Bulgaricus, Streptococcus thermophilus Lactobacillus lactis ssp. Lactis, Lactococcus lactis ssp. Lactis var. diacetylactis, L. lactis ssp. Cremoris, Leuconostoc mesenteroides ssp. Cremoris L. delbrueckii ssp. Bulgaricus, Streptococcus thermophilus L. casei, L. acidophilus, L. rhamnosus, L. johnsonii, Bifidobacterium lactis, B. bifidum, B. brevis L. kefir, L. kefiranofacies, L. brevis

High quality low fat cheese development using genetically modified starter cultures can open the doors to solve health problems create new opportunities to reduce disease.

Different species of Lactobacullus serving as starter culture for fermentation of plant products

Raw material	Dominating bacteria or starter cultures	Products
Cabbage	*Leuconostoc mesenteroides, Lb. plantarum* *Lb. curratus, lb. brevis, P. cerevisiae*	Sauerkraut
Cucumber	*Lb. brevis, P. cerevisiae, Lb. planatarumm* *Lb. pentosus, P. pentosaxeus, yeast*	Salted/pickled cucumber
Olives	*Lb. brevis, P. pentosaceus, Lb. planatarum* *Lb. pentosum, yeast*	Olives
Fruit juices	*Lb. caesi, Lb. planatarum, Lb. xylosus, Lb. sakei*	Fruit juice
Wheat and rye	*Lb. sanfranciscensis, Lb. brevis* *Lb. plantatarum, Lb. fermentum, Lb. fructivorans,* *Lb. delbrueckii*	Sourdough

Fermented products and microorganisms

Dairy product	
– Hard cheeses without eyes	L. lactis subsp. lactis, L. lactis subsp. cremoris
– Cheeses with small eyes	L. lactis subsp. lactis, L. lactis subsp. lactis var diacetylactis, L. lactis subsp. cremoris, Leuc. Menesteroides subsp. cremoris
– Swiss-and Italian-type cheeses	Lb. delbrueckii subsp. lactis, Lb. helveticus, Lb. casei, Lb. delbrueckii subsp. bulgaricus, S. thermophilus
– Butter and buttermilk	L. lactis subsp. lactis, L. lactis subsp. lactis var diacetylactis, L. lactis subsp. cremoris, Leuc. menesteroides subsp. cremoris

(continued)

– Yoghurt	Lb. delbrueckii subsp. bulgaricus, S. thermophilus
– Fermented, probiotic milk	Lb. casei, Lb. acidophilus, Lb. rhamnosus, Lb. johnsonii, B. lactis, B. bifidum, B. breve
– Kefir	Lb. kefir, Lb. kefiranofacies, Lb. brevis
Fermented meats	
– Fermented sausage (Europe)	Lb. sakei, Lb. curvatus
– Fermented sausage (USA)	P. acidilactici, P. pentosaceus
Fermented vegetables	
– Sauerkraut	Leuc. mesenteroides, Lb. plantarum, P. acidilactici
– Pickles	Leuc. mesenteroides, P. cerevisiae, Lb. brevis, Lb. plantarum, Leuc. mesenteroides, Lb. pentosus, Lb. plantarum
– Fermented olives	P. acidilactici, P. pentosaceus, Lb. plantarum,
– Fermented vegetables	Lb. fermentum
Fermented cereals	
– Sourdough	Lb. sanfransiscensis, Lb. farciminis, Lb. fermentum, Lb. brevis, Lb. plantarum Lb. amylovorus, Lb. reuteri, Lb. pontis Lb. panis, Lb. alimentarius, W. cibaria
Fermented fish products	Lb. alimentarius, C. piscicola

Future Trends

Starter culture application is in high demand in various fields like foods, beverages, fruit juices, dairy industry etc. Due to increased health awareness among the consumers to reduce the chemicals may be in the form of antibiotics/preservatives novel starter cultures are in demand. Development in molecular biology i.e. RDNA technology and microarrays techniques & with ease of genome sequencing, open up the possibilities to design and develop new starters such as disease resistance, production of antimicrobial which can target the development of unwanted organisms/modify the existing organisms with desired genes.

References

Buckenhuskes, H. J. (1993). Selection criteria for lactic acid bacteria to be used as starter cultures for various food commodities. *FEMS Microbiology Reviews, 12*, 253–271.

Hammes, W. P., & Knauf, H. J. (1994). Starters in the processing of meat products. *Meat Science, 36*, 155–168.

Hansen, E. B. (2002). Commercial bacterial starter cultures for fermentated foods of the future. *International Journal of Food Microbiology, 78*, 119–131.

Miralles, M. C., Flores, J., & Perez-Martinez, G. (1996). Biochemical tests for the selection of staphylococcus strains as potential meat starter cultures. *Food Microbiology, 13*, 227–236.

McFeeters, F. F. (2004). Fermentation microorganisms and flavor changes in fermented foods. *Journal of Food Science, 69*(1), 28–37.

Ling, M. Y., & Chou, C. C. (1996). Biochemical changes during the preparation of soy sauce koji with extruded and traditional raw materials. *International Journal of Food Science and Technology, 31*, 511–517.

Omafuvbe, B. O., Abiose, S. H., & Shonukan, O. O. (2002). Fermentation of soybean (Glycine max) for soy-daddawa production by starter cultures of Bacillus. *Food Microbiology, 19*, 561–566.

Uchida, M., Ou, J., Chen, B. W., Yuan, C. H., Zhang, X. H., Chen, S. S., Funatsu, Y., Kawasaki, K. I., Satomi, M., & Fukuda, Y. (2005). Effects os soy sauce koji and lactic acid bacteria on the fermentation of fish sauce from fresh water silver carp Hypophthalmichthys molitrix. *Fisheries Science, 71*(2), 422–430.

Holzapfel, W. H., Haberer, P., Snel, J., Schillinger, U., & Veld, J. H. (1998). Overview of gut flora and probiotics. *International Journal of Food Microbiology, 41*, 85–101.

Patel, A., Nihir, S., & Prajapati, J. B. (2013). Biosynthesis of vitamins and enzymes in fermented foods by lactic acid bacteria and related genera - A promising approach. *Journal of Food Science and Technology, 5*(2), 85–91.

Agarwal, S., & Prasad, R. (2013). Effect of stabilizer on sensory characteristics and microbial analysis of low fat frozen yoghurt incorporated with carrot pulp. *International Journal of Agri. Food Science Technology, 4*(8), 797–806.

Ayar, A., & Gurlin, E. (2014). Production and sensory, textural, physico-chemical properties of flavored spreadable yogurt. *Life Science Journal, 11*(4), 58–65.

Ali, A. A., & Mustafa, M. M. (2009). Use of starter cultures of lactic acid bacteria and yeasts in the preparation of kisra, a Sudanese fermented food. *Pakistan Journal of Nutrition, 8*(9), 1349–1353.

Stanley, E., Fitzgerald, G. F., Le Marrec, C., Fayarad, B., & Van Sinderen, D. (1997). Sequence analysis and characterization of f01205, a temperate bacteriophage infecting streptococcus thermophilus CNRZ1205. *Microbiology, 143*, 3417–3429.

Allison, G. E., & Klaenhammer, T. R. (1998). Phage resistance mechanisms in lactic acid bacteria. *International Dairy Journal, 8*, 207–226.

Sanders, M. E. (1998). Overview of functional foods: Emphasis on probiotic bacteria. *International Dairy Journal, 8*, 341–347.

Tannock, G. W. (1998). Study of intestinal microflora: A prerequisite for the development of probiotics. *International Dairy Journal, 8*, 527–533.

de Oliveira Leite, A. M., Migues, M. A., Peixoto, R. S., Rosado, A. S., Silva, J. T., & Paschoalin, V. M. (2013). Microbilogical, technological and therapeutic properties of kefir: A natural probiotic beverage. *Brazilian Journal of Microbiology, 44*(2), 341–349.

Syal P., & Vohra A (2013). Probiotic potential of yeast isolated from traditional Indian fermented foods. *International Journal of Microbiology Research, 5*(2), 390–398.

Nout, M. J. R. (1991). Ecology of accelerated natural lactic fermentation of sorghum based infant food formulas. *International Journal of Food Microbiology, 12*, 317–224.

Karimi, R., Sohrabvandi, S., & Mortazavian, A. M. (2012). Review article: Sensory characteristics of probiotic cheese. *Comprehensive Reviews in Food Science and Food Safety, 11*, 437–452.

Patil Pallavi, M.*, Kulkarni Avanti, A.**, & Kininge Pallavi, T.*** (2012). Production of milk clotting enzyme from Aspergillus oryzae under solid-state fermentation using mixture of wheat bran and rice bran. *International Journal of Scientific and Research Publications, 2*(10), 273–284.

Beresford*, T. P., Fitzsimons, N. A., Brennan, N. L., & Cogan, T. M. (2001). Recent advances in cheese microbiology. *International Dairy Journal, 11*, 259–274.

Leroy, F., Verluyten, J., & Vuyst, L. (2006). Functional meat starter cultures for improved sausage fermentation. *International Journal of Food Microbiology, 106*, 270–285.

Caplice, E., & Fitzerald, G. F. (1999). Food fermentation: Role of microorganisms in food production and preservation. *International Journal of Food Microbiology, 50*, 131–149.

Klaenhammer, T. R., & Kullen, M. J. (1999). Selection and design of probiotics. *International Journal of Food Microbiology, 50*, 45–57.

Holzapfel, W. (1997). Use of starter cultures on a household scale. *Food Control, 8*, 241–258.

Holzapfel, W. (2002). Appropriate starter culture technologies for a small scale fermentation in developing countries. *International Journal of Food Microbiology, 75*, 197–212.

Hansen, L. T., Allan-Wojtas, P. M., Jin, Y.-L., & Paulson, A. T. (2002). Survival of ca-alginate microencapsulated bifidobacterium spp. in milk and simulated gastrointestinal conditions. *Food Microbiology, 19*, 35–45.

Wakil, S. M., Ayenuro, O. T., & Abiola, K. (2014). Microbiological and nutritional assessment of starter developed fermented tigernut milk. *Food and Nutrition Sciences, 5*, 495–506.

Smid, E. J., Erkus, O., Spus, M., Wolkers-Rooijackers, J. C., Alexeeva, S., & Kleerebezem, M. (2014). Functional implications of the microbial community structure of undefined mesophilic starter cultures. *Microbial Cell Factories, 13*(Suppl 1), 52.

Hati, S., Mandal, S., & Prajapati, J. B. (2013). Novel starters for value added fermented dairy products. *Current Research in Nutrition and Food Science, 1*(1), 83–91.

Chapter 22
Ca-Alginate Liquid Core Capsule for Lactobacili Fermentation

Boon-Beng Lee, Pogaku Ravindra, and Eng-Seng Chan

Introduction

Lactic acid bacteria (LAB) have been long used for dairy and non-dairy based food fermentation since ancient time (Widyastuti et al. 2014; Rhee et al. 2001). Yogurt, cultured milk, cheese are dairy products that produced from LAB fermentation (Widyastuti et al. 2014). Some Asian traditional foods are examples of LAB fermented non-dairy products such as 'tempoyak' (acid fermented condiment) (Leisner et al. 2002), "tapai" (rice wine) (Chiang et al. 2006), 'kimchi' (fermented vegetables) (Rhee et al. 2001), "idli" in India and "puto" in Philippines (steamed bread) (Rhee et al. 2001), "sikhae" in Korea and "burong isda/dalag" in Philippine (fermented mixture of salted fish and cereals) (Rhee et al. 2001).

LAB are characterized by production of lactic acid as their major metabolic end product ($\geq$50 %) of a carbohydrate fermentation (Gereková et al. 2011; Somkuti 2000). Homofermentative LAB produce mainly lactic acid, and heterofermentative LAB produce lactic acid, carbon dioxide, and acetic acid or ethanol. Depending on the carbohydrate and the growth conditions (pH, nutrient density, and the number of bacterial cells in the inoculum), homofermentative LAB may change their metabolic pathway to produce end product of the fermentation of mixed acids such as formate and acetate, ethanol, and carbon dioxide (Kontula et al. 1999).

LAB consists of the Gram-positive genera: *lactobacillus, lactococcus, Carnobacterium, Entercoccus, leuconostoc, Oenococcus, Pediococcus, Streptococcus,*

B.-B. Lee (✉)
School of Bioprocess Engineering, Universiti Malaysia Perlis, Perlis, Arau 02600, Malaysia
e-mail: bblee@unimap.edu.my

P. Ravindra
Faculty of Engineering, Universiti Malaysia Sabah, Kota Kinabalu, Sabah 88999, Malaysia

E.-S. Chan
Chemical Engineering Discipline, School of Engineering, Monash University Malaysia, Jalan Lagoon Selatan, Bandar Sunway, Selangor 46150, Malaysia

P. Ravindra (ed.), *Advances in Bioprocess Technology*,
DOI 10.1007/978-3-319-17915-5_22

Tetragenococcus, Vagococcus, Bifidobacerium and Weissella (Gereková et al. 2011; Hofvendahl and Hahn-Hagerdal 2000; Klein et al. 1998). LAB are cocci, with exception of *lactobacillus* and *Carnobacterium* which are rods. LAB are nonmotile and nonspore forming (Gereková et al. 2011; Hofvendahl and Hahn-Hagerdal 2000). Most LAB are facultatively anaerobic and catalase negative. They have high acid tolerance (pH 5 and lower). The optimal temperature for growth varies between the genera from 20 to 45 °C. Most of the LAB is considered as Generally Regarded As Safe (GRAS), but some strains of *Streptococcus* is pathogenic (Hofvendahl and Hahn-Hagerdal 2000).

Lactobacillus is one of the important LAB that have been widely applied and studied because of their attractive features. In general, the members of the genus *Lactobacillus* can be isolated from gastrointestinal (GI) and genitourinary tract of human and animal, plants, raw milks, and insects (Giraffa et al. 2010; Markiewicz et al. 2008). They have been well known for their fermentative ability to enhance food safety and nutrients, to improve organoleptic quality, as well as to improve the health and functions of GI tract (Markiewicz et al. 2008; Chiang et al. 2006; Rhee et al. 2001). Moreover, they have been recognized as probiotics due to their potential therapeutic and prophylactic characteristics (Giraffa et al. 2010). Owing to these benefits, they have been added as enhancement to the dairy, food, pharmaceutical, animal feed, health care products (Giraffa et al. 2010; Markiewicz et al. 2008).

In recent years, production of lactic acid using lactobacilli has received much attention of researchers (Zhang et al. 2007; Ohkouchi and Inoue 2006; Oh et al. 2005; Sakai et al. 2004; Akerberg and Zacchi 2000). This is because lactic acid is one of the major materials used to produce bio-degradable plastic or poly (lactic acid) [PLA] bags and the use of biodegradable plastics has been popularized by many countries. As a result, the production of biodegradable plastics increased and hence remarkably increased the demand of lactic acid (Zhang et al. 2007).

Encapsulation

Many biological systems in their natural state are encapsulated. Most functions of living systems are based on the confinement of reactions in a limited space. Therefore, encapsulation is the basis of cell evolution and the beginning point of all life forms on earth. Encapsulation protects the internal materials from outer environment. Encapsulation is defined as the process by which a gaseous, liquid or solid ingredient is surrounded by a continuous film or coating of a gel matrix (Marison et al. 2004). The concept of encapsulation is becoming increasingly important in industrial applications and life sciences due to its unique features.

The fundamental principle of encapsulation has been applied to confine bioactive materials (e.g., microbial cells, enzymes, animal cells, plant cells, etc) using certain controlled chemical reaction to form the film or coating. At the end, the bioactive materials could be encapsulated in various shapes, namely spherical particle (i.e., bead, capsule), fiber, thread, cylinder, film etc. Among the shapes, spherical particle is the most easy to form with consistent size. Therefore, spherical beads and

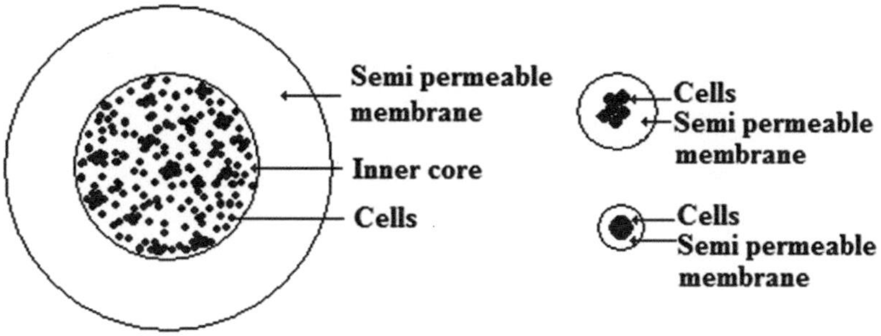

Fig. 22.1 Macrocapsule (*left*) and microcapsule (*right*)

capsules have been widely studied and used in research and industrial applications. They are differentiated based on their structure. Basically a capsule is defined as a hollow particle consisted of a defined external gel membrane (shell) and an inner core (solid or liquid) whereas the bead is defined as a solid dense gel particle.

Capsules are commonly categorised according to their size (macro, micro and nano-capsules), membrane polymer, core types and sometimes core material. Figure 22.1 illustrated the capsule in macro and micro size. Macrocapsules are formed by encapsulating multiple cells or cell aggregates in hollow semi-permeable materials and microcapsules are produced by enclosure individual cells or small cell aggregates in a semi-permeable membrane (Bhatia Surita et al. 2005). Macrocapsules are frequently used in Lactobacilli fermentation (Rodriguez-Huezo et al. 2011; Goderska et al. 2003; Dembczynski and Jankowski 2002; Yoo et al. 1996). This is because they can be handled easily and recovered from the fermentation medium without difficulty (centrifugation or filtration). Thus reducing the overall production cost.

The function of the semi-permeable membrane is to retain the cells but at the same time, allow nutrients and by-products to pass through it. The semi-permeable membrane of the capsules has been shown to offer many advantages to achieve high cell concentration in fermentation processes by increasing process stability (physical and biological) over long fermentation periods, increase the reuse number of biocatalysts, improve resistance of biomaterials to contamination, retain cells in continuous processes that operate at high dilution rate and provide protection to biomaterials from harsh environmental conditions such as pH, temperature, organic solvent and poison.

The inner core of the capsules could be made-up of solid or liquid. Besides, it is noticeable that the capsule could be formed with single core or multiple cores but the focus of this paper is only on the single liquid core capsule. The preference for the inner core type is dependent on the application, the ease of formation, the nature of the bioactive material that being encapsulated and so on. The features of these capsules were compared as shown in Table 22.1. As compared to liquid core capsules, solid core capsules are generally easier to be formed with uniform size and shape. The solid core capsules can withstand mechanical deformation and chemical chelating to a greater extent than the liquid core capsules. In contrast, the capsules with liquid in its core provide ample space for cellular growth and

Table 22.1 Features of solid core and liquid core capsules

Features	Solid core capsules	Liquid core capsules
Formation	Simple	Complicated
Mechanical and chemical stability	High	Low
Space for cell growth	Limited	Ample
Cell leakage	Moderate	Good
Mass transfer limitation	High	Low
Homogeneous reaction	Less	More

eliminate cell release to the fermentation medium (Rodriguez-Huezo et al. 2011; Koyama and Seki 2004a; Dembczynski and Jankowski 2002; Jankowski et al. 1997). Furthermore, the semi-porous membrane allows the solutes to diffuse into and out from the inner core of the capsule with minimum resistance. Since the capsules have low diffusion resistance, they have been used for medicinal drug delivery, nutritional supplements delivery and they also allow development of a new reactive perstraction system based on the principle of liquid–liquid extraction (Whelehan et al. 2010; Marison et al. 2004; Wyss et al. 2004, 2004, 2005; Stark et al. 2003).

Due to these attractive features, liquid core capsules have been applied to obtain high cell density cultivation in diverse industrial biotechnology processes. It has been well recognized that the cell density is of prime importance to attain high yield or productivity of bioproducts in microbial processes. Bioproducts of interest are of several types. These include the microbial cells themselves (e.g., starter culture, microflora) and the metabolites produced by cells (e.g., enzymes, pharmaceutical compounds, specialty chemicals, food additives, commodity chemicals etc).

Unfortunately, not all cells present a good growth when encapsulated (Moreno-Garrido et al. 2005). Furthermore, encapsulation may decrease the product output due to the interfering reactions between membrane polymer with the product and membrane polymers. This is due to the nature of product, variations in pH during formation and mass transfer limitation (Doleyres et al. 2002; Muralidhar et al. 2001). Membrane formed by certain polyelectrolytes is relatively compact and has low molecular weight cut off, which would limit its diffusion of substrates, essential nutrients and oxygen transfer through the membrane for sufficient cell growth. Meanwhile, metabolites or products might accumulate in the capsules to bring the product inhibition (Zhang et al. 2005; Gardin and Pauss 2001; Dias Joao Carlos et al. 2000). Therefore, the build-up materials of the capsules should be appropriately selected based on the intended functionality of capsules.

Alginate based capsules have stood out as one of the preferred encapsulation materials. Alginate (algin or alginic acid) is a gelling anionic algal polysaccharide that extracted mainly from brown seaweed. Alginate is a linear co-polymer composed of two monomeric units: D-mannuronic acid (M) and L-guluronic acid (G). Di- and multivalent cations, induce gelation by binding mainly to the G-blocks, like eggs in an egg box. Alginate gel has been widely used due to its unique characteristic such as bio-compatibility, inert, non-toxic, low shrinkage and high porosity.

It has been reported that alginate based capsules improved thermal stability of cells (Nussinovitch 2010; Yan and Lehe 2007), reduced cell leakage (Dembczynski and Jankowski 2002; Chang et al. 1996; Yoo et al. 1996), increased cell density (Chang et al. 1996), prolonged the activity of cells (Seong et al. 1997), allowed diffusion of substances of low molecular weight like glucose, fructose, sucrose, lactose, amino acids and vitamin B12 (Youngsukkasem et al. 2012; Chai et al. 2004; Dembczynski and Jankowski 2000), but restricted diffusion of large molecules like proteins through the capsular membrane (Nigam et al. 1988).

Ca-alginate Liquid Core Capsule Formation

In general, the (macro) liquid core capsules could be formed using single step or multiple steps methods. Some of the methods are modified from the typical procedure for Ca-alginate beads formation.

Single Step Methods

Extrusion-Dripping (Liquid Droplet Forming) Method

Extrusion-dripping or liquid droplet forming method is a common and simple method. This technique has been developed during the early 1980; capsules are produced by the reverse procedure of the formation of conventional Ca-alginate beads (Park and Chang 2000). It involves adding the cell containing cation ions solution through a syringe needle into a swirling alginate solution in a drop wise manner to produce spherical gel biocatalysts, as shown in Fig. 22.2. The drop (s) could be formed via simple gravitational force, vibration, pressurized air, and electrostatic force.

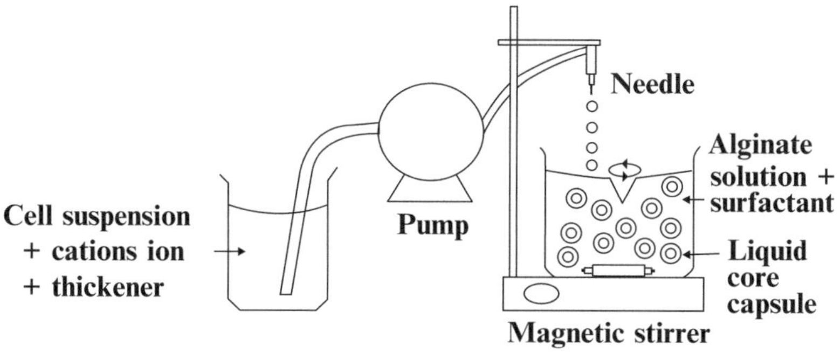

Fig. 22.2 Extrusion-dripping method (under simple gravity)

A calcium alginate membrane was formed immediately on the surface of the droplet by ionic interaction. Membrane thickness, size and mechanical strength can easily be controlled by alteration of the concentration of alginate and cations solution and hardening period (Koyama and Seki 2004a; Blandino et al. 1999; Yoo et al. 1996). The typical capsules diameter obtained by this technique is larger than 2 mm, even for small needle diameter (Dulieu et al. 1999). This method is easy, simple, low cost and required mild formation conditions. Generally, this method is difficult to be performed in a large industrial scale.

However, the application of extrusion-dripping method for liquid core capsule formation encountered several difficulties. First of all, the membrane thickness of the capsules was not uniform because it is dependent on the total quantity of capsules formed per batch, gelation time and concentration of the cation ion inside the liquid core (Koyama and Seki 2004a; Blandino et al. 1999). Secondly, the produced capsules may be aggregated. This is due to the diffusion of cation ions from the surface of one capsule to the surface of another nearby capsule. Thirdly, it is difficult to obtain spherical capsules because this method involves extrusion-dripping of low viscous solution (cations + cell suspension) into high viscous solution (alginate). The low viscous solution droplet is difficult to retain its spherical shape after the impact on the surface of alginate solution. In addition, the droplets also deformed during the process of penetration through the viscous alginate solution for gelation. Hence, less spherical capsules are commonly formed.

In order to overcome the problem, the mixture solution of cell suspension and cation ions is modified by addition of thickener to increase its viscosity such as sucrose (Zhang and Salsac 2012; Nussinovitch et al. 1996, 1997), dextran (Zhang and Salsac 2012; Nigam et al. 1988), starch (Goderska et al. 2003; Dembczynski and Jankowski 2002; Jankowshi et al. 1997), xanthan gum (Koyama and Seki 2004a; Seong et al. 1997; Yoo et al. 1996; Chang et al. 1996; Cheong et al. 1993), guar gum (Nussinovitch et al. 1996), polyethylene glycol (PEG) (Koyama and Seki 2004a), hydroxypropylmethylcellulose (Torre et al. 2000), chitosan (Mi et al. 2002) and carboxylmethylcellulose (CMC) (Westman et al. 2012; Youngsukkasem et al. 2012; Ylitervo et al. 2011; Ji et al. 2007; Yan and Lehe 2007; Talebnia et al. 2005; Chai et al. 2004; Blandino et al. 2000, 2001, 2002; Patel et al. 2000; Blandino et al. 1999; Nussinovitch et al. 1996). In some application, the viscous liquid inner core of the capsule may interrupt the functionality of the encapsulated biocatalyst. Therefore low molecular weight viscosifier like sucrose and dextran was used (Zhang and Salsac 2012; Nussinovitch et al. 1996, 1997; Nigam et al. 1988). After the capsules were formed, they were immersed in distilled water to allow these viscosifiers to diffuse out from the inner core of the capsule and an inviscid liquid core was obtained at the end.

Moreover, the sphericity of the capsules is further improved by reducing the surface tension of alginate solution with addition of surfactant [e.g., Tween 80 (Koyama and Seki 2004a, b; Dembczynski and Jankowski 2002) and Tween 20 (Youngsukkasem et al. 2012)]. The sphericity of the capsules is also be affected by the distance between nozzle tip and sodium alginate surface. Lastly, the alginate solution is required to be diluted to more than fourfold by a large amount of distilled water during harvesting; to avoid aggregation of capsules, where sphericity of the capsules is retained. As a result, wastage of alginate solution occurred (Koyama and Seki 2004a, b; Dembczynski and Jankowski 2002; Blandino et al. 1999; Jankowshi et al. 1997; Yoo et al. 1996).

Co-extrusion Dripping Method

The co-extrusion dripping method has been introduced to produce liquid core capsule for cells encapsulation. Cells suspension solution flows in the centric nozzle and the membrane forming alginate solution flows in the annulus, as shown in Fig. 22.3. Both solutions are co-extruded at the end of the concentric nozzle before dripped into a gelation bath (e.g. calcium or barium chloride solution). Although this technique is simple and suitable for fragile cells, several technological difficulties such as the flow rate of the solutions must be strictly controlled in order to obtain spherical capsules with uniform membrane and appropriate diameter (Dulieu et al. 1999) as well as to ensure the liquid inner core is positioned at the center of the capsule. Moreover, the alginate may be gelled and clogged at the orifice of the nozzle if the flow rate of the solutions was not properly determined. Even so, the clogging issue could be avoided if vibration force was applied to the nozzle (Whelehan et al. 2010; Marison et al. 2004; Wyss et al. 2004; Stark et al. 2003).

Sequestration Method

Calcium alginate gel is known to be sensitive to several chelators such as phosphate, citrate, lactate, sodium and magnesium (Strand et al. 2000). The gel became

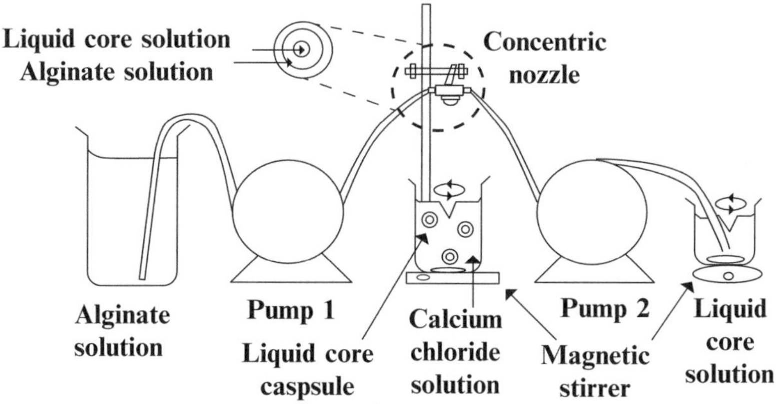

Fig. 22.3 Co-extrusion method

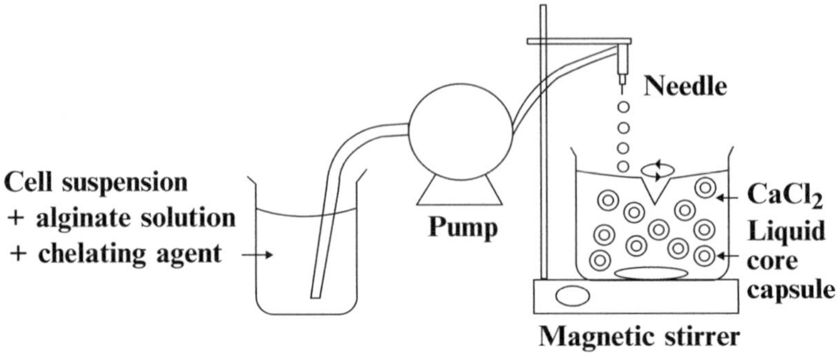

Fig. 22.4 Sequestration method

soften and eventually disintegrated when exposed to the chelators at large quantities over a period of time. Under proper control on the concentration of the chelator and gelation time in the capsule formation process, the conventional Ca-alginate bead setup could be used to form liquid core capsule (Rodriguez-Huezo et al. 2011). In which, the chelator is added in the alginate-cell suspension solution before allowing the solution to be extruded into the alginate solution (Rodriguez-Huezo et al. 2011), as shown in Fig. 22.4. The presence of chelator in the solution has minimized the cross-linking gelation between alginate and calcium cations (Rodriguez-Huezo et al. 2011) and hence liquid alginate solution remained in the inner core of the final capsule. The presence of chelator in the inner core of the capsule may prevent subsequent solidification of the core when the capsule is exposed to divalent or multivalent cation that may present in some working medium.

Multiple Steps Methods

Pre-gel Dissolving Method

The pregel dissolving (two-step) method involved formation of the calcium alginate beads containing cells at first using the typical Ca-alginate bead formation procedure (Nussinovitch et al. 1996; Wong and Chang 1991), as illustrated in Fig. 22.5. Then, the capsules are coated with other polymer like chitosan, poly-L-lysine and poly-L-ornithine which is chemically stable against the chelating agent like sodium citrate, phosphate and Ethylene diaminetetra acetic acid (EDTA) (Darrabie et al. 2005; Wen-tao et al. 2005; Park and Chang 2000; Nussinovitch et al. 1996; Wong and Chang 1991). The coated beads are retreated with sodium alginate to cross-link the remaining coating polymer on the surface of the beads. The liquid core capsules accomplished when the calcium alginate core is dissolved in a sodium citrate solution (Park and Chang 2000; Nussinovitch et al. 1996; Wong and Chang 1991). The inner core of the capsule formed via this method may be solidified easily

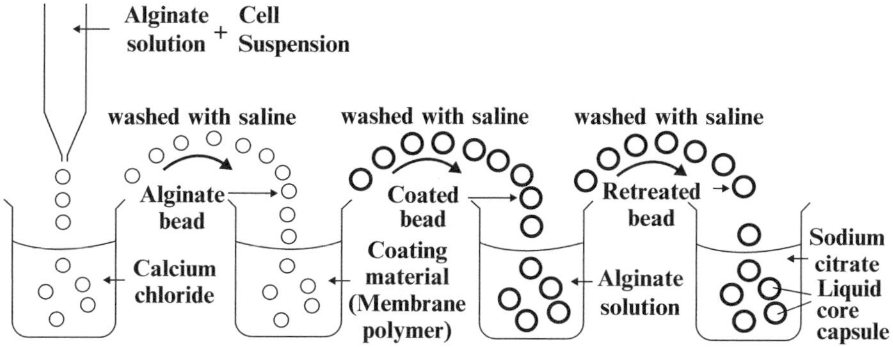

Fig. 22.5 Pregel dissolving method

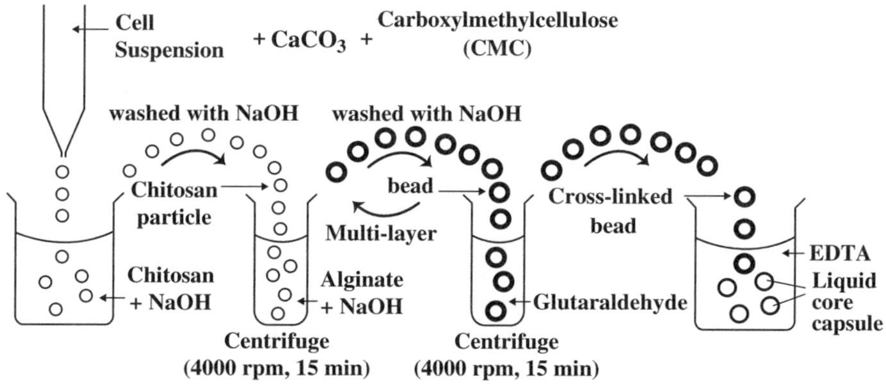

Fig. 22.6 Template assisted method

if the capsules were exposed to working medium which contained gelation agents such as divalent or multivalent cation. The method is not suitable for large scale production as it involved multiple steps.

Template Assisted Method

The template assisted method started with formation of a solid $CaCO_3$ particle which worked as a template to build-up the Ca-alginate membrane (Zhao et al. 2006, 2007), as illustrated in Fig. 22.6. The solid $CaCO_3$ particle was dissolved using EDTA after the Ca-alginate membrane is formed. As a result, liquid core capsule is produced. The method involved multiple steps and hence it is not suitable for industrial scale-up.

Stability Enhancement Methods

Ca-alginate based liquid core capsules are chemically less stable and sensitive to chelator like lactate, phosphate, and sodium that commonly available in the media of lactic acid fermentation. Therefore, the stability of the capsule is maintained by taking several preventive measures that have been developed and used for Ca-alginate beads. The simplest way is to add calcium chloride or calcium carbonate into the fermentation media (Dembczynski and Jankowski 2002; Koyama and Seki 2004a; Doleyres et al. 2002; Ivanova et al. 2002; Dias Joao Carlos et al. 2000; Yoo et al. 1996) to replace the calcium cation in the capsules that have been replaced by chelators during fermentation.

Moreover, it has been reported that the chemical stability of the Ca-alginate gel is influenced by the M/G ratio of the alginate used (Smidsrød and Haug 1972; Haug and Smidsrod 1965). In which M/G ratio is referred to the ratio of D-mannuronate to L-guluronate content that present in alginate. It is preferable to produce the capsules using alginate that contained high content of L-guluronate or low M/G ratio in order to produce more chemical stable capsules (Vorlop and Klein 1983). The capsules are expected to be stable as long as the molar ratio of electrolytes (e.g. Na^+, K^+, NH_4^+, Mg^{2+}) to Ca^{2+} cation in the fermentation medium is less than 20:1 (Smidsrod and Skjak-Braek 1990; Vorlop and Klein 1983).

The alginate capsules could be strengthen by using divalent or multi-valent cation such as barium, copper, manganese, strontium, iron, or aluminum ions (Vigo et al. 2005; Yoo et al. 1996; Smidsrod and Skjak-Braek 1990; Vorlop and Klein 1983; Hackel et al. 1975) that has high affinity to cross-link with alginate molecules, instead of Ca^{2+} cation. However, the use of strontium or barium ions is limited for biological applications due to their toxicity (Strand et al. 2000; Vorlop and Klein 1983).

On the other hand, the chemical stability of Ca-alginate capsules can be improved by coating the capsules with another polymer which is positively charged. The Ca-alginate capsules could be coated with polyamino acid such as Poly-L-Lysine (PLL) (Nussinovitch et al. 1996; Nigam et al. 1988), Poly-L-Ornithine (PLO) (Darrabie et al. 2005), and chitosan (Ylitervo et al. 2011; Krasaekoopt et al. 2003, 2006; Wen-tao et al. 2005; Ribeiro et al. 2005; Nussinovitch et al. 1996; Yoo et al. 1996). The coating polymer may alter the properties of the capsule like their permeability, biocompatibility and durability.

The coating process can be conducted by either one stage method or two stages method. In the one stage method using chitosan as coating material, the chitosan coated capsules were obtained by an uninterrupted way where chitosan solution was introduced into the oil dispersed Ca-alginate capsules (Krasaekoopt et al. 2003; Wen-tao et al. 2005). In the two stages method, the Ca-alginate capsules were isolated and then submerged into an aqueous chitosan solution, which gave more stable capsules than the one stage method (Westman et al. 2012; Ylitervo et al. 2011; Ribeiro et al. 2005; Nussinovitch et al. 1996; Yoo et al. 1996). The latter method is more commonly employed because the use of oil in the one stage method is not favorable for fermentation of lactobacilli.

Alternatively, the polymer could be added into the alginate solution to produce complex polymer composite capsules with improved stability. For example, sodium caseinate and Poly (methylene-co-guanidine) (PMCG) have been used to enhance the chemical stability of Ca-alginate capsules (Messaoud et al. 2015; Bucko et al. 2005; Zhang et al. 2002). However, the biocompatibility of the new polymer should be further examined (Zhang et al. 2002).

Lactobacili Fermentation

During LAB fermentation, it has been reported that the viability and growth of the lactobacilli is inhibited by its end product, i.e. lactic acid (or lactate). Consequently, the productivity of lactic acid by LAB is also decreased due to the inhibition effect. The end-product inhibition effect is commonly suppressed by using pH controller to neutralize the lactate, adding pH buffer in the fermentation media, employing membrane technology to separate lactate from the fermentation media, modifying the fermentation mode to dilute the concentration of lactate and encapsulating the bacteria for protection. As compared to the other approaches, encapsulation is attractive because it does not required complicated devices, it allows separation of LAB from the fermentation media with less cell damage, it enables the LAB can be reused easily and it simplifies the purification and downstream processes.

Table 22.2 shows the results of encapsulating lactobacilli in liquid core capsules for lactic acid fermentation as compared to that conducted by free cell and entrapment in bead (as control). Encapsulation of lactobacilli in liquid core capsules gave good protection and survivability to the cells from the end-product inhibition effect and harsh environments (Rodriguez-Huezo et al. 2011; Goderska et al. 2003; Dembczynski and Jankowski 2002; Yoo et al. 1996). In addition, the growth and productivity of cells were improved as the mass transfer resistance of the capsules was low where the diffusion of nutrients and product between the inner core and fermentation media was good (Rodriguez-Huezo et al. 2011; Koyama and Seki 2004b; Chai et al. 2004; Goderska et al. 2003; Dembczynski and Jankowski 2002; Yoo et al. 1996). It was observed that the leakage of cell from the immobilization matrix have been minimized because LAB was confined in the inner liquid core of the capsules.

The liquid core capsules are used to encapsulate the lactobacilli with the intention of protecting the cells from the harsh conditions (e.g. low pH, shear forces, inhibitors) during lactic acid fermentation, as shown in Table 22.2. Besides that, the liquid core capsules could be used to counteract the effect caused by the harsh conditions. The issue of low pH and present of inhibitor like lactate could be addressed by introducing encapsulated neutralizing agents in liquid core capsule in the fermentation media (Lee et al. 2013; Tik et al. 2001). The tested neutralizing agents are Alamine-336 in oleyl alcohol and calcium carbonate (Lee et al. 2013; Tik et al. 2001). Since the capsules have low mass transfer resistance, the solutes/ions could be freely diffused between the inner liquid core and the fermentation media through the capsular membrane to react with the inhibitors (Lee et al. 2013;

Table 22.2 Lactic acid fermentation by lactobacilli encapsulated in liquid core capsules

Lactobacillus sp.	Parameter	Comparison		Observation	Reference
		Free cell	Bead		
Lactobacillus rhamnosus LC705	Survivability in simulated gastrointestinal conditions	–	Higher	• Equal protection to the cells (homogeneous distribution in capsule) • Low cell leakage	Rodriguez-Huezo et al. (2011)
	Survivability during storage at 4 °C for 25 days	–	Higher		
Lactobacillus rhamnosus	Survivability in media of different pH (pH 2–8) at 37 °C	Higher	–	–	Goderska et al. (2003)
Lactobacillus rhamnosus	Cell viability	Higher (1 log)	–	Low cell leakage	Dembczynski and Jankowski (2002)
Lactobacillus casei	Cell viability Lactic acid production rate	–	Higher (1.5-fold) 2.7 g L^{-1} h^{-1}	• Low cell leakage • Homogenous growth (less mass transfer limitation) • Reusability up to five times	Yoo et al. (1996)

Whelehan et al. 2010; Marison et al. 2004; Koyama and Seki 2004b; Chai et al. 2004; Wyss et al. 2004; Stark et al. 2003). The neutralization could be achieved via liquid–liquid extraction (Whelehan et al. 2010; Marison et al. 2004; Wyss et al. 2004, 2004, 2005; Stark et al. 2003).

On the other hand, the liquid core capsule could allow co-immobilization of two (or more) biocatalysts for lactic acid fermentation using complex media like agricultural resources, food waste, agricultural waste, and industrial waste (Oh et al. 2005; Akerberg and Zacchi 2000), in similar manner that have been conducted using solid core capsules for the denitrification process (Dos Santos et al. 1996). There is an additional pre-treatment of the complex media is usually required to degrade impurities in the media and/or to produce fermentable sugar via saccharification for subsequent lactic acid fermentation (Zhang et al. 2007; Ohkouchi and Inoue 2006; Oh et al. 2005; Sakai et al. 2004; Akerberg and Zacchi 2000). Therefore, the enzyme required for saccharification can be encapsulated in the external capsular membrane while the fermentative microorganism can be encapsulated in the inner liquid core of the capsules. In this way, direct lactic acid fermentation can be carried out in one simple step in the complex fermentation media.

Summary

Lactic acid bacteria (LAB) have been widely used for production of various fermented dairy and non-dairy food products. LAB are known to enhance the organoleptic, nutritional, healthy, and functional quality of the food products. Encapsulation has been used to confine bioactive materials within a film or coating of a matrix for various purposes. Ca-alginate liquid core capsule is one of the common gel matrix used for encapsulation. The capsules could be formed using single or multiple steps method. The formation methods of the capsules were described and discussed. Since Ca-alginate liquid core capsules are chemically less stable, several approaches have been introduced to enhance the stability of the capsules. The approaches were discussed. On the other hand, the use of the capsules for LAB fermentation was discussed. The results of previous studies showed that the capsules have improved the survivability, viability, productivity and storage stability of LAB during fermentation.

References

Akerberg, C., & Zacchi, G. (2000). an economic evaluation of the fermentative production of lactic acid from wheat flour. *Bioresource Technology, 75*, 119–126.

Bhatia Surita, R., Khattak Sarwat, F., & Roberts Susan, C. (2005). Polyelectrolytes for cell encapsulation. *Current Opinion in Colloid & Interface Science, 10*, 45–51.

Blandino, A., Macias, M., & Cantero, D. (1999). Formation of calcium alginate capsules: Influence of sodium alginate and $CaCl_2$ concentration on gelation kinetics. *Journal of Bioscience and Bioengineering, 88*(6), 686–689.

Blandino, A., Macias, M., & Cantero, D. (2000). Glucose oxidase release from calcium alginate gel capsules. *Journal of Enzyme and Microbial Technology, 27*(3–5), 319–324.

Blandino, A., Macias, M., & Cantero, D. (2001). Immobilization of glucose oxidase within calcium alginate gel capsules. *Journal of Process Biochemistry, 36*, 601–606.

Blandino, A., Macias, M., & Cantero, D. (2002). Modelling and simulation of a bienzymatic reaction system co-immobilised within hydrogel-membrane liquid-core capsules. *Enzyme and Microbial Technology, 31*, 556–565.

Bucko, M., Vikartovska, A., Lacik, I., Kollarikova, G., Gemeiner, P., Patoprsty, V., & Brygin, M. (2005). Immobilization of a whole-cell epoxide-hydrolyzing biocatalyst in sodium alginate-cellulose sulphate-poly(methylene-co-guanidine) capsules using a controlled encapsulation process. *Enzyme and Microbial Technology, 36*, 118–126.

Chai, Y., Mei, L.-H., Lin, D.-Q., & Yao, S.-J. (2004). Diffusion coefficients in intrahollow calcium alginate microcapsules. *Journal of Chemical and Engineering Data, 49*, 475–478.

Chang, H. N., Seong, G. H., Yoo, I.-K., Park, J. K., & Seo, J.-H. (1996). Microencapsulation of recombinant *Saccharohyces cerevisiae* cells with lnvertase activity in liquid-core alginate capsules. *Biotechnology and Bioengineering, 51*, 157–162.

Cheong, S. H., Park, J. K., Kim, B. S., & Chang, H. N. (1993). Microencapsulation of yeast cells in the calcium alginate membrane. *Biotechnology Techniques, 7*(12), 879–884.

Chiang, Y. W., Chye, F. Y., & Mohd Ismail, A. (2006). Microbial diversity and proximate composition of Tapai, A Sabah's fermented beverage. *Malaysian Journal of Microbiology, 2*(1), 1–6.

Darrabie, M. D., Kendall, W. F., Jr., & Opara, E. C. (2005). Characteristics of poly-L-ornithine-coated alginate microcapsules. *Biomaterials, 26*, 6846–6852.
Dembczynski, R., Jankowski, T. (2000) Characterisation of small molecules diffusion in hydrogel-membrane liquid-core capsules, *Biochemical Engineering Journal 6*, 41–44.
Dembczynski, R., & Jankowski, T. (2002). Growth characteristics and acidifying activity of *Lactobacillus rhamnosus* in alginate/strach liquid-core capsules. *Enzyme and Microbial Technology, 31*(1–2), 111–115.
Dias Joao Carlos, T., Rezende Rachel, P., & Linardi Valter, R. (2000). Biodegradation of Acetonitrile by cells of Candida guilliermondii UFMG-Y65 immobilized in alginate, κ-carrageenan and citric pectin. *Brazilian Journal of Microbiology, 31*, 61–66.
Doleyres, Y., Paquin, C., LeRoy, M., & Lacroix, C. (2002). Bifidobacterium longum ATCC 15707 cell production during free- and immobilized-cell cultures in MRS-whey permeate medium. *Applied Microbiol Biotechnology, 60*, 168–173.
Dos Santos, V. A. P. M., Bruijnse, M., Tramper, J., & Wijffels, R. H. (1996). The magic-bead concept: An integrated approach to nitrogen removal with co-immobilized micro-organisms. *Applied Microbiology and Biotechnology, 45*, 447–453.
Dulieu, C., Poncelet, D., & Neufeld, R. J. (1999). Encapsulation and immobilization techniques. In W. M. Kuhtrieiber, R. P. Lanza, & W. L. Chick (Eds.), *Cell encapsulation technology and therapeutics* (pp. 3–17). Boston: Brikhauser.
Gardin, H., & Pauss, A. (2001). κ-Carrageenan/gelatin gel beads for the co-immobilization of aerobic and anaerobic microbial communities degrading 2,4,6-trichlorophenol under air-limited conditions. *Applied Microbiol Biotechnology, 56*, 517–523.
Gereková, P., Petruláková, Z., & Šturdík, E. (2011). Importance of lactobacilli for bread-making industry. *Acta Chimica Slovaca, 4*(2), 118–135.
Giraffa, G., Chanishvili, N., & Widyastuti, Y. (2010). Importance of lactobacilli in food and feed biotechnology. *Research in Microbiology, 161*, 480–487.
Goderska, K., Zybala, M., & Czarnecki, Z. (2003). Characterisation of microencapsulated *Lactobacillus Rhamnosus* LR7 strain. *Polish Journal of Food and Nutrition Sciences, 12*(3), 21–24.
Hackel, U., Klein, J., Megnet, R., & Wagner, F. (1975). Immobilisation of microbial cells in polymeric matrices. *European Journal of Applied Microbiology, 1*, 291–293.
Haug, E., & Smidsrod, O. (1965). The effect of divalent metals on the properties of alginate solutions ii, comparison of different metal ions. *Acta Chemica Scandinavica, 19*(2), 341–351.
Hofvendahl, K., & Hahn-Hagerdal, B. (2000). Factors affecting the fermentation lactic acid production from renewable resources. *Enzyme and Microbial Technology, 26*(2000), 87–107.
Ivanova, E., Chipeva, V., Ivanova, I., Dousset, X., & Poncelet, D. (2002). Encapsulation of lactic acid bacteria in calcium alginate beads for bacteruiocin production. *Journal of Culture Collections, 3*, 53–58.
Jankowshi, T., Zielinska, M., & Wysakowska, A. (1997). Encapsulation of lactic acid bacteria with alginate/starch capsules. *Biotechnology Techniques, 11*(1), 31–34.
Ji, C.-I., Cho, S.-M., Gu, Y.-S., & Kim, S.-B. (2007). The processing optimization of caviar analogs encapsulated by calcium-alginate gel membranes. *Food Science and Biotechnology, 16*(4), 557–564.
Klein, G., Pack, A., Bonaparte, C., & Reuter, G. (1998). Taxonomy and physiology of probiotic lactic acid bacteria. *International Journal of Food Microbiology, 41*, 103–125.
Kontula, P., Suihko, M. L., Von Wright, A., & Mattila-Sandholm, T. (1999). The effect of Lactose derivatives on intestinal Lactic acid bacteria. *Journal of Dairy Science, 82*(2), 249–256.
Koyama, K., & Seki, M. (2004a). Cultivation of yeast and plant cells entrapped in the low-viscous liquid-core of an alginate membrane capsule prepared by using polyethylene glycol. *Journal of Biotechnology and Bioengineering, 97*(2), 111–118.
Koyama, K., & Seki, M. (2004b). Evaluation of mass transfer characteristics in alginate-membrane liquid-core capsules prepared using polyethylene glycol. *Journal of Biotechnology and Bioengineering, 97*(2), 111–118.

Krasaekoopt, W., Bhandari, B., & Deeth, H. (2006). Survival of probiotics encapsulated in chitosan-coated alginate beads in yogurt from UHT- and conventionally-treated milks during storage. *Lebensmittel-Wissenschaft & Technologie, 39*, 177–183.

Krasaekoopt, W., Bhandari, B., & Deeth, H. (2003). Evaluation of encapsulation techniques of probiotics for yoghurt. *International Dairy Journal, 13*(1), 3–13.

Lee, B.-B., Zulkifli, N. A., & Ibrahim, R. (2013). *A pH regulating particles composition*. Malaysia: Universiti Malaysia Perlis. PI2013701999.

Leisner, J. J., Vancanneyt, M., Lefebvre, K., Vandemeulebroecke, K., Hoste, B., Euras Vilalta, N., Rusul, G., & Swings, J. (2002). Lactobacillus durianis sp. nov., isolated from an acid-fermented condiment (tempoyak) in Malaysia. *International Journal of Systematic and Evolutionary Microbiology, 52*, 927–931.

Yan, L., & Lehe, M. (2007). Production of Indigo by Immobilization of *E. coli* BL21 (DE3) cells in calcium-alginate gel capsules. *China Journal of Chemical Engineering, 15*(3), 387–390.

Marison, I., Peters, A., & Heinzen, C. (2004). Liquid core capsules for applications in biotechnology. In V. Nedovic & R. Willaert (Eds.), *Focus on biotechnology vol. 8A fundamental of cell immobilisation biotechnology* (pp. 185–204). The Netherlands: Kluwer Academic Publishers.

Markiewicz, L. H., Wasilewska, E., & Bielecka, M. (2008). Identification of lactobacillus strains present in fermented dairy products and their differentiation using molecular methods. *Polish Journal of Food and Nutrition Sciences, 58*(2), 251–256.

Messaoud, G. B., Sánchez-González, L., Jacquot, A., Probst, L., & Desobry, S. (2015). Alginate/sodium caseinate aqueous-core capsules: A pH-responsive matrix. *Journal of Colloid and Interface Science, 440*, 1–8.

Mi, F.-L., Sung, H.-W., & Shyu, S. S. (2002). Drug release from chitosan-alginate complex beads reinforced by a naturally occurring cross-linking agent. *Carbohydrate Polymers, 48*(1), 67–72.

Moreno-Garrido, I., Campana, O., Lubian, L. M., & Blasco, J. (2005). Calcium alginate immobilized marine microalgae: Experiments on growth and short-term heavy metal accumulation. *Marine Pollution Bulletin, 51*(8–12), 823–829.

Muralidhar, R. V., Jayachandran, G., & Singh, P. (2001). Development of high-density cultivation systems by bioencapsulation. *Current Science, 81*(3), 263–269.

Nigam, S. C., Tsao, I. F., Sakoda, A., & Wang, H. Y. (1988). Techniques for preparing hydrogel membrane capsules. *Biotechnology Techniques, 2*(4), 271–276.

Nussinovitch, A., Gershon, Z., & Nussinovitch, M. (1996). Liquid-core hydrocolloid capsules. *Food Hydrocolloids, 10*(1), 21–26.

Nussinovitch, A., Gershon, Z., & Nussinovitch, M. (1997). Temperature stable liquid core hydrocolloid capsules. *Food Hydrocolloids, 11*(2), 209–215.

Nussinovitch, A. (2010). Chapter 7 liquid core beads and their applications in food, biotechnology, and other fields. In A. Nussinovitch (Ed.), *Polymer macro- and micro-gel beads: Fundamentals and applications* (pp. 163–190). New York, USA: Springer Science + Business Media.

Oh, H., Wee, Y.-J., Yun, J.-S., Han, S. H., Jung, S., & Ryu, H.-W. (2005). Lactic acid production from agricultural resources as cheap raw materials. *Bioresource Technology, 96*, 1492–1498.

Ohkouchi, Y., & Inoue, Y. (2006). Direct production of L(+)-lactic acid from starch and food wastes using Lactobacillus manihotivorans LMG18011. *Bioresource Technology, 97*, 1554–1562.

Park, J. K., & Chang, H. N. (2000). Microencapsulation of microbial cells. *Biotechnology Advances, 18*, 303–319.

Patel, A. V., Pusch, I., Mix-Wagner, G., & Vorlop, K. D. (2000). A novel encapsulation technique for the production of artificial seeds. *Plant Cell Reports, 19*, 868–874.

Wen-tao, Q., Wei-ting, Y., Yu-bing, X., & Xiaojun, M. (2005). Optimization of Saccharomyces cerevisiae culture in alginate-chitosan-alginate microcapsule. *Biochemical Engineering Journal, 25*, 151–157.

Rhee, S. J., Lee, J. E., & Lee, C. H. (2001). Importance of lactic acid bacteria in Asian fermented foods. *Microbial Cell Factories, 10*(Suppl 1), S5.

Ribeiro, A. J., Silva, C., Ferreira, D., & Veiga, F. (2005). Chitosan-reinforced alginate microspheres obtained through the emulsification/internal gelation technique. *European Journal of Pharmaceutical Sciences, 25*, 31–40.

Rodriguez-Huezo, M. E., Lobati-Calleros, C., Reyes-Ocampo, J. G., Sandoval-Castilla, O., Perez-Alonso, C., & Pomentel-Gonzalez, D. J. (2011). Survivability of entrapped *Lactobacillus rhamnosus* in liquid- and gel- core alginate beads during storage and simulated gastrointestinal conditions. *Revista Mexicana de Ingenieria Quimica, 10*, 353–361.

Sakai, K., Taniguchi, M., Miura, S., Ohara, H., Matsumoto, T., & Shirai, Y. (2004). Making plastics from garbage. A novel process for poly-L-lactate production from municipal food waste. *Journal of Industrial Ecology, 7*(3–4), 63–74.

Seong, G. H., Han, S. J., Chang, H. N., & Lee, J. (1997). Whole cell enzyme microencapsulation of Escherichia coli with oxygen dependent inducible nar promoter. *Biotechnology Letters, 19*(9), 881–884.

Smidsrod, O., & Skjak-Braek, G. (1990). Alginate as immobilization matrix for cells. *Tibtech, 8*, 71–78.

Smidsrød, O., & Haug, A. (1972). Properties of poly(1,4-hexuronates) in the gel state. II. Comparison of gels of different chemical compositions. *Acta Chemica Scandinavica, 26*, 79–88.

Somkuti, G. A. (2000). Lactic acid bacteria. In J. Lederberg (Ed.), *Encyclopedia of microbiology* (2nd ed., Vol. 3). San Diego, CA, USA: Academic Press.

Stark, D., Kornmann, H., Munch, T., Sonnleitner, B., Marison, I. W., & von Stockar, U. (2003). Novel type of in situ extraction: Use of solvent containing microcapsules for the bioconversion of 2-phenylethanol from L-phenylalanine by *Saccharomyces cerevisiae*. *Biotechnology and Bioengineering, 83*(4), 376–385.

Strand, B. L., Morch, Y. A., & Skjak-Braek, G. (2000). Alginate as immobilization matrix for cells. *Minerva Biotechnologies, 12*, 223–233.

Talebnia, F., Niklasson, C., & Taherzadeh, M. J. (2005). Ethanol production from glucose and dilute-acid hydrolyzates by encapsulated *S. cerevisiae*. *Biotechnology and Bioengineering, 90* (3), 345–353.

Tik, N., Bayraktar, E., & Mehmetoglu, U. (2001). In situ reactive extraction of lactic acid from fermentation media. *Journal of Chemical Technology and Biotechnology, 76*, 764–768.

Torre, M. L., Maggi, L., Vigo, D., Galli, A., Bornaghi, V., Ma!eo, G., & Conte, U. (2000). Controlled release of swine semen encapsulated in calcium alginate beads. *Biomaterials, 21*, 1493–1498.

Vigo, D., Villani, S., Faustini, M., Accorsi, P. A., Galeati, G., Spinaci, M., Munari, E., Russo, V., Asti, A., Conte, U., & Torre, M. I. (2005). Follicle-like model by granulosa cell encapsulation in a barium alginate-protamine membrane. *Tissue Engineering, 11*(5/6), 709–714.

Vorlop, K. D., & Klein, J. (1983). New developments in the field of cell immobilization – Formation of biocatalysts by ionotropic gelation. In R. M. Lafferty (Ed.), *Enzyme technology* (pp. 219–235). Berlin, Heidelberg: Springer.

Westman, J. O., Taherzadeh, M. J., & Franzen, C. J. (2012). Proteomic analysis of the increased stress tolerance of *Saccharomyces cerevisiae* encapsulated in liquid core alginate-chitosan capsules. *PLoS One, 7*(11), e49335.

Whelehan, M., von Stockar, U., & Marison, I. W. (2010). Removal of pharmaceuticals from water: Using liquid core microcapsules as a novel approach. *Water Research, 44*, 2314–2324.

Widyastuti, Y., Rohmatussolihat, & Febrisiantosa, A. (2014). The role of lactic acid bacteria in milk fermentation. *Food and Nutrition Sciences, 5*, 435–442.

Wong, H., & Chang, T. M. S. (1991). A novel two step procedure for immobilizing living cells in microcapsuies for improving xenograft survival. *Artificial Cells in Immobilization Biotechnology, 19*(4), 687–697.

Wyss, A., Cordente, N., von Stockar, U., & Marison, I. W. (2004a). A novel approach for the extraction of herbicides and pesticides from water using liquid-core microcapsules. *Biotechnology and Bioengineering, 87*(6), 734–742.

Wyss, A., von Stockar, U., & Marison, I. W. (2004b). Production and characterization of liquid-core capsules made from cross-linked acrylamide copolymers for biotechnological applications. *Biotechnology and Bioengineering, 86*(5), 563–572.

Wyss, A., von Stockar, U., & Marison, I. W. (2005). A novel reactive perstraction system based on liquid-core microcapsules applied to lipase-catalyzed biotransformations. *Biotechnology and Bioengineering, 93*(1), 28–39.

Ylitervo, P., Franzén, C. J., & Taherzadeh, M. J. (2011). Ethanol production at elevated temperatures using encapsulation of yeast. *Journal of Biotechnology, 156*, 22–29.

Yoo, I. K., Seong, G. H., Chang, H. N., & Park, J. K. (1996). Encapsulation of *Lactobacillus casei* cells in liquid-core alginate capsules for lactic acid production. *Enzyme and Microbial Technology, 19*, 428–433.

Youngsukkasem, S., Rakshit, S. K., & Taherzadeh, M. J. (2012). Biogas production by encapsulated methane producing bacteria. *BioResources, 7*(1), 56–65.

Zhang, B., Cai, W.-M., & He, P.-J. (2007). Influence of lactic acid on the two-phase anaerobic digestion of kitchen wastes. *Journal of Environmental Science, 19*, 244–249.

Zhang, J., Yao, S.-J., & Guan, Y.-X. (2005). Preparation of macroporous sodium cellulose sulphate/poly (dimethldiallyammonium chloride) capsules and their characteristics. *Journal of Membrane Science, 255*, 89–98.

Zhang, L., & Salsac, A.-V. (2012). Can sonication enhance release from liquid-core capsules with a hydrogel membrane? *Journal of Colloid and Interface Science, 368*, 648–654.

Zhang, L. Y., Guan, Y. X., & Yao, S. J. (2002). Biocompatibility of SA/CS-CaCl2/PMCG Microcapsule in cell culture. *Sheng Wu Gong Cheng Xue Bao, 18*(6), 709–712.

Zhao, Q., Han, B., Wang, Z., Gao, C., Peng, C., & Shen, J. (2007). Hollow chitosan-alginate multilayer microcapsules as drug delivery vehicle: Doxorubicin loading and in vitro and in vivo studies. *Nanomedicine, 3*, 63–74.

Zhao, Q., Mao, Z., Gao, C., & Shen, J. (2006). Assembly of multilayer microcapsules on $CaCO_3$ particles from biocompatible polysaccharides. *Journal of Biomaterials Science Polymer Edition, 17*(9), 997–1014.

Part V
Policy and Regulations

Chapter 23
Intellectual Property Rights-Protection and Regulation

Sripathi Rao Kulkarni

Introduction

'Trial'—the word always carries the sense of inspiration and motivation for Scientists in their research endeavors, who have demonstrated with relentless efforts to gain knowledge and authority over mysteries held close to their hearts, thoughts, actions and deeds. The word 'trial' has been associated, at least in scientific pursuits, with another word 'error'. There were of course, mistakes and errors and numerous evidences in the past that support serendipity with scientific experiments on 'trial and error' basis. Development in research was and is possible by such experiments, more the number of attempts, more the clarity leading to go-no go decisions, as it associates with huge expenditure and costs. Historically, 'trial and error' had a purpose of being applied or demonstrated for the benefit of the society. On several occasions the said entity for the said purpose was tried successfully, more often served for another purpose down the line leading to 'evergreening' of the technology.

Bioprocess technology, in great detail, has been discussed in the earlier chapters pertaining to Metabolic and Biochemical Engineering, Food, Biomass & Bioenergy etc. and this chapter discusses the trends in IPRs, role of international and country-specific institutions in applying the regulations pertaining to various forms of protection of the innovations and inventions including commercialization exercises throughout the life cycle of the products, with the information gathered from various international agencies. Since IPRs pertain to the rights and exclusivities over intangible assets which could equally be traded as any other tangible assets, is dominated by the rules and regulations at each step of their existence and execution. Moreover, IPRs being territorial in nature, a huge conflict arose among the nations severing ties between them in trade and commerce. Attempts have been made to bring the nations

S.R. Kulkarni (✉)
Spangenberg Center for Law, Technology & the Arts, School of Law,
Case Western Reserve University, 11075 East Boulevard, Cleveland, OH 44106, USA
e-mail: sripathi_kulkarni@hotmail.com

P. Ravindra (ed.), *Advances in Bioprocess Technology*,
DOI 10.1007/978-3-319-17915-5_23

onto a common platform and work amicably towards smooth transition for the developmental activities. United Nations bodies have played a significant role in achieving the targets to most extent. Further, it deals with the prosecution of the inventions at the IP exclusivity issuing authorities.

The Paris Convention

The Paris Convention for the Protection of Industrial Property (the Paris Convention) was concluded in 1883 and has been revised several times. It applies to industrial property including patents, marks, industrial designs, utility models, trade names, geographical indications and the repression of unfair competition. It provides for national treatment, right of priority and common rules.

The principle of national treatment under the Paris Convention means that each contracting state must grant the same advantages to nationals of other contracting states as it grants to its own nationals with respect to the protection of industrial property. Nationals of non-contracting states are entitled to national treatment under certain conditions.

The right of priority means the following: on the basis of an earlier regular application filed in one of the contracting states, the applicant applies for protection of the same industrial property subject matter within a certain period of time (priority period) in any of the other contracting states. Then the later applications will not be affected by any event that may have taken place in the interval between the filing date of the first application (priority date) and the filing date of the later application, such as any publication of the invention claimed in a patent application or the sale of articles bearing the mark or incorporating an industrial design. The priority period under the Paris Convention lasts 12 months in the case of patents and utility models, and 6 months in the case of industrial designs and marks.

The common rules that must be followed by all contracting states include:

1. Patents granted in different contracting states for the same invention are independent of each other.
2. The grant of a patent may not be refused, and a patent may not be invalidated, just because the sale of the patented product, or of a product obtained by the patented process, is not allowed, restricted or is limited under national law.
3. Contracting states may take legislative measures providing for the grant of compulsory licenses, with certain limitations, to prevent the abuses which might result from the exclusive rights conferred.
4. The registration of a mark in a contracting state is independent of its possible registration in any other country, including the country of origin. Consequently, the lapse or annulment of the registration of a mark in one contracting state will not affect the validity of registration in other contracting states.
5. A contracting state must accept an application for a trademark which has been previously duly registered in another contracting state (the country of origin),

but it is allowed to refuse that application when it does not comply with the requirements under the national law.

6. Each contracting state must refuse registration and prohibit the use of marks which constitute a reproduction, imitation or translation, liable to create confusion, of a mark considered by the competent authority of that state to be well known in that state as being already the mark of a person entitled to the benefits of the Paris Convention and used for identical or similar goods.
7. Each contracting state must provide for effective protection against unfair competition (http://www.wipo.int/edocs/pubdocs/en/global_challenges/628/wipo_pub_628.pdf).

Innovations

Innovation could be any process of introducing a new product, technology, service, useful design, new process, artistic work, fabric, etc. to the market and the consumer. Every new idea transformed into a new product or service is considered to be an innovation—some innovations involve inventions, know-how and trade secrets, while others are based on new artistic perceptions, addressing emotions and senses. For example, innovations that result in a patent or trade secret, or the art and science of building a trademark are creative activities. There is, however, a great deal of creative activity that is not technology or brand related. Much of this creative activity is performed by individuals or small organizations and its results are often not as obvious to their creator or an observer or user as in the case of an invention (that may be protected by a patent) or a brand protected by product or service trademark (http://www.wipo.int/edocs/pubdocs/en/copyright/955/wipo_pub_955.pdf).

Although innovation is inherently a difficult process to define and quantify, rates of innovation can be measured by a number of variables including: patenting, licensing activities, royalties, technological development and absolute diffusion of new technologies and know-how. Increased levels of patenting suggest that individuals and companies see a clear value in their research and wish to protect and disseminate it (Pugatch et al. 2012).

Similarly, licensing activity (and accompanying royalty income) suggests the adoption, dissemination and use of technologies and processes otherwise not available or developed by a given entity or in a given country. International and trans-national licensing is of particular importance as it signifies the transfer of technologies from one country to another.

Trends in IP Activity

The patent application filings to the tune of hundreds of thousands speak the significance of the IPRs in the product/brand management of the companies from various IP offices. There has been a steep increase in the number of filings from China over the years whereas Japan and USA maintained more or less consistent trends. Finland tops the list of countries in the aspect of IP security followed by Sweden (Fig. 23.1).

Technology plays an important role in IP activity by major players across the globe, if the patenting activity in the field of Chemistry for the year 2011 is considered; most of the applications filed pertain to Pharmaceuticals followed by fine chemicals and basic materials chemistry. Though nanotechnology represents less activity, the domain has immense expectations in the future (Fig. 23.2).

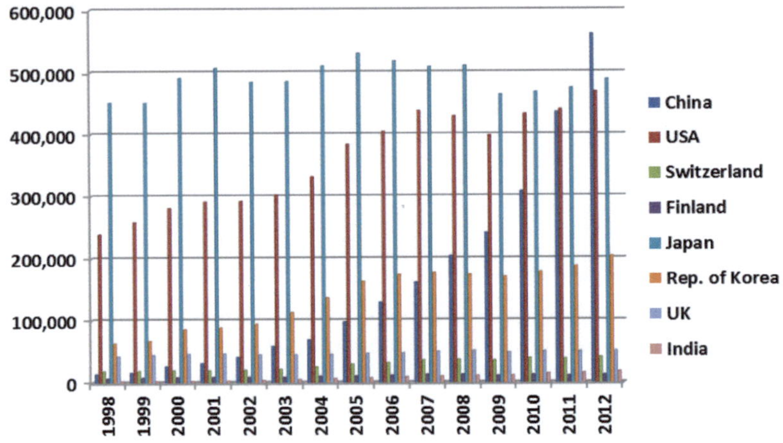

Fig. 23.1 Trends in patenting activity in major countries (1998–2012). (*Source*: www.Wipo.int)

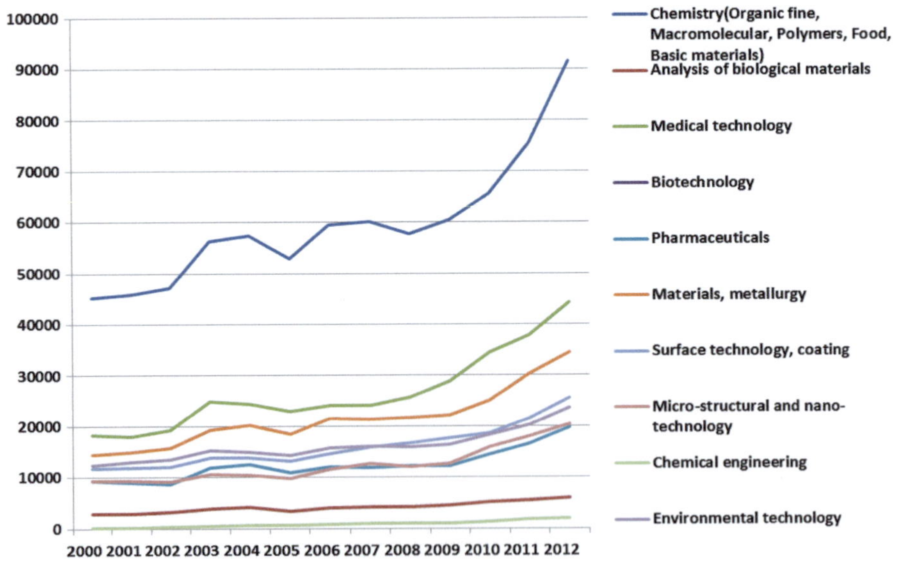

Fig. 23.2 Comparative patent applications in the various technology fields (2000–2012). (*Source*: wipo.int)

R&D Expenditure

Major chunk towards patenting at the global level goes from the governmental funding. Figure 23.3 below shows the relationship between R&D expenditure and resident patent applications for the top origins in terms of patent applications. By examining the data highlighted here, one can see that countries with a high R&D expenditure, such as China, Japan, Germany, the Republic of Korea and the US, are also associated with large numbers of resident patent applications.

China ranked number one in terms of resident patent applications, but its applications-to-GDP ratio is considerably lower than that of the Republic of Korea. Similarly, the US, which was ranked third for resident patent applications, had a lower resident applications-to-GDP ratio than Finland, Germany and Switzerland. The majority of these reported origins saw increases in their applications-to-population ratios between 2007 and 2012. China and the Republic of Korea saw the most notable increases, while Japan and the Netherlands reported the steepest decreases (http://www.wipo.int/export/sites/www/ipstats/en/wipi/2013/pdf/wipo_pub_941_2013_section_a.pdf).

The US was one of the pioneers, putting in place a legislative framework for promoting and encouraging technology transfer, partnerships and collaboration between industry, universities and federally funded institutions. Since the early to mid-1980s and the passage of the Patent and Trademark Amendment Act of 1980

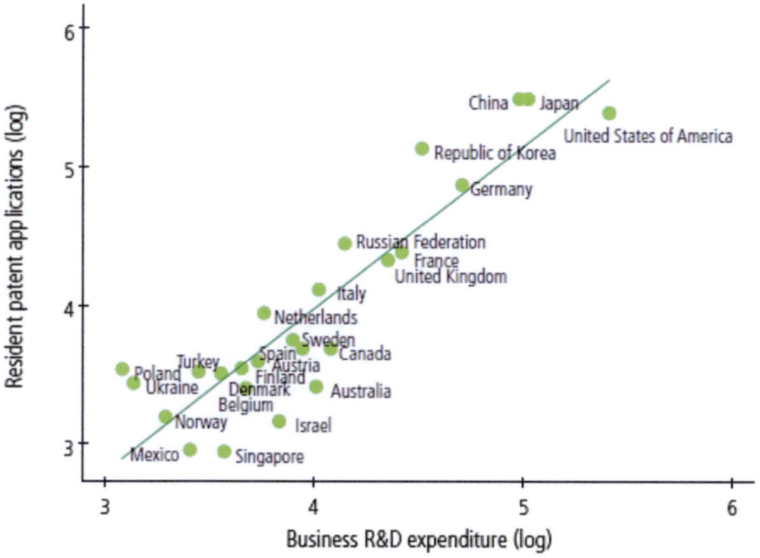

Fig. 23.3 Comparative R&D expenditures by various nations. *Note*: Business sector R&D expenditure is in constant 2005 purchasing power parity (PPP) US dollars, and R&D data lag by 1 year in order derive the patent-to-R&D dollar ratio. *Sources*: WIPO statistics Database and UNESCO Institute for Statistics, October 2013

(Bayh-Dole Act), the Stevenson-Wydler Technology Innovation Act and their subsequent amended acts (Federal Technology Transfer Act of 1986 and Technology Transfer Commercialization Act of 2000), American universities and federal research bodies have been allowed to commercialize and utilize the IP created through their research efforts (Pugatch et al. 2012).

Under the Bayh-Dole Act, universities and other centers funded by government grants engaged actively in the patenting and licensing process. Before 1980, academic institutions received fewer than 250 patents per year; by 2002, they were awarded more than 3,000 patents per year with licensing revenues surpassing $1.2 billion (AUTM Licensing Survey, FY 2002 Survey Summary-Available at: http://www.autm.net/surveys/02/2002spublic.pdf.) In FY2013, nearly 14 new commercial products were created each week—products based on university discoveries for which patents were typically filed 5–12 years prior. A study by the Biotechnology Industry Organization estimated the economic impact of university and nonprofit patent licensing from 1996 to 2010 was as much as $388 billion on the U.S. gross domestic product and $836 billion on the U.S. gross industrial output, while creating as many as three million jobs (http://www.autm.net/AM/Template.cfm?Section=FY_2013_Licensing_Activity_Survey&Template=/CM/ContentDisplay.cfm&ContentID=13870).

Much of the funds generated through university licensing and technology transfer activities are reinvested into the university and its research activities. For example, in the US the academic institutions that generate the largest amounts of licensing income have specific policies in place to reinvest the majority of this income into university and research activities. For instance, the University of California allocates all funds remaining after expenses and inventors' share to the campuses and research laboratories responsible for the licensed technologies (University of California 2011; Technology Transfer Annual Report 2011, p. 25).

In the report submitted to the President of USA on propelling innovation in drug discovery, development, and evaluation by the Executive Office of the President, President's Council of Advisors on Science and Technology in September 2012, emphasis was given to the fact that despite dramatic advances in biological knowledge, the rate of new drugs applications and new drug approvals has remained relatively constant for several decades. While the output of new drugs has remained constant, total R&D investment by industry in drug discovery and development have grown exponentially, in inflation-adjusted terms (see Fig. 23.4 below). As a result, the amortized R&D cost per newly approved drug has continued to grow.

Medical research may have once been defined by discrete discoveries in individual laboratories, but it currently more closely resembles a cumulative industry. Like building a computer, identifying targets and engineering compounds increasingly requires the interplay of many layers of transistors (genes) and circuits (enzyme pathways). By importing business models such as patent pools, biomedical companies along with government and academic research centers could develop more efficient ways of exchanging intellectual property and setting reimbursement without hindering progress or driving costs to unaffordable levels (Kesselheim and Avorn 2005).

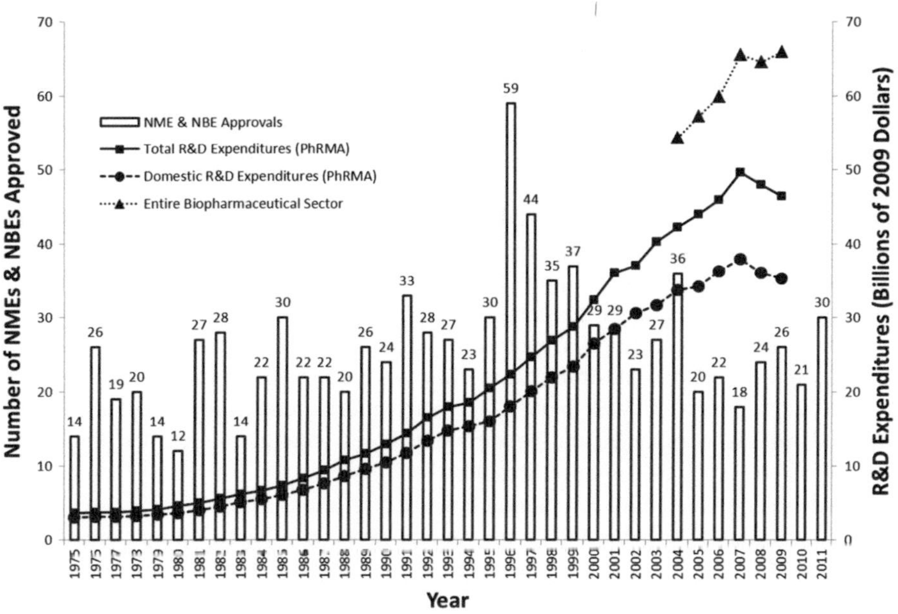

Fig. 23.4 Annual new molecular entity and new biologic entity approvals vs. R&D expenditures in 2009 dollars. (*Source*: www.whitehouse.gov/ostp/pcast)

World Trade Organization (WTO)

Intellectual property and conflicts over ownership of intangible assets is not a new subject, since early sixteenth century they have been practiced. Many organizations at national as well as international level have dedicated their efforts in bringing harmony towards settlement of issues relating to intangible assets management. The World Trade Organization (WTO), in this stream was born out of negotiations, and everything that WTO does now is the result of negotiations at diverse focal points. The WTO began life on 1 January 1995, but its trading system is half a century older. From 1948 to 1994, the General Agreement on Tariffs and Trade (GATT) provided the rules for much of world trade and presided over periods that saw some of the highest growth rates in international commerce. It seemed well-established, but throughout those 47 years, it was a provisional agreement and organization. The bulk of the WTO's current work comes from the 1986–94 negotiations called the Uruguay Round and led to the WTO's creation. Whereas, GATT mainly dealt with trade in goods, the WTO and its agreements now cover trade in services, and in traded inventions, creations and designs (intellectual property). The WTO is currently the host to new negotiations, under the "Doha Development Agenda" launched in 2001.

There are a number of ways of looking at the WTO. It's an organization for liberalizing trade, a forum for governments to negotiate trade agreements, a place for them to settle trade disputes. It operates a system of trade rules. Essentially, the

WTO is a place where member governments go, to try to sort out the trade problems they face with each other.

Where countries have faced trade barriers and wanted them lowered, the negotiations have helped to liberalize trade. At its heart are the WTO agreements, negotiated and signed by the bulk of the world's trading nations. These documents provide the legal ground-rules for international commerce. They are essentially contracts, binding governments to keep their trade policies within agreed limits. Although negotiated and signed by governments, the goal is to help producers of goods and services, exporters, and importers conduct their business, while allowing governments to meet social and environmental objectives.

The system's overriding purpose is to help trade flow as freely as possible—so long as there are no undesirable side-effects—because this is important for economic development and well-being. That partly means removing obstacles ensuring that individuals, companies and governments know what the trade rules around the world are "transparent" and predictable.

Trade relations often involve conflicting interests and the most harmonious way to settle the differences is through some neutral procedure based on an agreed legal foundation. That is the purpose behind the dispute settlement process written into the WTO agreements (http://www.wto.org/english/thewto_e/whatis_e/tif_e/fact1_e.htm).

The principles of WTO include-The trading system should be -without discrimination—a country should not discriminate between its trading partners (giving them equally "most-favored-nation" or MFN status); and it should not discriminate between its own and foreign products, services or nationals (giving them "national treatment"); freer—barriers coming down through negotiation; predictable—foreign companies, investors and governments should be confident that trade barriers (including tariffs and non-tariff barriers) should not be raised arbitrarily; tariff rates and market-opening commitments are "bound" in the WTO;-more competitive—discouraging "unfair" practices such as export subsidies and dumping products at below cost to gain market share; more beneficial for less developed countries—giving them more time to adjust, greater flexibility, and special privileges (http://www.wto.org/english/thewto_e/whatis_e/tif_e/fact2_e.htm).

Trade-Related Aspects of Intellectual Property Rights (TRIPS)

The WTO's agreement on Trade-Related Aspects of Intellectual Property Rights (TRIPS), negotiated in the 1986–1994 Uruguay Round, introduced intellectual property rules into the multilateral trading system for the first time. The TRIPS Agreement is an attempt to narrow the gaps in the way these rights are protected around the world, and to bring them under common international rules. It establishes minimum levels of protection that each government has to give to the intellectual property of fellow WTO members. In doing so, it strikes a balance between the long term benefits and possible short term costs to society. Society benefits in the long term when intellectual property protection encourages creation and invention, especially when the period of protection expires and the creations and inventions enter the public domain. Governments are allowed to reduce any short term costs through various exceptions, for example to tackle public health problems. And, when there are trade disputes over intellectual property rights, the WTO's dispute settlement system is now available.

The agreement covers five broad issues:

- How basic principles of the trading system and other international intellectual property agreements should be applied.
- How to give adequate protection to intellectual property rights.
- How countries should enforce those rights adequately in their own territories.
- How to settle disputes on intellectual property between members of the WTO.
- Special transitional arrangements during the period when the new system is being introduced.

The TRIPS Agreement provides flexibility for governments to fine tune the protection granted in order to meet social goals. For patents, it allows governments to make exceptions to patent holders' rights such as in national emergencies, anti-competitive practices, or if the right-holder does not supply the invention, provided certain conditions are fulfilled. For pharmaceutical patents, the flexibility has been clarified and enhanced by the 2001 Doha Declaration on TRIPS and Public Health. The enhancement was put into practice in 2003 with a decision enabling countries that cannot make medicines themselves, to import pharmaceuticals made under compulsory license. In 2005, members agreed to make this decision a permanent amendment to the TRIPS Agreement, which will take effect when two thirds of members accept it (http://www.wto.org/english/tratop_e/trips_e/tripsfactsheet_pharma_2006_e.pdf).

Types of Intellectual Property

There are several types of intellectual properties as they deal with inventions, artistic and music related works and many new things that were non-existent earlier. The areas covered by the TRIPS Agreement include:

Copyright and Related Rights

Copyrights include literary and artistic works such as novels, poems and plays, films, musical works, artistic works such as drawings, paintings, photographs and sculptures, and architectural designs. Rights related to copyright include those of performing artists in their performances, producers of phonograms in their recordings, and those of broadcasters in their radio and television programs (http://www.wipo.int/export/sites/www/freepublications/en/statistics/943/wipo_pub_943_2013.pdf).

The field of copyright and related rights has expanded enormously with the technological progress of the last several decades, which has brought new ways of spreading creations by such forms of worldwide communication as satellite broadcast and compact discs. WIPO is deeply involved in the ongoing international debate to shape new standards for copyright protection in cyberspace. The Organization administers the WIPO Copyright Treaty and the WIPO Performances and Phonogram Treaty (often known together as the "Internet Treaties"), which sets down international norms aimed at preventing unauthorized access to and use of creative works on the Internet or other digital networks (http://www.wipo.int/edocs/pubdocs/en/copyright/450/wipo_pub_l450cr.pdf).

The importance of protecting intellectual property was first recognized in the Paris Convention for the Protection of Industrial Property in 1883 and the Berne Convention for the Protection of Literary and Artistic Works in 1886. Both treaties are administered by the World Intellectual Property Organization (WIPO). Countries generally have laws to protect intellectual property for two main reasons. One is to give statutory expression to the moral and economic rights of creators in their creations and to the rights of the public in accessing those creations. The second is to promote creativity, and the dissemination and application of its results, and to encourage fair trade, which would contribute to economic and social development (http://www.wipo.int/edocs/pubdocs/en/intproperty/909/wipo_pub_909.pdf).

Copyright law protects only the form of expression of ideas, not the ideas themselves. The creativity protected by copyright law is creativity in the choice and arrangement of words, musical notes, colors and shapes. So copyright law protects the owner of property rights against those who copy or otherwise take and use the form in which the original work was expressed by the author (http://www.wipo.int/edocs/pubdocs/en/intproperty/909/wipo_pub_909.pdf).

Case Example

In his will, the nineteenth century English author Charles Dickens left "all my private papers whatsoever and wheresoever" to his sister-in-law Georgina Hogarth. He left the remainder of his "real and personal estate (including my copyrights)" to his children. Included in the estate was the manuscript of 'The Life of Christ', which Dickens had handwritten for the use of his children, and which he did not intend to publish. Some years after Dickens's death, however, a trustee acting for the surviving family sold the copyright of the manuscript to a publisher and also gave him a photocopy of the original manuscript to be used as the basis of the book to be published.

The question arose whether Hogarth's heirs, Dickens's children's heirs, or both, were entitled to the money under the publishing contract. If both were entitled, then how would they share?

The Court decided:

1. The bequest of "private papers" passed only the property in the physical manuscript of 'The Life of Christ', not any copyright in it.
2. The bequest of Dickens's "personal estate" passed all the author's copyrights, and would have done so even if the word "copyright" had not been specifically mentioned.
3. The consent of both estates would be needed before the manuscript of The Life of Christ could practicably be published: In Re Dickens [1935] Ch. 267 (U.K.: High Court & Court of Appeal).

MR JUSTICE BENNETT in the High Court:
The common law has this conception with regard to rights of property in a literary work written, marked or impressed or otherwise recorded upon some material thing namely, that the material thing might, as a subject of property, be separated from the literary work recorded on it and that the literary work might be regarded as an incorporeal subject of property and be owned separately from the material thing upon which it was recorded. ... The question to be answered is: Did Georgina Hogarth take the two species of property, the physical thing, the manuscript, and also the incorporeal property, the copyright, or did she only take the physical thing, the manuscript?

LORD JUSTICE MAUGHAM in the Court of Appeal:
This work was written by the author not for publication but for the benefit of his family, and he retained it for their private study. ... Some copies were made with the author's permission, one, for instance, for his eldest son. . .; but the work was unpublished in the author's lifetime and was not intended for publication. In these circumstances, I am of opinion that, like letters, diaries and memoranda, it comes fairly within the description of private papers. The manuscript, that is the pieces of paper with the writing on them, therefore passed on the death of the author to Miss Hogarth.

If Charles Dickens had finished some novel or story but had not published it at the date of his death, it would have passed under the bequest of his copyrights. Nor can I see that any different conclusion should be arrived at in the present case because of the bequest to Miss Hogarth of "all his private papers whatsoever." This phrase, standing alone, obviously does not properly or naturally carry with it a right to publish private papers. ... In my opinion, on the true construction of the will, the exclusive right to publish the Life of Christ passed to [the trustees of Dickens's children], whilst the original manuscript became the personal property of Miss Hogarth. If the trustees and Miss Hogarth could not agree there would be a sort of deadlock. The trustees might, perhaps, have published from a copy, but Miss Hogarth could not properly have published from the original without the consent of the trustees.

LORD JUSTICE ROMER:
If one person had been the owner of the copyright and another the owner of the manuscript, neither one singly could have conferred upon the purchasers the rights conferred by the deed of assignment. The copyright was practically worthless if the purchaser could not make use of the manuscript. The right to examine and copy and photograph the manuscript was practically worthless if the purchaser could not

obtain the copyright. The concurrence of both owners in the sale could alone have resulted in the payment of a substantial sum, and in the circumstances it seems reasonably clear that neither owner would join in the sale except on the terms that such sum should be equally divided between them.

The Court therefore ordered that the monies be divided equally between both estates (http://www.wipo.int/edocs/pubdocs/en/copyright/844/wipo_pub_844.pdf).

Trademarks, Including Service Marks

A trademark is a distinctive sign that identifies certain goods or services as those produced or provided by a specific person or enterprise. The holder of a registered trademark has the legal right to exclusive use of the mark in relation to the products or services for which it is registered. The owner can prevent unauthorized use of the trademark, or a confusingly similar mark, so as to prevent consumers and the public in general from being misled. Trademarks can be maintained indefinitely by paying renewal fees. The procedures for registering trademarks are governed by the rules and regulations of national and regional IP offices. Trademark rights are limited to the jurisdiction of the authority that registers the trademark. Trademarks can be registered by filing an application at the relevant national or regional office(s), or by filing an international application through the Madrid System (http://www.wipo.int/export/sites/www/freepublications/en/statistics/943/wipo_pub_943_2013.pdf).

The system of international registration of marks is governed by two treaties: the Madrid Agreement Concerning the International Registration of Marks, which dates from 1891, and the Protocol Relating to the Madrid Agreement, which was adopted in 1989, entered into force on December 1, 1995, and came into operation on April 1, 1996. Common Regulations under the Agreement and Protocol also came into force on that date. The system is administered by the International Bureau of WIPO, which maintains the International Register and publishes the WIPO Gazette of International Marks (http://www.wipo.int/edocs/pubdocs/en/marks/418/wipo_pub_418.pdf).

The objectives of the system are two fold. Firstly, it facilitates the obtaining of protection for marks (trademarks and service marks). The registration of a mark in the International Register produces, in the Contracting Parties designated by the applicant. Further Contracting Parties may be designated subsequently. Secondly, since an international registration is equivalent to a bundle of national registrations, the subsequent management of that protection is made much easier. There is only one registration to renew, and changes such as a change in ownership or in the name or address of the holder, or a limitation of the list of goods and services, can be recorded in the International Register through a single simple procedural step. On the other hand, if it is desired to transfer the registration for only some of the designated Contracting Parties, or for only some of the goods or services, or to limit the list of goods and services with respect to only some of the designated Contracting Parties, the system is flexible enough to accommodate this (http://www.wipo.int/edocs/pubdocs/en/marks/418/wipo_pub_418.pdf).

The Madrid system: A one-stop shop

- File one international application instead of multiple national applications.
- File in one language.
- Pay one set of fees in one currency.
- Obtain an international registration covering multiple territories.
- Expand protection in new territories.
- Renew every 10 years with one simple procedure.
- Manage portfolio of marks through one centralized system (http://www.wipo.int/edocs/pubdocs/en/marks/1039/wipo_pub_1039.pdf).

Case Example

Though there are several Trademark infringement cases, a case pertaining Pharmaceuticals and Biotechnology is discussed here:

Ciba-Geigy Canada Ltd. v. Apotex Inc. and Novopharm Limited, [1992]

Ciba-Geigy is a pharmaceutical laboratory which has manufactured and sold metoprolol tablets in Canada under the trade name "Lopresor" since 1977. Metoprolol is a prescription drug generally prescribed for hypertension. The case involved a complaint by Ciba-Geigy Canada Ltd. against two other pharmaceutical companies, Apotex and Novopharm, which had mimicked the size, shape, and colour of metoprolol. In terms of their effectiveness, the Apotex and Novopharm had already been deemed interchangeable with Metoprolol according to Ontario law. The sole issue was whether the appearance of the competing products infringed upon Ciba-Geigy's trademark (http://www.trademarkcases.ca/ciba-geigy-canada-ltd-v-apotex-inc-1992-3-s-c-r-120/).

At the trial level, Ciba-Geigy failed to establish that the customers, namely physicians and pharmacists that prescribe or dispense metoprolol, were confused in choosing the brand of metoprolol to give to patients due to the similar appearance of the tablets. For this reason, the Supreme Court of Ontario refused to issue an interlocutory injunction because Ciba-Geigy failed to show that there was a "serious issue" to be tried. At the Court of Appeal, Ciba-Geigy argued that the customers affected by the passing-off should include ultimate consumer of the prescribed drug as they are likely to be confused by the similar appearance of the products in question. The court unanimously allowed Ciba-Geigy's appeal, finding that the appearance of the Apotex and Novopharm produces constituted a misrepresentation to patients. The Court of Appeal rejected its argument and dismissed the appeal (http://en.wikipedia.org/wiki/Ciba-Geigy_Canada_Ltd._v._Apotex_Inc).

Geographical Indications

A geographical indication is a sign used on products that have a specific geographical origin and possess qualities or a reputation that are due to that origin.

In order to work as a GI, a sign must identify a product as originating in a given place, in addition, the qualities or reputation of the product should be essentially due to the place of origin. Since the qualities depend on the geographical place of production, there is a link between the product and its original place of production.

Most commonly, a GI consists of the name of the place of origin of the good, such as "Jamaica Blue Mountain" or "Darjeeling". But non-geographical names, such as "Vinho Verde", "Cava" or "Argan Oil", or symbols commonly associated with a place, can also constitute a GI. In essence, whether a sign functions as a GI is a matter of national law and consumer perception.

Since the adoption of the Agreement on Trade-Related Aspects of Intellectual Property Rights (the TRIPS Agreement) in 1994, which contains a section on geographical indications (GIs), this form of intellectual property (IP) has attracted increasing attention from policymakers and trade negotiators, as well as producers (mostly of agricultural products), lawyers and economists across the world. It is undoubtedly because of the TRIPS Agreement section on GIs that the issue now appeals to more and more nations beyond the rather restricted list of countries that have traditionally pursued active GI policies.

The period following the conclusion of the Paris Convention saw numerous efforts aimed at increasing the level of multilateral protection afforded to indications of source and appellations of origin, which led, among others things, to the adoption of the Madrid Agreement for the Repression of False or Deceptive Indications of Source on Goods of 1891, and the Lisbon Agreement for the Protection of Appellations of Origin and their International Registration of 1958 (Lisbon Agreement), and to the inclusion, in the TRIPS Agreement, of a special section on GIs.

Geographical indications are distinctive signs used to differentiate competing goods. They are collectively owned with a strong inherent origin-base, namely the geographical origin to which they refer. The reference to geographical origin—most regularly for agricultural products—combined with the use of traditional extraction and processing methods, presents an interesting marketing potential in terms of product branding.

However, the use of geographical origin brands also presents a number of challenges. Owing to their collective nature, those who produce and market GIs must engage in collective action with regard to production methods, quality standards and control, as well as product distribution and marketing (http://www.wipo.int/edocs/pubdocs/en/geographical/952/wipo_pub_952.pdf).

Case Example

Prior to 1999 there was no specific legislation to regulate geographical indication, TRIPS Agreement enacted the Geographical Indications of Goods (Registration and Protection) Act, 1999, and this act excludes unauthorized persons from misusing GIs. This would protect the interest of producers, manufacturers and thereby consumer from being deceived by the falsity of geographical origin to economic prosperity of the producer of such goods and promote goods bearing GIs in export market.

RiceTec Inc., had been trying to enter the international Basmati market with brands like "Kasmati" and "Texmati". Ultimately, the company claimed to have developed a new strain of aromatic rice by interbreeding basmati with another variety. They sought to call the allegedly new variety as Texmati or American Basmati. RiceTecInc, was issued the Patent number 5663484 on Basmati rice lines and grains on September 2, 1997.

Centre for Food Safety, Research Foundation for Science, Technology and Ecology, filed legal petitions in the United States. The Council of Scientific and Industrial Research also objected to it. RiceTec has been forced to give up the title of its patent, it has been forced to give up 15 of its 20 claims, including those with the most far-reaching implications related to biopiracy. This has enabled for plaintiffs to project Basmati as the rice variety grown in the sub-Himalayan regions of India and Pakistan proclaiming its regional importance (Mukherjee 2008).

Industrial Designs

An industrial design constitutes the ornamental or aesthetic aspect of an article. A design may consist of three-dimensional features, such as the shape or surface of an article, or of two-dimensional features, such as patterns, lines or color.

Industrial designs are applied to a wide variety of products of industry and handicraft: from technical and medical instruments to watches, jewelry, and other luxury items; from housewares and electrical appliances to vehicles and architectural structures; and from textile to leisure goods.

Industrial design protection helps to ensure a fair return on investment. An effective system of protection also benefits consumers and the public at large, by promoting fair competition and honest trade practices (http://www.wipo.int/designs/en/).

In addition to Paris Convention, The Hague Agreement Concerning the International Registration of Industrial Designs and The Locarno Agreement Establishing an International Classification for Industrial Designs are the international treaties governing the policies and practices relating to Industrial Designs.

Case Example

Bodum USA, Inc. And Pi Design Ag. V. Trudeau Corporation (1889) Inc.

The plaintiffs commenced an action against the company Trudeau Corporation (1889) Inc. and sought relief in application of the Industrial Design Act, on the grounds of infringement of two Canadian industrial designs registered under numbers 107,736 and 114,070, which correspond to Bodum double wall glasses (Fig. 23.5 below).

Fig. 23.5 Bodum's double walled glasses. (*Source*: http://decisions.fct-cf.gc.ca/fc-cf/decisions/en/item/61469/index.do)

Fig. 23.6 Trudeau's double walled glasses. (*Source*: http://decisions.fct-cf.gc.ca/fc-cf/decisions/en/item/61469/index.do)

Trudeau denied acting in violation of the industrial designs in question, and also denied directing public attention to its wares in such a way as to cause or be likely to cause confusion between its wares and the wares of Bodum (Fig. 23.6 below). Furthermore, Trudeau sought a declaration that the industrial designs in question are and have always been invalid.

The industrial designs of Bodum (107,736 and 114,070) were registered with the Office of the Commissioner of Patents of the Canadian Intellectual Property Office on February 1, 2006. The priority date for the industrial designs in question is February 18, 2004. The industrial designs have no registered variants. Trudeau's double wall glasses were introduced to the Canadian market in the fall of 2006. At the time, Trudeau was aware of the double wall glasses marketed by Bodum.

The issues raised in this case are the following:

1. Was there infringement of industrial designs 107,736 and 114,070?
2. Is the registration of industrial designs 107,736 and 114,070 invalid?

3. Does Trudeau's marketing of double wall glasses constitute unfair competition (offence of confusion)?

The utilitarian part was that the insulating quality of the double wall glasses doesn't just keep hot drinks hot for a longer period of time, it also keeps cold drinks cold longer. Another nice thing about them—there is no condensation water when you serve cold drinks, therefore no messy rings on your table. And by the way, they're great for ice cream as well. Double wall glasses are truly multifunctional. They are made from borosilicate glass and are dishwasher safe.

After weighing the testimony of the expert witness, the prior art, and the parties' arguments, the Court finds that the Trudeau glasses do not have the features attributed to them by the plaintiffs and that the Trudeau glasses are not infringing products, consequently, the industrial designs in question must be expunged from the register (http://decisions.fct-cf.gc.ca/fc-cf/decisions/en/item/61469/index.do).

Patents

A set of exclusive rights granted by law to applicants for inventions that are new, non-obvious and commercially applicable. Patents are valid for a limited period of time (generally 20 years), during which patent holders can commercially exploit their inventions on an exclusive basis. In return, applicants are obliged to disclose their inventions to the public in a manner that enables others, skilled in the art, to replicate the invention. The patent System is designed to encourage innovation by providing innovators with time-limited exclusive legal rights, thus enabling innovators to appropriate a return on their innovative activity (http://www.wipo.int/export/sites/www/freepublications/en/statistics/943/wipo_pub_943_2013.pdf).

A patent only gives an inventor the right to prevent others from using the patented invention. It says nothing about whether the product is safe for consumers and whether it can be supplied. Patented pharmaceuticals still have to go through rigorous testing and approval before they can be put on the market (http://www.wto.org/english/tratop_e/trips_e/tripsfactsheet_pharma_2006_e.pdf).

PCT system: The Patent Cooperation Treaty (PCT), an international treaty administered by WIPO, facilitates the acquisition of patent rights in a large number of jurisdictions. The PCT system simplifies the process of multiple national patent filings by reducing the requirement to file a separate application in each jurisdiction. However, the decision on whether or not to grant patent rights remains in the hands of national and regional patent offices, and patent rights remain limited to the jurisdiction of the patent-granting authority. The PCT international application process starts with the international phase, during which an international search and possibly a preliminary examination are performed, and concludes with the national phase, during which national and regional patent offices decide on the patentability of an invention according to national law (http://www.wipo.int/export/sites/www/freepublications/en/statistics/943/wipo_pub_943_2013.pdf).

PCT Timeline (Fig. 23.7 Below)

After 30 months, each PCT application enters the national phase wherein the application is reviewed for the grant of a patent by the individual countries' national IP offices. There is no binding for the member state to abide by the search and examination reports generated by the PCT-IB. Each national IP office has its own domestic rules and Acts pertaining to prosecution of patent applications received through PCT or directly from the applicants at domestic level.

A general procedure of patenting is discussed however, it may change from country to country according to its legal framework, since IPRs are territorial in nature.

35 U.S.C. 101 states that "whoever invents or discovers any new and useful process, machine, manufacture, or composition of matter, or any new and useful improvement thereof may obtain a patent therefor, subject to the conditions and requirements of this title".

It has been interpreted as imposing three requirements.

First, whoever invents or discovers an eligible invention may obtain only ONE patent therefor. Secondly, a claimed invention must fall within one of the four eligible categories of invention, i.e., process, machine, manufacture, or composition of matter, as these categories have been interpreted by the courts. Thirdly, a claimed invention must be useful or have a utility that is specific, substantial and credible (http://www.uspto.gov/web/offices/pac/mpep/s2104.html).

An invention is any art or process (way of doing or making things), machine, manufacture, design, or composition of matter, or any new and useful improvement thereof, or any variety of plant, which is or may be patentable under the patent laws of the United States (35 U.S.C. 100).

Inventors typically execute the deed of assignment wherein they assign the rights associated with the invention to the assignee. The assignee is the actual owner of the

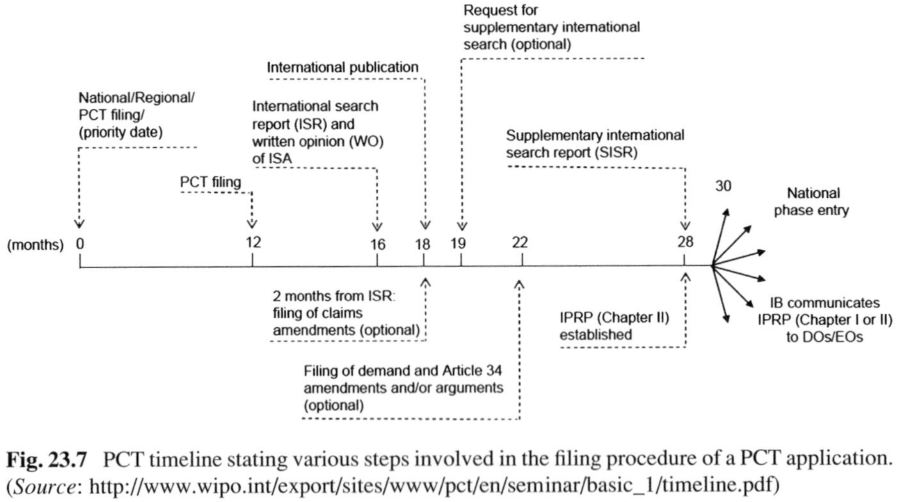

Fig. 23.7 PCT timeline stating various steps involved in the filing procedure of a PCT application. (*Source*: http://www.wipo.int/export/sites/www/pct/en/seminar/basic_1/timeline.pdf)

patent and enjoys the rights associated with the commercialization, licensing etc. of the patent.

The Patent application could be filed as a provisional application if the inventors are not ready with the final specification, wherein the priority is obtained. A provisional application is not required to have a formal patent claim or an oath or declaration. The inventors have a non-extendable 12 month window to submit a non-provisional specification after filing a provisional specification. This forms significant on the patent term, by filing a provisional application first, and then filing a corresponding non-provisional application that references the provisional application within the 12-month provisional application pendency period, a patent term endpoint may be extended by as much as 12 months. Further, the non-provisional application must have at least one inventor in common with the inventor(s) named in the provisional application to claim benefit of the provisional application filing date.

Each application for a patent is published promptly after the expiration of a period of 18 months from the earliest filing date. A meticulous examination of the patent application is carried out by the examiner evaluating the merits of the invention over the existing art taking into consideration the novelty, non-obviousness and the utility aspects of the proposed invention. A series of communication exchange commences between the examiner and the inventors towards making the application qualify for the grant of a patent.

A patent is personal property and may be sold to others or mortgaged; it may be bequeathed by a will; and it may pass to the heirs of a deceased patentee. The patent law provides for the transfer or sale of a patent, or of an application for patent, by an instrument in writing. Such an instrument is referred to as an assignment and may transfer the entire interest in the patent. The assignee, when the patent is assigned to him or her, becomes the owner of the patent and has the same rights that the original patentee had (http://www.uspto.gov/patents/resources/general_info_concerning_patents.jsp#heading-27).

Patent Infringement

Infringement of a patent consists of the unauthorized making, using, offering for sale, selling or importing any patented invention during the term of the patent. If a patent is infringed, the patentee may sue for relief in the appropriate court. The patentee may ask the court for an injunction to prevent the continuation of the infringement and may also ask for an award of damages because of the infringement. In such an infringement suit, the defendant may raise the question of the validity of the patent, which is then decided by the court. The defendant may also aver that what is being done does not constitute infringement. Infringement is determined primarily by the language of the claims of the patent and, if what the defendant is making does not fall within the language of any of the claims of the patent, there is no literal infringement.

In U.S. suits for infringement of patents follow the rules of procedure of the federal courts. From the decision of the district court, there is an appeal to the Court of Appeals for the Federal Circuit. The Supreme Court may thereafter take a case by writ of certiorari. If the United States Government infringes a patent, the patentee has a remedy for damages in the United States Court of Federal Claims. The government may use any patented invention without permission of the patentee, but the patentee is entitled to obtain compensation for the use by or for the government. The USPTO has no jurisdiction over questions relating to infringement of patents. In examining applications for patent, no determination is made as to whether the invention sought to be patented infringes any prior patent. An improvement invention may be patentable, but it might infringe a prior unexpired patent for the invention improved upon, if there is one (http://www.uspto.gov/patents/resources/general_info_concerning_patents.jsp#heading-27).

Case Example

There are several patent infringement cases however, the case of Novartis v. Union of India is discussed here as it forms the classic case of the territorial nature of patent laws and implications on the trade and commerce between the nations.

On July 17, 1998, Novartis, a multinational company headquartered in Basel, Switzerland, filed a patent application in India. At that time the Act allowed acceptance of product patent applications as per "mail box" process and the same was contemplated to be examined post January 1, 2005 once India introduced product patent regime. This was in line with TRIPS requirement. Meanwhile, Novartis applied for and was granted exclusive marketing rights (EMR) in relation to Product, which also was in line with TRIPS.

The bone of contention was the Section 3[d] of the new patent regime which reads as:

> *[(d) the mere discovery of a new form of a known substance which does not result in the enhancement of the known efficacy of that substance or the mere discovery of any new property or new use for a known substance or of the mere use of a known process, machine or apparatus unless such known process results in a new product or employs at least one new reactant.*
>
> *Explanation: For the purposes of this clause, salts, esters, ethers, polymorphs, metabolites, pure form, particle size, isomers, mixtures of isomers, complexes, combinations and other derivatives of known substance shall be considered to be the same substance, unless they differ significantly in properties with regard to efficacy;]*

Novartis received pre-grant opposition filed by Cancer Patients Aid Association, NATCO Pharma, Cipla, Ranbaxy Laboratories and Hetero Drugs before the examination of its application that was in 'mail-box'.

In view of the Act, in January, 2006, the Assistant Controller of Patents upholding the pre-grant oppositions, on the grounds of novelty, non-obviousness and moreover on incremental improvement aspects, rejected the Novartis' application.

In May 2006, Novartis filed a petition in the Madras High Court on the premise of Sections 3[d] and 3[b] stating that:

- Section 3(d) is unconstitutional as it violates the provision of the TRIPS agreement.
- The Indian patent act doesn't define the term 'efficacy' and provides unguided power on the Controller. Hence it is arbitrary, illogical and vague.

IPAB was constituted and in April, 2007, and the Madras High Court transferred the appeal from the Controller's order rejecting patent to the IPAB. However, the Madras High Court, reserving the right to pronounce had issued a judgment of the constitutional validity of Section 3(d) of the Act. In June, 2009 the IPAB rejected the Novartis' patent.

Later, in August 2009 Novartis filed a Special Leave Petition with the Indian Supreme Court in 2009 challenging the denial of the Glivec beta crystal form patent on two grounds, based on Sections 3(d) and 3(b) of the Indian patent law. However, on April 1, 2013, the Supreme Court, in its landmark decision concluded that Section 3(d) does not violate the TRIPS mandate rather prevents frivolous patenting without neglecting valuable incremental innovations in pharmaceuticals and is very well compatible with TRIPS agreement (http://documents.lexology.com/2284cfd8-021d-40a8-be6e-875943f0507e.pdf) and (http://www.lexology.com/library/detail.aspx?g=3f92413f-107c-4886-aca7-24633a341e22).

In response, Novartis contends that a decision issued by the Indian Supreme Court regarding the Novartis breakthrough medicine Glivec® (imatinibmesylate) provides clarification on Indian patent law and discourages innovative drug discovery essential to advancing medical science for patients. Novartis has never been granted an original patent for Glivec in India. The Court denied an appeal challenging the rejection of a patent for Glivec, a life-saving medicine for certain forms of cancer, patented in nearly 40 countries including China, Russia, and Taiwan (http://www.novartisoncology.com/).

Layout-Designs (Topographies) of Integrated Circuits

Article 35 of the TRIPS Agreement requires Member countries to protect the layout-designs of integrated circuits in accordance with the provisions of the IPIC Treaty (the Treaty on Intellectual Property in Respect of Integrated Circuits), negotiated under the auspices of WIPO in 1989. These provisions deal with; inter alia, the definitions of "integrated circuit" and "layout-design (topography)", requirements for protection, exclusive rights, and limitations, as well as exploitation, registration and disclosure.

An "integrated circuit" means a product, in its final form or an intermediate form, in which the elements, at least one of which is an active element, and some or all of the interconnections are integrally formed in and/or on a piece of material and which is intended to perform an electronic function.

A "layout-design (topography)" is defined as the three-dimensional disposition, however expressed, of the elements, at least one of which is an active element, and of some or all of the interconnections of an integrated circuit, or such a three-dimensional disposition prepared for an integrated circuit intended for manufacture. The obligation to protect layout-designs applies to such layout-designs that are original in the sense that they are the result of their creators' own intellectual effort and are not commonplace among creators of layout-designs and manufacturers of integrated circuits at the time of their creation. The exclusive rights include the right of reproduction and the right of importation, sale and other distribution for commercial purposes (http://www.wto.org/english/tratop_e/trips_e/intel2_e.htm#layoutdesigns).

A diplomatic conference was held at Washington, D.C., in 1989, which adopted a Treaty on Intellectual Property in Respect of Integrated Circuits, also called the Washington Treaty or IPIC Treaty. The Treaty, signed at Washington on May 26, 1989, is open to States Members of WIPO or the United Nations and to intergovernmental organizations meeting certain criteria. The Treaty has been incorporated by reference into the TRIPS Agreement of the World Trade Organization (WTO), subject to the following modifications: the term of protection is at least 10 (rather than 8) years from the date of filing an application or of the first commercial exploitation in the world, but Members may provide a term of protection of 15 years from the creation of the layout-design; the exclusive right of the right-holder extends also to articles incorporating integrated circuits in which a protected layout-design is incorporated, in so far as it continues to contain an unlawfully reproduced layout-design; the circumstances in which layout-designs may be used without the consent of right-holders are more restricted; certain acts engaged in unknowingly will not constitute infringement (http://www.wto.org/english/docs_e/legal_e/27-trips.pdf).

Undisclosed Information, Including Trade Secrets

A trade secret is "information, including a formula, pattern, compilation, program, device, method, technique, or process, that:

1. Derives independent economic value, actual or potential, from not being generally known to the public or to other persons who can obtain economic value from its disclosure or use.
2. Is the subject of efforts that are reasonable under the circumstances to maintain its secrecy."

Depending on the legal system, the protection of trade secrets forms part of the general concept of protection against unfair competition or is based on specific provisions or case law on the protection of confidential information.

The subject matter of trade secrets is usually defined in broad terms and includes sales methods, distribution methods, consumer profiles, advertising strategies, lists of suppliers and clients, and manufacturing processes. While a final determination of what information constitutes a trade secret will depend on the circumstances of each individual case, clearly unfair practices in respect of secret information include industrial or commercial espionage, breach of contract and breach of confidence (http://www.wipo.int/sme/en/ip_business/trade_secrets/trade_secrets.htm).

Article 10bis of the Paris Convention (1967), Members shall protect undisclosed information in accordance with paragraph 2 and data submitted to governments or governmental agencies in accordance with paragraph 3.

Natural and legal persons shall have the possibility of preventing information lawfully within their control from being disclosed to, acquired by, or used by others without their consent in a manner contrary to honest commercial practices so long as such information:

1. Is secret in the sense that it is not, as a body or in the precise configuration and assembly of its components, generally known among or readily accessible to persons within the circles that normally deal with the kind of information in question.
2. Has commercial value because it is secret.
3. Has been subject to reasonable steps under the circumstances, by the person lawfully in control of the information, to keep it secret.

Members, when requiring, as a condition of approving the marketing of pharmaceutical or of agricultural chemical products which utilize new chemical entities, the submission of undisclosed test or other data, the origination of which involves a considerable effort, shall protect such data against unfair commercial use. In addition, Members shall protect such data against disclosure, except where necessary to protect the public, or unless steps are taken to ensure that the data are protected against unfair commercial use (http://www.wto.org/english/docs_e/legal_e/27-trips.pdf).

Precautions to be taken for Trade secrets:
Trade secrets are widely used by SMEs. In fact, many SMEs rely almost exclusively on trade secrets for the protection of their IP (although in many cases they may not even be aware that trade secrets are legally protected). It is important, therefore, to make sure that enterprises take all necessary measures to protect their trade secrets effectively. This includes:

- Firstly, considering whether the secret is patentable and, if so, whether it would not be better protected by a patent.
- Secondly, making sure that a limited number of people know the secret and that all those who do are well aware that it is confidential information.
- Thirdly, including confidentiality agreements within employees' contracts. Under the law of many countries, however, employees owe confidentiality to their employer even without such agreements. The duty to maintain confidentiality on the employer's secrets generally remains, at least for a certain period of time, even after the employee has left the employment.
- Fourthly, signing confidentiality agreements with business partners whenever disclosing confidential information (http://www.wipo.int/sme/en/ip_business/trade_secrets/trade_secrets.htm).

DELL

Dell, the computer company, has a number of patents and some pending applications in the United States on its unique business models. A patent may reveal a lot of valuable information, but at the same time, it provides exclusivity in the marketplace. In 1999, Dell used its patent portfolio as collateral in a $16 billion cross-licensing deal with IBM that provided Dell with lower cost computer components. This freed Dell from having to pay IBM several millions of dollars in royalties and further reduced Dell's cost of doing business.

WAL-MART

Wal-Mart, on the other hand, appears to rely more on the protection afforded by the law of trade secrets for protection of its business model, regardless of the fact that the law protecting secret information is often regarded as a relatively ineffective mechanism for protection against theft of proprietary information from past key employees to competitors (http://www.wipo.int/export/sites/www/sme/en/documents/wipo_magazine/04_2002.pdf).

There are several cases wherein Trade Secret was an integral part of the lawsuit between the individuals and/or companies. One such case is discussed below:

Vaughan Co. v. Global Bio–Fuels Tech., LLC, (N.D.N.Y. Nov. 15, 2012).

Vaughan Company filed suit against Global Bio–Fuels Technology, LLC and Richard Behnke (collectively "Defendants"). Vaughan alleged, amongst other claims, a claim for misappropriation of trade secrets related to Vaughan contact list and other confidential information. Defendants filed a motion to dismiss. Behnke was an employee of Vaughan during which time he had access to extensive information. As part of his responsibilities, Behnke was assigned to take pictures for use by Vaughan. During his employment with Vaughan, Behnke formed Global a company with similar business as Vaughan. Defendant's used said pictures on Global's website and in a trademark application without Vaughan's consent. Behnke subsequently resigned from his position at Vaughan. However, he refused to return Vaughan's laptop containing confidential company information that Vaughan alleges Defendants used to compete against Vaughan in contract bids. The court denied Defendants motion. In regards to the misappropriation claim, the court concluded that Vaughan adequately alleged that Defendants possessed trade secrets and used such information in breach of their fiduciary duties to maintain the confidentiality (Trade Secret Case Law – 2013; www.apps.americanbar.org).

World Intellectual Property Organization (WIPO)

WIPO is the global forum for intellectual property services, policy, information and cooperation, with 188 member countries.

Mission: To lead the development of a balanced and effective international intellectual property (IP) system that enables innovation and creativity for the benefit of all.

One of the oldest specialized agencies of the United Nations, the World Intellectual Property Organization (WIPO) has a long and interesting past. The Paris Convention was the first major step taken to help creators ensure that their intellectual works protected in other countries. The need for international protection of intellectual property (IP) became evident when foreign exhibitors refused to attend the International Exhibition of Inventions in Vienna, Austria in 1873 because they were afraid their ideas would be stolen and exploited commercially in other countries.

In 1893, the two secretariats were set up to administer the Paris and Berne Conventions combine to form WIPO's immediate predecessor, the United International Bureaux for the Protection of Intellectual Property—best known by its French acronym, BIRPI. The organization, with a staff of seven, is based in Berne, Switzerland and in 1970 the Convention establishing the World Intellectual Property Organization (WIPO) came into force and BIRPI was thus transformed to become WIPO. The newly established WIPO is a member state-led, intergovernmental organization, with its headquarters in Geneva, Switzerland. In 1974 WIPO joined the United Nations (UN) family of organizations, became a specialized agency of the UN. All member states of the UN are entitled, though not obliged, to become members of the specialized agencies.

WIPO provides:

- A policy forum to shape balanced international IP rules for a changing world.
- Global services to protect IP across borders and to resolve disputes.
- Technical infrastructure to connect IP systems and share knowledge.
  Cooperation and capacity-building programs to enable all countries to use IP for economic, social and cultural development.
- A world reference source for IP information (http://www.wipo.int/about-wipo/en/).

United States Food And Drug Administration (USFDA): IND, NDA and ANDA

The FDA's Drug Review Process: Ensuring Drugs Are Safe and Effective

Most drugs that undergo preclinical (animal) testing never even make it to human testing and review by the FDA. The drugs that do must undergo the agency's rigorous evaluation process, which scrutinizes everything about the drug-from the design of clinical trials to the severity of side effects to the conditions under which the drug is manufactured.

Drug Review Steps Simplified

1. Preclinical (animal) testing.
2. An investigational new drug application (IND) outlines what the sponsor of a new drug proposes for human testing in clinical trials.
3. Phase 1 studies (typically involve 20–80 people).
4. Phase 2 studies (typically involve a few dozen to about 300 people).
5. Phase 3 studies (typically involve several hundred to about 3,000 people).
6. The pre-NDA period, just before a new drug application (NDA) is submitted. A common time for the FDA and drug sponsors to meet.
7. Submission of an NDA is the formal step asking the FDA to consider a drug for marketing approval.
8. After an NDA is received, the FDA has 60 days to decide whether to file it so it can be reviewed.
9. If the FDA files the NDA, an FDA review team is assigned to evaluate the sponsor's research on the drug's safety and effectiveness.
10. The FDA reviews information that goes on a drug's professional labeling (information on how to use the drug).
11. The FDA inspects the facilities where the drug will be manufactured as part of the approval process.
12. FDA reviewers will approve the application or issue a complete response letter.

Reviewing Applications

Once a new drug application is filed, an FDA review team—medical doctors, chemists, statisticians, microbiologists, pharmacologists, and other experts—evaluates whether the studies the sponsor submitted show that the drug is safe and effective for its proposed use. No drug is absolutely safe; all drugs have side effects. "Safe" in this sense means that the benefits of the drug appear to outweigh the known risks.

The review team analyzes study results and looks for possible issues with the application, such as weaknesses of the study design or analyses. Reviewers determine whether they agree with the sponsor's results and conclusions, or whether they need any additional information to make a decision.

Each reviewer prepares a written evaluation containing conclusions and recommendations about the application. These evaluations are then considered by team leaders, division directors, and office directors, depending on the type of application.

Reviewers receive training that fosters consistency in drug reviews, and good review practices remain a high priority for the agency.

Sometimes, the FDA calls on advisory committees, who provide FDA with independent opinions and recommendations from outside experts on applications to market new drugs, and on FDA policies. Whether an advisory committee is needed depends on many things.

Accelerated Approval

Traditional approval requires that clinical benefit be shown before approval can be granted. Accelerated approval is given to some new drugs for serious and life-threatening illnesses that lack satisfactory treatments. This allows an NDA to be approved before measures of effectiveness that would usually be required for approval are available.

Instead, less traditional measures called surrogate endpoints are used to evaluate effectiveness. These are laboratory findings or signs that may not be a direct measurement of how a patient feels, functions, or survives, but are considered likely to predict benefit. For example, a surrogate endpoint could be the lowering of HIV blood levels for short periods of time with anti-retroviral drugs.

Gleevec (imatinib mesylate), an oral treatment for patients with a life-threatening form of cancer called chronic myeloid leukemia (CML), received accelerated approval. The drug was also approved under the FDA's orphan drug program, which gives financial incentives to sponsors for manufacturing drugs that treat rare diseases. Gleevec blocks enzymes that play a role in cancer growth. The approval was based on results of three large Phase 2 studies, which showed the drug could substantially reduce the level of cancerous cells in the bone marrow and blood.

Most drugs to treat HIV have been approved under accelerated approval provisions, with the company required to continue its studies after the drug is on the market to confirm that its effects on virus levels are maintained and that it ultimately benefits the patient. Under accelerated approval rules, if studies don't confirm the initial results, the FDA can withdraw the approval.

Because premarket review can't catch all potential problems with a drug, the FDA continues to track approved drugs for adverse events through a post-marketing surveillance program.

Bumps in the Road

If the FDA decides that the benefits of a drug outweigh the known risks, the drug will receive approval and can be marketed in the United States. But if there are problems with an NDA or if more information is necessary to make that determination, the FDA may issue a complete response letter.

Common problems include unexpected safety issues that crop up or failure to demonstrate a drug's effectiveness. A sponsor may need to conduct additional studies—perhaps studies of more people, different types of people, or for a longer period of time.

Manufacturing issues are also among the reasons that approval may be delayed or denied.

The FDA outlines the justification for its decision in a complete response letter to the drug sponsor and CDER gives the sponsor a chance to meet with agency officials to discuss the deficiencies. At that point, the sponsor can ask for a hearing, correct any deficiencies and submit new information, or withdraw the application.

The Role of User Fees

On September 27, 2007, President Bush signed into law the Food and Drug Administration Amendments Act of 2007 which includes the reauthorization and expansion of the Prescription Drug User Fee Act. The reauthorization of PDUFA will significantly broaden and upgrade the agency's drug safety program, and facilitate more efficient development of safe and effective new medications for the American public.

In addition to setting time frames for review of applications, PDUFA sets goals to improve communication and sets goals for specific kinds of meetings between the FDA and drug sponsors. It also outlines how fast the FDA must respond to requests from sponsors.

The Quality of Clinical Data

The Food and Drug Administration relies on data that sponsors submit to decide whether a drug should be approved. To protect the rights and welfare of people in clinical trials, and to verify the quality and integrity of data submitted, the FDA's Division of Scientific Investigations (DSI) conducts inspections of clinical investigators' study sites. DSI also reviews the records of institutional review boards to be sure they are fulfilling their role in patient protection.

DSI seeks to determine such things as whether the study was conducted according to the investigational plan, whether all adverse events were recorded, and whether the subjects met the inclusion/exclusion criteria outlined in the study protocol.

At the conclusion of each inspection, FDA investigators prepare a report summarizing any deficiencies. In cases where they observe numerous or serious deviations, such as falsification of data, DSI classifies the inspection as "official action indicated" and sends a warning letter or Notice of Initiation of Disqualification Proceedings and Opportunity to Explain (NIDPOE) to the clinical investigator, specifying the deviations that were found. The NIDPOE begins an administrative process to determine whether the clinical investigator should remain eligible to receive investigational products and conduct clinical studies.

The FDA has established an independent Drug Safety Oversight Board (DSOB) to oversee the management of drug safety issues (http://www.fda.gov/Drugs/ResourcesForYou/Consumers/ucm289601.htm).

Center for Drug Evaluation and Research (CDER)

The center's best-known job is to evaluate new drugs before they can be sold. CDER's evaluation not only prevents quackery, but also provides doctors and patients the information they need to use medicines wisely. The center ensures that drugs, both brand-name and generic, work correctly and that their health benefits outweigh their known risks.

Drug companies seeking to sell a drug in the United States must first test it. The company then sends CDER the evidence from these tests to prove the drug is safe and effective for its intended use. A team of CDER physicians, statisticians, chemists, pharmacologists, and other scientists reviews the company's data and proposed labeling. If this independent and unbiased review establishes that a drug's health benefits outweigh its known risks, the drug is approved for sale. The center doesn't actually test drugs itself, although it does conduct limited research in the areas of drug quality, safety, and effectiveness standards (http://www.fda.gov/Drugs/DevelopmentApprovalProcess/HowDrugsareDevelopedandApproved/default.htm).

Application Types

Investigational New Drug (IND) Application
New Drug Application (NDA)
Abbreviated New Drug Application (ANDA): Generics
Therapeutic Biologic Applications (BLA)
Drug Applications for Over-the-Counter (OTC) Drugs

Investigational New Drug (IND) Application

The IND is the means through which the sponsor technically obtains this exemption from the FDA.

During a new drug's early preclinical development, the sponsor's primary goal is to determine if the product is reasonably safe for initial use in humans, and if the compound exhibits pharmacological activity that justifies commercial development. When a product is identified as a viable candidate for further development, the sponsor then focuses on collecting the data and information necessary to establish that the product will not expose humans to unreasonable risks when used in limited, early-stage clinical studies.

FDA's role in the development of a new drug begins when the drug's sponsor (usually the manufacturer or potential marketer), having screened the new molecule for pharmacological activity and acute toxicity potential in animals, wants to test its diagnostic or therapeutic potential in humans. At that point, the molecule changes in legal status under the Federal Food, Drug, and Cosmetic Act and becomes a new drug subject to specific requirements of the drug regulatory system.

There are three IND types:

- An Investigator IND is submitted by a physician who both initiates and conducts an investigation, and under whose immediate direction the investigational drug is administered or dispensed. A physician might submit a research IND to propose studying an unapproved drug, or an approved product for a new indication or in a new patient population.
- Emergency Use IND allows the FDA to authorize use of an experimental drug in an emergency situation that does not allow time for submission of an IND in accordance with 21CFR, Sec. 312.23 or Sec. 312.34. It is also used for patients who do not meet the criteria of an existing study protocol, or if an approved study protocol does not exist.
- Treatment IND is submitted for experimental drugs showing promise in clinical testing for serious or immediately life-threatening conditions while the final clinical work is conducted and the FDA review takes place.

There are two IND categories:

- Commercial.
- Research (non-commercial).

The IND application must contain information in three broad areas:

- Animal Pharmacology and Toxicology Studies—Preclinical data to permit an assessment as to whether the product is reasonably safe for initial testing in humans. Also included is any previous experience with the drug in humans (often foreign use).
- Manufacturing Information—Information pertaining to the composition, manufacturer, stability, and controls used for manufacturing the drug substance and the drug product. This information is assessed to ensure that the company can adequately produce and supply consistent batches of the drug.
- Clinical Protocols and Investigator Information—Detailed protocols for proposed clinical studies to assess whether the initial-phase trials will expose subjects to unnecessary risks. Also, information on the qualifications of clinical investigators—professionals (generally physicians) who oversee the administration of the experimental compound—to assess whether they are qualified to fulfill their clinical trial duties. Finally, commitments to obtain informed consent

from the research subjects, to obtain review of the study by an institutional review board (IRB), and to adhere to the investigational new drug regulations.

Once the IND is submitted, the sponsor must wait 30 calendar days before initiating any clinical trials. During this time, FDA has an opportunity to review the IND for safety to assure that research subjects will not be subjected to unreasonable risk (http://www.fda.gov/Drugs/DevelopmentApprovalProcess/HowDrugsareDevelopedandApproved/ApprovalApplications/InvestigationalNewDrugINDApplication/default.htm).

New Drug Application (NDA)

The NDA application is the vehicle through which drug sponsors formally propose that the FDA approve a new pharmaceutical for sale and marketing in the U.S. The data gathered during the animal studies and human clinical trials of an Investigational New Drug (IND) become part of the NDA.

The goals of the NDA are to provide enough information to permit FDA reviewer to reach the following key decisions:

- Whether the drug is safe and effective in its proposed use(s), and whether the benefits of the drug outweigh the risks.
- Whether the drug's proposed labeling (package insert) is appropriate, and what it should contain.
- Whether the methods used in manufacturing the drug and the controls used to maintain the drug's quality are adequate to preserve the drug's identity, strength, quality, and purity.

The documentation required in an NDA is supposed to tell the drug's whole story, including what happened during the clinical tests, what the ingredients of the drug are, the results of the animal studies, how the drug behaves in the body, and how it is manufactured, processed and packaged (http://www.fda.gov/Drugs/DevelopmentApprovalProcess/HowDrugsareDevelopedandApproved/ApprovalApplications/NewDrugApplicationNDA/default.htm).

Out of around 1,000 NDA/BLAs approved by the USFDA during 2004–2013, 820 were for the new chemical entities and the rest could be categorized as found in the Fig. 23.8 below:

Hatch-Waxman Act and Generics

An Abbreviated New Drug Application (ANDA) contains data which when submitted to FDA's Center for Drug Evaluation and Research, Office of Generic Drugs, provides for the review and ultimate approval of a generic drug product.

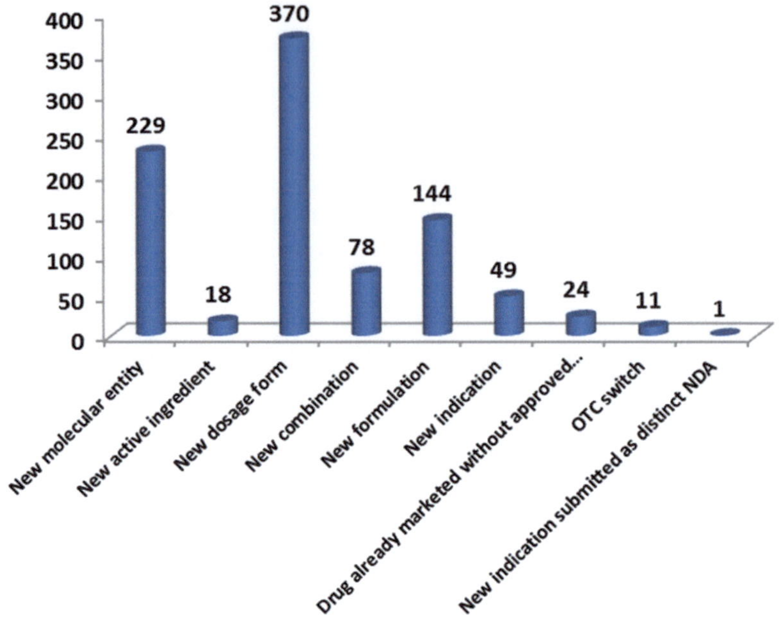

Fig. 23.8 Categories of NDA/BLAs approved by USFDA during 2004–2013. Out of around 1,000 NDA/BLAs approved by the USFDA during 2004–2013, following figure (Fig. 23.9) gives the glimpse of categories of applications with biological entities

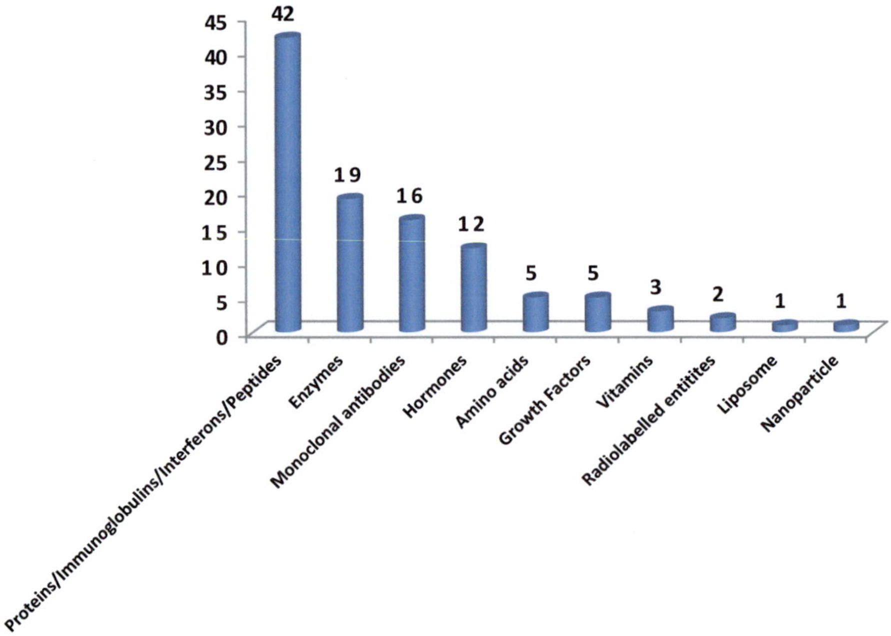

Fig. 23.9 Categories of biological entities approved by USFDA during 2004–2013

Once approved, an applicant may manufacture and market the generic drug product to provide a safe, effective, low cost alternative to the American public.

A generic drug product is one that is comparable to an innovator drug product in dosage form, strength, route of administration, quality, performance characteristics and intended use. All approved products, both innovator and generic, are listed in FDA's Approved Drug Products with Therapeutic Equivalence Evaluations (Orange Book).

Generic drug applications are termed "abbreviated" because they are generally not required to include preclinical (animal) and clinical (human) data to establish safety and effectiveness. Instead, generic applicants must scientifically demonstrate that their product is bioequivalent (i.e., performs in the same manner as the innovator drug). One way scientists demonstrate bioequivalence is to measure the time it takes the generic drug to reach the bloodstream in 24–36 healthy, volunteers. This gives them the rate of absorption, or bioavailability, of the generic drug, which they can then compare to that of the innovator drug. The generic version must deliver the same amount of active ingredients into a patient's bloodstream in the same amount of time as the innovator drug.

Using bioequivalence as the basis for approving generic copies of drug products was established by the "Drug Price Competition and Patent Term Restoration Act of 1984," also known as the Waxman-Hatch Act. This Act expedites the availability of less costly generic drugs by permitting FDA to approve applications to market generic versions of brand-name drugs without conducting costly and duplicative clinical trials. At the same time, the brand-name companies can apply for up to 5 additional years longer patent protection for the new medicines they developed to make up for time lost while their products were going through FDA's approval process. Brand-name drugs are subject to the same bioequivalence tests as generics upon reformulation (http://www.fda.gov/Drugs/DevelopmentApprovalProcess/HowDrugsareDevelopedandApproved/ApprovalApplications/AbbreviatedNewDrugApplicationANDAGenerics/default.htm).

The FDA guidance also is intended to provide recommendations to sponsors and/or applicants planning to include bioavailability (BA) and bioequivalence (BE) information for orally administered drug products in investigational new drug applications (INDs), new drug applications (NDAs), abbreviated new drug applications (ANDAs), and their supplements.

Studies to measure BA and/or establish BE of a product are important elements in support of INDs, NDAs, ANDAs, and their supplements. As part of INDs and NDAs for orally administered drug products, BA studies focus on determining the process by which a drug is released from the oral dosage form and moves to the site of action. BA data provide an estimate of the fraction of the drug absorbed, as well as its subsequent distribution and elimination. BA can be generally documented by a systemic exposure profile obtained by measuring drug and/or metabolite concentration in the systemic circulation over time. The systemic exposure profile determined during clinical trials in the IND period can serve as a benchmark for subsequent BE studies.

Studies to establish BE between two products are important for certain changes before approval for a pioneer product in NDA and ANDA submissions and in the presence of certain post-approval changes in NDAs and ANDAs. In BE studies, an applicant compares the systemic exposure profile of a test drug product to that of a reference drug product (RLD). For two orally administered drug products to be bioequivalent, the active drug ingredient or active moiety in the test product must exhibit the same rate and extent of absorption as the reference drug product.

Both BA and BE studies are required by regulations, depending on the type of application being submitted. BE information is required to ensure therapeutic equivalence between a pharmaceutically equivalent test drug product and a reference listed drug (http://www.fda.gov/downloads/Drugs/GuidanceComplianceRegulatoryInformation/Guidances/UCM070124.pdf) and (http://thomas.loc.gov/cgi-bin/bdquery/z?d098:SN01538:@@@D&summ2=m&|TOM:/bss/d098query.html).

505(b) 2 v. ANDA

The 505(b)(1) process is what the industry is familiar with; it is executed for new drugs like those discovered by big pharma versus the 505(b)(2) process, which can take an existing drug and makes small modifications, often significantly advancing the medication for the patients' benefit (http://www.camargopharma.com/Userfiles/white-paper/Cmrgo_WhitePaperApprovalPthwy_VFb.pdf).

Federal Food, Drug, and Cosmetic Act defines a 505(b)2 application as an application that contains full reports of investigations of safety and effectiveness but where at least some of the information required for approval comes from studies not conducted by or for the applicant and for which the applicant has not obtained a right of reference.

505(b)2 applications or the 'paper NDAs' are unique in nature wherein, the 505 (b)(2) applicant may qualify for 3 or 5 years of market exclusivity, depending on the extent of the change to the previously approved drug and the type of clinical data included in the NDA. This distinguishes a 505(b)(2) from an ANDA, where exclusivity can be held for only 180 days. A 505(b)(2) application may also be eligible for orphan drug or pediatric exclusivity.

The 505(b)(2) approval route can be utilized for a wide range of products, especially for those that represent a limited change from a previously approved drug. The following are examples of changes to approved drugs which would be appropriate to submit as 505(b)(2) applications:

- Changes in dosage form, strength, route of administration, formulation, dosing regimen, or indication.
- A new combination product where the active ingredients have been previously approved.
- Change to an active ingredient (e.g., different salt, ester complex, chelate, etc.).

- New molecular entity when studies have been conducted by other sponsors and published information is pertinent to the application (e.g., a pro-drug or active metabolite of an approved drug.).
- Change from an Rx indication to an OTC indication.
- Change to an OTC monograph drug (e.g., non-monograph indication, new dosage form).
- Drugs with naturally derived or recombinant (i.e., biological) active ingredients where additional limited clinical data is necessary to show the ingredient is the same as the ingredient in the reference drug.
- Bio-inequivalence for drug products where the rate and or extent of absorption exceed or are otherwise different from the standards for bioequivalence compared to a listed drug. Additional studies might be required to document the safety and efficacy at the different rate and extend of delivery.

The 505(b)(2) applications are not appropriate for products:

- That are covered under Section 505(j).
- For which the only difference is lower extent of absorption than reference drug.
- For which the only difference is an unintended lower rate of absorption than reference drug (21 CFR 314.54 – 505(b)(2) applications and www.regprofessional.com).

ANDA: Application containing information to show that the proposed product is identical in active ingredient, dosage form, strength, route of administration, labeling, quality, performance characteristics, and intended use, among other things, to a previously approved product (section 505(j)).

The process starts with the filing of an Abbreviated New Drug Application (ANDA) by a generic manufacturer with one of the four Paragraph certifications:

A Paragraph I certification is one for which the originator firm has not filed patent information for its branded product.

Paragraph II certification relates to when the branded product's patent has already expired (i.e., the end of market exclusivity).

Paragraph III certification relates to cases when the generic manufacturer notes that the patent on the branded product will expire on a certain date and that it seeks to enter only after patent expiry or end of market exclusivity.

Paragraph IV, argues that the generic manufacturer does not infringe on a branded product's patents or that those patents are invalid.

As an incentive to encourage generics to file Paragraph IV certifications, the Act provided a 180-day period of marketing exclusivity to the first to file an ANDA with a Paragraph IV certification.

There are, however, some regulatory challenges that are unique to 505(b)(2) applications. Unlike a 505(b)(1) NDA, wherein the sponsor owns all the data necessary for approval (or has obtained the right to reference), the filing or approval of a 505(b)(2) application may be delayed due to patent or exclusivity protection on the reference drug. Sponsors filing 505(b)(2) applications must include patent certifications in their applications and must also provide notice of certain patient certifications to the NDA and patent holders of the reference drug.

A major challenge with 505(b)(2) applications is determining what additional information is needed to support the proposed change of the previously approved drug. As noted in 21 CFR 314.54, the "application need contain only that information needed to support the proposed modification(s) of the listed drug." This will usually be a case-by-case determination (21 CFR 314.54–505(b)(2) applications and www.regprofessional.com).

Biosimilars

A biosimilar is a biological product that is highly similar to a U.S.-licensed reference biological product notwithstanding minor differences in clinically inactive components, and for which there are no clinically meaningful differences between the biological product and the reference product in terms of the safety, purity, and potency of the product.

The Patient Protection and Affordable Care Act (Affordable Care Act), signed into law by President Obama on March 23, 2010, amends the Public Health Service Act (PHS Act) to create an abbreviated licensure pathway for biological products that are demonstrated to be "biosimilar" to or "interchangeable" with an FDA-licensed biological product. Under this new law, a biological product may be demonstrated to be "biosimilar" if data show that, among other things, the product is "highly similar" to an already-approved biological product (http://www.fda.gov/Drugs/DevelopmentApprovalProcess/HowDrugsareDevelopedandApproved/ApprovalApplications/TherapeuticBiologicApplications/Biosimilars/ucm241718.htm).

As the first wave of biopharmaceuticals has expired or is about to approach expiry, the opportunity for the development of subsequent biosimilar or follow-on protein product (FOPP) versions of these products has emerged. However, second entry biopharmaceuticals differ significantly from traditional chemical generics. Biosimilars, unlike conventional generic drugs, require more quality data and therefore must demonstrate full comparability (including immunogenicity data), to the reference product. This of course begs the question: what is the incentive to develop a biosimilar? Unlike with traditional generic products, there is the potential in Europe to extrapolate to all indications of the reference product, which may go part of the way answering this question (Saenger 2012).

Sandoz' somatropin biosimilar was first given the go-ahead in Europe and the U.S. in 2006, where it is marketed as Omnitrope®. It represented the first biosimilar product to be approved in both these markets. Omnitrope was also sanctioned in Canada earlier this year (http://www.genengnews.com/gen-news-highlights/sandoz-launches-somatropin-biosimilar-in-japan/64624116/).

Celltrion, a global biopharmaceutical company located in Incheon, Republic of Korea, announced that on August 8, 2014, it completed the filing procedure to obtain US FDA approval for its infliximab biosimilar-Remsima® through USFDA's biosimilar regulatory approval pathway, the Biologics Price Competition and Innovation Act of 2009 (BPCIA). This marks the first 351(k) biosimilarmAb application to be filed in the U.S.A. and the second application for a biosimilar after Sandoz.

After prior consultation with the US FDA, Celltrion conducted additional clinical trials (starting on October 2013 and lasting 6 months) to determine the bioequivalency of the originator products with Remsima®. Specifically, Celltrion tested for Pharmacokinetic/Pharmacodynamic (PK/PD) equivalency and safety equivalency for the three distinct products, the originator products sold in the US, the originator products sold in Europe, and Remsima®. These additional clinical trial data, along with Celltrion's established global clinical trial data, were submitted to the US FDA by Celltrion as part of its application.

The patent for the originator drug is set to expire at the end of 2018 however, in support of its US FDA application; Celltrion has currently filed a lawsuit in the federal court of Massachusetts seeking a declaratory judgment that Janssen Biotech's remaining patents on the original reference drug Remicade® (infliximab) are invalid and unenforceable.

Celltrion has already obtained approval for its biosimilar infliximab product, which uses the registered brand name of Remsima®, from over 50 countries worldwide, including most of advanced regulatory agencies including Europe, Canada and Japan. In fact, Remsima® remains the world's first and only biosimilar mAb to be approved by the European EMA, Japan PMDA, and Health Canada (http://www.celltrion.com).

Access to Medicines

Prior patent-rule pluralism in both the developed and developing world had allowed discrimination between fields of invention, for example by excluding medicines, but TRIPS expressly outlawed such discrimination. Similarly, it was no longer permissible to discriminate against imports in favor of local products, thus allowing major pharmaceutical companies to control the place of production. Because of TRIPS, the major pharmaceutical producers succeeded in consolidating their monopoly power internationally—they have exclusive rights under TRIPS to exclude others from "making, using, offering for sale, selling, or importing" patented pharmaceutical products or "products made with a patented process" (http://www.wto.org/english/docs_e/legal_e/27-trips.pdf). If drug prices increase, in addition to the obvious implications for public health, this could be potentially politically disastrous for many politicians in developing states who are already under pressure from their constituents to improve access to medicines and lower pharmaceutical prices (Cohen and Illingworth 2003).

When a patent holder can exclude others, it frequently charges monopoly prices, and its profit-maximizing strategy in developing countries is typically to sell medicines at high prices to the rich even if that price excludes purchase by or for the vast majority of a country's population. A more pragmatic solution, currently pursued by health activists internationally, is the promotion of robust generic pharmaceutical production, operating at efficient economies-of-scale so that medicines can be made available at the lowest possible cost. To make these drugs available to all, activists have succeeded in establishing funding structures such as the Global Fund to Fight AIDS, Tuberculosis, and Malaria and the U.S. President's Emergency Plan for AIDS Relief, and in agitating for greatly enhanced bilateral and multilateral donations so that there are reliable and sustainable reservoirs of purchasing power to support a market in generic pharmaceuticals and finance purchase of large quantities of medicine (Baker 2009).

Recently the concept of 'Patent pool' is adopted to improve access to appropriate, affordable HIV medicines and technologies for people living with HIV in developing countries (http://www.medicinespatentpool.org/about/). Patent pools can be defined as an agreement between two or more patent owners to license one or more of their patents to one another or to third parties. Often, patent pools are associated with complex technologies that require complementary patents in order to provide efficient technical solutions. Generally, these patent pools cover mature technologies. Pools also frequently represent the basis for industry standards that supply firms with the necessary technologies to develop compatible products and services. In that case, they rather concern technologies that are yet to be fully developed.

Patent pools have been the subject of ongoing discussions from both a legal and an economic perspective. On the one hand, patent pools may have positive effects on competition and innovation. By sharing intellectual property assets, companies may develop new products and reduce their transaction costs. On the other hand, under specific circumstances, patent pools may provide an opportunity for a possible anti-competitive behavior: like any cooperation among competitors, they involve an inherent risk of collusive behavior. In other words, a patent pool may be regarded as a cartel. In addition, there may be competition-related concerns regarding the licensing practices and restrictions they entail (http://www.wipo.int/export/sites/www/ip-competition/en/studies/patent_pools_report.pdf).

Médecins Sans Frontières (MSF), or 'Doctors Without Borders' initiated the Campaign for Access to Essential Medicines in 1999 to increase access to essential medicines in developing countries. "Essential medicines" are those drugs that are needed in sufficient supply to treat a disease common to a population. However, most diseases common to populations in developing countries are no longer common to populations in developed countries; therefore, pharmaceutical companies find that producing these drugs is no longer profitable and may raise the price per treatment, decrease development of the drug (and new treatments) or even stop production of the drug. MSF started the campaign to put pressure on governments and pharmaceutical companies to increase funding for essential medicines. MSF played a major role in urging the drug maker Novartis to drop its case against India's patent law that prevents Novartis from patenting its drugs in India (http://en.wikipedia.org/wiki/M%C3%A9decins_Sans_Fronti%C3%A8res).

Compulsory Licensing

Compulsory licensing is when a government allows someone else to produce the patented product or process without the consent of the patent owner (however, if a compulsory license is issued, adequate remuneration must still be paid to the patent holder). In current public discussion, this is usually associated with pharmaceuticals, but it could also apply to patents in any field.

The agreement allows compulsory licensing as part of the agreement's overall attempt to strike a balance between promoting access to existing drugs and promoting research and development into new drugs. But the term "compulsory licensing" does not appear in the TRIPS Agreement. Instead, the phrase "other use without authorization of the right holder" appears in the title of Article 31. Compulsory licensing is only part of this since "other use" includes use by governments for their own purposes.

Compulsory licensing and government use of a patent without the authorization of its owner can only be done under a number of conditions aimed at protecting the legitimate interests of the patent holder, such as national emergencies, extreme urgency or public noncommercial use. Compulsory licensing must meet certain additional requirements. In particular, it cannot be given exclusively to licensees (e.g. the patent-holder can continue to produce), and usually it must be granted mainly to supply the domestic market (http://www.wto.org/english/news_e/pres03_e/pr350_e.htm).

Case Example

Bayer Corporation, USA ("Bayer") had developed Sorafenib Tosylate, marketed as Nexavar, and obtained a patent from United States Patent Office in 1999. The drug is a life extending drug which is used for treating patients suffering from advanced stages of kidney cancer (Renal Cell Carcinoma) and liver cancer (Hepatocellular Carcinoma).

Bayer was granted a patent for the Drug in India in March 2008. On December 6, 2010, Natco Pharma Ltd approached Bayer for grant of a voluntary license. Bayer rejected Natco's request for grant of voluntary license and requested Natco to approach within 14 days in case Natco had anything further to add (http://www.lexology.com/library/detail.aspx?g=5dc7c10a-7a1b-4b09-a387-7aa7a09b5c95).

After the expiration of 3 years from the date of grant of the Indian patent to the Drug, Natco applied to the Controller General of Patents, India for a compulsory license under Section 84 (1) of the Patents Act 1970 proposing to manufacture and sell the drug at a price of Rs. 8,800 (approx. USD 160) for a month's dose—a fraction of Bayer's price of Rs. 280,000 (approx. 5,098 USD). Bayer opposed this application on various grounds, however in March 2012, the Controller granted the first Compulsory License to Natco to manufacture and sell the drug.

Bayer promptly filed an appeal against the compulsory license order before the Intellectual Property Appellate Board in Chennai. Meanwhile, the compulsory license had a dramatic effect on the drug's price—bringing it down to 8,800 rupees. Bayer got a 6 % royalty on sales by Natco.

The mechanism of compulsory license worked the following grounds: the reasonable requirements of the public with respect to the patented invention have not been satisfied; the patented invention is not available to the public at a reasonably affordable price; or the patented invention has not been worked in the territory of India.

Granting the compulsory license to Natco reinvigorated the old debate about patents versus patients in India and turned the spotlight on the escalating battle between multinational pharmaceutical companies and the country's generic drug manufacturers (http://www.ip-watch.org/2013/03/04/indias-first-compulsory-licence-upheld-but-legal-fights-likely-to-continue/).

Parallel Imports, Grey Imports and 'Exhaustion' of Rights

Parallel or grey-market imports are not imports of counterfeit products or illegal copies. These are products marketed by the patent owner (or trademark- or copyright-owner, etc.) or with the patent owner's permission in one country and imported into another country without the approval of the patent owner.

For example, suppose company 'A' has a drug patented in the country 'X' and the country 'Y', which it sells at a lower price in the country 'Y'. If a company 'B' buys the drug in the country 'Y' and imports it into the country 'X' at a price that is lower than company A's price, that would be a parallel or grey import.

The legal principle here is "exhaustion", the idea that once company 'A' has sold a batch of its product (in this case, in country 'Y'), its patent rights are exhausted on that batch and it no longer has any rights over what happens to that batch.

The TRIPS Agreement simply says that none of its provisions, except those dealing with nondiscrimination ("national treatment" and "most-favored-nation treatment"), can be used to address the issue of exhaustion of intellectual property rights in a WTO dispute. In other words, even if a country allows parallel imports in a way that another country might think violates the TRIPS Agreement; this cannot be raised as a dispute in the WTO unless fundamental principles of non-discrimination are involved. The Doha Declaration clarifies that this means that members can choose how to deal with exhaustion in a way that best fits their domestic policy objectives.

Importing Under Compulsory Licensing

Article 31(f) of the TRIPS Agreement says products made under compulsory licensing must be "predominantly for the supply of the domestic market". This applies to countries that can manufacture drugs—it limits the amount they can export when the drug is made under compulsory license. And it has an impact on countries unable to make medicines and therefore wanting to import generics. They would find it difficult to find countries that can supply them with drugs made under compulsory licensing.

The legal problem for exporting countries was resolved on 30 August 2003 when WTO members agreed on legal changes to make it easier for countries to import cheaper generics made under compulsory licensing if they are unable to manufacture the medicines themselves. When members agreed on the decision, the General Council chairperson also read out a statement setting out members' shared understandings on how the decision would be interpreted and implemented. This was designed to assure governments that the decision will not be abused.

The decision actually contains three waivers:

- Exporting countries' obligations under Article 31(f) are waived—any member country can export generic pharmaceutical products made under compulsory licenses to meet the needs of importing countries.
- Importing countries' obligations on remuneration to the patent holder under compulsory licensing are waived to avoid double payment. Remuneration is only required on the export side.
- Exporting constraints are waived for developing and least-developed countries so that they can export within a regional trade agreement, when at least half of the members were categorized as least developed countries at the time of the decision. That way, developing countries can make use of economies of scale.

Carefully negotiated conditions apply to pharmaceutical products imported under the system. These conditions aim to ensure that beneficiary countries can import the generics without undermining patent systems, particularly in rich countries. They include measures to prevent the medicines from being diverted to the wrong markets. And they require governments using the system to keep all other members informed, although WTO approval is not required. At the same time phrases such as "reasonable measures within their means" and "proportionate to their administrative capacities" are included to prevent the conditions becoming burdensome and impractical for the importing countries. All WTO member countries are eligible to import under this decision. Several potential exporting countries changed their laws and regulations in order to implement the waivers and to allow production exclusively for export under compulsory license (http://www.wto.org/english/news_e/pres03_e/pr350_e.htm).

Research Exceptions

Many countries use this provision to advance science and technology. They allow researchers to use a patented invention for research, in order to understand the invention more fully. In addition, some countries allow manufacturers of generic drugs to use the patented invention to obtain marketing approval—for example from public health authorities—without the patent owner's permission and before the patent protection expire. The generic producers can then market their versions as soon as the patent expires.

In Canada, this exemption is known as the Bolar provision or Roche-Bolar provision, named after the case Roche Products v. Bolar Pharmaceutical. Bolar was a generic drug manufacturer. Roche was a brand-name pharmaceutical company which made and sold Dalmane, the active ingredient of which was protected by patent. Before patent expiration, Bolar used the patented chemical in experiments to determine if its generic product was bioequivalent to Dalmane in order to obtain FDA approval for its generic version of Dalmane. Bolar argued that its use of the patented product was not infringement under the experimental use exception to the patent law.

The Court of Appeals for the Federal Circuit rejected Bolar's contention holding that the experimental use exception did not apply because Bolar intended to sell its generic product in competition with Roche's Dalmane after patent expiration and, therefore, Bolar's experiments had a business purpose.

Bolar also argued that public policy in favor of availability of generic drugs immediately following patent expiration justified the experimental use of the patented chemical because denying such use would extend Roche's monopoly beyond the date of patent expiration. The court rejected this argument, stating that such policy decisions should be made by Congress. Likewise, the court decided that apparent policy conflicts between statutes such as the Food and Drug Act and the Patent Act should be decided by Congress and not the courts.

Shortly after Roche v Bolar was decided, Congress passed a law permitting use of patented products in experiments for the purpose of obtaining FDA approval (section 271-e-1 of the Drug Price Competition and Patent Term Restoration Act, informally known as the "Hatch-Waxman Act" [Public Law 98-417], which established the modern system for FDA approval of generic drugs). In 2005, the U.S. Supreme Court considered the scope of the Hatch-Waxman exemption in Merck v. Integra. The Supreme Court held that the statute exempts from infringement all uses of compounds that are reasonably related to submission of information to the government under any law regulating the manufacture, use or distribution of drugs. The TRIPS Agreement says governments can also act to prevent patent owners and other holders of intellectual property rights from abusing intellectual property rights, "unreasonably" restraining trade, or hampering the international transfer of technology (http://en.wikipedia.org/wiki/Roche_Products_v._Bolar_Pharmaceutical).

Conclusion

Various aspects of intellectual property rights have been discussed in this chapter along with the international treaties associated with different types of intangible assets. Several other aspects of importance relating to effectively bringing the molecules from the bench to the market place under the current era of competition among the players are dealt with. Still the changing environments leave the scope for re-visiting the older laws and bringing in new laws accordingly with crucial amendments suitable for today's scenario. We know that timely protection of innovations/inventions impart exclusivity to the owners thereby giving sizeable return on investments similarly, timely amendment of older laws if commenced, would be beneficial for the expedited development of nations.

Acknowledgements The author is thankful to DBT-India towards DBT-CREST Fellowship and to CSIR-CDRI, India for all support.

References

http://www.wipo.int/edocs/pubdocs/en/global_challenges/628/wipo_pub_628.pdf. Retrieved October 9, 2014.

http://www.wipo.int/edocs/pubdocs/en/copyright/955/wipo_pub_955.pdf. Retrieved October 10, 2014.

Pugatch, M. P., et al. (2012). *How global biotechnology benefits from intellectual property rights*. New York, NY: Pugatch Consilium.

http://www.wipo.int/export/sites/www/ipstats/en/wipi/2013/pdf/wipo_pub_941_2013_section_a.pdf. Retrieved October 13, 2014.

AUTM Licensing Survey, FY 2002 survey summary. Retrieved October 15, 2014, from http://www.autm.net/surveys/02/2002spublic.pdf.

http://www.autm.net/AM/Template.cfm?Section=FY_2013_Licensing_Activity_Survey&Template=/CM/ContentDisplay.cfm&ContentID=13870. Retrieved October 15, 2014.

University of California. (2011). *Technology transfer annual report 2011*, p. 25.

http://www.whitehouse.gov/ostp/pcast. Retrieved October 15, 2014.

Kesselheim, A. S., & Avorn, J. (2005). University-based science and biotechnology products: Defining the boundaries of intellectual property. *JAMA, 293*, 850–854.

http://www.wto.org/english/thewto_e/whatis_e/tif_e/fact1_e.htm. Retrieved October 16, 2014.

http://www.wto.org/english/tratop_e/trips_e/tripsfactsheet_pharma_2006_e.pdf. Retrieved October 16, 2014.

http://www.wipo.int/export/sites/www/freepublications/en/statistics/943/wipo_pub_943_2013.pdf. Retrieved October 16, 2014.

http://www.wipo.int/edocs/pubdocs/en/copyright/450/wipo_pub_l450cr.pdf. Retrieved October 16, 2014.

http://www.wipo.int/edocs/pubdocs/en/intproperty/909/wipo_pub_909.pdf. Retrieved October 16, 2014.

http://www.wipo.int/edocs/pubdocs/en/copyright/844/wipo_pub_844.pdf. Retrieved October 17, 2014.

http://www.wipo.int/export/sites/www/freepublications/en/statistics/943/wipo_pub_943_2013.pdf. Retrieved October 17, 2014.

http://www.wipo.int/edocs/pubdocs/en/marks/418/wipo_pub_418.pdf. Retrieved October 17, 2014.

http://www.wipo.int/edocs/pubdocs/en/marks/1039/wipo_pub_1039.pdf. Retrieved October 17, 2014.

http://www.trademarkcases.ca/ciba-geigy-canada-ltd-v-apotex-inc-1992-3-s-c-r-120/. Retrieved October 17, 2014.

http://en.wikipedia.org/wiki/Ciba-Geigy_Canada_Ltd._v._Apotex_Inc. Retrieved October 17, 2014.
http://www.wipo.int/edocs/pubdocs/en/geographical/952/wipo_pub_952.pdf. Retrieved October 17, 2014.
Mukherjee, U. (2008). A study of the basmati case (India-US basmati rice dispute): The geographical indication perspective. SSRN. http://ssrn.com/abstract=1143209.
http://www.wipo.int/designs/en/. Retrieved October 17, 2014.
http://decisions.fct-cf.gc.ca/fc-cf/decisions/en/item/61469/index.do. Retrieved October 17, 2014.
http://www.wto.org/english/tratop_e/trips_e/tripsfactsheet_pharma_2006_e.pdf. Retrieved October 17, 2014.
http://www.wipo.int/export/sites/www/pct/en/seminar/basic_1/timeline.pdf. Retrieved October 17, 2014.
http://www.uspto.gov/web/offices/pac/mpep/s2104.html. Retrieved October 17, 2014.
http://www.uspto.gov/patents/resources/general_info_concerning_patents.jsp#heading-27. Retrieved October 17, 2014.
http://documents.lexology.com/2284cfd8-021d-40a8-be6e-875943f0507e.pdf. Retrieved October 17, 2014.
http://www.lexology.com/library/detail.aspx?g=3f92413f-107c-4886-aca7-24633a341e22. Retrieved October 17, 2014.
http://www.novartisoncology.com/. Retrieved October 17, 2014.
http://www.wto.org/english/tratop_e/trips_e/intel2_e.htm#layoutdesigns. Retrieved October 20, 2014.
http://www.wto.org/english/docs_e/legal_e/27-trips.pdf. Retrieved October 10, 2014.
http://www.wipo.int/sme/en/ip_business/trade_secrets/trade_secrets.htm. Retrieved October 10, 2014.
http://www.wipo.int/export/sites/www/sme/en/documents/wipo_magazine/04_2002.pdf. Retrieved October 10, 2014.
Trade Secret Case Law – 2013; Retrieved October 16, 2014, from www.apps.americanbar.org.
http://www.wipo.int/about-wipo/en/. Retrieved October 9, 2014.
http://www.fda.gov/Drugs/ResourcesForYou/Consumers/ucm289601.htm. Retrieved May 22, 2014.
http://www.fda.gov/Drugs/DevelopmentApprovalProcess/HowDrugsareDevelopedandApproved/default.htm. Retrieved April 7, 2014.
http://www.fda.gov/Drugs/DevelopmentApprovalProcess/HowDrugsareDevelopedandApproved/ApprovalApplications/InvestigationalNewDrugINDApplication/default.htm. Retrieved April 14, 2014.
http://www.fda.gov/downloads/Drugs/GuidanceComplianceRegulatoryInformation/Guidances/UCM070124.pdf. Retrieved April 18, 2014.
http://thomas.loc.gov/cgi-bin/bdquery/z?d098:SN01538:@@@D&summ2=m&|TOM:/bss/d098query.html. Retrieved September 12, 2014.
http://www.camargopharma.com/Userfiles/whitepaper/Cmrgo_WhitePaperApprovalPthwy_VFb.pdf. Retrieved July 2, 2014.
21 CFR 314.54 – 505(b)(2) applications. Retrieved October 13, 2014.
www.regprofessional.com. Retrieved October 13, 2014.
http://www.fda.gov/Drugs/DevelopmentApprovalProcess/HowDrugsareDevelopedandApproved/ApprovalApplications/TherapeuticBiologicApplications/Biosimilars/ucm241718.htm. Retrieved October 13, 2014.
Saenger, P. (2012). Current status of biosimilar growth hormone. *The Indian Journal of Pediatrics, 79*(1), 92–98.
http://www.genengnews.com/gen-news-highlights/sandoz-launches-somatropin-biosimilar-in-japan/64624116/. Retrieved October 13, 2014.
http://www.celltrion.com/en/COMPANY/notice_view.asp?idx=456&code=ennews&intNowPage=1&menu_num=&align_year=all. Retrieved October 13, 2014.

Cohen, J. C., & Illingworth, P. (2003). The dilemma of intellectual property rights for pharmaceuticals: The tension between ensuring access of the poor to medicines and committing to international agreements. *Developing World Bioethics, 3*(1), 27–48.

Baker, B. K. (2009). Patents, pricing, and access to essential medicines in developing countries, virtual mentor. *American Medical Association Journal of Ethics, 11*(7), 527–532.

http://www.medicinespatentpool.org/about/. Retrieved October 13, 2014.

http://www.wipo.int/export/sites/www/ip-competition/en/studies/patent_pools_report.pdf. Retrieved 13 Oct 2014.

http://en.wikipedia.org/wiki/M%C3%A9decins_Sans_Fronti%C3%A8res. Retrieved October 13, 2014.

http://www.wto.org/english/news_e/pres03_e/pr350_e.htm. Retrieved October 9, 2014.

http://www.lexology.com/library/detail.aspx?g=5dc7c10a-7a1b-4b09-a387-7aa7a09b5c95. Retrieved October 17, 2014.

http://www.ip-watch.org/2013/03/04/indias-first-compulsory-licence-upheld-but-legal-fights-likely-to-continue/. Retrieved October 20, 2014.

http://en.wikipedia.org/wiki/Roche_Products_v._Bolar_Pharmaceutical. Retrieved October 16, 2014.

Index

P. Ravindra (ed.), *Advances in Bioprocess Technology*,
DOI 10.1007/978-3-319-17915-5

D

R

MIX
Papier aus verantwortungsvollen Quellen
Paper from responsible sources
FSC® C105338

If you have any concerns about our products,
you can contact us on
ProductSafety@springernature.com

In case Publisher is established outside the EU,
the EU authorized representative is:
Springer Nature Customer Service Center GmbH
Europaplatz 3, 69115 Heidelberg, Germany

Printed by Libri Plureos GmbH
in Hamburg, Germany